Power Electronics and RF Power Systems Analysis

Program Examples in BASIC and C

Carl Eichenauer

P T R Prentice Hall
Englewood Cliffs, New Jersey 07632

Eichenauer, Carl,
    Power electronics and RF power systems analysis : program examples
in BASIC and C / by Carl Eichenauer.
       p.   cm.
    Includes bibliographical references and index.
    ISBN 0-13-689910-2
     1. Power electronics—Data processing.   2. Electric network
analysis—Data processing. I. Title.
    TK7881.15.E43   1993
    621.381'54—dc20   92-15065
                                     CIP

Editorial/production supervision
  and interior design: *Ann Sullivan*
Cover design: *Wanda Lubelska*
Buyer: *Mary Elizabeth McCartney*
Acquisitions editor: *Bernard M. Goodwin*
Editorial assistant: *Diane Spina*

© 1993 by P T R Prentice-Hall, Inc.
A Simon & Schuster Company
Englewood Cliffs, New Jersey 07632

The publisher offers discounts on this book when ordered
in bulk quantities. For more information, write:

Special Sales/Professional Marketing
Prentice-Hall, Inc.
Professional Technical Reference Division
Englewood Cliffs, New Jersey 07632

All rights reserved. No part of this book may be
reproduced, in any form or by any means,
without permission in writing from the publisher.

Printed in the United States of America
10 9 8 7 6 5 4 3 2 1

ISBN 0-13-689910-2

Prentice-Hall International (UK) Limited, *London*
Prentice-Hall of Australia Pty. Limited, *Sydney*
Prentice-Hall Canada Inc., *Toronto*
Prentice-Hall Hispanoamericana, S.A., *Mexico*
Prentice-Hall of India Private Limited, *New Delhi*
Prentice-Hall of Japan, Inc., *Tokyo*
Simon & Schuster Asia Pte. Ltd., *Singapore*
Editora Prentice-Hall do Brasil, Ltda., *Rio de Janeiro*

Books are to be returned on or before the last date below.

DUE 21 NOV 2006

To

Jennifer, Joyce, and Evelyn

Contents

Preface xi

Chapter 1 **Introduction** 1

    1-1 Energy and Power 2

    1-2 Power Electronics 4

    1-3 Power Sources 5

Chapter 2 **Single-Phase Power Circuits** 7

    2-1 Prime Power Interface Terminology 7

    2-2 Single-Phase Power Network Configuration 9

    2-3 Power Transformer Characteristics 10

    2-4 Preliminary Distribution Center Analysis 12

    2-5 General-Purpose Steady-State Analysis Program 17

    2-6 Power Systems With Two Sources 21

Chapter 3 Three-Phase Power Circuits   28

   3-1 Balanced Three-Phase Systems   29

   3-2 Conversion of Three-Phase Networks   32

   3-3 Unbalanced Three-Phase Circuits   34

   3-4 General-Purpose Three-Phase Analysis Program   36

   3-5 Analysis of an Unbalanced WYE-Connected Load   40

   3-6 Analysis of an Unbalanced Delta-Connected Load   42

   3-7 Analysis of an Unbalanced Composite Load   42

   3-8 High-Voltage Substitution Analysis   45

Chapter 4 Long Transmission Lines   52

   4-1 Long-Line Equation Development   53

   4-2 Prime Power Line System Analysis   59

   4-3 RF Transmission Line System Analysis   61

   4-4 Types of Transmission Lines   64

   4-5 Transmission Line Auxiliary Elements   73

Chapter 5 Fourier Transformations   79

   5-1 Fourier Series Principles   80

   5-2 Tabular Fourier Analysis   85

   5-3 DC Power Supply System Analysis   92

Chapter 6 RF Power-Generating Systems   99

   6-1 Definitive Documents and Modulating Waveforms   100

   6-2 System Functional Block Diagrams   105

   6-3 System Power Amplification Devices   107

   6-4 Practical System Applications   119

Chapter 7 **Transient Analysis of Power Electronics Systems** **128**

7-1 Transient Analysis Approaches 128

7-2 Network Building Block Definitions 129

7-3 Numerical Integration Program 134

7-4 Simplified DC Power Supply Analysis 140

7-5 Operational Amplifier Transient Analysis 142

7-6 Bandpass Amplifier Transient Response 144

Chapter 8 **DC Power Sources and Inverters** **147**

8-1 Single-Phase Center-Tap Full-Wave DC Power Supply 149

8-2 Power Supply Commutation Considerations 157

8-3 Single-Phase Full-Wave Bridge DC Power Supply 159

8-4 Three-Phase Full-Wave Bridge DC Power Supply 161

8-5 Twelve-Pulse Bridge DC Power Supply Analysis 169

8-6 Regulators and Switching DC Power Supplies 179

8-7 Fifty-Kilohertz DC-to-DC Converter Power Supply System 180

8-8 Controlled Three-Phase Bridge DC Power Supplies/Inverters 185

8-9 Analysis of AC-to-DC Converter for a Shunt Motor 190

8-10 Analysis of a DC-to-AC Inverter Using a Shunt Generator 192

8-11 Analysis of an Uninterruptible AC Power Source 199

Chapter 9 **Low-Level Pulsing Techniques** **206**

9-1 Phased Array Pulsing System 207

9-2 Grid Pulsed Traveling-Wave Amplifier Tube System 216

Chapter 10 High-Level Pulsing Techniques 224

10-1 Hard Switch Pulser for a Magnetron 226

10-2 Line Pulser System Elements 236

10-3 Line Pulser with Pulse Transformer and Nonlinear RF Amplifier Load 242

10-4 MTI Performance Versus DC Charging Stability 255

Chapter 11 Analysis of an MTI System's Performance Capabilities 259

11-1 Radar Clutter and System Improvement Factors 259

11-2 Design Approach for a High-Stability RF Amplifier 262

11-3 DC-to-DC Power Supply with a DC Voltage Source 264

11-4 DC-to-DC Power Supply with a Current Source 272

11-5 Steady-State DC-to-DC Power Supply Analysis 277

Chapter 12 BASIC-to-C Conversion 282

12-1 BASIC Programming 282

12-2 C Language Programming 283

12-3 Brief Reviews of BASIC-to-C Conversions 283

12-4 Concluding Thought 298

Bibliography 299

Appendix A BASIC and C Software Operation 301

Appendix B BASIC and C Listings

1. Parallel and Series Conversions of Circuit Elements 309
2. Steady State Network Analyses 309
3. Delta to Y or Y to Delta Circuit Conversion 316
4. Three-Phase System Analysis from READ Statements 318
5. Transmission Line Analysis 323
6. Rectangular Pulse Spectrum Analysis 332
7. Graphical Fourier Analysis 335
9. Bessel Function of the First Kind 340
11. DC Motor Acceleration Analysis Program 341
12. Transient Numerical Integration Program 343
13. Simplified Power Supply Transient Analysis 343
14. Operational Amplifier Transient Analysis 345
15. Band Pass Amplifier Transient Analysis 347
16. DC Power Supply Transient Analysis One-Phase C.T. 349
17. DC Power Supply Transient Analysis One-Phase Bridge 351
18. DC Power Supply Transient Analysis Three-Phase Bridge 354
18C. Three-Phase F.W.P.S., RC Snubber 358
18D. Controlled Three-Phase Bridge Converter For Shunt Motor 362
18E. Controlled Three-Phase Bridge Inverter For Shunt Generator 367
18G. Twelve Pulse F.W.P.S. 372
18U. Nonsynchronous Three-Phase Y-Delta UPS Driving a Resistance Load 377
19. DC to DC Converter Power Supply Transient Analysis 381

21 Phased Array Pulsing System Analysis   384

22 Phased Array Pulsing System Analysis (Modified)   386

24 Grid Pulser Analysis Program   389

25 Hard Switched Analysis Program   392

26 Pulse Forming Network Analysis Program   394

27 Line Type Pulser Analysis Program   396

28 DC Resonant Charging/Inverse Diode Analysis   399

29 Three-Phase F.W.P.S. with DC Resonant Charging and Inverse Diode   400

31 MTI Stability of TWT RF Power Generator (Voltage Source)   405

33 MTI Stability of TWT RF Power Generator (Current Source)   408

FOURIER.C   Graphical Fourier Analysis Program   411

TRANS_LI.C   Steady-State Circuit Analysis Program   412

SSANAL.C   Steady-State Circuit Analysis Program   413

SSANAL2.C   Steady-State Bandpass Amplifier Analysis   416

OPAMP.C   Operational Amplifier Tabular Transient Response Program   419

OPAMPP.C   Operational Amplifier Transient Response Plotting Program   420

3PHFWPS.C   Three-Phase Full Wave Power Supply Transient Analysis   421

12PULSE.C   Twelve Pulse DC Power Supply Transient Analysis   422

Index   **427**

Preface

When an electronic system's concept is initially drawn in block diagram form, two fundamental technical areas need to be defined. First, the input interface between the active source and the system network, and second, the output interface between the system network and the system load. The objective of this book is to provide technical background information in representative problem areas of concern, and to provide computer programs that permit quantitative problem solutions to be obtained in an expedient manner.

Who can benefit from a book that broadens their power electronics system insight/technical analysis capabilities/background? First, power electronics design engineers—they can more effectively analyze the areas that degrade design performance resulting from unknown input and output interface properties. Second, power system engineers—they can quantitatively define/analyze how waveforms existing at the prime power/power electronics interface adversely affect the systems. Third, computer analysts and programmers—they can use the program as a point of departure to expand their present expertise/capabilities by developing new, improved, and far more comprehensive analysis programs/techniques. Fourth, engineering students—they can develop background/analysis capabilities in areas that are not addressed in normal courses of study.

From a steady-state point of view, in the first five chapters of the book we develop power electronics system input and output interface background/computer analysis programs. From a transient point of view, in Chapter 6 we describe/illustrate the use of an elementary transient analysis program, and in the remainder of the book, develop a broad range of power electronics transient analyses procedures/programs.

The personal computer programs used in this book are of a very fundamental nature. The steady-state circuit analysis programs behave in a manner similar to commercially available analysis programs. When transient analysis programs are investigated, the system principles and equations are defined and developed in considerable detail. Individual analysis programs are then prepared. This approach helps the reader develop both programming skills and the background required for correcting programs that hang up.

Both BASIC and C programming have been used in this book. BASIC is a universally available language. Almost every technically oriented person can read a BASIC program and, if required, rapidly modify it to perform a revised analysis. However, in its interpreter format, BASIC has the fundamental problem of executing at a relatively slow rate.

On the other hand, C is a compiler language. C programming has, within the last ten years, become almost the standard language in many technical areas of activity. One reason for this is that the execution time required to perform analyses with compiled C programs (i.e., based on comparisons between several of the book's programs written in both C and BASIC) is typically less than half that of BASIC programs. However, C requires a higher degree of programming skill than BASIC, C code requires more time to modify than BASIC, and C programming requires that the programmer acquire/purchase additional software unique to the C compiling process.

The computer programs used in Chapters 2 through 11 of the book are written in BASIC. The rationale: The relative simplicity of BASIC programming permits readers to devote the major portion of their study activities toward electrical system analysis problems rather than toward the computer programming aspects of the problems. However, because of its superior execution capabilities, ANSI Standard C language capabilities are reviewed briefly in Chapter 12. The reviews illustrate how to convert eight of the book's commonly used BASIC programs into C source code. The manner in which C execute programs function is also illustrated.

All BASIC and C programs used in the book are listed in the appendixes. The floppy disk provided contains thirty-two BASIC programs (written in Microsoft BASIC). Thirty-four C source code programs, and one C execute code program (compiled with Zortech files). All programs are for use on IBM or IBM-compatible computers.

Chapter 1

Introduction

This is a book about electrical *interfaces*. What do we mean by "interfaces"? According to the dictionary an interface is the place at which independent systems meet and interact on or communicate with each other. The places we will be concerned about are those at which electrical systems meet and interact with each other. Why should we be concerned about these interfaces? Simply because these interactions influence the *performance capabilities* of both systems.

All electronic system designs require quantitative definitions of their *input* and *output* interface capabilities. The input interface definition establishes the mutual relationships that must exist between the external power source and the electronics system. The output interface definition establishes the performance capabilities that must exist between the electronic system's internally generated energy and the external system load.

To illustrate this rationale, assume that we are reviewing the design of a new dc power supply. Ac power is delivered to the power supply's input interface from a 60-Hz source. All practical 60-Hz sources provide service to a number of other public utility customers, so the amplitude of the source voltage can be expected to fluctuate as a function of time. Therefore, we require an accurate *definition* of power source performance capabilities at the input interface.

Energy from the ac source is processed by the electronics system. All practical power supply systems deliver less-than-ideal levels of dc voltage at the system's output interface. For example, the power supply's dc output voltage will change as a function of both input interface voltage fluctuations and variations in the value of the power supply's external load. Therefore, it is essential that accurate definitions exist at the output interface relative to the electronic system's *voltage regulation* performance capabilities.

This very simplified example illustrates only two of a large number of variables that contribute to less-than-ideal overall electronic system performance capabilities. A major objective of this book is to illustrate how electronic system performance capabilities can be determined *quantitatively*.

1-1 ENERGY AND POWER

What is the source of the prime energy that makes a power electronic systems function? A variety of prime energy sources exist: fossil fuel, nuclear, hydroelectric, and so on. In addition, a number of energy *conversions* are required before a source's potential energy is delivered to the input interface of a power electronics system. For example, consider a typical waterwheel power-generating system. Potential energy is stored in an elevated reservoir of water behind a dam. An enclosed conduit (a penstock) connects the water reservoir to a waterwheel turbine in a powerhouse near the base of the dam. The elevation difference between the ends of the penstock is called the *head*. Energy conversion occurs when water flow through the penstock delivers kinetic energy to the turbine blades and causes the turbine's drive shaft to rotate. The energy delivered by the turbine is given by the hydraulic expression

$$W = kQHEt \tag{1-1}$$

where W is the energy in joules (watt-seconds), k is a numerical coefficient, Q is the quantity of water flowing through the penstock per second, H is the head, E is the turbine efficiency (typically, 0.80 to 0.85), and t is time in seconds.

If the solution for Eq. (1-1) is to be in *System International Units (SI)* [1], then $k = 9806$, Q is in cubic meters/second, and H is in meters. In the United States an alternative set of units is also defined by the National Bureau of Standards. For a quantitative solution using these *NBS* units, k has a value of 84.63, Q is in cubic feet/second, and H is in feet. The system analyses presented in this book are compatible with both SI and NBS units. Table 1-1 compares frequently encountered SI and NBS units. Approximate conversion factors are also shown.

In the waterwheel system's powerhouse, a second energy conversion occurs: Electrical energy is generated by mechanically coupling the turbine shaft to the shaft of an ac generator. This interaction converts mechanical energy of rotation into electrical energy. If the generator's load has a constant value, the quantities in Eq. (1-1) can be changed to their electrical counterparts as follows:

$$W = VIt = I^2Rt = \frac{V^2}{R}t \tag{1-2}$$

where V is the potential generated, in volts; I is load current, in amperes; and R is load resistance, in ohms.

The following two examples illustrate use of the foregoing principles.

TABLE 1-1 COMPARISON OF ALTERNATIVE SYSTEM UNITS AND QUANTITIES

Quantity	SI unit Name	SI unit Symbol	NBS unit Name	NBS unit Symbol
1. Length	meter	m	foot	ft
2. Mass	kilogram	kg	slug	
3. Time	second	s	second	s
4. Acceleration	m/s/s		ft/s/s	
5. Velocity	m/s		ft/s	
6. Force	newton = kg·m/s/s	N	pound = slug-ft/s/s	lb
7. Energy/work	joule = N·m	J	ft lb	
8. Power	watt = J/s	W	watt 0.7376 ft-lb/s	W
9. Current	ampere	A	ampere	A
10. Potential	volt	V	volt	V
11. Resistance	ohm	Ω	ohm	Ω
12. Inductance	henry	H	henry	H
13. Capacitance	farad	F	farad	F
14. Frequency	hertz	Hz	hertz	Hz

APPROXIMATE CONVERSION FACTORS

1 m = 3.281 ft	1 ft = 0.3048 m
1 J = 0.7376 ft-lb	1 ft-lb = 1.356 J
1 kg = 0.06852 slug	1 slug = 14.59 kg
1 N = 0.2248 lb	1 lb = 4.448 N
1 km = 0.6214 mile (mi)	1 mi = 1.609 km
1 cm = 0.3937 inch (in.)	1 in. = 2.54 cm (exact)

Acceleration of gravity at 45° latitude at sea level
9.807 m/s/s   32.174 ft/s/s

Example 1-1

A small hydroelectric plant has a penstock elevation differential of 45.72 m (150 ft), a maximum flow rate of 5.633 m³/s (200 ft³/s), a turbine efficiency of 82%, and a generator efficiency of 95%. What is the plant's maximum energy-generating capacity per second?

Solution If we substitute the hydraulic terms into Eq. (1-1) and multiply the result by electrical efficiency, we obtain

$$W = (2071 \times 10^6) \, 0.95 = 1.967 \times 10^6 \text{ J} \quad \text{(watt-seconds)}$$

When electrical energy is utilized, the order of conversion is reversed. For example, the heating element of an electric range converts all the energy supplied into heat. On the other hand, the filament of an incandescent lamp converts over 90% of the energy supplied into radiated heat and perhaps 2% into light. A storage battery converts approximately 90% of its charging energy into chemical energy stored in its cells. Electric motors convert 65 to 95% of their input energy into the mechanical energy of rotation. Electronic ac-to-dc power supplies convert 50 to 90% of the input interface ac energy into output interface dc energy delivered to the system's external load.

Power is defined as the *rate* of energy transformation. Hence for steady-state system loading, power levels can be obtained simply by dividing the energy terms of Eq. (1-2) by time in seconds, to give

$$P = VI = I^2R = \frac{V^2}{R} \tag{1-3}$$

where P is in watts.

Example 1-2

Returning to the hydroelectric system of Example 1-1, assume that the generator has a rated output of 7200 V. What full-capacity load current must the generator deliver to establish steady-state operating conditions?
Solution From Eq. (1-3) we can write

$$I = \frac{1.967 \times 10^6}{7200} = 273 \text{ A}$$

If the quantities involved in Eqs. (1-2) and (1-3) are not constant, a representative power expression is

$$p = vi \tag{1-4}$$

where p, v, and i now define the *transient* power, voltage, and current functions. Similarly, a representative energy expression (over an analysis period from 0 to t seconds) is as follows:

$$w = \int_0^t p \, dt \tag{1-5}$$

Example 1-3

To illustrate transient behavior, assume that we have a 10-F capacitor initially charged to 4 V. At $t = 0$, a 0.1-Ω resistor is connected across the capacitor instantaneously to form a transient electrical system. The system equations are $v = 4e^{-t}$ and $i = 40e^{-t}$.
 (a) What is the expression for the power dissipated as a function of time, and how much power is being dissipated at $t = 1$ s?
 (b) What is the expression for the energy delivered by the system as a function of time, and how much energy has been supplied at $t = 2$ s?
Solution (a) By substituting in Eq. (1-4) we have

$$p = 160e^{-2t}$$

When $t = 1$, $p = 21.65$ W.
(b) From Eq. (1-5) we can write/integrate as follows:

$$w = \int_0^t (160e^{-2t}) \, dt = 80(1 - e^{-2t})$$

When $t = 2$, $w = 78.53$ J.

1-2 POWER ELECTRONICS

Integrated circuit (IC) chips that contain 1 billion electronic circuit components may be with us by the year 2000. The enormous system capabilities of these devices are difficult to conceive. One thing is certain, however: The power output interface levels of these

ICs will be quite small. When higher-level requirements and functions exist external to the IC, auxiliary power electronic circuits and hardware must be provided to deliver levels and functions that are compatible with overall system performance capabilities.

What is *power electronics*? Typically, this term defines the area of electronic circuit technology where system operating power levels range from a few watts to a few megawatts. Some examples are audio, video, and RF amplifiers; power supplies; motor drives; communications systems; and high-voltage dc power transmission systems.

What is unique about power electronics? At high power levels, energy conversion must be made *efficiently*. For example, if we consider electronic systems that operate at power levels of a few watts, high levels of performance can be achieved by connecting an electronically controlled variable resistance [2] in series with the output interface. The power loss in the series resistance is relatively high; hence system efficiency is relatively low. On the other hand, when we consider systems operating in the kilowatt-to-megawatt region, low-efficiency variable-resistance techniques are no longer acceptable from an *operating cost* point of view. Instead, power electronics systems use on–off switching techniques. System losses are then low in both the on and off states, and system efficiency is relatively high.

1-3 POWER SOURCES

What is a *prime power* system? It is a system, typically designed and owned by a public utility, that delivers prime power service to private and industrial customers. For customers with high-level electronic system loads (falling in the region from hundreds of kilowatts to megawatts) the utility often delivers energy to the user's facility directly from the nearest utility substation.

Analyses of high-level input interface problems require a broad scope of technical capabilities. Clearly, power electronics system personnel require backgrounds in both electronic and power system technology, and similarly, prime power personnel should have backgrounds in both technologies.

In reduced-power-level systems the energy from the utility often passes through one or more lower-level power distribution systems. It is, nevertheless, important for power electronic system personnel to be aware of the characteristics of the overall system and its *other* attached loads. Otherwise, significantly impaired performance capabilities can be experienced when the power electronics system hardware is evaluated. For example, the prime power load frequently consists of a large number of relatively low-level power electronic systems, acting jointly, to produce deteriorated performance capabilities to the overall group of systems. Similarly, power system engineers should be aware of the prime line consequences that a large group of small power electronic system loads introduce on utility lines.

In some cases engine-generator sets provide prime power to the system. If the power electronics system and its associated external load present other than absolutely constant loading on the generator, as before, it is essential that close liaison exist between the power source personnel and the power electronics personnel. This is because engine-generators are typically *less stable* than public utility sources.

In Chapters 2 and 3 we review the *steady-state* characteristics associated with typical prime source/power electronics interfaces. In Chapters 4 and 5 we review the steady-state characteristics associated with typical power electronics/system load interfaces. With this input and output interface background established, in the remainder of the book we address *overall* power electronics system performance in terms of both steady-state and transient analyses.

Chapter 2

Single-Phase Power Circuits

The objectives of this chapter are (1) to review single-phase prime power circuit terminology and some of the major elements associated with a single-phase system, (2) to illustrate how to obtain power system problem solutions by setting up and solving system equations, and (3) to demonstrate three computer programs that simplify the process of obtaining problem solutions.

2-1 PRIME POWER INTERFACE TERMINOLOGY

The significance of the interface between a single-phase prime power system and a power electronics system was outlined in Chapter 1. Five fundamental interface terms and a brief definition of each follow.

Voltage

Voltage is the line-to-line root-mean-square (rms) voltage at the interface.

Current

Current is the rms line current at the interface.

Volt-amperes

This term is defined as the product of the rms line-to-line voltage and the rms line current at the interface.

Power

Power is the product of the volt-amperes and the power factor present at the interface.

Power Factor

This term is defined as power at the interface divided by volt-amperes at the interface. Power factor for a circuit is also defined as the cosine of the angle between circuit voltage and circuit current. These decimal values are often multiplied by 100 and expressed as percent power factor.

Several less easily defined terms that are also of importance in system analyses include the following.

Frequency

Frequency is the prime power source's nominal frequency in hertz and its tolerance as a function of time. These values are of particular significance when engine–generator sources are employed.

Source Impedance

The source impedance is the internal impedance of the prime power source. This may be expressed in terms of actual resistive and reactive values, or it may be given in terms of a source transformer's nameplate ratings.

Voltage Tolerance

The definition of this value is complex. It includes normal line voltage regulation. This regulation is a function of the application and removal of loads on the utility system. Voltage tolerance also includes transient voltage fluctuations such as system load faults, load switching functions, and lightning strokes.

Waveform

The definition of waveform characteristics is also complex. There are two points of view involved: that of the utility and that of the customer. From the customer's perspective, the source may not deliver a pure sine waveform of constant amplitude, due to both the prime generator's characteristics and/or the rapidly fluctuating nature of other utility customer loads. From the utility's perspective, prime line waveform distortion occurs due to rectification and switching functions within the electronic system.

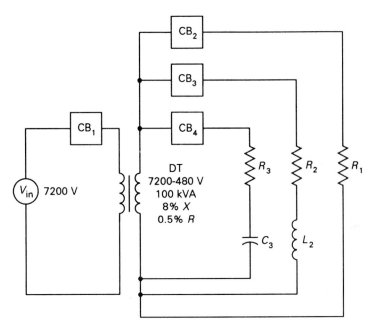

Figure 2-1 Elementary single-phase power network.

We do not address all of the foregoing areas in this chapter. Some of them require transient system analyses—a subject addressed in later chapters.

2-2 SINGLE-PHASE POWER NETWORK CONFIGURATION

Figure 2-1 illustrates an elementary single-phase power system with three load circuits. Distribution transformer DT, in conjunction with its associated high- and low-voltage circuit switching elements, forms a distribution center. The center functions to *control* energy flow to the three load circuits. It must be capable of energizing or deenergizing any of the system circuits—under *normal or abnormal* operating conditions. The following expands on some of the major system elements and prime power source's characteristics.

Circuit Breaker CB 1

Circuit breakers are mechanical devices designed to close or open electrical contact members. These contact members in turn function to close or open their associated electrical circuit. Circuit breakers are equipped with a trip coil that opens the circuit automatically under abnormal (e.g., circuit fault) operating conditions. It is essential to know the worst-case fault current levels, initially to specify, and later to adjust a breaker. Worst-case currents are determined by using the source impedance of the system. In subsequent problems we show how these levels are determined.

CB_1 provides the main incoming prime line circuit protection function for the system (e.g., it protects the prime power line from faults in the incoming feeder line, faults in transformer DT primary circuit, etc.).

Distribution Transformer DT

Through the action of the transformer's step-down turns ratio, the center supplies the required 480-V operating level for the system's three load branches. The transformer's internal impedance is the fundamental quantity that determines 480-V fault current levels.

Circuit Breakers CB_2–CB_4

The center provides individual opening and closing functions, under normal or abnormal operating conditions, for each of the three branch load connection paths. Note that with the three breaker 480-V circuit arrangement shown, any path can be *isolated* from any other path for servicing. This is a fundamental switching requirement for multiple-path power system designs. In addition, the transformer can be isolated for service by opening all breakers.

Prime Power Source

The distribution center is fed from one phase of a public utility substation's 12,470-V 60-Hz three-phase line. Three-phase circuits are described in Chapter 3; however, in this case a 7200-V source is derived simply by a connection between one of the high-voltage lines and the substation's neutral.

Initially, let us assume that the center's location is physically close to the utility substation. Also assume that the substation's installed 12,470-V service capacity is orders of magnitude greater than the load of the 480-V service center; hence the source can be considered to have zero impedance. Based on these simplifications, we will not attempt to define the fault current in CB_1 in our initial single-phase analyses.

2-3 POWER TRANSFORMER CHARACTERISTICS

Before proceeding with system analyses, we review some background on transformers and their nameplate ratings (i.e., as shown beside DT in Fig. 2-1). Electrical performance capabilities of transformers can be derived from the value of *mutual coupling* that exists between the primary and secondary windings. The following expression defines the parameters involved:

$$M = k(L_1 L_2)^{1/2} \qquad (2\text{-}1)$$

where M is the mutual inductance between the two windings, L_1 the *decoupled* circuit inductance of the electrically excited winding, L_2 the *decoupled* circuit inductance of the secondary winding, and k the *coefficient of coupling* ($k = 0$ for uncoupled circuits and $k = 1$ for totally coupled circuits).

To increase the mutual coupling to levels suitable for prime power–coupled circuits,

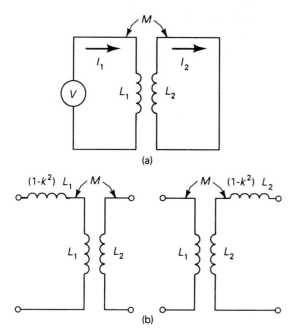

Figure 2-2 Power transformer testing and equivalent circuit: (a) transformer short-circuit test; (b) alternate transformer circuits.

the coils are wound on appropriate types of metallic cores. Almost all of the magnetic flux lines generated by winding L_1 will then pass through the core. This produces nearly the maximum attainable value of mutual inductance and also a value of k nearly equal to 1. The limited number of flux lines that do not pass through the core produce *leakage inductance*.

Leakage inductance in a transformer is determined quantitatively by means of a short-circuit test. A simplified diagram of the test arrangement is shown in Fig. 2-2a. In this test the terminals of the low-voltage winding are usually short circuited. A lower-that-rated value of voltage is applied to the high-voltage winding, and this voltage is adjusted until rated current flows in the shorted winding.

It can be seen from Fig. 2-2a that the value of current in the source mesh will consist of two components: (1) it will include the current through L_1 due to the source voltage V alone, and (2) it will include the current in L_1 due to mutual coupling from shorted winding L_2. Hence we can write a Kirchhoff voltage loop equation around the source mesh as follows:

$$j\omega L_1 I_1 + j\omega M I_2 - V = 0 \tag{2-2}$$

where source voltage V is a sine wave of f hertz, the *radian frequency* $\omega = 2\pi f$, the primary circuit reactance is $j\omega L$, and the secondary circuit-coupled reactance is $j\omega M$. Similarly, for the other mesh equation we have

$$j\omega L_2 I_2 + j\omega M I_1 = 0 \tag{2-3}$$

When Eqs. (2-2) and (2-3) are solved simultaneously,

$$V = j\omega I_1(L_1 - \frac{M^2}{L_2}) \tag{2-4}$$

If the value of M from Eq. (2-1) is substituted into Eq. (2-4), we have

$$V = j\omega L_1 I_1 (1 - k^2) \tag{2-5}$$

The input impedance of the test setup can be obtained simply by dividing the source voltage by the source current. This gives

$$Z_{in} = j\omega L1(1 - k^2) \tag{2-6}$$

The leakage inductance referred to the primary winding is then

$$L_{Lp} = L_1(1 - k^2) \tag{2-7}$$

and referred to the secondary winding it is

$$L_{Ls} = L_2(1 - k^2) \tag{2-8}$$

Through the use of Eq. (2-7) or (2-8) the power transformer can also be drawn as shown in Fig. 2-2b. In these configurations the coefficient of coupling between L_1 and L_2 is unity.

In Fig. 2-1 leakage reactance value X (i.e., ω times the leakage inductance) and series resistance R are expressed as percentages of the a full-load impedance of the transformer DT. In power transformers there is always a small exciting current present (its customary value is 1 or 2% of the full-load current). If we neglect this current, the full-load impedance value is, in terms of the transformer's ratings,

$$Z = \frac{(\text{rated voltage})^2}{\text{rated volt-amperes}} \tag{2-9}$$

Hence the quantitative values of series reactance and series resistance are

$$X_S = (\%X) Z \qquad R_S = (\%R) Z \tag{2-10}$$

These figures can be referred to either the primary or the secondary winding, depending on the area of analysis desired.

2-4 PRELIMINARY DISTRIBUTION CENTER ANALYSIS

With the background areas noted above established, we can proceed with an analysis of the network configuration of Fig. 2-1. If we refer the analysis the secondary winding of DT, our diagram can be drawn as shown in Fig. 2-3. Equivalent values of load impedance, leakage reactance, and series resistance become, using Eqs. (2-9) and (2-10),

$$Z = \frac{480^2}{100 \times 10^3} = 2.304 \ \Omega$$

$$X_4 = 0.08 \times 2.304 = 0.18432 \ \Omega$$

$$R_4 = 0.005 \times 2.304 = 0.01152 \ \Omega$$

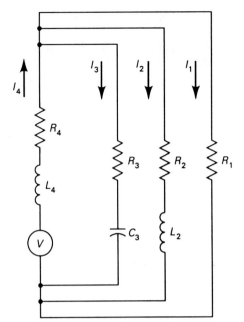

Figure 2-3 Analysis diagram for single-phase power network.

We are now in a position to determine the fault current in the 480-V circuits. The obvious worst case is a zero-impedance short in the secondary circuit of DT. Hence

$$\text{fault current} = \frac{480}{(0.18432^2 + 0.01152^2)^{1/2}}$$
$$= 2598 \text{ A}$$

Circuit breakers CB_2–CB_4 must be rated to clear at least this level of fault current.

Values of load circuit inductance and capacitance are usually expressed as their equivalent reactive values (i.e., $X = j\omega L$ or $X = -1/j\omega C$). The distribution center loads then have the following values:

$$R_1 = 9.216 \text{ }\Omega$$
$$R_2 = 3.6864 \text{ }\Omega$$
$$X_2 = jwL_2 = 2.7648 \text{ }\Omega$$
$$R_3 = 7.3728 \text{ }\Omega$$
$$X_3 = \frac{-1}{j\omega C_3} = -5.5296 \text{ }\Omega$$

We now have all of the necessary information available. In the following we first analyze the performance of the individual load paths. In Section 2-5 we perform an overall analysis of the distribution center.

Analysis of Branch 1

The simplest distribution center load occurs when only CB_1 and CB_2 in Figure 2-1 are closed and the transformer is loaded by R_1 alone. A practical load branch of this nature might consist of the incandescent lighting and resistance heating equipment used for a facility.

By referring to Figure 2-3 and using Kirchhoff's voltage law, we can write the loop equation as follows:

$$[(R_4 + R_1) + j\omega L_4]I_1 = 480 + j0$$

where, for ease of reference, the source voltage has a reference angle of zero degrees. The term in square brackets has a value of

$$Z = [(0.01152 + 9.216)^2 + (0.18432)^2]^{1/2} = 9.2294 \ \Omega$$

and its phase angle is

$$\theta = \arctan \frac{0.18432}{0.01152 + 9.612} = 1.1443°$$

We can now proceed to define the path's interface capabilities as follows:

$$\text{load current} = I_1 = \frac{480 \text{ at } 0°}{9.2294 \text{ at } 1.1443°}$$
$$= 52.0079 \text{ A at } -1.1443°$$

$$\text{power factor} = \cos(-1.1443°) = 0.9998 \quad \text{or} \quad 99.98\%$$

$$\text{load voltage} = I_1 R_1 = (52.0079 \text{ at } -1.11443°)(9.216)$$
$$= 479.305 \text{ V at } -1.1443°$$

$$\text{utility volt-amperes} = 52.0079(480) = 24{,}964 \text{ VA}$$

$$\text{utility power} = \text{VA (PF)} = 24{,}964(0.9998) = 24{,}959 \text{ W}$$

$$\text{load voltage regulation} = \frac{100(480 - 479.305)}{479.305} = 0.1450\%$$

Analysis of Branch 2

In Fig. 2-1 if we now close breakers CB_1 and CB_3, we encounter a considerably less desirable system load. For a representative electronic system installation, this might be typical of the 60-Hz loading from a group of dc power supplies. Proceeding as before, the loop equation is

$$[(0.01152 + 3.6864) + j(0.18432 + 2.7648)]I_2 = 480 + j0$$

The form of solution noted above yields the following interface capabilities:

$$\text{load current} = 101.4821 \text{ A at } -38.5726°$$

$$\text{power factor} = \cos(-38.5726°) = 0.7818 \text{ or } 78.18\%$$

$$\text{load voltage} = 467.6295 \text{ V at } -1.7027°$$

$$\text{utility volt-amperes} = 48{,}711 \text{ VA}$$

$$\text{utility power} = 38{,}083 \text{ W}$$

$$\text{load voltage regulation} = 2.64367\%$$

The public utility cost of generating and transmitting power is a function of the current level involved. With an inductive load such as this, only about 78% of the volt-amperes transmitted is converted into power. The line current is therefore about 28% higher than it would be with a unity power factor load. As a result, public utilities assess a *power service cost* multiplier when the customer's load power factor is lower than perhaps 95%. A typical power factor correction technique is reviewed in the next section.

Analysis of Branch 3

This branch consists of resistance R_3 in series with capacitance C_3. The R_3–C_3 series circuit was used to replace a lagging power factor load in parallel with a power factor correction element. Circuit background is as follows.

Initial branch properties. Initially, the branch consisted of a resistance of 5.76 Ω connected in series with an inductive reactance of 5.76 Ω—elements with an atrocious power factor of 70.7%.

Correction procedure. The normal procedure for correcting lagging power factors is to connect capacitance across the power line. In this case the owner made a correction sufficient to raise the *overall* system power factor above 95%. He lumped a reactance of $-j6.582857$ Ω (402.953μF) at this location.

Example 2-1

Based on the initial series values of $5.76 + j5.76$ connected in parallel with correction capacitor $-j6.582857$, how were the original series equivalent values of R_3 and X_3 determined?

Solution One approach is to use a simple series–parallel conversion computer program and proceed as follows:

1. Load Appendix 1. The upper diagram of Fig. 2-4 shows the circuit. Our approach is first to convert the series L and R branch to its parallel equivalent. Issue a <RUN> command to the computer. Respond to monitor prompt 1 with the letter N followed by a <RETURN> key operation. For prompt 2 enter letters SP and <RETURN>. For prompts 3 and 4 enter the series resistive and reactive values, each followed by <RETURN>. The converted equivalent parallel circuit values are then displayed on lines 5 and 6.

2. We now have three elements in parallel, as shown in the second figure. Our next step is to combine the two parallel reactances. On the second run respond to prompt 1 with letter Y and <RETURN>, prompt 2 with P and <RETURN>, prompt 3 with the inductive reactance and <RETURN>, and prompt 4 with the capacitive reactance and <RETURN>. The resultant parallel reactance is then printed out on line 5.

3. We now have two elements in parallel. Our next step is to convert these two values into their series equivalent circuit. In the third run simply respond as shown. Lines 5 and 6 show the desired set of series values that we have assigned for use in our analysis. This easy-

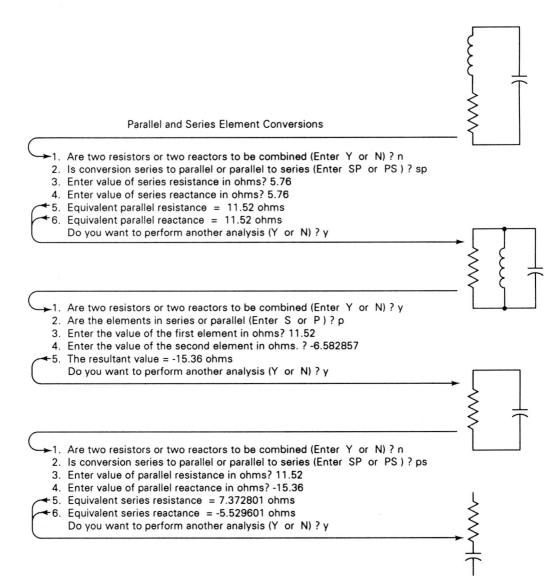

```
        Parallel and Series Element Conversions
 ┌─
 └→1. Are two resistors or two reactors to be combined (Enter  Y  or  N) ? n
    2. Is conversion series to parallel or parallel to series (Enter  SP  or  PS ) ? sp
    3. Enter value of series resistance in ohms? 5.76
    4. Enter value of series reactance in ohms? 5.76
  ←5. Equivalent parallel resistance  =  11.52 ohms
  ←6. Equivalent parallel reactance  =  11.52 ohms
       Do you want to perform another analysis (Y  or  N) ? y
 ─┘


 ┌─
 └→1. Are two resistors or two reactors to be combined (Enter  Y  or  N) ? y
    2. Are the elements in series or parallel (Enter  S  or  P ) ? p
    3. Enter the value of the first element in ohms? 11.52
    4. Enter the value of the second element in ohms. ? -6.582857
  ←5. The resultant value = -15.36 ohms
       Do you want to perform another analysis (Y  or  N) ? y
 ─┘


 ┌─
 └→1. Are two resistors or two reactors to be combined (Enter  Y  or  N) ? n
    2. Is conversion series to parallel or parallel to series (Enter  SP  or  PS ) ? ps
    3. Enter value of parallel resistance in ohms? 11.52
    4. Enter value of parallel reactance in ohms? -15.36
  ←5. Equivalent series resistance  =  7.372801 ohms
  ←6. Equivalent series reactance  =  -5.529601 ohms
       Do you want to perform another analysis (Y  or  N) ? y
 ─┘
```

Figure 2-4 Example of parallel and series circuit conversions with Appendix 1.

to-use program has been found to be an effective analysis tool for numerous other purposes in addition to power factor analyses.

Now, returning to the analysis, we can write the circuit equation for the composite branch as

$$[(0.01152 + 7.3728) + j(0.18432 - 5.5296)]I_3 = 480 + j0$$

and the resultant interface capabilities are:

$$\text{line current} = 52.6550 \text{ A at } 35.8996°$$

$$\text{power factor} = \cos(35.8996°) = 0.8100 = 81.00\%$$
$$\text{load voltage} = 485.269 \text{ V at } -0.9703°$$
$$\text{utility volt-amperes} = 25{,}274\text{VA}$$
$$\text{utility power} = 20{,}473\text{W}$$
$$\text{load voltage regulation} = -1.0858\%$$

Two things are of interest here: (1) for this branch alone the power factor is vastly overcorrected, and (2) a negative voltage regulation results; that is, the voltage rises above the no-load line voltage. The composite results of these two effects are shown in the following section.

2-5 GENERAL-PURPOSE STEADY-STATE ANALYSIS PROGRAM

If we used a traditional overall analysis approach, we would now document a set of system equations and solve them simultaneously. A much more expedient approach is to perform the analysis on a personal computer. The computer program we will use was developed by Bert Erickson [3]. Appendix 2 is a variation/extension of his published program.

A significant feature of the program from a user's point of view is that it is not necessary to set up system equations. This straightforward program's rationale and operation can be summarized as follows. *Source:* A zero-impedance single-phase generator functions to drive the system. One side of the generator attaches to the system's reference node (number 0) and the other side attaches to system node 1. *Diagram:* Before using the program, sketch the circuit diagram. Number all nodes sequentially, starting with node 1. *Input loading*: The program is then loaded into the computer and a <RUN> command is issued. In accordance with computer prompts, all technical information (element symbols, node numbers, and values) is entered into the program from the computer keyboard. *Running*: One of three forms of analysis is then selected: (1) an output voltage/gain versus frequency, (2) an impedance versus frequency, or (3) power system performance at a fixed frequency. Finally, program output response details are supplied in response to further computer screen prompts. The system then proceeds with the analysis, and requested results are displayed.

Appendix 2 Analysis

To illustrate how Appendix 2 functions, let us proceed to solve the distribution center problem. Figure 2-5a shows how to perform the first step—the diagram is sketched. Next, the nodes are labeled as shown by the encircled numbers. The zero-voltage return bus to the generator has its node labeled 0. Then the source output is labeled node 1, and so on. The line currents shown are for reference and are not required from a programming point of view.

Load the program and enter a <RUN> command. You have a choice of a computer screen printout or a combined screen and printer presentation of the solution by responding to the computer prompts. While progressing through the program operation steps that follow, refer to Fig. 2-5b.

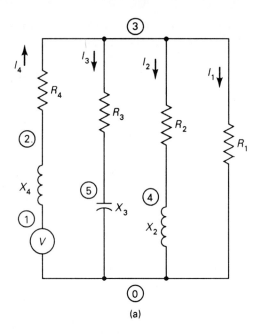

The input information for this analysis is as follows:

(*Note:* Separate all multiple-entry information with commas.)

1. Number of network nodes =? 5
2. Enter element symbol ? X4
   Nodes and ohms are? 1,2,.18432
   At what frequency was reactance measured? 60
3. Enter element symbol ? R4
   Nodes and ohms are? 2,3,.01152
4. Enter element symbol ? R1
   Nodes and ohms are? 3,0,9.216
5. Enter element symbol ? R2
   Nodes and ohms are? 3,4,3.6864
6. enter element symbol ? X2
   Nodes and ohms are? 4,0,2.7648
   At what frequency was reactance measured? 60
7. Enter element symbol ? R3
   Nodes and ohms are? 3,5,7.3728
8. Enter element symbol ? X3
   Nodes and ohms are? 5,0,-5.5296
   At what frequency was reactance measured? 60
9, Enter element symbol ? E

(b)

Figure 2-5 Procedure for using Appendix 2 analysis program: (a) analysis diagram element and node labeling; (b) entry of circuit parameters into computer.

Number of nodes. Your response to "number of nodes" should not include zero-voltage reference node 0—for this analysis the node number is 5. Entering too low a number will cause the program to cease when a higher node number is encountered. Entering too high a number will increase the operating time for program execution. Follow the number with a <RETURN> key operation.

Entering elements. Element entry information is either a two- or three-step process. For the X_4 case, this symbol is entered followed by a <RETURN> key operation. Then the nodes and the element value are entered (with appropriate commas) followed by a <RETURN> key operation. Finally, in the case of X entries only, the frequency at which the X value is defined is entered followed by a <RETURN> key operation. All other elements from 3 to 8 are entered in similar fashion. In response to 9, the letter E is entered, followed by a <RETURN> key operation. This signifies that the data-entry procedure is complete.

Type of analysis. A series of program screen prompts request the type of analysis to be performed. Respond to each of the first two with the letter N and <RETURN> key operations. Respond to the third with the letter Y and a <RETURN> key operation. If you are using a printer display, the heading and the list of elements shown in Fig. 2-6 will be printed out as you enter the previously described element input information.

Source voltage. In response to (1) in Fig. 2-6, enter the analysis voltage of 480 and a <RETURN> key operation.

Frequency. In response to prompt (2), enter the frequency at which the analysis is to be performed followed by a <RETURN> key operation. The frequency does not have to correspond with the value at which any X elements were entered since X values are converted to equivalent L and C elements by the program.

First node. In response to (3), respond with the first node number of the particular branch of the circuit to be analyzed. In this case the transformer reactance/resistance branch is going to be analyzed; hence 1 is entered as the first node, followed by a <RETURN> key operation. There will then be a time pause as the program solves for the voltage between nodes 1 and 0.

Second node. In similar fashion, (4) requests the second node number of the branch. Enter 3 followed by a <RETURN> key operation. The program calculates the voltage between node 3 and 0 and the difference in voltage between the two nodes. Then (5) automatically prints the voltage drop across the branch and the phase angle between this voltage and current.

Branch elements. Prompts (6) and (7) request the element values of the branch. Enter the information and follow by <RETURN> key operations. Solutions appear in (8)–(11). By taking the cosine of line current phase in (8), we can establish the overall system power factor as 97.12%.

```
The Input Information For This Analysis Is As Follows:
    1. Number of network nodes = 5
    2. X4 = 1 , 2 , .18432
    3. R4 = 2 , 3 , .01152
    4. R1 = 3 , 0 , 9.215999
    5. R2 = 3 , 4 , 3.6864
    6. X2 = 4 , 0 , 2.7648
    7. R3 = 3 , 5 , 7.3728
    8. X3 = 5 , 0 ,-5.5296
    9. End of list
```

```
This Is A Branch Power Analysis Where:
   (1) Source voltage = 480
   (2) Frequency = 60 Hz
   (3) The first branch node is 1
   (4) The second branch node is 3
   (5) Branch voltage = 32.60728 @ 72.62985 degrees
   (6) Branch resistance = .01152 Ohms
   (7) Branch reactance = .18432 Ohms
   (8) Branch current = 176.5613 @-13.79371 degrees
   (9) Branch volt-amperes = 5757.183
  (10) Branch power = 359.123 W
  (11) Branch power factor = 6.237826E-02
```

```
This Is A Revised Branch Power Analysis Where:
   (3) The first branch node is 3
   (4) The second branch node is 0
   (5) Branch voltage = 471.294 @-3.78608 degrees
   (6) Branch resistance = 9.215999 Ohms
   (7) Branch reactance = 0 Ohms
   (8) Branch current = 51.13867 @-3.78608 degrees
   (9) Branch volt-amperes = 24101.35
  (10) Branch power = 24101.35 W
  (11) Branch power factor = 1
```

```
This Is A Revised Branch Power Analysis Where:
   (3) The first branch node is 3
   (4) The second branch node is 0
   (5) Branch voltage = 471.294 @-3.78608 degrees
   (6) Branch resistance = 3.6864 Ohms
   (7) Branch reactance = 2.7648 Ohms
   (8) Branch current = 102.2773 @-40.65593 degrees
   (9) Branch volt-amperes = 48202.69
  (10) Branch power = 38562.15 W
  (11) Branch power factor = .8
```

```
This Is A Revised Branch Power Analysis Where:
   (3) The first branch node is 3
   (4) The second branch node is 0
   (5) Branch voltage = 471.294 @-3.78608 degrees
   (6) Branch resistance = 7.3728 Ohms
   (7) Branch reactance =-5.5296 Ohms
   (8) Branch current = 51.13867 @ 33.08378 degrees
   (9) Branch volt-amperes = 24101.35
  (10) Branch power = 19281.07 W
  (11) Branch power factor = .8
```

Figure 2-6 Appendix 2 solution for 480-V distribution system.

The program then requests if another solution is desired. A letter Y response, followed by a <RETURN> key operation, followed by responses to additional prompts will result in the second solution, shown for the R_1 branch. Similar procedures will result in the third set of solutions for the R_2–X_2 branch, and for the fourth set of solutions for the R_3–X_3 branch. These corrected overall system branch currents depart slightly from the previous individual branch values.

2-6 POWER SYSTEMS WITH TWO SOURCES

The failure of a key element in a single-source power distribution center usually results in power loss for an unacceptably long period. One solution to this problem is to use a system design that provides a *backup* source of power. To illustrate the backup source approach, suppose that the distribution system's owner plans to add a second load center to the system and that the new center will be located 304.8 m (1000 ft) from the original center. The design approach for the system is outlined below.

Principle

Both centers will be effectively connected in parallel.

Mode of Operation

In the event of a 480-V power outage at the first center, the second center will provide an uninterrupted source of power to both locations. Note that in this approach it is assumed that a failure in the 7200-V source is not the reason for the outage.

Original Center Diagram

Figure 2-7 shows the arrangement to be investigated. The original center is at the left. It consists of the 7200/480-V 100-kVA transformer, its leakage reactance X_1, and its winding resistance R_1 (both referred to the secondary winding). The composite of the three original loads are represented by X_2 and R_2.

New Center Diagram

The new load center is shown at the right. It consists of a 7200/480-V 250-kVA transformer (i.e., rating increased to handle the load of both centers), leakage reactance X_3, winding resistance R_3 (again referred to the secondary), and composite load X_4 and R_4.

Transmission Line

Because of the significant distance between the two centers it is necessary to take the system's transmission line characteristics into consideration. The 480-V transmission line's electrical characteristics are denoted by X_5 and R_5. The 7200-V line's characteristics are denoted by R_H and X_H.

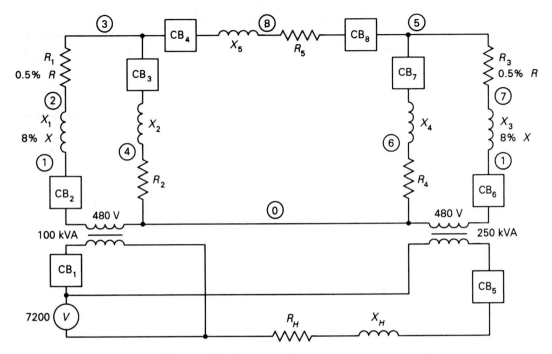

Figure 2-7 Diagram for two 480-V centers with backup line.

Definition of System Element Values

The 100-kVA transformer characteristics have previously been established as

$$X_1 = 0.18432 \, \Omega$$
$$R_1 = 0.01152 \, \Omega$$

We can readily sum up the parallel equivalent value of the three system loads (e.g., use Appendix 1). The second distribution center loads with the same values. They are

$$X_2 = X_4 = 0.464 \, \Omega$$
$$R_2 = R_4 = 2.629 \, \Omega$$

Now consider the 250-kVA transformer. From Eq. (2-9) the full-load impedance is

$$Z = \frac{480^2}{250 \times 10^3} = 0.9216 \, \Omega$$

and from Eq. (2-10) we have

$$X_3 = 0.08 \times 0.9216 = 0.073728 \, \Omega$$
$$R_3 = 0.005 \times 0.9216 = 0.004608 \, \Omega$$

For short 60-Hz transmission lines (e.g., with lengths less than perhaps 50 km) it is considered acceptable to use only the line inductance and resistance in analyses such as this. The inductance of each conductor in the line is

$$L = k(\ln\frac{D}{r} + 0.25) \times 10^{-8} \text{ H per unit length} \qquad (2\text{-}11)$$

where $k = 20$ for meter units (6.096 for feet), ln is the natural log, D the center-to-center spacing of the parallel conductors, and r the conductor radius (both in the same units of measure). To simplify analysis, say that both the 7200-and 480-V lines have conductor spacings of 30.48 cm (12 in.), and use size 4/0 conductor of radius 0.5842 cm (0.230 in.). Then

$$L = 304.8 \times 20(\ln\frac{30.48}{0.5842} + 0.25) \times 10^{-8} = 0.25631 \times 10^{-3} \text{ H}$$

and accounting for the two conductors of each line, we have

$$X_S = X_H = 2wL = 4\pi(60)\, 0.25631 \times 10^{-3} = 0.193 \,\Omega$$

From wire tables, copper 4/0 conductors have a resistance of 0.1851 Ω/km (0.298 Ω/mi). Hence for these cases,

$$R_S = R_H = 2\frac{304.8}{1000}\, 0.1851 = 0.113 \,\Omega$$

If we assume that the 7200-V source is at the original distribution center, we can reflect the high-voltage line's inductance and resistance through the distribution transformer at the new distribution center. These equivalent low-voltage-side values can be added to X_3 and R_3. The reflected values are obtained from

$$X_R = \frac{X_H}{N^2} \qquad R_R = \frac{R_H}{N^2} \qquad (2\text{-}12)$$

Based on the transformer's 15:1 turn ratio, quantitative figures are

$$X_R = \frac{0.193}{15^2} = 0.000858 \,\Omega$$

$$R_R = \frac{0.113}{15^2} = 0.000502 \,\Omega$$

and the corrected values of X_3 and R_3 become

$$X_3 = 0.074596 \,\Omega$$

$$R_3 = 0.00511 \,\Omega$$

Based on the values above, the fault current is

$$\text{fault current} = \frac{480}{(0.074596^2 + 0.00511^2)^{1/2}}$$
$$= 6420 \text{ A}$$

```
The Input Information For This Analysis Is As Follows:
--- ----- ----------- --- ---- -------- -- -- --------
         1. Number of network nodes = 8
         2. X1 = 1 , 2 , .18432
         3. R1 = 2 , 3 , .01152
         4. X2 = 3 , 4 , .464
         5. R2 = 4 , 0 , 2.629
         6. X3 = 1 , 7 , .074596
         7. R3 = 7 , 5 , .00511
         8. X4 = 5 , 6 , .464
         9. R4 = 6 , 0 , 2.629
        10. R5 = 5 , 8 , .113
        11. X5 = 8 , 3 , .193
        12. End of list
-------------------------------------------------------
This Is A Branch Power Analysis Where:
    (1) Source voltage = 480
    (2) Frequency = 60 Hz
    (3) The first branch node is 5
    (4) The second branch node is 3
    (5) Branch voltage = 9.183969 @ 58.75055 degrees
    (6) Branch resistance = .113 Ohms
    (7) Branch reactance = .193 Ohms
    (8) Branch current = 41.06457 @-.9007369 degrees
    (9) Branch volt-amperes = 377.1358
   (10) Branch power = 190.5518 W
   (11) Branch power factor = .5052604
-------------------------------------------
This Is A Revised Branch Power Analysis Where:
    (3) The first branch node is 3
    (4) The second branch node is 0
    (5) Branch voltage = 471.934 @-2.884225 degrees
    (6) Branch resistance = 2.629 Ohms
    (7) Branch reactance = .464 Ohms
    (8) Branch current = 176.7786 @-12.89344 degrees
    (9) Branch volt-amperes = 83427.83
   (10) Branch power = 82158.04 W
   (11) Branch power factor = .9847798
-------------------------------------------
This Is A Revised Branch Power Analysis Where:
    (3) The first branch node is 1
    (4) The second branch node is 3
    (5) Branch voltage = 25.27791 @ 69.95571 degrees
    (6) Branch resistance = .01152 Ohms
    (7) Branch reactance = .18432 Ohms
    (8) Branch current = 136.8744 @-16.46785 degrees
    (9) Branch volt-amperes = 3459.897
   (10) Branch power = 215.8223 W
   (11) Branch power factor = 6.237826E-02
-------------------------------------------
This Is A Revised Branch Power Analysis Where:
    (3) The first branch node is 5
    (4) The second branch node is 0
    (5) Branch voltage = 476.3657 @-1.912183 degrees
    (6) Branch resistance = 2.629 Ohms
    (7) Branch reactance = .464 Ohms
    (8) Branch current = 178.4387 @-11.92139 degrees
    (9) Branch volt-amperes = 85002.07
   (10) Branch power = 83708.33 W
   (11) Branch power factor = .9847798
-------------------------------------------
```

```
This Is A Revised Branch Power Analysis Where:
   (3) The first branch node is 1
   (4) The second branch node is 5
   (5) Branch voltage = 16.3666 @ 76.21582 degrees
   (6) Branch resistance = .00511 Ohms
   (7) Branch reactance = .074596 Ohms
   (8) Branch current = 218.8902 @-9.865301 degrees
   (9) Branch volt-amperes = 3582.488
  (10) Branch power = 244.8349 W
  (11) Branch power factor = 6.834215E-02
-------------------------------------------------
```

Figure 2-8 Analysis of backup power system with both sources activated.

This value establishes the minimum interrupting rating of the 480-V circuit breakers used in the new load center design.

Analysis of Normal System Operation with Both Centers Supplying Load Power

To perform this analysis load (Appendix 2), issue a <RUN> command and enter the calculated element values. The input display at the top of Figure 2-8 will then appear. Select the power analysis option and respond to the branch power analyses prompts. The remainder of the run should result. A review of the major analysis results shows the following.

First analysis. The 480-V transmission line assumes a current value of 41.06 A. Since the 250-kVA transformer has lower impedance (compared to the 100-kVA unit), some current is fed from the new center to the original center. This presents no problem insofar as system operation is concerned.

Second analysis. The voltage across the load at the original center has risen a few tenths of a volt, due to the transmission line feedback from the new center. This is a secondary advantage of the new approach.

Third analysis. The current delivered by the 100-kVA transformer has dropped off significantly, due to the feedback from the new center. This is an additional advantage of the new approach—the transformer is less heavily loaded.

Fourth analysis. The line voltage at the new center is somewhat higher than it was at the original center. The lower impedance of the 250-kVA transformer provides this further advantage for the approach.

Fifth analysis. The current delivered by the 250-kVA transformer is well within its ratings.

Summary. The runs show that for dual-source operation the backup power system is operating satisfactorily.

```
The Input Information For This Analysis Is As Follows:
    1. Number of network nodes = 7
    2. X2 = 3 , 4 , .464
    3. R2 = 4 , 0 , 2.629
    4. X3 = 1 , 2 , .074596
    5. R3 = 2 , 5 , .00511
    6. X4 = 5 , 6 , .464
    7. R4 = 6 , 0 , 2.629
    8. R5 = 5 , 7 , .113
    9. X5 = 7 , 3 , .193
   10. End of list
```

```
This Is A Branch Power Analysis Where:
    (1) Source voltage = 480
    (2) Frequency = 60 Hz
    (3) The first branch node is 5
    (4) The second branch node is 3
    (5) Branch voltage = 37.47233 @ 43.2156 degrees
    (6) Branch resistance = .113 Ohms
    (7) Branch reactance = .193 Ohms
    (8) Branch current = 167.5512 @-16.4357 degrees
    (9) Branch volt-amperes = 6278.532
   (10) Branch power = 3172.294 W
   (11) Branch power factor = .5052604
```

```
This Is A Revised Branch Power Analysis Where:
    (3) The first branch node is 3
    (4) The second branch node is 0
    (5) Branch voltage = 447.2998 @-6.426515 degrees
    (6) Branch resistance = 2.629 Ohms
    (7) Branch reactance = .464 Ohms
    (8) Branch current = 167.5511 @-16.43573 degrees
    (9) Branch volt-amperes = 74945.58
   (10) Branch power = 73804.9 W
   (11) Branch power factor = .9847798
```

```
This Is A Revised Branch Power Analysis Where:
    (3) The first branch node is 5
    (4) The second branch node is 0
    (5) Branch voltage = 472.4291 @-2.961346 degrees
    (6) Branch resistance = 2.629 Ohms
    (7) Branch reactance = .464 Ohms
    (8) Branch current = 176.9641 @-12.97056 degrees
    (9) Branch volt-amperes = 83603
   (10) Branch power = 82330.55 W
   (11) Branch power factor = .9847798
```

```
This Is A Revised Branch Power Analysis Where:
    (3) The first branch node is 1
    (4) The second branch node is 5
    (5) Branch voltage = 25.748 @ 71.42522 degrees
    (6) Branch resistance = .00511 Ohms
    (7) Branch reactance = .074596 Ohms
    (8) Branch current = 344.3589 @-14.6559 degrees
    (9) Branch volt-amperes = 8866.552
   (10) Branch power = 605.9593 W
   (11) Branch power factor = 6.834215E-02
```

Figure 2-9 Analysis of backup power system with 250-kVA transformer alone.

Backup Analysis with Only the 250-kVA Center Activated

Our primary objective is to determine how the system functions in its backup role. We can simulate this situation if circuit breaker CB_2 at the output node of the 100-kVA transformer is opened. To run this analysis, remove X_1 and R_1 from the diagram and make a few minor node changes. When Appendix 2 is run again, the revised input parameters should look like the display at the top of Fig. 2-9. The revised set of runs show the following.

First analysis. The 480-V transmission line current has assumed a substantial value of 167.6 A and there is a 3.172-kW loss in the line.

Second analysis. The voltage at the original center has dropped to 447.3 V. It must be kept in mind that this is a backup system and that some losses of voltage and power in the transmission line are inevitable.

Third analysis. The line voltage at the new center has retained an acceptable value of 472.4 V.

Fourth analysis. The line current of the 250-kVA transformer is 344.4 A, well within ratings.

Summary. Except for the voltage decrease of about 5% at the original load center, the design is a complete success. As a temporary measure while the necessary repairs are in progress at the original distribution center, the taps at the 250-kVA transformer could be set to increase its output voltage. This would result in somewhat higher-than-normal voltage at the new center and more nearly normal voltage at the original center. An alternative (and more costly) approach would be to redesign the 480-V transmission line for lower losses.

Chapter 3

Three-Phase Power Circuits

Why do we use three-phase power circuits? They are *cost-effective*. To illustrate, late in the nineteenth century, power industry analysts investigated alternative approaches for transmitting electrical energy over long distances. Their quantitative analyses showed that for equal values of transmission line length, equal transmitted power levels, equal line-to-line voltages, and equal power losses, three-phase lines require one-third *less copper* (cubic volume) than do equivalent single-phase lines.

In addition, the early analysts demonstrated the cost-effectiveness of three-phase generator and motor designs. For a given physical size, three-phase machines have 150% *higher output capability* than single-phase machines. In addition, three-phase motors develop almost *constant torque*—like a steam turbine. Single-phase motors deliver pulsating torque—like a reciprocating engine. In power electronics an equivalent of the three-phase constant-torque effect is present. With three-phase ac energy sources, the unfiltered output of power rectifier circuits maintains a relatively constant dc voltage level. With single-phase ac energy sources, wide fluctuations occur in the unfiltered dc output voltage. Hence based on equivalent performance, the most cost-effective designs result from systems using three-phase energy sources. In subsequent chapters we demonstrate a number of additional electrical and mechanical advantages that occur when three-phase power electronics systems are used.

In this chapter we review (1) balanced three-phase systems with delta- and wye-connected load circuits; (2) balanced three-phase circuit voltages, currents, and power relationships; (3) delta-to-wye and wye-to-delta circuit conversion; (4) unbalanced three-phase circuit analysis; (5) computer programs that significantly reduce analysis complexity, and (6) major system element characteristics.

3-1 BALANCED THREE-PHASE SYSTEMS

A three-phase system is simply several single-phase systems that are displaced in time phase from one another. Initially, we review the fundamental delta and wye interconnections associated with these single-phase elements.

System with Wye Source and Wye Load

Figure 3-1 shows the most fundamental three-phase system arrangement. The three-phase source—almost always represented as three wye-connected single-phase generators—is shown at the left of the figure. The center of the wye source is referred to as its *neutral* connection.

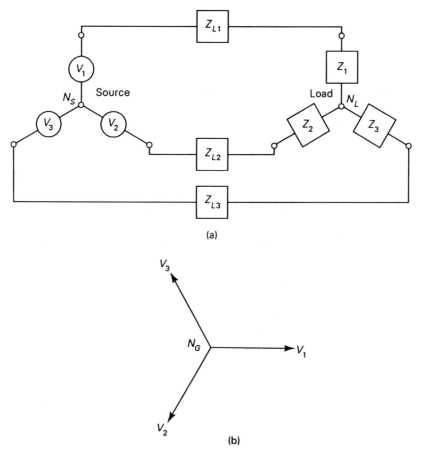

Figure 3-1 Three-phase balanced wye network and phasor diagram: (a) balanced three-phase network with wye source and wye load; (b) phasor diagram of balanced three-phase source voltages.

Sec. 3-1   Balanced Three–Phase Systems   29

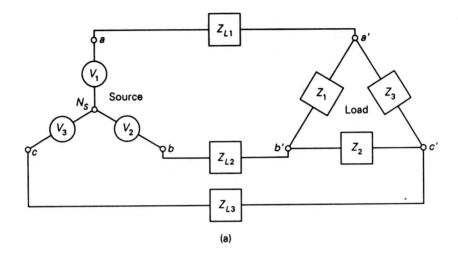

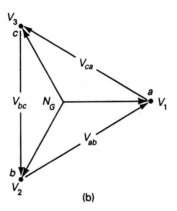

Figure 3-2 Balanced wye-to-delta network and voltage diagram: (a) balanced three-phase network with wye source and delta load; (b) wye-generated voltages and delta line voltages at source.

The *phasor* diagram of Fig. 3-1b shows instantaneous values of the generator voltage relationships (their peak scalar values are defined as $V_1 = V_2 = V_3 = V$). At the specific point in time shown in the diagram, the complex values of these three voltages are

$$V_1 = V + j0 \tag{3-1}$$

$$V_2 = -0.5\,V - j\left(\frac{3^{1/2}}{2}\right) V \tag{3-2}$$

$$V_3 = -0.5\,V + j\left(\frac{3^{1/2}}{2}\right) V \tag{3-3}$$

As an ac generator's windings are rotated by the system's prime mover, the phasors rotate in like manner. The machine develops new complex values of voltage as a function of time, but its peak voltages and phase angles relative to one another remain the same.

In the analyses performed in this chapter, the phase relationships indicated in Fig. 3-1b will be used as the reference for problem solutions.

Now consider the load and line elements shown in Fig. 3-1a. The load impedances for a balanced system are $Z_1 = Z_2 = Z_3 = Z_Y$, and the line impedances are $Z_{L1} = Z_{L2} = Z_{L3} = Z_L$. In a balanced wye system the generated currents delivered by V_1, V_2, and V_3, respectively, are clearly equal to load currents in Z_1, Z_2, and Z_3, respectively. If the neutral point N_S of the generator is connected to the neutral point N_L of the load, elementary system calculations will show that the current in the conductor is zero. Hence we have a classical case where the system acts exactly like three independent single-phase generators and loads. If we use the rms values and phase-angle symbology for the generated voltages of Figs. 3-1, 3-2, and 3-3, the rms line currents are

$$I_1 = \frac{V_1 \text{ at } 0°}{Z_L + Z_Y} \tag{3-4}$$

$$I_2 = \frac{V_2 \text{ at } 240°}{Z_L + Z_Y} \tag{3-5}$$

$$I_3 = \frac{V_3 \text{ at } 120°}{Z_L + Z_Y} \tag{3-6}$$

in their phasor form.

Phase volt-amperes, watts, power factor, and voltage regulation can be determined with the same single-phase solution methods that we used in Chapter 2. Three-phase volt-amperes and watts are simply three times the values obtained for one phase.

System with Wye Source and Delta Load

Figure 3-2a illustrates a delta load arrangement. All elements are identical to the wye system except the load. It is now necessary to deal with line-to-line voltages since these are the values that appear across load impedances $Z_1 = Z_2 = Z_3 = Z_D$. We define the node-to-node voltages by the lowercase letters a, b, and c. In the voltage diagram of Fig. 3-2b, the line-to-line voltages are defined as V_{ab} (i.e., the voltage at a relative to the voltage at b), V_{bc}, and V_{ca}. These voltages have the following values:

$$V_{ab} = 1.5 \text{ V} + j\left(\frac{3^{1/2}}{2}\right) \text{V} = 3^{1/2} \text{ V at } 30° \tag{3-7}$$

$$V_{bc} = 0 - j3^{1/2} \text{ V} = 3^{1/2} \text{ V at } 270° \tag{3-8}$$

$$V_{ca} = -1.5 \text{ V} + j\left(\frac{3^{1/2}}{2}\right) \text{V} = 3^{1/2} \text{ V at } 150° \tag{3-9}$$

Delta circuits are somewhat more involved from an analytical point of view because of the dual currents present at each of the load nodes. For that reason they are often converted into their equivalent wye network. In the next section, the conversion of delta-to-wye and wye-to-delta circuits is reviewed. Based on that information, for the special case of balanced circuits, a set of delta load impedances can be converted to the equivalent wye set simply by dividing the delta values by 3. From a single-phase circuit background, the line currents for this system are, then,

$$I_1 = \frac{V_1 \text{ at } 0°}{Z_L + Z_D/3} \tag{3-10}$$

$$I_2 = \frac{V_2 \text{ at } 240°}{Z_L + Z_D/3} \tag{3-11}$$

$$I_3 = \frac{V_3 \text{ at } 120°}{Z_L + Z_D/3} \tag{3-12}$$

System volt-amperes, watts, power factor, and voltage regulation can now be determined through use of conventional single-phase analysis procedures.

The currents in the delta impedances can be determined from the following loop equations:

$$I_{a'b'} = \frac{(V_{ab} - I_1 Z_L + I_2 Z_L)}{Z_D} \tag{3-13}$$

$$I_{b'c'} = \frac{(V_{bc} - I_2 Z_L + I_3 Z_L)}{Z_D} \tag{3-14}$$

$$I_{c'a'} = \frac{(V_{ca} - I_3 Z_L + I_2 Z_L)}{Z_D} \tag{3-15}$$

3-2 CONVERSION OF THREE-PHASE NETWORKS

Why do we deal with both delta and wye networks? An extensive array of technical and physical considerations are involved. Two simplified explanations follow.

Delta Networks

Delta networks (e.g., delta-connected transformer banks) are used primarily with three-wire transmission circuits where a ground reference is not associated with the system. The closed primary winding loop of the delta provides harmonic attenuation. All prime power service contains some harmonic components (i.e., resulting from less than ideal power generator characteristics, electronic loads, nonsinusoidal transformer exciting currents, etc.). In a balanced three-phase delta system, fundamental, third, fifth, and so on, harmonics cannot appear in the mesh voltages.

Wye Networks

Wye networks are used primarily for distribution circuits. A low-impedance ground system (attached to the neutral) is a fundamental part of almost all wye-connected configurations. Wye-connected systems are vulnerable to all harmonic components. For that reason, typical three-phase transformers have their primary windings connected in delta and their secondary windings connected in wye. Transformers connected wye-to-wye are sometimes

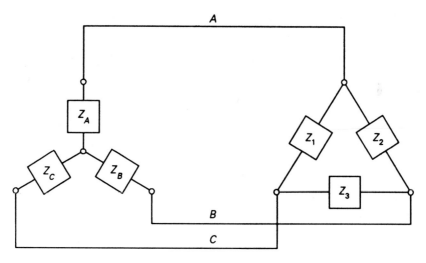

Figure 3-3 Delta-to-wye and wye-to-delta conversion nomenclature.

used with transmission systems, but they are typically equipped with a tertiary delta winding to provide harmonic suppression.

Network Conversions

It is frequently necessary to convert one type of system to another, as we did in the preceding section. A set of conversion expressions can be developed based on the nomenclature shown in Fig. 3-3.

The rationale for the wye-to-delta conversions is that the systems are equivalent if the impedance between any pair of lines A, B, and C for the delta is the same as that between the corresponding pair of lines for the wye if the third line is broken. We can describe this situation mathematically with the following three equations:

$$\text{Line } A \text{ open:} \quad Z_A + Z_B = \frac{Z_3(Z_1 + Z_2)}{Z_1 + Z_2 + Z_3}$$

$$\text{Line } B \text{ open:} \quad Z_A + Z_C = \frac{Z_1(Z_2 + Z_3)}{Z_1 + Z_2 + Z_3}$$

$$\text{Line } C \text{ open:} \quad Z_A + Z_B = \frac{Z_2(Z_1 + Z_2)}{Z_1 + Z_2 + Z_3}$$

If these equations are solved simultaneously for Z_A, Z_B, and Z_C, we obtain the following delta-to-wye conversions:

$$Z_A = \frac{Z_1 Z_2}{Z_1 + Z_2 + Z_3} \tag{3-16}$$

$$Z_B = \frac{Z_2 Z_3}{Z_1 + Z_2 + Z_3} \tag{3-17}$$

Sec. 3-2 Conversion of Three–Phase Networks

$$Z_C = \frac{Z_1 Z_3}{Z_1 + Z_2 + Z_3} \tag{3-18}$$

For the alternate case, these three equations can be solved for Z_1, Z_2, and Z_3 to give the following wye-to-delta conversions:

$$Z_1 = \frac{Z_A Z_B + Z_B Z_C + Z_C Z_A}{Z_B} \tag{3-19}$$

$$Z_2 = \frac{Z_A Z_B + Z_B Z_C + Z_C Z_A}{Z_C} \tag{3-20}$$

$$Z_3 = \frac{Z_A Z_B + Z_B Z_C + Z_C Z_A}{Z_A} \tag{3-21}$$

The six transformations above can be performed very rapidly if the program of Appendix 3 is used. The following example illustrates its application.

Example 3-1

Consider a delta network operating at a frequency of 60 Hz, with ohmic values of $40 + j60$, $100 + j0$, and $50 - j20$ for the three branch impedances. We want to convert this delta network to its equivalent wye.

Solution To obtain solutions, we load Appendix 3, issue a <RUN> command, and respond to the screen prompts as follows: first prompt, enter the letter M for a monitor display or a letter P for a printer display; second prompt, enter the letter DY for the delta-to-wye conversion analysis; third prompt, enter 60 for the line frequency; fourth prompt, enter 40 for the branch 1 resistance in ohms; fifth prompt, enter three items: the letter X (i.e., L for inductance, C for capacitance, or X for reactance), then a comma, and finally, 60 (i.e., the inductance in henries, the capacitance in farads, or the sign and value of the reactance in ohms); and sixth through ninth prompts, enter the values for the other two branches in the same manner.

The system then proceeds to print the solutions as shown in Fig. 3-4. Note that the program also provides equivalent element values (i.e., it converts reactive values to their circuit element equivalent, or vice versa). The second run uses the wye-to-delta conversion program. It demonstrates that when the first run's wye solution values are entered, they convert back to the original delta values.

3-3 UNBALANCED THREE-PHASE CIRCUITS

The goal in a three-phase system design is to achieve balanced three-phase sources and three-phase loads. However, physically realizable systems always have some degree of unbalance, and unfortunately, this complicates the mathematical analysis procedures.

The conversion expressions of Section 3-2 are as valid for unbalanced loads as they are for balanced loads. As a result, we are now able to represent unbalanced three-phase systems by means of a wye–wye diagram as shown in Fig. 3-5.

Traditionally, an analysis would start by writing the simplest set of system line current equations possible. In the direction of simplification, note that by using Kirchhoff's

Three Phase Circuit Conversion

1. This is a delta to Y circuit conversion.

2. The power source frequency is 60 Hz.

3. The delta circuit elements are as follows:
   Branch 1 resistance = 40 Ohms
   Branch 1 reactance = 60 Ohms (.1591549 H)
   Branch 2 resistance = 100 Ohms
   Branch 2 reactance = 0 Ohms (0 H)
   Branch 3 resistance = 50 Ohms
   Branch 3 reactance =-20 Ohms (1.326291E-04 F)

4. The equivalent Y circuit elements are as follows:
   R1 = 18.46154
   X1 = 7.692308 (2.040448E-02 H)
   R2 = 26.5252
   X2 = 25.9947 (6.895308E-02 H)
   R3 = 23.07692
   X3 =-15.38462 (1.724178E-04 F)

Three Phase Circuit Conversion

1. This is a Y to delta circuit conversion.

2. The power source frequency is 60 Hz.

3. The Y circuit elements are as follows:
   Branch 1 resistance = 18.46154 Ohms
   Branch 1 reactance = 7.692308 Ohms (2.040448E-02 H)
   Branch 2 resistance = 26.5252 Ohms
   Branch 2 reactance = 25.9947 Ohms (6.895308E-02 H)
   Branch 3 resistance = 23.07692 Ohms
   Branch 3 reactance =-15.38462 Ohms (1.724178E-04 F)

4. The equivalent delta circuit elements are as follows:
   R1 = 39.99999
   X1 = 60.00001 (.159155 H)
   R2 = 99.99999
   X2 = 0 (0 H)
   R3 = 49.99999
   X3 =-20.00001 (1.326291E-04 F)

Figure 3-4 Examples of delta-to-wye and wye-to-delta conversions using Appendix 3.

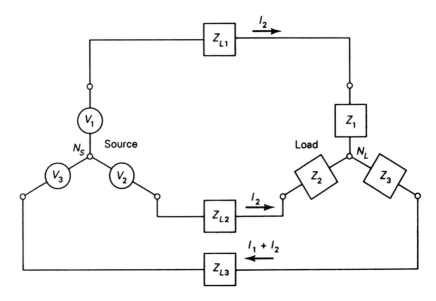

Figure 3-5 Three-phase unbalanced wye network analysis diagram.

node equation at load neutral N_L, line current 3 is directly related to line currents $I_1 + I_2$. As a result, we can then define the system with the following two Kirchhoff loop equations:

$$(Z_{L1} + Z_1)I_1 - (Z_{L2} + Z_2)I_2 = V_1 - V_2 \qquad (3\text{-}22)$$
$$(Z_{L2} + Z_2)I_2 + (Z_{L3} + Z_3)(I_1 + I_2) = V_2 - V_3$$

which can be rearranged as

$$(Z_{L3} + Z_3)I_1 + (Z_{L2} + Z_{L3} + Z_2 + Z_3)I_2 = V_2 - V_3 \qquad (3\text{-}23)$$

The line currents can then be determined by first solving these two equations simultaneously for I_1 and I_2. Then from a node summation we have

$$I_1 + I_2 + I_3 = 0 \quad \text{or} \quad I_3 = -(I_1 + I_2) \qquad (3\text{-}24)$$

This provides the first set of solutions for an unbalanced three-phase system. Determination of the system's other unknown voltages and currents typically involves a number of time-consuming exercises in complex number manipulation.

3-4 GENERAL-PURPOSE THREE-PHASE ANALYSIS PROGRAM

As a means of saving analysis time and improving solution accuracy, a three-phase modification of Appendix 2 was developed. The new program is listed in Appendix 4. Figure 3-6 shows the delta load analysis diagram used with the program. Instead of entering system information from the computer keyboard, the program uses READ statements. This approach typically results in a more accurate data input procedure and reduces

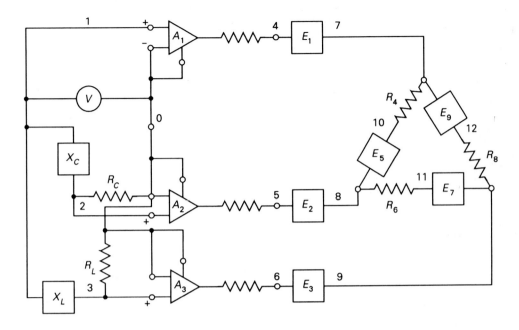

Note: Diagram E symbols have the following prefixes and values:
1. For inductances, prefix is L and value is in henries.
2. For capacitances, prefix is C and value is in farads.
3. For reactances, prefix is X and value is in + or − ohms.

Figure 3-6 Balanced three-phase source and unbalanced delta load.

the time required to enter revised programs. The general structure/modification approach for the program's DATA statements is summarized below.

Delta System Structure

Lines 1950–2010. These DATA statements define the three-phase source generators. Three operational amplifier models (part of Appendix 4) are used. A_1 is driven directly from program voltage generator V, while A_2 and A_3 are driven from V via $\pm 60°$ phase-shift networks consisting of X_C/R_C and X_L/R_L. The relative gain levels assigned to the generators are set by the fifth portion of each op amp entry, and the internal resistance of each generator phase is the sixth entry.

When entering balanced generator analysis data, only the resistance value terms in lines 1950–1970 require modification. To simulate unbalanced phase voltages, appropriate multipliers can be applied to the gain-level terms in these lines.

Lines 2020–2040. These are DATA statements that define the reactive components of source impedance. Figure 3-6 indicates all reactive terms with an E symbol, since the program will accept input information on either inductance, capacitance, or reactance values. The note below the diagram explains the input format. When revising the statements for a new analysis, the value term and possibly the symbol letter in each of the lines above require modification.

Lines 2050–2100. These are DATA statements that define the series-connected load elements of a delta network. As noted in the preceding item, the reactive terms on the diagram are denoted by E prefixes and the rules outlined apply here as well.

When revising the statements for a new analysis, the value terms of all the lines above and possibly the symbol letter in lines 2060, 2080, and 2100 require modification. For a delta load the second node numbers for lines 2060, 2080, and 2100 must be 8, 9, and 7, respectively.

Wye System Structure

Lines 1950–2140. These are DATA statement lines used to define the source generators for the wye-loaded network shown in Fig. 3-7. The same principles as noted in the delta diagram description apply here. Insert revised wye analysis network values as required.

Lines 2050, 2070, and 2090. These lines define the wye analysis network resistance values as required.

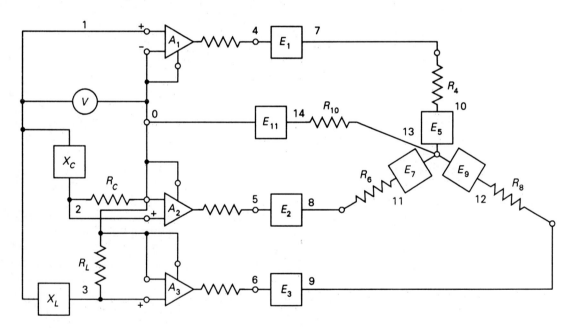

Note: Diagram E symbols have the following prefixes and values:
1. For inductances, prefix is L and value is in henries.
2. For capacitances, prefix is C and value is in farads.
3. For reactances, prefix is X and value is in + or − ohms.

Figure 3-7 Balanced three-phase source and unbalanced delta-wye load.

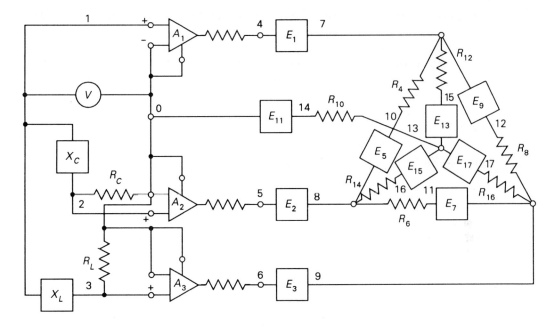

Note: Diagram E symbols have the following prefixes and values:
1. For inductances, prefix is L and value is in henries.
2. For capacitances, prefix is C and value is in farads.
3. For reactances, prefix is X and value is in + or − ohms.

Figure 3-8 Balanced three-phase source and unbalanced delta-wye load.

Lines 2060, 2080, and 2100. For a wye load the second node numbers must be 13 for all three lines. Insert reactance values as required.

Lines 2110 and 2120. These are DATA statements that define the generator neutral-to-load neutral conductor when a four-wire wye-connected system is used. When revising the statements for a new analysis, insert revised value terms as required.

Composite System Structure

Lines 1950–2010. These DATA statements define the three-phase source generators as described under delta system structure above.

Lines 2020–2180. These are DATA statements that define series-connected elements for the combined delta and wye loaded network shown in Fig. 3-8.

When revising data for a new analysis, check/modify all value terms and the symbols in lines 2020, 2030, 2040, 2060, 2080, 2100, 2120, 2140, 2160, and 2180, as required.

Appendix 4 functions in a manner similar to its single-phase counterpart. The voltages at the significant circuit nodes are calculated by means of successive runs. The voltage differences between nodes are calculated and used to determine key system operating values, as in Appendix 2. The programs differ in that you must define if a delta circuit, a wye circuit, or a combined delta and wye circuit is to be analyzed.

The background information above may make it appear that the use of this program is a formidable task. To illustrate that this is not the case, three example runs will be presented to demonstrate the operating principles and the analysis information obtained from the program.

3-5 ANALYSIS OF AN UNBALANCED WYE-CONNECTED LOAD

This analysis will use as its source a 12,470–480/277-V 60-Hz three-phase 500-kVA delta–wye-connected transformer that has 6% leakage reactance and 0.5% series resistance. On a per phase basis, the full-load impedance is

$$Z = \frac{(480/3^{1/2})^2}{500 \times 10^3/3} = 0.4608 \; \Omega$$

The leakage reactance and series resistance values are

$$X = 0.06 \times 0.4608 = 0.027648 \; \Omega$$

$$R = 0.005 \times 0.4608 = 0.002304 \; \Omega$$

From a circuit breaker specification point of view, the line-to-neutral fault current is

$$I_{\text{fault}} = \frac{480/3^{1/2}}{(0.027648^2 + 0.002304^2)^{1/2}}$$
$$= 9989 \; \text{A}$$

The unbalanced wye-connected load has a full-voltage calculated value of 390 kVA and a power factor of 80%. Its per phase values are

$$Z_1 = 0.4096 + j\,0.3072 \; \Omega$$

$$Z_2 = 0.4726 + j\,0.3545 \; \Omega$$

$$Z_3 = 0.5585 + j\,0.4189 \; \Omega$$

The element values above are inserted in Appendix 4 as follows.

Lines 1950, 1960, and 1970. These per phase series resistance R values (0.002304 Ω) are the final terms of generators A_1, A_2, and A_3.

Lines 2020, 2030, and 2040. These per phase leakage reactance X values (0.027648 Ω) are the final terms of lines 2020, 2030, and 2040.

Lines 2050 and 2060. The phase 1 load impedance values (0.4096 and 0.3072 Ω) are the final terms of lines 2050 and 2060, respectively. The second node number in line 2060 must be 13.

```
Line-Line V = 480 : Line-Neut. V = 277.1282 : F = 60 Hz
-- System Elements Are As Follows: --
A1,  1,   0,   4,   0,  -1,  .002304
A2,  2,   0,   5,   0,   2,  .002304
A3,  3,   0,   6,   0,   2,  .002304
XC,  1,   2,  -1.732051
RC,  2,   0,   1
XL,  1,   3,   1.732051
RL,  3,   0,   1
X1,  4,   7,   .027648
X2,  5,   8,   .027648
X3,  6,   9,   .027648
R4,  7,  10,   .4096
X5, 10,  13,   .3072
R6,  8,  11,   .4726
X7, 11,  13,   .3545
R8,  9,  12,   .5585
X9, 12,  13,   .4189
R10, 13, 14,   1000000
X11, 14,  0,   .027648
```

Input and output nodes are 1 , 7

FREQ	REAL	IMAG	AMPL	ANGLE
6.00E+01	268.303	-10.138	268.495	357.836

New input and output nodes are 1 , 8

FREQ	REAL	IMAG	AMPL	ANGLE
6.00E+01	-141.073	-227.631	267.801	238.212

New input and output nodes are 1 , 9

FREQ	REAL	IMAG	AMPL	ANGLE
6.00E+01	-127.230	237.768	269.669	118.151

New input and output nodes are 1 , 13

FREQ	REAL	IMAG	AMPL	ANGLE
6.00E+01	20.279	-12.703	23.929	327.938

```
Line To Line Voltage 1-2 = 463.5649 @ 27.98079 Degrees
Line To Line Voltage 2-3 = 465.605 @ 268.2963 Degrees
Line To Line Voltage 3-1 = 466.8019 @ 147.922 Degrees
         Line Current 1 = 484.451 @-36.27654 Degrees
         Line Current 2 = 454.9165 @-163.7694 Degrees
         Line Current 3 = 416.365 @ 83.62449 Degrees
Line 1 To Neutral Voltage = 248.0371 @ .5925719 Degrees
Line 2 To Neutral Voltage = 268.7541 @ 233.1034 Degrees
Line 3 To Neutral Voltage = 290.68 @ 120.495 Degrees
Neutral To Ground Voltage = 23.92936 @-32.06246 Degrees
Neutral To Ground Current = 2.392937E-05 @-32.06246 Degrees
```

Figure 3-9 Appendix 4 analysis of unbalanced three-phase wye load.

Lines 2070 and 2080. The phase 2 load impedance values (0.4726 and 0.3545 Ω) are the final terms of lines 2070 and 2080, respectively. The second node number in line 2080 must be 13.

Lines 2090 and 2100. The phase 3 load impedance values (0.5585 and 0.4189 Ω) are the final terms of lines 2090 and 2100, respectively. The second node number in line 2100 must be 13.

For this analysis, simply enter a <RUN> command. Then respond to the prompts by, first, requesting a wye analysis; second, specifying 480-V line to line; and third, specifying 60-Hz line frequency.

The description of the program's input parameters will then be printed out as shown at the top of Fig. 3-9. Refer to Fig. 3-6 and note that the voltage at nodes 7, 8, 9, and 13 (relative to node 0) are then calculated automatically and printed out in Fig. 3-9 on four successive analyses. These values are then further processed by the program to provide the detailed system performance printout shown at the bottom of the run.

In this run the transformer-to-load neutral connection was effectively eliminated by inserting a 1-MΩ value for R_{10} in line 2110. Hence this is a floating wye analysis.

3-6 ANALYSIS OF AN UNBALANCED DELTA-CONNECTED LOAD

In the next analysis we use the configuration of Fig. 3-6. To give the program a check point, the same transformer source will be used. The delta load will use values that are equivalent to those used in the wye analysis above. Using Appendix 3, these values are.

$$Z_1 = 1.2288 + j0.92167 \ \Omega$$

$$Z_2 = 1.6755 + j1.2568 \ \Omega$$

$$Z_3 = 1.4521 + j1.0891 \ \Omega$$

It is then necessary to modify lines 2050–2100. The ohmic terms above are inserted as the value in each succeeding line. Note that in order to define the delta circuit connections, the second node number in line 2060 is 8, in line 2080 it is 9, and in line 2100 it is 7.

Follow the same procedure as that noted for the preceding run except in this case request a delta analysis. Figure 3-10 shows the completed run. It can be seen that good correlation exists between the solutions for the delta analysis and those of the equivalent wye analysis.

3-7 ANALYSIS OF AN UNBALANCED COMPOSITE LOAD

Figure 3-8 shows the form of circuit configuration to be analyzed in this section. In this analysis the delta load loop uses the same elements that we established for the last analysis, and the wye load uses the same elements that we used in our first analysis with this program. Since this analysis uses a four-wire three-phase configuration, insert a practical value of 0.002304 in line 2110 for R_{10}. Retain the delta listing from the previous run.

```
Three phase system analysis with a delta connected load.

Line-Line V = 480 ; Line-Neut. V = 277.1282 ; F = 60 Hz

-- System Elements Are As Follows: --

A1, 1, 0, 4, 0, -1, .002304
A2, 2, 0, 5, 0, 2, .002304
A3, 3, 0, 6, 0, 2, .002304
XC, 1, 2, -1.732051
RC, 2, 0, 1
XL, 1, 3, 1.732051
RL, 3, 0, 1
X1, 4, 7, .027648
X2, 5, 8, .027648
X3, 6, 9, .027648
R4, 7, 10, 1.2288
X5, 10, 8, .92167
R6, 8, 11, 1.6755
X7, 11, 9, 1.2568
R8, 9, 12, 1.4521
X9, 12, 7, 1.0891

Input and output nodes are 1 , 7
```

FREQ	REAL	IMAG	AMPL	ANGLE
6.00E+01	268.303	-10.137	268.494	357.836

New input and output nodes are 1 , 8

FREQ	REAL	IMAG	AMPL	ANGLE
6.00E+01	-141.073	-227.631	267.801	238.212

New input and output nodes are 1 , 9

FREQ	REAL	IMAG	AMPL	ANGLE
6.00E+01	-127.230	237.768	269.669	118.151

```
Line To Line Voltage 1-2 = 463.5647 @ 27.98086 Degrees
Line To Line Voltage 2-3 = 465.6048 @ 268.2963 Degrees
Line To Line Voltage 3-1 = 466.8013 @ 147.9221 Degrees
       Line Current 1 = 484.4454 @-36.27832 Degrees
       Line Current 2 = 454.9174 @-163.7702 Degrees
       Line Current 3 = 416.3698 @ 83.6237 Degrees
    Delta Current 1-2 = 301.7917 @-8.891129 Degrees
    Delta Current 2-3 = 222.3009 @ 231.4225 Degrees
    Delta Current 3-1 = 257.171 @ 111.0515 Degrees
```

Figure 3-10 Appendix 4 analysis of unbalanced three-phase delta load.

Three phase system analysis with a delta and Y connected load.

Line-Line V = 480 ; Line-Neut. V = 277.1282 ; F = 60 Hz

-- System Elements Are As Follows: --

```
A1,  1 ,  0 ,  4 ,  0 , -1 , .002304
A2,  2 ,  0 ,  5 ,  0 ,  2 , .002304
A3,  3 ,  0 ,  6 ,  0 ,  2 , .002304
XC,  1 ,  2 , -1.732051
RC,  2 ,  0 ,  1
XL,  1 ,  3 ,  1.732051
RL,  3 ,  0 ,  1
X1,  4 ,  7 , .027648
X2,  5 ,  8 , .027648
X3,  6 ,  9 , .027648
R4,  7 , 10 , 1.2288
X5, 10 ,  8 , .92167
R6,  8 , 11 , 1.6755
X7, 11 ,  9 , 1.2568
R8,  9 , 12 , 1.4521
X9, 12 ,  7 , 1.0891
R10, 13 , 14 , .002304
X11, 14 ,  0 , .027648
R12,  7 , 15 , .4096
X13, 15 , 13 , .3072
R14,  8 , 16 , .4726
X15, 16 , 13 , .3545
R16,  9 , 17 , .5585
X17, 17 , 13 , .4189
```

Input and output nodes are 1 , 7

FREQ	REAL	IMAG	AMPL	ANGLE
6.00E+01	258.262	-19.143	258.971	355.761

New input and output nodes are 1 , 8

FREQ	REAL	IMAG	AMPL	ANGLE
6.00E+01	-143.658	-216.087	259.483	236.383

New input and output nodes are 1 , 9

FREQ	REAL	IMAG	AMPL	ANGLE
6.00E+01	-117.471	234.868	262.606	116.572

New input and output nodes are 1 , 13

FREQ	REAL	IMAG	AMPL	ANGLE
6.00E+01	2.867	0.362	2.890	7.205

Line To Line Voltage 1-2 = 447.5789 @ 26.10515 Degrees
Line To Line Voltage 2-3 = 451.7141 @ 266.6765 Degrees
Line To Line Voltage 3-1 = 453.5379 @ 145.9397 Degrees

```
              Line Current 1 = 968.7655 @-39.81852 Degrees
              Line Current 2 = 881.2699 @-163.2104 Degrees
              Line Current 3 = 782.4741 @ 81.08798 Degrees
           Delta Current 1-2 = 291.3846 @-10.76683 Degrees
           Delta Current 2-3 = 215.6689 @ 229.8028 Degrees
           Delta Current 3-1 = 249.8639 @ 109.0692 Degrees
     Line 1 To Neutral Voltage = 256.1391 @-4.367442 Degrees
     Line 2 To Neutral Voltage = 261.3808 @ 235.904 Degrees
     Line 3 To Neutral Voltage = 263.5788 @ 117.1649 Degrees
     Neutral To Ground Voltage = 2.889701 @ 7.204971 Degrees
             Y Current In Leg 1 = 500.2716 @-41.23734 Degrees
             Y Current In Leg 2 = 442.4334 @ 199.0302 Degrees
             Y Current In Leg 3 = 377.5443 @ 80.29332 Degrees
     Neutral To Ground Current = 104.1565 @-78.03139 Degrees
```

Figure 3-11 Appendix 4 analysis of unbalanced three-phase composite load.

(*Note*: The original DATA listings from lines 2130 to 2180 of Appendix 4 define the composite wye elements). The transformer used in the previous analysis is also used, so the system arrangement represents a three-phase low-voltage power distribution center operating under temporarily overloaded conditions.

A <RUN> command followed by the previously noted prompt responses (except in this case a composite analysis must be requested) will result in the printout shown in Fig. 3-11. Because the four successive program analyses involve 17 circuit nodes, the matrices used to obtain the problem solutions become significantly larger, and the solution time for each analysis is somewhat longer. Nevertheless, the analysis time is orders of magnitude reduced from that required for a manual solution.

Two points of interest are present in the printout. First, despite the overloaded transformer and the objectionably low power factor, the worst-case voltage regulation is only about 3%. Second, the unbalanced load causes a substantial amount of neutral current to flow—approximately 20% of the average line current.

3-8 A HIGH-VOLTAGE SUBSTATION ANALYSIS

Transformer and Switching Elements

When multiple megawatt levels of power are involved, high-source voltage levels are required. The power distribution center is then typically referred to as a *substation*. The system elements we will analyze are derived from a group of 10- to 14-MVA electronic system substations that operate from 115-kV three-phase source lines [4].

Substation capabilities are determined by the main step-down transformer's ratings. The 10-MVA figure indicates that the transformer can operate continuously with convection cooling with a maximum 55°C rise. At 14 MVA, forced air cooling is used and there will be a 65°C rise.

The line-to-line primary voltage is 115-kV delta connected, and the secondary line-to-line voltage is 12.47 kV wye connected. The reactance and resistance values are 8.5% and 0.25% based on the 10-MVA rating. A substation includes both high- and low-voltage switching. We will investigate the interrupting capabilities required at both the high- and low-voltage levels.

High-Voltage Transmission Lines

For three-phase line applications, conductor-to-neutral capacitance is given by

$$C = \frac{K \times 10^{-8}}{\log_{10}(D/r)} \tag{3-25}$$

where C is in farads per unit length of line, $k = 2.413$ for kilometer unit lengths (3.883 for mile unit lengths), D the center-to-center spacing of conductors, and r the conductor radius (D and r must be expressed in the same units).

For three-phase lines the inductance of each conductor is

$$L = (k_1 \log_{10}\frac{D}{r} + k_2) \times 10^{-6} \tag{3-26}$$

where L is in henries per unit length of line, $k_1 = 460.54$ and $k_2 = 50$ for kilometer unit lengths (741.13 and 80.47 for mile unit lengths). D and r are the same as in Eq. (3-25).

The 10/14-MVA substation's 115-kV source line is 120.7 km (75 mi) long. The line uses 500-kcm conductors [i.e., the 500-kilocircular mil conductor diameter is about 2.068 cm (0.814 in.), and for copper the line resistance is about 0.07087 Ω/km (0.114 Ω/mi)] arranged in equilateral format as shown in Fig. 3-12a. Line-to-line spacing is 304.6 cm (120 in.). The values for each conductor are

$$C = 120.7\left[\frac{2.413 \times 10^{-8}}{\log_{10}(304.7/1.034)}\right] = 1.179 \times 10^{-6} \text{F}$$

$$L = 120.7(460.54 \log_{10}\frac{304.7}{1.034} + 50) \times 10^{-6} = 0.1433 \text{ H}$$

$$R = 120.54 \times 0.07087 = 8.553 \text{ }\Omega$$

The 115-kV line-to-neutral fault current is determined by the series reactance of L and the series resistance of R. The switching element interrupting capability must be at least

$$\text{I fault} = \frac{115 \times 10^3/3^{1/2}}{\{[2\pi (60)0.1433]^2 + 8.553^2\}^{1/2}}$$
$$= 1214 \text{ A}$$

Performance Analysis Using Appendix 4

We will refer the system analysis to the 12.47-kV level. Figure 3-13 shows the circuit configuration. Analysis arrangements/combinations of system elements follow.

To simulate the transmission line as shown in Fig. 3-12b, the internal resistances of A_1, A_2, and A_3 are equal to one-half of the transformed value of R [i.e., assume the effective wye-to-wye transformation ratio value to be $(12.47/115)^2$]. This gives

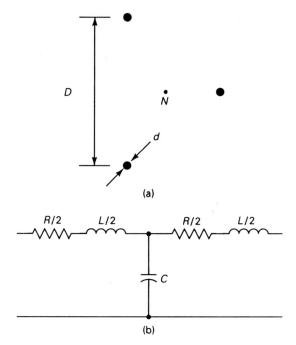

Figure 3-12 Long three-phase transmission line layout and analysis: (a) equilateral triangular three-phase conductor arrangement; (b) circuit for long transmission line simplified analysis.

$$R = \frac{8.553}{2} \frac{12.47^2}{115} = 0.0503 \ \Omega$$

The converted values for E_1, E_2, and E_3 are obtained in a similar manner, to give

$$X_1 = X_2 = X_3 = [0.1433\pi(60)] \left(\frac{12.47}{115}\right)^2 = 0.3176 \ \Omega$$

The per phase 12.47-kV characteristics of the 10/14-MVA transformer are

$$Z_t = \frac{(12{,}470/3^{1/2})^2}{10 \times 10^6/3} = 15.55 \ \Omega$$

$$X_t = 0.085(15.55) = 1.3218 \ \Omega$$

$$R_t = 0.0025(15.55) = 0.03888 \ \Omega$$

If the remaining half of the transmission line L and R elements are combined with the transformer elements of the last step, we obtain

$$X_{13} = X_{15} = X_{17} = 0.3176 + 1.3218 = 1.6394 \ \Omega$$

$$R_{12} = R_{14} = R_{16} = 0.0503 + 0.03888 = 0.08918 \ \Omega$$

From a system design point of view, the line-to-neutral fault-interrupting rating of the 12.47-kV incoming line circuit breaker can now be determined. Add the other half of

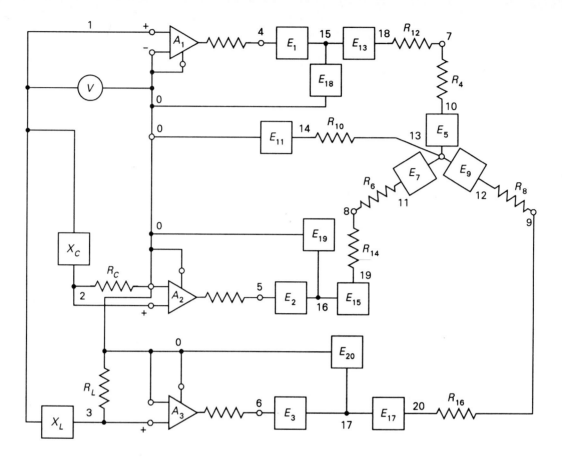

Note: Diagram E symbols have the following prefixes and values:
1. For inductances, prefix is L and value is in henries.
2. For capacitances, prefix is C and value is in farads.
3. For reactances, prefix is X and value is in + or − ohms.

Figure 3-13 115-kV power system referenced to 12.47-kV level.

the transformed line reactance and line resistance to the two terms above, and solve for the current as in previous fault calculations:

$$\text{I fault} = \frac{(12{,}470/3^{1/2})}{[(1.6394 + 0.3167)^2 + (0.08918 + 0.0503)^2]^{1/2}}$$
$$= 3670 \text{ A}$$

The transformed values of the line capacitance elements become

$$X_{18} = X_{19} = X_{20} = \frac{-1}{2\pi(60)1.179 \times 10^{-6}} \left(\frac{12.47}{115}\right)^2$$
$$= -26.454 \; \Omega$$

The site's phase load values can be combined to give the following six equivalent 90% power factor load elements:

$$R_4 = 31.104 \ \Omega$$
$$X_5 = 15.064 \ \Omega$$
$$R_6 = 29.549 \ \Omega$$
$$X_7 = 14.311 \ \Omega$$
$$R_8 = 32.659 \ \Omega$$
$$X_9 = 15.817 \ \Omega$$

A substation transformer normally has the neutral of its secondary connected, via a grounding resistor, to the substation's buried ground grid. Some inductance is also present in the loop. The values used in this analysis are

$$R_{10} = 10 \ \Omega$$
$$X_{11} = 1 \ \Omega$$

```
Three phase system analysis with a Y connected load.
Line-Line V = 12470 : Line-Neut. V = 7199.558 : F = 60 Hz
-- System Elements Are As Follows: --
A1,  1 ,  0 ,  4 ,  0 , -1 ,  .0503
A2,  2 ,  0 ,  5 ,  0 ,  2 ,  .0503
A3,  3 ,  0 ,  6 ,  0 ,  2 ,  .0503
XC,  1 ,  2 , -1.732051
RC,  2 ,  0 ,  1
XL,  1 ,  3 ,  1.732051
RL,  3 ,  0 ,  1
X1,  4 , 15 ,  .3176
X2,  5 , 16 ,  .3176
X3,  6 , 17 ,  .3176
R4,  7 , 10 , 31.104
X5, 10 , 13 , 15.064
R6,  8 , 11 , 29.549
X7, 11 , 13 , 14.311
R8,  9 , 12 , 32.659
X9, 12 , 13 , 15.817
R10, 13 , 14 , 10
X11, 14 ,  0 ,  1
R12,  7 , 18 ,  .08918
X13, 15 , 18 , 1.6394
R14,  8 , 19 ,  .08918
X15, 16 , 19 , 1.6394
R16,  9 , 20 ,  .08918
X17, 17 , 20 , 1.6394
X18, 15 ,  0 , -26.454
X19, 16 ,  0 , -26.454
X20, 17 ,  0 , -26.454
Input and output nodes are 1 , 7
```

FREQ	REAL	IMAG	AMPL	ANGLE
6.00E+01	7072.362	-356.452	7081.339	357.115

Figure 3-14 Analysis of 115/12.47-kV 10-MVA substation.

Sec. 3-8   A High–Voltage Substation Analysis

Figure 3-14 *(cont.)*

New input and output nodes are 1 , 8

FREQ	REAL	IMAG	AMPL	ANGLE
6.00E+01	-3844.542	-5931.176	7068.193	237.049

New input and output nodes are 1 , 9

FREQ	REAL	IMAG	AMPL	ANGLE
6.00E+01	-3244.533	6296.024	7082.860	117.263

New input and output nodes are 1 , 13

FREQ	REAL	IMAG	AMPL	ANGLE
6.00E+01	-27.739	-91.543	95.653	253.142

```
Line To Line Voltage 1-2 = 12257.91 @ 27.05116 Degrees
Line To Line Voltage 2-3 = 12241.91 @ 267.1907 Degrees
Line To Line Voltage 3-1 = 12275.74 @ 147.1855 Degrees
Line 1 To Neutral Voltage = 7105.04 @-2.136752 Degrees
Line 2 To Neutral Voltage = 6976.339 @ 236.8313 Degrees
Line 3 To Neutral Voltage = 7151.837 @ 116.7299 Degrees
Neutral To Ground Voltage = 95.65311 @ 253.1425 Degrees
         Y Current In Leg 1 = 205.5866 @-27.97816 Degrees
         Y Current In Leg 2 = 212.4852 @ 210.9897 Degrees
         Y Current In Leg 3 = 197.0878 @ 90.88868 Degrees
Neutral To Ground Current = 9.517841 @ 247.4319 Degrees
```

This completes the system element list. However, Appendix 4 must have several bookkeeping statements modified before the program can accommodate the number of the system elements used. The specific changes are:

In line 130, change "Y = 20",and "UL = 24."
In line 560, change "AND Y = 20."
In line 670, change "Y = 20," change to 635, delete 670.
In line 2400, change "Y = 20," "R = R(4)," and "X = X(5)."
In line 2410, change "Y = 20," "R = R(6)," and "X = X(7)."
In line 2420, change "Y = 20," "R = R(8)," and "X = X(9)."

The foregoing statement changes correct only for a wye circuit analysis. Now change the DATA statements and add three new statements for E_{18}, E_{19}, and E_{20}. Enter a <RUN> command, and respond to prompts in the usual fashion. The 20 nodes slow the program execution significantly, so be patient. The printed output is shown in Fig. 3-14.

The printout presents no particular surprises. If 7200-V line to neutral is taken as the no-load value, a worst-case voltage regulation of about 1.83% is present, and the neutral current is about 4.63% of normal line current for the loading profile analyzed.

When dealing with long lines it is wise to run a check and determine the actual value of no-load voltage. If we assume that there is always at least 1% loading on the system, a check can easily be performed simply by running the decimal point on the values of R_4, X_5, R_6, X_7, R_8, and X_9 two places to the right and repeating the analysis. It will show that the average line-to-neutral voltage is approximately 7285 V, slightly over 1% higher than the nominal value. For this length line the increase is not really significant, but on longer lines the voltage regulation may show a substantial increase. This is shown by example in Section 4-2.

Chapter 4

Long Transmission Lines

A transmission line is a physical device that conveys electrical energy from one location (typically, from an energy source interface) to another location (typically, to a system load interface). *Parallel wire* lines for prime power systems were fundamental parts of several analyses that we performed in Chapters 2 and 3. In these problems we were able to use simplified line models because the lines were physically short in terms of the wavelength of a 60-Hz wave. When more exact calculations are required for a prime power system, or when analyzing RF transmission lines, it is necessary to use *long-line* mathematical procedures.

In the next section we derive the fundamental expressions that define the performance of long parallel wire transmission lines. These expressions provide the technical background for a comprehensive transmission line computer program. Two elementary parallel wire line system applications will be used initially to demonstrate this program.

The background developed for parallel wire lines is also directly applicable to a number of types of RF transmission line systems. We will review the capabilities of several widely used types of RF lines and their associated line elements. Following this, the computer program is used to analyze several additional power and RF transmission line applications.

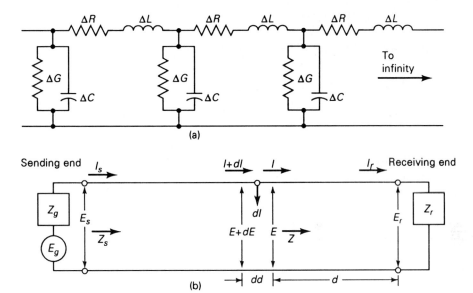

Figure 4-1 System models for transmission line analysis: (a) incremental element representation of a transmission line; (b) notation used in transmission line equation development.

4-1 LONG-LINE EQUATION DEVELOPMENT

This analysis of transmission line operation is based on use of the model shown in Fig. 4-1a. It is composed of the following building blocks.

Series Resistance

R is the resistance in ohms per unit length of line.

Series Inductance

L is the inductance in henries per unit length of line.

Series Impedance

z is the impedance in ohms per unit length of line, $z = R + j\omega L$.

Shunt Capacitance

C is the capacitance in farads per unit length of line.

Shunt Conductance

G is the conductance in siemens per unit length of line.

Shunt Admittance

y is the shunt admittance in siemens per unit length of line, $y = G + j\omega C$.

Unit length is usually defined in measurement values that are appropriate to the analysis under consideration. For example, L is typically defined in henries per kilometer (or per mile) when dealing with long prime power transmission lines. In RF applications, L is often defined in microhenries per 100 m (or per 100 ft).

In the following analysis, we consider the line to be composed of an *infinite* number of the building blocks shown. Then from a length point of view, each line segment will have a wavelength that is infinitesimally short. The infinitesimal length, shown in Fig. 4-1b, is defined as dd.

Energy is supplied to the line by source E_g through sending end impedance Z_g. At its receiving end, the line is terminated in load Z_r. The fundamental principles used in this analysis are as follows:

Shunt Voltage

The shunt voltage between conductors, E, varies from point to point as we progress along the line. This variation is, in part, a function of series current I flowing through series impedance z.

Series Current

In addition, the variation in series current I is a function of the change in shunt current in admittance y. This change is in turn a function of the shunt voltage.

Incremental Variations

The first objective in this analysis is to define the incremental voltage and current variations that occur on the infinitesimal length of line dd. We can write the voltage relationship as follows:

$$dE = I(z\,dd) \qquad (4\text{-}1)$$

where this expression states that the incremental change in voltage is equal to the line current times the line impedance per unit length times the incremental length dd.

We can write the current relationship as follows:

$$dI = E(y\,dd) \qquad (4\text{-}2)$$

where this expression states that the incremental change in current is equal to the line shunt voltage times the line admittance per unit length times the analysis increment dd. If we divide both expressions by dd we have

$$\frac{dE}{dd} = Iz \tag{4-3}$$

$$\frac{dI}{dd} = Ey \tag{4-4}$$

Both equations contain the unknowns E and I. To find E and I as functions of any position on the line, the equations must be solved (i.e., we need an equation containing only one unknown). To eliminate one of the unknowns, say I, we can take the derivative of Eq. (4-3) with respect to d:

$$\frac{d^2E}{dd^2} = z\frac{dI}{dd} \tag{4-5}$$

If we substitute Eq. (4-4) into Eq. (4-5), we obtain

$$\frac{d^2E}{dd^2} = zyE \tag{4-6}$$

The same approach can be used to eliminate E in Eq. (4-4). It gives

$$\frac{d^2I}{dd^2} = zyI \tag{4-7}$$

We now require a solution technique for linear differential equations with constant coefficients. Most texts on the subject give a set of rules that can be applied here. An example follows.

Equation format. Set up the auxiliary equation. The auxiliary equation is simply an algebraic expression where the differential terms are replaced by algebraic terms with the same exponents (using any convenient letter, such as m).

Equation roots. Solve the algebraic equation for its roots. Hence for Eq. (4-6) we have

$$m^2 = zy$$

and the roots are $m_1 = (zy)^{0.5}$ and $m_2 = -(zy)^{0.5}$.

Solution form. For Eq. (4-6) the solution has the form

$$E = C_1 \exp(m_1 d) + C_2 \exp(m_2 d)$$

where C_1 and C_2 are constants that require further evaluation, exp is the base of natural logarithms, and d the line length. When the roots are substituted into the equation, we obtain

$$E = C_1 \exp[(zy)^{0.5}d] + C_2 \exp[-(zy)^{0.5}d] \tag{4-8}$$

To evaluate I, first write Eq. (4-3) as follows:

$$I = \frac{1}{z}\frac{dE}{dd} \tag{4-9}$$

Next, we can differentiate Eq. (4-8) and substitute it into Eq. (4-9) as follows:

$$I = \frac{1}{z}\{(zy)^{0.5} C_1 \exp[(zy)^{0.5}d] - (zy)^{0.5} C_2 \exp[-(zy)^{0.5}d]\} \quad (4\text{-}10)$$
$$= \frac{y}{z} C_1 \exp[(zy)^{0.5}d] - \frac{y}{z} C_2 \exp[-(zy)^{0.5}d]$$

To determine the constants, we will evaluate Eq. (4-8) and (4-10) at the load end of the line. The evaluation conditions at the load are

$$d = 0 \quad E = E_r \quad I = I_r \quad (4\text{-}11)$$

These conditions are then substituted into Eqs. (4-8) and (4-10) as follows:

$$E_r = C_1 + C_2 \quad (4\text{-}12)$$

$$I_r = \left(\frac{y}{z}\right)^{0.5}(C_1 - C_2) \quad (4\text{-}13)$$

If these two equations are solved simultaneously, we obtain

$$C_1 = \frac{1}{2}\left[E_r + \left(\frac{z}{y}\right)^{0.5} I_r\right]$$

$$C_2 = \frac{1}{2}\left[E_r - \left(\frac{z}{y}\right)^{0.5} I_r\right]$$

The transmission line's *characteristic impedance* is defined as

$$Z_0 = \left(\frac{z}{y}\right)^{0.5} \quad (4\text{-}14)$$

and the transmission line's *propagation constant* is defined as

$$P = (zy)^{0.5} \quad (4\text{-}15)$$

Further particulars on both Z_0 and P will be presented shortly. At this juncture let us conclude this derivation by substituting C_1, C_2, Z_0, and P into Eqs. (4-8) and (4-10) and document the famous long-line equations as follows:

$$E = \frac{1}{2}\left[(E_r + Z_0 I_r)e^{pd} + (E_r - Z_0 I_r)e^{-pd}\right] \quad (4\text{-}16)$$

$$I = \frac{1}{2}\left[(I_r + \frac{E_r}{Z_0})e^{pd} + (I_r - \frac{E_r}{Z_0})e^{-pd}\right] \quad (4\text{-}17)$$

It is important to note that each of the equations above contains two similarly structured terms. The first term of each equation defines the *incident wave* and the second term defines the *reflected wave*. The rationale for this can be developed as follows. First, terminate the line with an impedance that has a value equal to Z_0. Second, when this is done, we see that the coefficient of the second term of each equation becomes zero. Third, as a result, for this matched load operating condition, the only wave present on the line will be the incident wave—due to the first term of each equation. The propagation direction of the incident wave is always from the source toward the load. Fourth, if other than a matched load condition is present, reflected energy will propagate on the line. Reflected wave propagation is defined by the second term of each equation. The propagation direction of the reflected wave is always from the load toward the source.

With this background, let us return to our definition of Z_0. Equation (4-14) can be expanded in terms of our original unit-length building blocks as follows:

$$Z_0 = \left(\frac{z}{y}\right)^{0.5} = \left(\frac{R + j\omega L}{G + j\omega C}\right)^{0.5} \qquad (4\text{-}18)$$

where the quantitative value can be determined from the line's physical dimensions and electrical properties. In Eqs. (3-25) and (3-26) we found that a power line's L and C per unit values are direct functions of the line's structural dimensions. Similarly, a line's R and G values per unit length are simply functions of the line's structural materials (i.e., the line conductor series loss resistance per unit length and the reciprocal of the line insulation shunt loss resistance per unit length).

If a transmission line of infinite length is considered, waves will propagate along the line in only one direction (i.e., from the source toward the load), and Z_0 is the exact value of impedance the wave will encounter as it propagates along the line. This fact is, of course, completely consistent with the incident wave (matched load) propagation rationale noted earlier.

The transmission line's propagation constant can be expanded using the unit-length building block terms as follows:

$$P = (zy)^{0.5} = [(R + j\omega L)(G + j\omega C)]^{0.5} \qquad (4\text{-}19)$$

The significance of P lies in the fact that all practical lines have losses. Hence waves of voltage and current become smaller as they propagate—as either incident or reflected waves—along a line. The ratio of voltage and current magnitudes at two points on the line (point a and point b separated by a distance Δd) can be expressed as follows [by using Eq. (4-16) and (4-17)]:

$$\frac{E_a}{E_b} = \frac{I_a}{I_b} = e^{P\Delta d} \qquad (4\text{-}20)$$

Since both quantities in the voltage or current ratios are complex numbers, we can write

$$P = A + jB \qquad (4\text{-}21)$$

The attenuation in distance Δd is said to be $A\,\Delta d$ nepers, where A is defined as the *attenuation constant* (measured in nepers per unit length of line) and B is the *phase constant* (measured in radians per unit length of line). Nepers can be defined from Eq. (4-20) as follows:

$$A\,\Delta d = \ln\frac{E_a}{E_b} = \ln\frac{I_a}{I_b} \qquad (4\text{-}22)$$

The propagation constant in terms of the voltage and current ratios can be expanded as follows:

$$\frac{E_a}{E_b} = \frac{I_a}{I_b} = e^p = e^{A+jB} = e^A \times e^{jB} = e^A(\cos B + j \sin B) \quad (4\text{-}23)$$

The final form shown denotes clearly both the attenuation and phase angle as the wave propagates down the line.

The imaginary part of the propagation constant, B, for ideal air insulated line is

$$B = \frac{2\pi f}{v} \quad (4\text{-}24)$$

radians/unit length, where f is frequency and v is its *velocity of propagation* along the line. For example, consider a 60-Hz power line where the velocity of propagation is approximately the same as the velocity of light: 3×10^8 m/s (9.84×10^8 ft/s). Then B has a value of approximately $377/3 \times 10^5 = 0.001256$ rad/km (0.00202 rad/mi) or 0.0721 deg/km (0.116 deg/mi).

We now have all of the quantities that are required to perform an analysis. The steady-state voltage and current at any point on the line can be determined by inserting the distance d (from the load to the point of interest on the line) and solving the equations. If a number of calculations are to be performed, this may become quite time consuming. A computer program that can reduce typical analysis calculation times by perhaps an order of magnitude is described in a next section.

For RF lines in air, the following parallel wire line equation can be used to determine attenuation constant A per unit length of line. It is valid for a line constructed from circular copper conductors,

$$A = \frac{kF^{0.5}}{d \log_{10}(2D/d)} \quad (4\text{-}25)$$

where A is in nepers/unit length, $k = 3.02 \times 10^{-3}$ for 100-m unit lengths (3.62×10^{-4} for 100-ft unit lengths), F is in megahertz, d is conductor diameter, and D is the center-to-center spacing [d and D measured in centimeters (inches)].

Although nepers per unit of line length express the mathematical logarithmic attenuation in transmission lines, the following conversion factors are often useful:

$$\begin{aligned} A &= 0.1151 \times (\text{dB/unit length}) \quad \text{nepers per unit length} \\ A &= 8.686 \times (\text{nepers/unit length}) \quad \text{dB per unit length} \end{aligned} \quad (4\text{-}26)$$

Phase constant B for RF lines is lower (often much lower) than the value given by Eq. (4-24). Practical two-conductor lines have, to some degree, insulation dielectric between their conductors. The dielectric constant of this intraconductor material retards the velocity of wave propagation down the line. Hence in line analyses, practical line values of B are corrected relative to air-insulated line values by means of a propagation constant.

4-2 PRIME POWER LINE SYSTEM ANALYSIS

The fundamental equations derived in Section 4-1 can be used to perform system analyses of any type of transmission line system. The type of information available to us for a prime power line system analysis is usually presented in somewhat different form than the information presented for an RF transmission line analysis. The computer analysis program that we will use in this chapter is compatible with both types of systems. In this section we review prime power system applications of the program.

Prime Line Parameters

In Section 3-9, per unit values of series impedance z and shunt admittance y were used in conjunction with a T-section model. In the T model, we did not include the line's shunt conductance term because it typically introduces an insignificant effect on the calculations. Otherwise, we were dealing there with the identical quantitites that we will use here (i.e., the quantities described in Section 4-1).

Prime power analysts ordinarily use kilometers or miles to express their per unit lengths. For example, if a line is 100 km long, the parameters are often expressed in terms of a 100-km per unit line length. However, any desired per unit length can be used in the power line portion of our analysis program. In addition, the program can provide analyses based on physical line parameters such as line spacings, conductor resistance per unit length, and the shunt conductance per unit length.

Prime power personnel also ordinarily analyze their three-phase lines in terms of balanced three-phase line loading. As we know from our investigations in Chapter 3, this significantly reduces any problem's complexity because only single-phase analyses need to be performed.

Full-Load Prime Line Analysis Using Appendix 5

With the prime line background above established, let us now analyze a representative three-phase prime line system. In Appendix 5 we provide solutions in terms of both SI and NBS units.

For our first illustration we use SI units and analyze a 60-Hz three-phase line with a balanced 0.80 lagging power factor load. The load's total power dissipation is 18 MW, and the balanced three-phase line-to-line voltage at the load is 110 kV. This problem will be solved in terms of the line's physical parameters.

The line is constructed with 0000 copper conductors (diameter = 1.168 cm and resistance = 0.1684 Ω/km). It is 321.8 km long and has equilateral triangular construction with 304.8-cm conductor-to-conductor spacing. Assume that the line conductance is 10^{-12} S/km, that the source series resistance and reactance both have negligible values of 10^{-6} Ω, and that induction and radiation losses from the line are negligible. What steady-state conditions will prevail at each end of the line?

Before starting the solution we must determine the impedance of the wye-connected load. We know that V_{ln}, the line-to-neutral voltage, = $110 \times 10^3/3^{0.5}$ = 63,508.53 V.

```
This Is A Steady State Analysis Of A Transmission Line With Losses
---- -- - ------ ----- -------- -- - ------------ ---- ---- ------
    Is this a Utility line or an RF line application (U or R)? U
    Is line information physical or electrical (P or E)? P
    Are line dimensions in SI or NBS units (S or N) ? S
    What is the diameter of the conductors in cm? 1.168
    What is the center to center spacing of conductors in cm ? 304.8
    What is the series resistance of one conductor per km? .1684
    What is the shunt conductance of one conductor to neutral per km? 1E-12
 1. The line's frequency of operation in Hz =  60
 2. The line's wave velocity propagation factor = .967534
 3. The line's characteristic impedance =  393.6795
 4. The line's attenuation factor in nepers / km = 5.616021E-04
 5. The line's length in km =  321.8
    Do you know the value of load impedance? Y
 6. The load's series resistance value in ohms =  430.222
 7. The loads series reactance value in ohms = 322.66
    The value of ZL =  537.7736  @  36.86934 Deg.
 8. The source's series resistance value in ohms =  .000001
 9. The source's series reactance value in ohms  =  .000001
    Will analysis be referred to load voltage or source voltage (S or L)? L
10. The load rms voltage =  63508.83
-------------------------------------------------------------------
The Solutions At The Line's Sending And Receiving Ends Are:
--- --------- -- --- ------ ------- --- ----
Input Current =  88.002 @   2.771 Deg.: Input Power = 6551518.000 Watts
Input Voltage = 75005.530 @    9.766 Deg.
Input Impedance: RI =  845.97: XI =  103.80: ZI =  852.31@   6.996 Deg.
Output Current = 118.096 @  -36.869 Deg.: Output Power = 6000146.000 Watts
Output Voltage =  63508.830 @   0.000 Deg.
Power Attenuation =  -0.382 dB: Line Efficiency =  91.584 %
-------------------------------------------------------------------
```

Figure 4-2 Appendix 5 analysis of 321.8-km 60-Hz three-phase prime line.

From trigonometry, the voltage across the resistance is 80% of V_{ln} and the voltage across the reactance is 60% of V_{ln}. Therefore, on a per phase basis,

$$R = \frac{(0.8V_{ln})^2}{P/3}$$

$$= \frac{(0.8 \times 63{,}508.53)^2}{6 \times 10^6} = 430.222 \ \Omega$$

$$X = \frac{6}{8} 430.222 = 322.667 \ \Omega$$

Now load the program of Appendix 5 and issue a <RUN> command. Then, using Fig. 4-2 as a reference, simply respond to the first eight computer prompts with the information we have defined above. The program proceeds to calculate and display the line characteristics shown in display items 1–4. Next enter the line length in response to the prompt for item 5, answer Y to the prompt regarding the load, and then enter the foregoing values of load and source resistances/reactances in response to the prompts for items 6–9. Finally, refer the analysis to the load by entering L, and enter the load voltage in response to prompt 10. The system analysis solutions at the sending and the receiving end interfaces of the line will then develop rapidly on your printer. Two interesting things shown in the long line analysis are the following.

System efficiency. Power transmission has been accomplished over the 304.8-km line length with 91.585% efficiency.

Input and output interface line currents. The current at the source is significantly less than that at the load. When we look into load mismatching (and voltage/current standing-wave ratios) in the next section, the reason for this situation will become clear.

Voltage Regulation Analysis Using Appendix 5

How can we obtain the voltage regulation of the utility line above at its load end? From Fig. 4-2 the full-load line-to-neutral voltage is 63,508 V. To obtain the open-circuit voltage, let us assume that (1) all line parameters remain the same, (2) the source voltage remains at its calculated value, and (3) the load is an open circuit. When you enter these data into the program, your new run should show that the open-circuit line-to-neutral voltage is 81,842 V. These two voltage values give

$$\text{voltage regulation} = \frac{100(81{,}842 - 63{,}508)}{63{,}508} = 28.73\%$$

Solutions Using Electrical Line Parameters

Recall that we used physical line parameters to obtain solutions for the two problems above. The program would provide the same solutions if an electrical solution had been requested. To obtain that solution, an appropriate set of per unit values of the line's R, X, G, and C parameters would be entered in response to an alternate set of program prompts.

4-3 RF TRANSMISSION LINE SYSTEM ANALYSIS

In Section 4-1 we noted that fundamental long-line equations (4-16) and (4-17) describe incident and reflected waves that propagate along transmission lines. In this section we define these waves in terms of the *standing-wave ratios* that they produce along lines. We will also show how standing waves influence a number of significant transmission line performance capabilities. Following this, we use Appendix 5 to analyze typical RF transmission line systems.

Reflection Coefficients and VSWR

If a transmission line's load end resistance Z_r is not equal to the value of the line's characteristic impedance Z_0, the voltage across the load will be the result of the summation of two traveling waves. One is the incident wave that is sent down the line by the source generator. The second is the wave reflected from line termination Z_r. The value of the reflected wave is equal to the incident wave times the *reflection coefficient* Γ. For low-loss lines,

$$\Gamma = \frac{Z_r - Z_0}{Z_r + Z_0} \qquad (4\text{-}27)$$

At any point along a transmission line the voltage and current amplitudes and phases can be summed up just as at the load. For all cases other than that of a matched load, amplitude variations as a function of distance from the load will be present. The curves defined by these amplitude variations are called *standing waves*. If a voltage wave is under observation, the ratio of a maximum value to one of its two adjacent minimum values is called the *voltage standing-wave ratio* (VSWR). If the load's reflection coefficient is known, VSWR can be determined in a low-loss line without actually measuring the wave amplitudes. It is given by

$$\text{VSWR} = \frac{1 + |\Gamma|}{1 - |\Gamma|} \qquad (4\text{-}28)$$

Significance of Load Impedance and VSWR

To conduct a transmission line analysis Eqs. (16) and (17) require values of characteristic impedance and load impedance. Characteristic impedance is a fundamental quantity that can be established from physical measurements on a transmission line. As the frequency of operation of an RF line increases, load impedance is often an unknown and difficult to measure quantity. On the other hand, the VSWR of a transmission line can be determined with relative ease by line measurement—by using any of a number of techniques described in Section 4-6.

If a transmission line's load Z_r value is unknown, it can be determined as follows. First, measure the VSWR at the load end of the line. Second, determine the reflection coefficient by using Eq. (4-28). Third, determine the reflection coefficient's phase angle by measuring the electrical angle between the end of the line and the point (toward the source end) where the next voltage maximum occurs (e.g., by using slotted-line techniques described in Section 4-6). The phase angle of Γ is equal to twice the measured angle. Fourth, determine load impedance by rearranging Eq. (4-27) and solving for the complex value of Z_r.

Why are we concerned with the value of VSWR on a line? Several of the major advantages of a low VSWR are as follows.

Higher power-handling capability A line's insulation breakdown voltage is determined by its peak VSWR voltage. Low VSWR results in low peaks.

Higher transmission efficiency. Low VSWR results in nearly uniform losses along a line/lower overall losses.

Nearly uniform input impedance. Line input power as a function of frequency is nearly constant with low VSWR. It varies widely with high VSWR.

High system stability. Many RF power amplifiers specify the maximum VSWR (often less than 1.1:1) into which they can operate in a stable manner.

RF System Analysis Using Appendix 5

For our second example using Appendix 5 we analyze a parallel wire RF line using NBS units. The system operates at a frequency of 9.348 MHz. The line has a characteristic impedance of 600 Ω, the velocity propagation factor is 0.95, the attenuation is 0.25 dB per 100 ft, and the length of the line is 50 ft. The load impedance is $300 + j0$ ohms, the source impedance is $1200 + j0$ ohms, and the load voltage is 450 V. Assume that induction and radiation losses from the line are negligible.

What are the steady-state conditions that prevail at each end of the line? What is the voltage and current distribution along the line?

To run this exercise issue a <RUN> command. Note that in this case our prompt responses are in terms of electrical parameters. Simply enter responses to prompts from the information above as shown in Fig. 4-3. A display of the system's fundamental performance capabilities will develop rapidly, as shown in Fig. 4-3. They show the following.

Efficiency. The line propagates energy with an efficiency of 96.489%.

Sending and receiving end performance. From the sending end/receiving end display one might be led to believe that the individual values of voltage and current decay at a relatively uniform rate between the two ends of the line. The reason this appears to be the case is that the 50-ft line length chosen at this operating frequency is equal to one-half wavelength (i.e., any low-loss transmission line has approximately the same voltage and current values at half-wavelength increments along its length).

To display the voltage/current standing waves along the line, next respond to the prompts with N and Y to initiate a dual VSWR plot. To obtain the plots in 1-ft increments along the entire line, next respond to the prompts with 0, 50, 1, and 0.4 (i.e., voltage scale maximum = source voltage times factor). Note that from this computer run we can determine the following.

Standing waves. The figure both plots and tabulates the wave values that exist along the line. The load is producing a 2:1 standing-wave ratio; hence the voltages, currents, resistances, and reactances are the result of the combined action of the steady-state incident and reflected waves that exist along the line. A tabulated value check shows that peak-to-valley ratios of approximately 2:1 exist for the voltage and current waves along the line.

```
This Is A Steady State Analysis Of A Transmission Line With Losses
---- -- - ------ ----- -------- -- - ------------ ---- ---- ------
    Is this a Utility line or an RF line application (U or R)? R
    Is this a parallel wire, coaxial, or waveguide line? P
    Are line dimensions in SI or NBS units (S or N) ? N
    Do you know the line's electrical characteristics? Y
 1. The line's frequency of operation in MHz =    9.3480
 2. The line's characteristic impedance value in ohms =  600
 3. The line's velocity of propagation factor value = .95
 4. The line's attenuation in dB / 100 ft = .25
 5. The line's length value in ft =  50
    Do you know the value of load impedance? Y
 6. The load's series resistance value in ohms =  300
 7. The loads series reactance value in ohms = 0
    The value of ZL = 300 @ 0 Deg.
 8. The source's series resistance value in ohms =  1200
 9. The source's series reactance value in ohms = 0
    Will analysis be referred to load voltage or source voltage (S or L)? L
10. The load rms voltage = 450
-----------------------------------------------------------------
    The Solutions At The Line's Sending And Receiving Ends Are:
    --- -------- -- --- ------ ------- --- --------- ---- ----
    Input Current =    1.511 @ 180.000 Deg.: Input Power =    699.562 Watts
    Input Voltage =    462.996 @ 180.000 Deg.
    Input Impedance: RI = 306.43: XI =  -0.00: ZI = 306.43@ -0.000 Deg.
    Output Current =   1.500 @   0.000 Deg.: Output Power =   675.000 Watts
    Output Voltage =   450.000 @   0.000 Deg.
    Power Attenuation = -0.155 dB: Line Efficiency = 96.489 %
-----------------------------------------------------------------
    Is a plot of intermediate line voltage and current solutions required (Y,N)? Y
    How far from the load (in ft) is the initial solution required?  0
    How far from the load (in ft) is the final solution required?  50
    What point to point  increments (in ft) are required?  2
    What scale multiplying factor is required? = .4
```

Figure 4-3 Appendix 5 analysis of parallel wire line characteristics.

Line impedance. If we divide the voltage value by the current value at any point along the line, a line impedance value will result. A corollary of this statement is that if we cut a line at any point along its length and terminate the source section in the impedance at the cut point—the active line will retain the same V and I values shown.

4-4 TYPES OF TRANSMISSION LINES

Four representative types of RF transmission lines are illustrated in Fig. 4-4. They are all evolutionary developments of the parallel wire line. We examine their capabilities in the following.

Parallel Wire Lines

This chapter has thus far been devoted completely to parallel wire lines. We summarize their predominant features below.

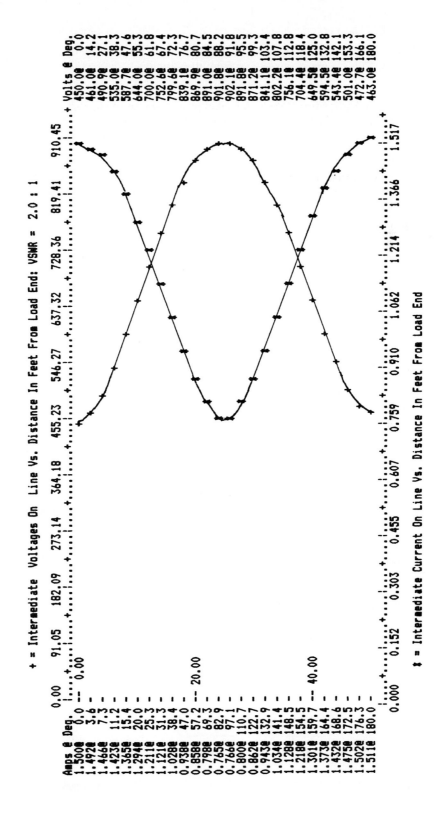

External fields. Electric and magnetic fields extend appreciable distances beyond the conductors as shown in Fig. 4-5. These fields induce currents in adjacent metal objects. This results in power loss from the system.

Line radiation. Radiation takes place from the conductors. This results in power loss from the system. Line radiation can also induce interference with nearby communication/service systems.

Structure. Parallel wire lines are simple and inexpensive to construct. They are not widely used for high-power transmission applications (fields, radiation), but they are used extensively as television receiving lines because of their low cost. The lines are not appropriate for many common applications: underground lines, underwater lines, and so on.

Circuit elements. Short parallel wire lines are frequently used as electronic circuit elements for RF systems housed in shielded enclosures. These parallel wire sections are typically used to replace conventional *LC* resonator circuits in VHF and UHF RF power generators.

Coaxial Lines

These are shielded lines consisting of an outer conducting cylinder and a centered inner conducting cylinder. A variety of forms of insulating support structures are used to secure the inner conductor at the center of the line's cross section. Coaxial lines provide high efficiency and high power-handling capabilities across the HF and UHF regions.

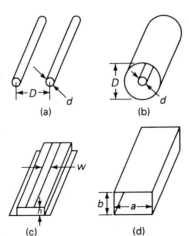

Figure 4-4 Representative types of transmission lines: (a) parallel wire line; (b) coaxial line; (c) microstrip line; (d) waveguide line.

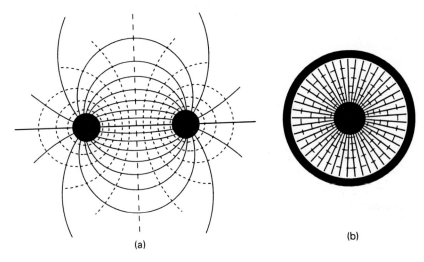

Figure 4-5 Representative electric and magnetic field patterns: (a) parallel wire; (b) coaxial transmission lines. [After E. W. Kimbark et al., *Principles of Radar* (Cambridge, Mass.: MIT Press, 1944).]

For an air-insulated line using copper as its inner and outer conductors, coaxial line attenuation is

$$A = \frac{kF^{0.5}(1 + D/d)}{D \log_{10}(D/d)} \quad (4\text{-}29)$$

nepers per unit length where $k = 3.02 \times 10^{-3}$ for 100-m unit lengths (3.62×10^{-4} for 100-ft unit lengths), F is in megahertz, d is the outer diameter of the inner conductor, and D is the inner diameter of the outer conductor [d and D in centimeters (inches)].

For center conductor support, rigid lines use insulated support rings spaced appreciable distances apart. Flexible lines use dielectrics such as solid polyethelene, dense insulating foam, and continuous helical strips as support members. The dielectric must be taken into account when defining the phase constant and losses of a line.

Example 4-1

The performance capabilities of a 60.96-m length of coaxial line at 98.4 MHz need to be determined. The electrical characteristics of the line are unknown. How can we determine if the line will have an efficiency of greater than 80%? Consider performing some cable tests—then use Appendix 5 in its physical mode.

Solution Lab tests showed that first, the inner and outer conductors were copper; second, the inner conductor had an outside diameter of 1.27 cm; third, the outer conductor had an inside diameter of 2.921 cm; fourth, the power factor of the insulation was 0.0005; fifth, the average dielectric constant was 1.05; sixth, the line's load impedance was $45 + j10$ ohms; and seventh, the generator's source impedance was effectively 0. Load the information above into the program—it shows that the line efficiency is 81.895%.

Some of the capabilities and shortcomings of coaxial lines are the following.

Induction and radiation. Induction and radiation power losses do not occur because of the line's confined electric and magnetic field lines as shown in Fig. 4-5.

Structure. Both rigid and flexible lines are available. Flexible lines permit a broad range of installation capabilities. Some forms of coaxial lines can be enclosed and pressurized—thus permitting installation in severe environmental circumstances. Coaxial lines are more complex and costly than parallel wire lines.

Circuit elements. Sections of coaxial line can be fabricated as high-power resonant circuit elements. They are self-shielding; hence additional shield enclosures are usually not required.

Shortcomings. First, as with all transmission lines, the line losses increase as a function of frequency. Second, when the spacing between conductors exceeds approximately one-half wavelength, the line's upper operating frequency is reached. Third, for higher frequencies of operation, it is necessary to decrease the structural dimensions. This results in lower power-handling capabilities and higher losses per unit length.

Microstrip Lines

Microstrip lines are widely used in conjunction with VHF, UHF, and SHF solid-state RF amplifier structures. The physical configuration of a microstrip line is shown in Fig. 4-4c. The conductors and dielectric are typically formed from copper-clad circuit board material.

A number of experimentally determined coefficients are needed before quantitative definitions of microstrip capabilities can be established. Two expressions help define their general nature of line attenuation. The conductor loss expression has the form

$$A_C = \frac{k\,(F\,e)^{1/2}}{H} \tag{4-30}$$

where attenuation is in nepers per unit length, k is a coefficient depending on the line's structure, F is the operating frequency, e the dielectric constant, and H the dielectric thickness. Correction factors are required in the value of k to account for variations in conductor dimensions and dielectric thickness.

The dielectric attenuation expression has the form

$$A_D = k_1 F(\text{FP})e^{1/2} \tag{4-31}$$

where attenuation is in nepers per unit length, k_1 is a second coefficient depending on the line's structure, and FP is the power factor or loss tangent of the dielectric. As with A_C, correction factors are required in the value of k_1 with this expression.

The significance of the correction factors required stems from the fact that microstrip line is, in effect, a more complex form of parallel wire line. Flux line fringing is therefore more complex. From a practical point of view, a microstrip design project typically starts from a well-stocked file of empirical data. An initial design can then be fabricated for laboratory evaluation and modification.

In the range 2 to 10 GHz representative power levels of several hundred watts can be handled before conductor heating becomes excessive. Corona discharges, due to sharp metal to insulation transitions in microstrip, typically limit peak pulse power levels to under 10 kW.

The capabilities and shortcomings of microstrip are as follows.

Induction and radiation.   Fringing of the electric and magnetic fields from the microstrip conductors results in coupling with external circuits. This, in turn, results in power loss from the system.

Structure.   Microstrip lines are relatively simple and inexpensive to construct. Their losses are relatively high and their power-handling capabilities are relatively low. Since their prime area of application is associated with relatively low-power RF systems, these limitations do not seriously affect their wide range of applications. It is necessary to keep line runs as short as practical.

Circuit elements.   The physical format of microstrip makes it highly adaptable to co-mounted resonant circuit/solid-state amplifier structures on a printed circuit board.

Example 4-2

Line-end matching stub(s) function to minimize VSWR and maximize energy transfer to the load. We can use Appendix 5, first, to determine the physical distance from the load to the VSWR stub attachment point, and second, to determine the type and length of the stub. (*Note*: The stub is a section of the same type of line as the main line.) From the point of stub attachment to the source, the line then operates as though it were terminated in a matched load.

Consider a 5400-MHz RF source driving a microstrip line with $Z_0 = 50\ \Omega$, propagation factor $= 0.65$, attenuation $= 40$ dB per 100 ft, and length $= 6.5$ ft. The load impedance is $40 + j30\ \Omega$. The source's internally generated voltage is 50 V rms and its impedance of $50 + j15\ \Omega$. What are the load and source stub locations, lengths, and operating modes?

Solution   Load Appendix 5, issue a <RUN> command, and then enter P, N, and Y followed by the information above. Following the initial printout, continue by responding N, N, and Y to the prompts that follow. The resulting analysis shows that the stub attachment point is 0.0328 ft from the load, that its length is 0.018 ft, and that its unattached end is short circuited. Then enter Y to obtain source stub information. Its attachment point is 0.0296 ft from the source, its length is 0.024 ft, and that its unattached end is short circuited.

What do the stubs accomplish for us? The initial analysis shows that with no stubs the system transfers 5.91 W of the generated power to the load. Under stub-matched operating conditions the system power transfer will be

$$P = a \log (40 \times \frac{6.5}{100} \frac{25^2}{50}) = 6.87\ W$$

Waveguide Lines

A waveguide is a hollow pipe that can provide low-loss power transfer capabilities between an RF generator and an RF load. It can provide high-power propagation capabilities over the UHF and SHF frequency regions. Most waveguides use a rectangular cross section, but circular cross sections are occasionally used as well.

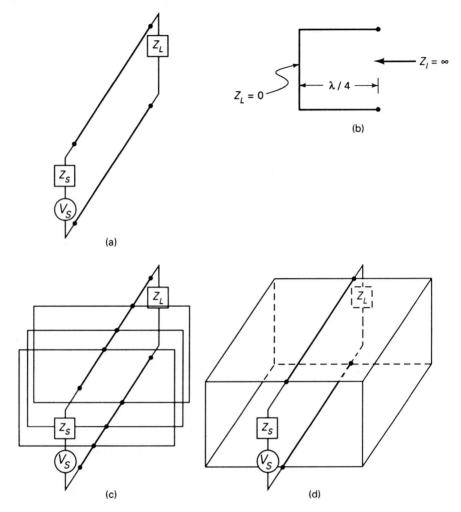

Figure 4-6 Waveguide rationale based on parallel wire line considerations: (a) open wire line system; (b) shorted quarter-wave stub; (c) open wire line with λ/4 stubs; (d) "equivalent" waveguide system.

The rationale for waveguide propagation can be visualized from Fig. 4-6. In the first sketch a conventional two-wire line has an RF source at one end and an RF load at the other. Figure 4-6b shows an open-wire segment one-quarter wavelength long at the operating frequency of the system (i.e., the input impedance of a quarter-wave shorted stub is infinite). In Fig. 4-6c a number of these stubs have been placed along the two-wire line, and since their input impedance is infinite, they do not affect energy propagation. In Fig. 4-6d an infinite number of the stubs has been added to form a pipe. Energy is therefore clearly propagating down a rectangular waveguide.

To form a frame of reference, let us divert from waveguides for a moment and consider the fields associated with the two-wire lines shown in Fig. 4-5a. The parallel wire line exhibits the unconfined nature of the solid electric field lines and the dotted

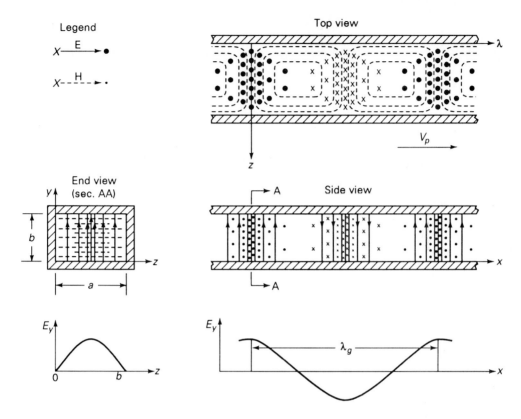

Figure 4-7 Field configuration for the dominant mode in rectangular waveguide. [After E. W. Kimbark et al., *Principles of Radar* (Cambridge, Mass.: MIT Press, 1944).]

magnetic field lines—the basis of the induction and radiation problems associated with this type of line. On the other hand, Fig. 4-5b shows the well-defined and shielded field associated with a coaxial line. If a ground plane were constructed perpendicular to the paper on the vertical magnetic field line in Fig. 4-5a, we would approximate the field of a microstrip line (use only the right or left half of the illustration). These three types of lines have a common characteristic. Waves propagate down them through use of the transverse electromagnetic mode (TEM). This means that energy from dc up to their maximum useful RF frequency can propagate down them due to their two-conductor structure.

Figure 4-7 illustrates the dominant mode fields in a rectangular waveguide. The end view shows that the electric field has a distribution in accordance with our previous simplified rationale. Similarly, the top and side views show the nature of the E and H fields as they propagate along the guide longitudinally. The E waveform shown below these two cross sections indicates the sinusoidal-shaped distribution of this field.

The simplest field pattern that will satisfy waveguide boundary conditions is illustrated in Fig. 4-7. An infinite number of patterns are possible, but most applications use this particular field pattern. It is referred to as the transverse electric 1, 0, or $TE_{1,0}$ mode of operation. The first subscript denotes the number of half-wave space variations in the field in going across the guide in the *a* direction. The second subscript denotes the number of variations across the guide in the *b* direction.

For a rectangular guide, operating in its dominant mode, the *cutoff* wavelength and frequency are (using SI units)

$$\lambda_c = 2a$$
$$F_c = \frac{30,000}{\lambda_c}$$
(4-32)

where F_c is in megahertz and a is the longer guide dimension in centimeters.

A waveguide acts as a low-pass filter. If frequencies lower than cutoff are applied, propagation down the line effectively ceases. Over the $TE_{1,0}$-mode frequency range, attenuation is

$$A = k\frac{(a/2b)\,(F/F_c)^{3/2} + (F/F_c)^{-1/2}}{a^{3/2}\,[(F/F_c)^2 - 1]^{1/2}}$$
(4-33)

nepers/unit length where $k = 1.71$ for 100-m unit lengths (0.129 for 100-ft unit lengths), a and b are guide width and height in centimeters (inches), and F is in megahertz.

Even though waveguides have air dielectric, the *wavelength in the guide* is a function of the frequency of operation. This relation is

$$\lambda' = \frac{\lambda}{[1 - (\lambda/\lambda_c)^2]^{1/2}}$$
(4-34)

where λ' is the wavelength within the guide and λ is the wavelength external to the guide.

The phase-shift constant is given by

$$B = \frac{2\pi}{\lambda'}$$
(4-35)

in radians.

The characteristic impedance in ohms is

$$Z_0 = \frac{120\pi\lambda' a}{b\lambda}$$
(4-36)

Waveguide line characteristics are typically expressed in terms of the electric and magnetic fields that exist as energy is propagated. VSWR-related measurement procedures provide a practical means of determining energy propagation characteristics in a waveguide.

Waveguide capabilities/shortcomings are as follows.

Induction and radiation. Induction and radiation losses do not occur because of the confined electromagnetic field.

Structure. The solid metal duct structure is simple and relatively economical to manufacture. The dielectric loss is negligible (air dielectric). Wall loss is present but it is less than with a comparable coaxial line (i.e., no center conductor). The physical size of guides is impractical for use below perhaps 400 MHz.

Circuit elements. Waveguide RF resonant structures are widely used in both transmission line and RF power generation circuits. These elements have the advantage of a self-shielding structure.

4-5 TRANSMISSION LINE AUXILIARY ELEMENTS

Transmission line monitoring, measuring, and wave conversion elements form a part of most practical RF systems. Some representative forms of devices/applications follow.

Wave Monitoring Equipment

A well-established method of measuring VSWR is by means of mechanically adjustable equipment. The information is obtained by moving a monitor/detector device along the line and documenting the maximum values, the minimum values, and the physical distance between them. If done with sufficient precision, the VSWR and the RF wavelength of operation of the system can be determined.

Figure 4-8a shows how this can be done with a parallel wire line system. In the arrangement shown, a quarter-wave stub is moved along the line, and a low-impedance detector provides readings at the load end of the stub. The rationale is that the low-impedance RF ammeter at the end of the stub effectively acts as a short circuit; hence the quarter-wave stub provides the dual advantages of, first, not loading the line since it appears as a very high impedance at the point of contact with the line, and second, the monitoring equipment is remote from the high-level fields that occur close to the conductors in an open wire line.

Figure 4-8b shows how a slotted-line measurement is made on a coaxial line. Within the slot, a probe (that acts like a short antenna) relays the signal it picks up to the external detector. The detector may consist of an RF diode detector circuit or a *bolometer* (a temperature-sensitive fine wire whose resistance changes as a function of the current through it) and associated low-frequency/dc readout circuits.

Figure 4-8c illustrates the slotted line principle applied to a waveguide system. The slot is placed on the longer dimension of the guide. Probes and detectors similar to the coaxial case can be used.

These three approaches can provide VSWR and/or the value of an unknown load impedance as described in Section 4-4. However, they sometimes exhibit limited accuracy due to nonlinear characteristics in the measuring equipment.

A more recent form of coaxial line VSWR measurement is shown in Fig. 4-8d. This technique uses a pair of short auxiliary lines placed internal to the main line, parallel to the center conductor, but close to the outer conductor. The lines have a detector at one end and a matching resistor (for their own characteristic impedance) at the opposite end. One of the lines monitors the incident wave and the other monitors the reflected wave. With this very simple device (a dual *directional coupler*) and associated readout elements, the user can monitor VSWR directly on an indicator.

Figure 4-8e shows how this technique is applied to waveguide lines. In this arrangement the incident and reflected waves are coupled out of the main guide by means of a pair of holes that pass into the auxiliary guide. The triangular structure at the end of each

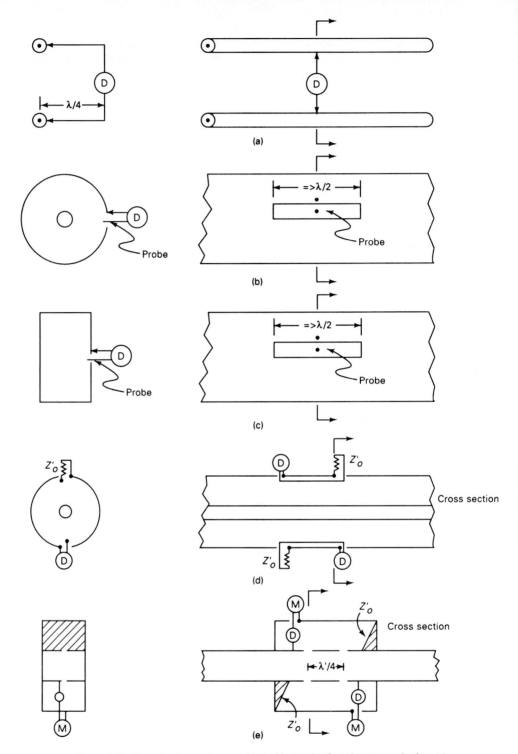

Figure 4-8 Two-wire line and waveguide incident and reflected wave monitoring: (a) parallel wire line USWR measurement; (b) slotted coaxial line USWR measurement; (c) slotted waveguide USWR measurement; (d) coaxial directional coupler USWR measurement; (e) waveguide directional coupler measurement.

auxiliary guide is a matching resistor. A detector is placed at the opposite end of each auxiliary guide. Otherwise, the principles are the same as with the coaxial dual directional coupler. Many variations from the techniques shown are used in commercially available coaxial and waveguide couplers.

Example 4-3

This 3000-MHz waveguide transmission line example illustrates an application of some of the foregoing auxiliary line devices. The line has inner dimensions of 7.123 and 3.404 cm, it is constructed of copper, and it is 32.81 m long. It is terminated with a load of unknown impedance. Directional coupler measurements show that the load has a VSWR of 1.5:1. Slotted-line measurements show that the first voltage maximum occurs 0.55 rad from the load. What is the load impedance? Approximately what transfer efficiency should be expected from this system?

Solution The use of Appendix 5 in its unknown load impedance mode shows that

$$Z_L = 143.51 + j53.29 \ \Omega \qquad \text{transfer efficiency} = 84.927\%$$

Waveguide Circulators

A circulator can be defined as a waveguide network, consisting of a wye symmetrical junction, loaded at its center with ferrite. The ferrite is biased with a dc magnetic field. In its ideal form, a circulator is a nonreciprocal device that can transfer microwave energy from one port to an adjacent port, fully decoupling or isolating all other ports. Figure 4-9 is a sectional concept of a junction circulator [5].

To illustrate, assume that an RF power generator system has a maximum VSWR at its output port specified as 1.25:1 and that the radiator for the system exhibits a VSWR

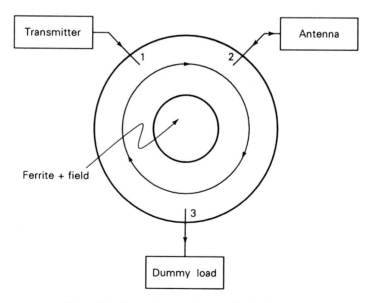

Figure 4-9 Single-junction circulator block diagram.

of up to 3:1 across its wide RF operating range. If we connect the RF power generator's output transmission line to an ideal junction circulator's port 1, the circulator transfers all of its port 1 energy in the direction of the arrowhead to port 2, where the transmission line to the radiator is connected. From RF line theory, the radiator will then reflect some of its energy back to port 2 due to its VSWR. This reflected energy will in turn all be transferred in the direction of the arrowhead to port 3 that is terminated in a matched dummy load. The reflected energy from the antenna is then all absorbed in the dummy load. Since no energy is left to transfer in the arrowhead direction to port 1, the RF power generator will operate into a matched load. Practical circulators are available that can easily provide 1.25:1 VSWR (or lower) capabilities at the RF power generator's output port.

Other configurations, known as differential phase-shift circulators, are capable of handling much higher power levels than junction circulators. Some differential units have RF capabilities in excess of 10 MW peak and 100 KW average. Under high-VSWR conditions, external cooling via heat exchangers may be required for high power units.

Circulators are feasible over a frequency range of about 1.5 to 30 GHz. They typically have insertion losses in the range of 0.15 to 0.5 dB. Their power-handling capabilities decrease as a function of frequency.

Miscellaneous Auxiliary Devices

RF power splitters and combiners are frequently required in RF power-generating systems. The coaxial four-port device (see Fig. 6-13) is one effective approach that can be used. Waveguide four-port devices use the same general principles.

A second common waveguide approach uses the *magic T* shown in Fig. 4-10a. When this arrangement is used as an RF power splitter, RF power enters the unit at port 1, and the divided RF output appears at ports 2 and 3. As an RF combiner, ports 2 and 3 would be used for the two RF inputs and port 1 would deliver the RF output. Port 4 would be terminated in a matched load. In an ideal case no power would be absorbed in the dummy load. For situations where source impedances, phase relationships, and load matches are not at their ideal values, the load would absorb some power.

There are two supplementary elements in the magic T. One is the diaphragm in the port 4 line, and the other is the cylindrical post at the junction of all 4 lines. These elements perform the line matching/resonating functions required for combiner/splitter action.

A transition from parallel wire line to coaxial line is often required in RF power generator designs. A device that transforms from a balanced to unbalanced line is usually referred to as a *balun*. A variety of approaches are used, but a simple transmission line approach is shown in Fig. 4-10b. Assume that the parallel line's conductors are balanced with respect to ground but that the outer surface of the coaxial line can have any status with respect to ground. An auxiliary quarter-wave sleeve placed around this conductor and shorted at the end remote from the junction provides an open circuit between the outer coaxial surface and the wire that attaches to that surface—thereby providing a substantial degree of RF isolation.

Coaxial line to waveguide *transformations* are frequently required. One approach is shown in Fig. 4-10c. The coaxial line is assumed to be operating in its TEM mode and

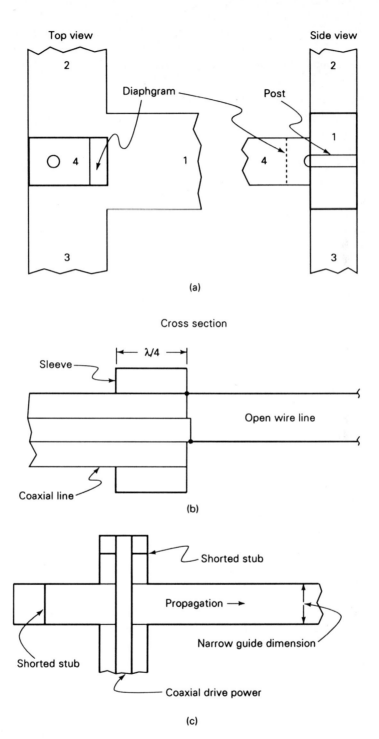

Figure 4-10 Some miscellaneous auxilliary transmission line devices: (a) waveguide magic T hybrid junction; (b) coaxial/open wire balun; (c) cross section of coaxial line-to-waveguide transistion.

the waveguide in its $TE_{1,0}$ mode. The coaxial line enters the long side of the waveguide close to its end. By adjusting the length of the two stubs the center conductor of the coaxial line is made to function like an antenna, and it propagates energy down the guide. A simpler system deletes the coaxial stub and accomplishes matching by adjusting the end stub on the guide and the length of the unterminated coaxial center conductor's projection into the guide.

Chapter 5

Fourier Transformations

Why do we need *transforms* to conduct power electronics analyses? Primarily because transforms permit us to obtain steady-state analysis solutions for systems that have nonsinusoidal waveforms. What system status must exist for us to perform a transform? The system must have reached steady-state operating conditions—on a cycle-to-cycle basis. The following two system descriptions will illustrate how we proceed with transform analyses.

Pulse Spectrum Systems

The first analyses we perform concern pulse generating systems. These systems typically function to modulate RF oscillators or RF amplifiers in radar transmitter systems. The waveforms generated by the pulsers are normally defined as functions of time (i.e., pulse duration, pulse rise and fall time, interpulse time, etc.), but the RF systems are defined in terms of their frequency of operation. How do we correlate these areas?

A radar system typically has only one RF carrier frequency. The process of RF pulse modulation, reviewed in Chapter 6, shows that the system delivers RF output over a *frequency spectrum*—a band of output frequencies. Transforms allow us to determine the pulser's steady-state spectrum in terms of a series of sinusoidal waveforms. If we correlate the pulse spectrum with the principles of RF modulation, we can determine the system's RF pulse spectrum.

Prime Line Interaction Systems

Our second set of analyses concern a system whose 60-Hz prime power source delivers an exactly sinusoidal voltage waveform. When this waveform is applied to the ac input port of a steady-state dc power supply, nonsinusoidal waveforms will be present throughout the power supply system. Transient analyses can be used to determine the waveforms of power supply systems. How can we obtain and then utilize this transient information?

The techniques used to perform transient analyses are described in Chapter 8. Assume that we have this power supply information available. Then the transient information (in the time domain) can be transformed into a *series* of sinusoidal voltages or currents (in the frequency domain). Following this, through the application of several straightforward mathematical principles, the transform terms can provide us with the steady-state solutions we need for system analysis.

5-1 FOURIER SERIES PRINCIPLES

We will conduct our transforms of periodic waves by using a *Fourier series*. The series can be defined as follows:

$$y = f(x) = A_0 + A_1 \sin x + B_1 \cos x + A_2 \sin 2x + B_2 \cos 2x + \\ A_3 \sin 3x + B_3 \cos 3x + \cdots + A_N \sin Nx + B_N \sin Nx \quad (5\text{-}1)$$

where y is a periodic transient function of time defined over 2π radians, and $N = 4, 5, 6, \ldots$. Except for constant A_0, each term in this infinite series is the product of a unique *coefficient* and the instantaneous value of a sine or cosine wave. The fundamental wave frequency equals 1 over the period of the transient in seconds. Other waves are harmonics of this frequency.

One approach used to determine the series coefficients is to perform appropriate mathematical operations on Eq. (5-1)—operations that will eliminate all terms except the desired term. Thereby, the value of each coefficient can be established.

To illustrate, say that we want to determine the A_0 term of a wave that is defined between 0 and 2π. The mathematical operation used in this case is to multiply Eq. (5-1) by dx and then integrate each term as follows:

$$\int_0^{2\pi} y\, dx = \int_0^{2\pi} A_0\, dx + \int_0^{2\pi} A_1 \sin dx + \int_0^{2\pi} B_1 \cos dx + \int_0^{2\pi} A_2 \sin 2x\, dx \\ + \int_0^{2\pi} B_2 \cos dx \cdots + \int_0^{2\pi} A_N \sin Nx\, dx + \int_0^{2\pi} B_N \cos Nx\, dx \quad (5\text{-}2)$$

When we integrate, all of the terms above except the first two are equal to zero, so we have

$$\int_0^{2\pi} y\, dx = A_0 \int_0^{2\pi} dx = 2\pi A_0 \quad A_0 = \frac{1}{2\pi} \int_0^{2\pi} y\, dx \quad (5\text{-}3)$$

Through the use of similar multiplication, integration, and trigonometric procedures, all of the Fourier coefficients can be determined. They are:

$$A_1 = \frac{1}{\pi} \int_0^{2\pi} y \sin x \, dx \qquad (5\text{-}4)$$

$$B_1 = \frac{1}{\pi} \int_0^{2\pi} y \cos x \, dx \qquad (5\text{-}5)$$

$$\vdots$$

$$A_n = \frac{1}{\pi} \int_0^{2\pi} y \sin Nx \, dx \qquad (5\text{-}6)$$

$$B_n = \frac{1}{\pi} \int_0^{2\pi} y \cos Nx \, dx \qquad (5\text{-}7)$$

Analytical Evaluation of Fourier Coefficients

Except for the most elementary transient waveforms, analytical evaluation of Fourier coefficients necessitates the use of relatively complex mathematical procedures. As a means of demonstrating the approach, we analyze one of the most frequently mentioned waveforms in power electronics—a rectangular pulse. For this analysis the pulse is located on its time scale as shown in Fig. 5-1. T_P is the pulse duration in seconds, T_R the pulse repetition rate (PRR) period in seconds, and A the pulse amplitude (typically, expressed in volts or amperes). When the pulse duration is equally divided between the start and the conclusion of period T_R, the use of odd and even functions simplifies the mathematical development.

Odd and even functions are defined below. They simplify the equations as stated following each definition:

If $y = f(x) = -f(x)$, the series is odd and has sine terms only (5-8)

If $y = f(x) = f(-x)$, the series is even and has cosine terms only (5-9)

For the wave form of Fig. 5-1 we will have cosine terms only.
For the A_0 coefficient, $N = 0$ and $f(x) = 1$. Therefore,

$$\begin{aligned}
A_0 &= \frac{1}{\pi} \left[\int_0^{\pi(T_P/T_R)} 1 \, dx + \int_{\pi[2-(T_P/T_R)]}^{2\pi} 1 \, dx \right] \\
&= \frac{1}{\pi} \left[\left. x \right|_0^{\pi(T_P/T_R)} + \left. x \right|_{\pi[2-(T_P/T_R)]}^{2\pi} \right] \\
&= \frac{1}{\pi} \left(\pi \frac{T_P}{T_R} - 0 + 2\pi - 2\pi + \pi \frac{T_P}{T_R} \right) \\
&= 1 \left(\frac{T_P}{T_R} \right)
\end{aligned} \qquad (5\text{-}10)$$

Sec. 5-1   Fourier Series Principles

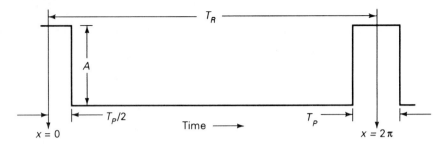

Figure 5-1 Parameters of a representative rectangular pulse.

For the harmonic coefficients,

$$A_N = 1/\pi \left[\int_0^{\pi(T_P/T_R)} 1[\cos(Nx)]\, dx + \int_{\pi[2-(T_P/T_R)]}^{2\pi} 1[\cos(Nx)]\, dx \right]$$

$$= \frac{1}{N\pi} \left[|\sin(Nx)|_0^{\pi(T_P/T_R)} + |\sin(Nx)|_{\pi[2-(T_P/T_R)]}^{2\pi} \right] \quad (5\text{-}11)$$

$$= \frac{1}{N\pi} [\sin(N\pi \frac{T_P}{T_R}) - 0 + 0 + \sin(N\pi \frac{T_P}{T_R})]$$

$$= \frac{2\sin[N\pi\,(T_P T_R)]}{N\pi}$$

Since $T_P/T_R = $ DU, where DU is the duty factor, we can write Eq. (5-11) as

$$A_N = \frac{2\mathrm{DU}\,\sin(N\pi\cdot\mathrm{DU})}{N\pi\cdot\mathrm{DU}} \quad (5\text{-}12)$$

If we define the analysis angle as $\theta = 2\pi\cdot\mathrm{PRR}\cdot t$, where t is the analysis time, the complete series can be written as follows:

$$f(x) = 2\mathrm{DU}\,(0.5 + \frac{\sin \pi\cdot\mathrm{DU}}{\pi\cdot\mathrm{DU}}\cos\theta$$
$$+ \frac{\sin 2\pi\cdot\mathrm{DU}}{2\pi\cdot\mathrm{DU}}\cos 2\theta \quad (5\text{-}13)$$
$$+ \frac{\sin 3\pi\cdot\mathrm{DU}}{3\pi\cdot\mathrm{DU}}\cos 3\theta + \cdots)$$

A number of the terms in the series may have a value of zero. These zero values typically indicate an axis crossing (where the coefficients reverse polarity).

Rectangular Pulse Analysis Computer Program

To illustrate the use of Eq. (5-13), consider a radar system that uses a rectangular pulse train with $A = 1$ V, $T_P = 10 \times 10^{-6}$ s, PRR $= 10{,}000$ Hz, and DU $= 0.1$. If these values were substituted into Eq. (5-13), a table of coefficients could be calculated laboriously. To simplify the process, we can simply load a computer program listed in Appendix 6, issue a <RUN> command, and respond to prompts 1 and 2 with the pulse duration and repetition rate listed above. To complete the input, we request 40 harmonics in response to prompt 3, a coefficient versus frequency display in response to prompt 4, and a coefficient versus time display (to be defined later) in response to prompt 5. The printer outputs shown in Fig. 5-2 will result.

Figure 5-2a provides both graphic and numeric response values for the dc term and each harmonic. The following information can be determined from the display: First, the

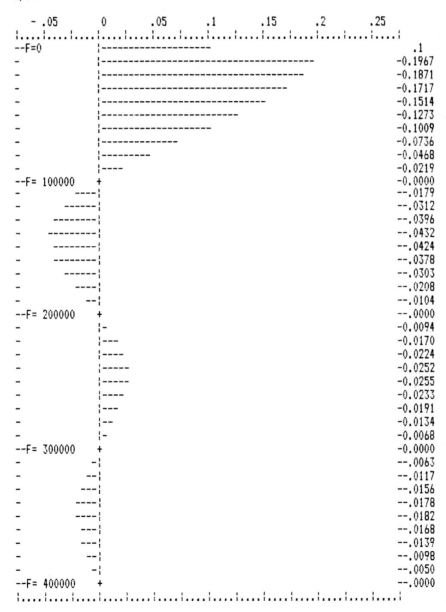

Figure 5-2 Appendix 6 analysis of a rectangular pulse.

Sec. 5-1   Fourier Series Principles

Figure 5-2 (*cont.*)

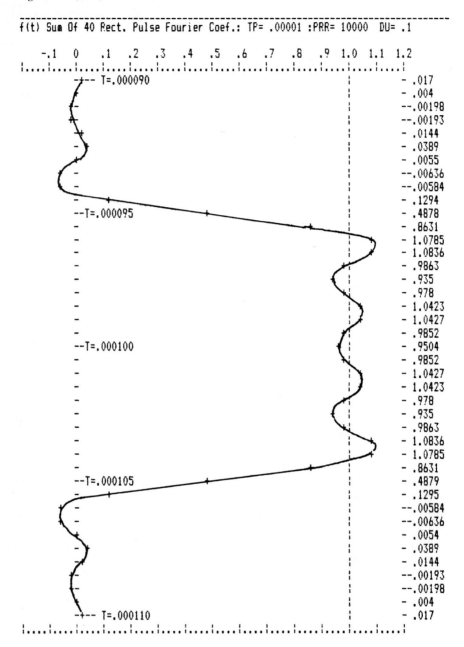

major lobe of the spectrum contains 1/DU frequency lines whose frequency increments = PRR. Second, succeeding lobes also have 1/DU lines. The lobes have alternate positive and negative values. Third, if the pulse is used to modulate an RF power generator, the first axis crossover (at $1/T_P$) provides a useful design guide. The crossover frequency can be used to establish the approximate bandwidth required (e.g., in the radar system's receiver). In a simple system, the receiver's 3-dB down bandwidth should be approximately $2/T_P$ or 200 kHz. Fourth, a power summation for all the spectrum lines will show that over 90% of the pulse energy is contained in the lines that constitute the major lobe.

In addition to Fourier transforms, it is often informative to run inverse transforms—transforms that sum up all the Fourier terms and return us to the time domain. For example, the question might be asked: What degree of distortion will be introduced if we pass the spectrum of an ideal rectangular pulse train through a limited bandwidth channel (e.g., such as a receiver)? When we answered the fifth prompt with a Y, the inverse transform of Fig. 5-2b resulted.

Figure 5-2b shows that if we passed a complete set of Fourier terms through an ideal low-pass filter with a 400-kHz bandwidth: First, significant rise and fall time slopes develop on the originally rectangular pulse; and second, significant ripple is present on the originally flat top of the pulse.

For a simple radar system the response shown is perhaps adequate, but the equivalent rise time of the pulse may not be satisfactory for more sophisticated applications. If we passed more Fourier coefficients (e.g., if we responded to computer prompt 3 with a request for 160 harmonics), the response would be improved. However, in a radar case where both signal and noise are present, the wider bandwidth for this case would also allow more noise to be passed. Therefore, the capability of receiving a weak signal in the presence of an increased receiver noise level is clearly a factor that should be investigated further during the radar system's design phase.

5-2 TABULAR FOURIER ANALYSES

It is often not feasible to perform an analysis that is completely mathematical (i.e., definition of the mathematical procedures become too complex). When this is the case, a tabular ordinate graphical approach is used. In the analyses that follow tabular approaches will be used exclusively.

Tabular Analysis Program

Typical transient analyses on a computer are performed by means of numerical integration. The solutions are usually obtained through the use of uniformly spaced incremental time steps. The transform program to be described, Appendix 7, uses the same (or related) analysis time steps as the foundation for its solution approach.

The fundamental purpose of the program is to obtain the Fourier coefficients and their phase angles at each of the harmonically related frequencies. However, as noted previously, the end outputs we are concerned about are system voltage, current, volt-amperes, and power. The program also automatically solves for these values if we request them.

For any nonsinusoidal current, the program determines the transient wave's rms value by using the following relationship:

$$I = \left[\frac{I_o + (I_{m1}^2 + I_{m2}^2 + \cdots + I_{mn}^2)}{2}\right]^{1/2} \quad (5\text{-}14)$$

where, in terms of subscripts, the letter m defines the peak value of the wave, and the following digit defines the harmonic number of that term in the series. The program obtains all necessary information from the coefficient and phase-angle solutions. It then proceeds to square each term, sum the squared terms, divide by 2, and extract the square root of the result.

The rms value of nonsinusoidal voltage waves are determined in the same fashion. Volt-amperes are determined by the rms product,

$$\text{VA} = VI \quad (5\text{-}15)$$

and the program will display this value if requested.

Power is determined by using the following relationship:

$$P = 0.5\,[V_{m1}I_{m1}\cos(A_{v1} - A_{I1}) + V_{m2}I_{m2}\cos(A_{v2} - A_{I2}) + \cdots \\ + V_{mn}I_{mn}\cos(A_{vn} - A_{In})] \quad (5\text{-}16)$$

If requested, the program will calculate and display this value.

The general procedure for using the program is as follows,

Define the fundamental frequency. For example, consider a train of rectangular pulses that are uniformly spaced in the time domain. The fundamental frequency can be defined as $F = 1/T_R$, where T_R is the repetition period of the transient waveform.

Define the analysis time step. Short time steps yield higher analysis accuracy; however, excessively short time steps unduly increase the amount of time required to perform the transform.

Define the wave. Either a mathematical or graphical waveform description can be used. If a mathematical waveform description is feasible, follow the instructions given in the next section. The graphical input description follows.

Define the graphical data input mode. Two approaches are possible with this program. In the first approach, each transient data ordinate is entered in response to a request from the program. In the second approach, the ordinates are entered into the program in the form of DATA statements. These statements are READ as the program executes.

Define the printout type. Either a monitor display or a line printer display of the DATA listing and/or the transform results can be provided.

Define if a power analysis is required. To obtain a meaningful power analysis, a specific arrangement of data input is required.

Analysis of a Train of Equally Spaced Rectangular Pulses

For our first example, the pulse characteristics and the analysis procedure are defined as follows. First, all pulses will have a duration of 0.5 s and an amplitude of 100 V. Second, the pulse-to-pulse spacing will be 1 s. Third, the pulses will deliver power into a 1-Ω load. Fourth, manual data input from the keyboard will be used. Fifth, the analysis increment will equal the pulse-to-pulse spacing divided by 64.

All Fourier transforms are composed of an infinite number of terms. With many waveforms, only a limited number of terms are required to provide sufficient accuracy for analysis work, but with rectangular waveforms, the terms decay at a slow rate. Hence a fairly large number is required for this analyses. In this example the pulse waveform is used to modulate an RF amplifier; hence the analysis is for pulse spectrum investigation purposes.

From a practical point of view, a rectangular pulse is never achieved (due to the presence of parasitic circuit elements such as connecting lead inductances and capacitances). As a result, rise and fall times are always present. Figure 5-1 illustrates the time-domain display for a single pulse in the system outlined above. The rise and fall times are physically represented by the linear slopes shown for data points 15, 16, 17, and 47, 48, 49. These slopes are positioned so that the area under the voltage pulse is the same as that of an equivalent rectangular voltage pulse. If the pulse to be analyzed is positioned on the time scale as shown in the figure, half of the transform terms will be eliminated (when symmetrical pulses are analyzed).

With this background established, proceed as follows to perform the analysis by first loading the program and then entering a <RUN> command. Then proceed as follows. First prompt: Answer the question presented with the digit 1 (to indicate that tabular input is to be used). Press <RETURN>. Second prompt: Enter the number 65 (to indicate that a total of 65 data points will be entered). Press <RETURN>. Third prompt: Enter the digit 1 (to indicate the fundamental frequency will be the first term of the analysis). Press <RETURN>. Fourth prompt: Enter the number 17 (to indicate that 17 harmonics will be calculated). Press <RETURN>. Fifth prompt: Enter the letter M (to indicate that you will enter data ordinates manually). Press <RETURN>. Ordinate prompts (65): Enter each sequential ordinate from Fig. 5-3. Press <RETURN> after each. Next prompt: Enter the digit 1 (to indicate a 1 s pulse to pulse period). Press <RETURN>. Initial program execution: Periodic calculations will appear on the screen

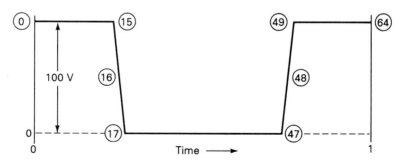

Figure 5-3 Practical rectangular pulse in the time domain.

as each harmonic is calculated. When they are complete, respond to the next prompt with the letter M or P (to define the display mode you want). Press <RETURN> (if P is selected, make sure that your printer is on), respond to the next prompt with the letter Y, and again press <RETURN>. Next prompt: Enter the letter Y (to indicate that a data point listing is required). Press <RETURN>. Figure 5-4 shows the display. Note that

```
-----------------------------------------------------------------------
            ---Numerical Integration Evaluation Of Fourier Coefficients---
                         Number Of Ordinates =  65
                              D-C Term, A(0) =  50
                             -- Data Ordinates --
  Point #  Ordinate   Point #  Ordinate   Point #  Ordinate   Point #  Ordinate   Point #  Ordinate
  -------  --------   -------  --------   -------  --------   -------  --------   -------  --------
       0   100.000         1   100.000         2   100.000         3   100.000         4   100.000
       5   100.000         6   100.000         7   100.000         8   100.000         9   100.000
      10   100.000        11   100.000        12   100.000        13   100.000        14   100.000
      15   100.000        16    50.000        48    50.000        49   100.000        50   100.000
      51   100.000        52   100.000        53   100.000        54   100.000        55   100.000
      56   100.000        57   100.000        58   100.000        59   100.000        60   100.000
      61   100.000        62   100.000        63   100.000        64   100.000

           HARMONIC      COS TERM      SIN TERM     MAGNITUDE       PHASE        PERCENT
           --------      --- ----      --- ----     ---------       -----        -------
              1           63.66          0.00         63.66          0.00        100.00
              2            0.00          0.00          0.00          0.00          0.00
              3          -21.22          0.00         21.22        180.00         33.33
              4            0.00          0.00          0.00          0.00          0.00
              5           12.74          0.00         12.74          0.00         20.01
              6            0.00          0.00          0.00          0.00          0.00
              7           -9.11          0.00          9.11        180.00         14.30
              8            0.00          0.00          0.00          0.00          0.00
              9            7.10          0.00          7.10          0.00         11.15
             10            0.00          0.00          0.00          0.00          0.00
             11           -5.84          0.00          5.84        180.00          9.17
             12            0.00          0.00          0.00          0.00          0.00
             13            4.99          0.00          4.99          0.00          7.83
             14            0.00          0.00          0.00          0.00          0.00
             15           -4.39          0.00          4.39        180.00          6.90
             16            0.00          0.00          0.00          0.00          0.00
             17            3.98          0.00          3.98          0.00          6.25

          Fundamental Frequency =  1   Hertz

          Volt-Ampere & Power Analysis
          ----------------------------

Harmonic  Voltage  V. Angle  Amperes  A. Angle   Power
--------  -------  --------  -------  --------   -----
    1      63.66     0.00     63.66     0.00    2026.43
    2       0.00     0.00      0.00     0.00       0.00
    3      21.22   180.00     21.22   180.00     225.18
    4       0.00     0.00      0.00     0.00       0.00
    5      12.74     0.00     12.74     0.00      81.11
    6       0.00     0.00      0.00     0.00       0.00
    7       9.11   180.00      9.11   180.00      41.46
    8       0.00     0.00      0.00     0.00       0.00
    9       7.10     0.00      7.10     0.00      25.20
   10       0.00     0.00      0.00     0.00       0.00
   11       5.84   180.00      5.84   180.00      17.04
   12       0.00     0.00      0.00     0.00       0.00
   13       4.99     0.00      4.99     0.00      12.43
   14       0.00     0.00      0.00     0.00       0.00
   15       4.39   180.00      4.39   180.00       9.64
   16       0.00     0.00      0.00     0.00       0.00
   17       3.98     0.00      3.98     0.00       7.93

   RMS Voltage = 70.33085
   RMS Current = 70.33085
   Volt Amperes = 4946.427
         Power = 4946.428
   Power Factor = 1
```

Figure 5-4 Fourier coefficient/power analysis for tabular rectangular pulse using Appendix 7.

data points with a value of zero are not listed. This list is valuable on new programs because it allows you to check that the data have been entered in the manner you intended. The listing is shown at the top of Fig. 5-4. Solution display: The ordinate list is followed automatically by the solution display shown in Fig. 5-4. For each harmonic it defines the sine and cosine coefficients, the overall magnitude, the phase angle, and the percentage amplitude (relative to the fundamental frequency). Note that as a result of our time-scale arrangement, only cosine terms are present in the transform. Next prompt: Enter the letter N (to indicate that no further inputs to the analyses are to be supplied at this time). Press <RETURN>. Next prompt: Because of its linearity and its symmetry, this system has a very simple power analysis. Without further discussion at this juncture, answer the next question Y (to indicate that power calculations are to be performed). The power analysis chart will then be displayed.

The analysis shows that even with the limited number of terms considered, both the volt-amperes and power calculations have values of approximately 5000 (5000 is the exact solution for an ideal rectangular pulse). This display occurs because the program has automatically loaded the current transform columns with the voltage data (i.e., we did not enter any current information, so the program again loaded the only data available for the second part of the power analysis). For the special case of a 1-Ω load resistance, the current values are of course equal to the voltage values. A more elaborate power solution is examined in Fig. 5-5.

Analysis of Pulse Train with Cosine-Squared Rise and Fall Times

For this analysis the waveform is defined as follows. First, the pulses will use shaped rise and fall times. Shaped rise and fall times function to attenuate the higher-frequency harmonics at an increased rate (i.e., relative to the harmonics of a rectangular pulse). The arrangement chosen for this example uses cosine-squared curved slopes that merge with a flat pulse top. Second, the effective pulse duration is 0.1 s, based on the pulse's average voltage value. Third, the pulse-to-pulse spacing will be 1 s. Fourth, the pulses will be delivering power into a 1-Ω load. Fifth, the pulse shape will be described mathematically. Sixth, the analysis time increment will equal the pulse-to-pulse spacing divided by 320.

Program lines 400–600 of Appendix 7 list the subroutine for the shaped pulse. When this subroutine is called by the program, it defines the pulse by using the same general amplitude versus time positioning as in the previous example. From an incremental point of view, the pulse's initial flat-top region occurs during the first 10 time-increment periods, the cosine-squared fall time of the pulse occurs during the next of 12 time increments, the pulse's zero-output region over the next 277 time increments, and the pulse's rise time and concluding flat top (in a manner similar to the initial flat top and fall time) during the final 22 time increments.

All pulse increments and ordinates are established automatically by a relatively straightforward set of equations in the FOR–NEXT loop in the subroutine. This analysis approach greatly simplifies loading data into the program. Entering manual response inputs from the keyboard (as in the previous example) or entering DATA statements into the program from the keyboard (as in the next exercise) is not required. However, for

```
---Numerical Integration Evaluation Of Fourier Coefficients---
                    Number Of Ordinates = 321
                    D-C Term, A(0) = 10
                          -- Data Ordinates --
Point #  Ordinate   Point #  Ordinate   Point #  Ordinate   Point #  Ordinate   Point #  Ordinate
-------  --------   -------  --------   -------  --------   -------  --------   -------  --------
   0     100.000       1     100.000       2     100.000       3     100.000       4     100.000
   5     100.000       6     100.000       7     100.000       8     100.000       9     100.000
  10     100.000      11      98.296      12      93.301      13      85.355      14      75.000
  15      62.941      16      50.000      17      37.059      18      25.000      19      14.645
  20       6.699      21       1.704     299       1.704     300       6.699     301      14.645
 302      25.000     303      37.059     304      50.000     305      62.941     306      75.000
 307      85.355     308      93.301     309      98.296     310     100.000     311     100.000
 312     100.000    313     100.000      314     100.000     315     100.000     316     100.000
 317     100.000    318     100.000      319     100.000     320     100.000
```

HARMONIC	COS TERM	SIN TERM	MAGNITUDE	PHASE	PERCENT
1	19.65	0.00	19.65	0.00	100.00
2	18.61	0.00	18.61	0.00	94.73
3	16.97	0.00	16.97	0.00	86.35
4	14.82	0.00	14.82	0.00	75.44
5	12.32	0.00	12.32	0.00	62.70
6	9.62	0.00	9.62	0.00	48.97
7	6.90	0.00	6.90	0.00	35.10
8	4.30	0.00	4.30	0.00	21.87
9	1.96	0.00	1.96	0.00	9.99
10	0.00	0.00	0.00	0.00	0.00
11	-1.52	0.00	1.52	180.00	7.73
12	-2.57	-0.00	2.57	-180.00	13.07
13	-3.15	-0.00	3.15	-180.00	16.03
14	-3.31	-0.00	3.31	-180.00	16.85
15	-3.12	-0.00	3.12	-180.00	15.86
16	-2.66	-0.00	2.66	-180.00	13.53
17	-2.03	-0.00	2.03	-180.00	10.32
18	-1.32	-0.00	1.32	-180.00	6.72
19	-0.62	0.00	0.62	180.00	3.17
20	0.00	0.00	0.00	0.00	0.00
21	0.50	0.00	0.50	0.00	2.53
22	0.84	0.00	0.84	0.00	4.29
23	1.03	0.00	1.03	0.00	5.24
24	1.07	0.00	1.07	0.01	5.46
25	0.99	0.00	0.99	0.00	5.06
26	0.83	0.00	0.83	0.00	4.22
27	0.62	0.00	0.62	0.00	3.13
28	0.39	0.00	0.39	0.00	1.97
29	0.18	0.00	0.18	0.00	0.89
30	0.00	0.00	0.00	0.00	0.00
31	-0.13	0.00	0.13	180.00	0.64
32	-0.20	0.00	0.20	180.00	1.01
33	-0.22	0.00	0.22	180.00	1.14
34	-0.21	0.00	0.21	180.00	1.07
35	-0.17	0.00	0.17	180.00	0.87
36	-0.12	0.00	0.12	180.00	0.62
37	-0.07	0.00	0.07	180.00	0.37
38	-0.03	0.00	0.03	180.00	0.16
39	-0.01	0.00	0.01	180.00	0.04
40	0.00	0.00	0.00	0.00	0.00

Fundamental Frequency = 1 Hertz

Figure 5-5 Fourier coefficients for shaped pulse using Appendix 7.

```
Volt-Ampere & Power Analysis
---------------------------

Harmonic  Voltage  V. Angle  Amperes  A. Angle  Power
--------  -------  --------  -------  --------  -----
   1       19.65     0.00     19.65     0.00    193.00
   2       18.61     0.00     18.61     0.00    173.20
   3       16.97     0.00     16.97     0.00    143.92
   4       14.82     0.00     14.82     0.00    109.83
   5       12.32     0.00     12.32     0.00     75.88
   6        9.62     0.00      9.62     0.00     46.29
   7        6.90     0.00      6.90     0.00     23.77
   8        4.30     0.00      4.30     0.00      9.23
   9        1.96     0.00      1.96     0.00      1.93
  10        0.00     0.00      0.00     0.00      0.00
  11        1.52   180.00      1.52   180.00      1.15
  12        2.57  -180.00      2.57  -180.00      3.29
  13        3.15  -180.00      3.15  -180.00      4.96
  14        3.31  -180.00      3.31  -180.00      5.48
  15        3.12  -180.00      3.12  -180.00      4.86
  16        2.66  -180.00      2.66  -180.00      3.53
  17        2.03  -180.00      2.03  -180.00      2.06
  18        1.32  -180.00      1.32  -180.00      0.87
  19        0.62   180.00      0.62   180.00      0.19
  20        0.00     0.00      0.00     0.00      0.00
  21        0.50     0.00      0.50     0.00      0.12
  22        0.84     0.00      0.84     0.00      0.35
  23        1.03     0.00      1.03     0.00      0.53
  24        1.07     0.01      1.07     0.01      0.57
  25        0.99     0.00      0.99     0.00      0.49
  26        0.83     0.00      0.83     0.00      0.34
  27        0.62     0.00      0.62     0.00      0.19
  28        0.39     0.00      0.39     0.00      0.08
  29        0.18     0.00      0.18     0.00      0.02
  30        0.00     0.00      0.00     0.00      0.00
  31        0.13   180.00      0.13   180.00      0.01
  32        0.20   180.00      0.20   180.00      0.02
  33        0.22   180.00      0.22   180.00      0.03
  34        0.21   180.00      0.21   180.00      0.02
  35        0.17   180.00      0.17   180.00      0.01
  36        0.12   180.00      0.12   180.00      0.01
  37        0.07   180.00      0.07   180.00      0.00
  38        0.03   180.00      0.03   180.00      0.00
  39        0.01   180.00      0.01   180.00      0.00
  40        0.00     0.00      0.00     0.00      0.00

   RMS Voltage = 30.10371
   RMS Current = 30.10371
  Volt Amperes = 906.2333
         Power = 906.2333
  Power Factor = 1
```

Figure 5-6 Volt-ampere and power analysis for shaped pulse using Appendix 7.

anything other than the relatively simple pulse described in this example, the mathematical procedures may become overwhelming.

With this background established, a computer analysis of the system can be initiated by using the same starting procedure as in the last analysis. Then proceed as follows.

Analysis definition prompt: Enter the digit 2 (this indicates that a mathematical pulse definition will be used). Press <RETURN>. The ordinate values calculated by the subroutine will then be displayed rapidly on the screen in sequential fashion. Next prompt: Enter the digit 1 (this indicates that the analysis will start with the fundamental frequency). Press <RETURN>. Next prompt: Enter the number 40 (this indicates 40 harmonics will be calculated). Press <RETURN>. Next prompt: Enter the digit 1 (to define the pulse to pulse period in seconds). Press <RETURN>. Subsequent prompts: Execute the remainder of the program by using the same operations as in the previous analysis. The resultant display of this analysis is shown in Figs. 5-5 and 5-6.

Two significant results of this analysis are as follows. First, the shaped pulse harmonic energy attenuation is significantly greater than with a rectangular pulse. For example, if the maximum amplitudes in the first sideband lobes are compared, the rectangular pulse voltage peak is 33.33% of the fundamental, while the shaped pulse peak 16.85%. The peak third sideband lobe figures are 14.30 and 1.14% respectively. Hence analyses of this type readily permit the performance of quantitative evaluations of alternate pulse systems. Second, from the volt-ampere and power analysis, the rms voltage has a value of approximately 30.1 V as compared to a value of approximately 31.6 V for an equivalent (based on voltage waveform area) rectangular pulse. Similarly, the power levels are approximately 906 and 1000 W, respectively. For these two relatively simple waveforms, the ratios above can be calculated without a Fourier analysis. However, when dealing with more complex waveforms, the Fourier approach can often save the analyst a great deal of time.

5-3 DC POWER SUPPLY SYSTEM ANALYSIS

In Chapter 2 we noted that electronic loads could produce line voltage and line current distortion in the public utility system. In this analysis the transient waveforms we use were obtained by using power supply system analysis principles developed in Chapter 7. Figure 5-7 shows the network diagram used for the transient analysis. Its major features are: first, a three-phase wye-connected prime power generator [including its internal transient inductances (L_G)]; second, a three-phase linear auxiliary load with an 85% power factor (designated as three sets of elements L_A and R_L that are connected across each of the delta input legs of the dc power supply); third, a delta–wye-connected rectifier transformer for the dc power supply; and fourth, the remainder of the three-phase full-wave power supply, including the rectifier transformer's leakage inductances, six rectifiers connected as a full-wave circuit, a transient suppressor, a dc ripple filter, and a dc load resistor.

The transient analysis showed that all steady-state voltage and current waveforms in this system were nonsinusoidal. Our objective is to use Fourier transforms to obtain the following quantitative solutions: first, the rms voltage across one delta primary leg;

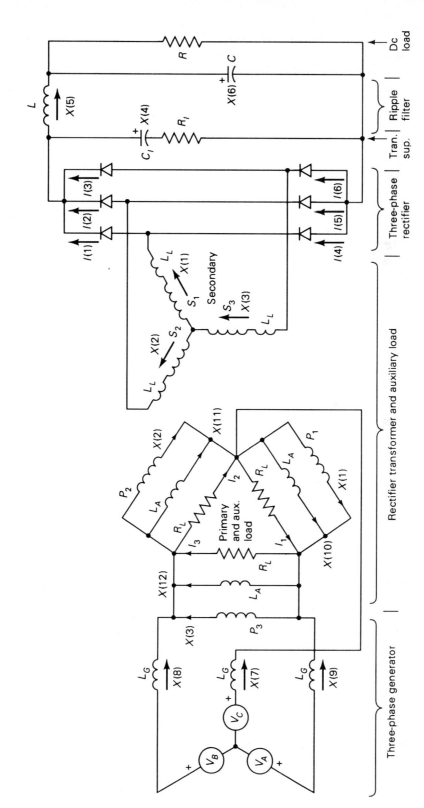

Figure 5-7 Dc power supply system transient analysis network diagram.

93

second, the total rms current taken by the primary leg elements; third, the volt-amperes supplied by one generator phase; and fourth, the power delivered by one generator phase.

The solution procedure for the analysis is as follows.

Source of Input Data

The data from the transient analysis program (gathered from the twenty-fourth cycle of the transient analysis—after steady-state conditions prevailed) have been used to provide the necessary inputs for the program. Phase 3 information was used for the analysis (with a balanced three-phase system, under steady-state conditions the other two phases will have nearly identical solutions). In Fig. 5-7 this includes V_3, the generator line-to-line voltage across primary winding P_3, and SUM_3, the sum of the currents in the three elements associated with delta path 3.

The 161 data points associated with each of the three quantities above have been listed in the form of DATA statements in the program (V_3 data are listed in the program on lines 2100–2220, and SUM_3 data are listed on lines 2300–2430. By proper calls of the DATA statements, we can obtain analysis solutions for the required quantities.

Analysis Program Procedures

With this background established, initiate the computer analysis by first performing the preliminary steps outlined in the analysis of Section 5.2. Then proceed as follows. Next prompt: Enter the digits 161 (the number of data points). Press <RETURN>. Next two prompts: Enter 1 and 25, respectively (to obtain solutions for harmonics 1 through 25). Press <RETURN> after each entry. Next prompt: Enter the letter R (this will READ and store the V_3 inputs for the remainder of the program's execution). Press <RETURN>. Next prompt: Enter .016 (to define the waveform's period). Press <RETURN>. Next prompt: Enter M or P (to define your printing output mode). Press <RETURN>. Next prompt and printout: Enter the letter Y (to check the DATA points). Press <RETURN>. Following the list, the program will automatically display the V_3 Fourier output. These analyses displays are shown in Fig. 5-8. Next prompt: Enter the letter Y (to continue the analysis). Press <RETURN>. Next prompt: Enter the letter R (to READ the SUM_3 inputs). Press <RETURN>. Next prompt: Enter .016 (to define the waveform period). Press <RETURN>. Next prompt: Enter the letter M or P (to define the output printing mode). Press <RETURN>. Next prompt and printout: Enter the letter Y (to check DATA points). Press <RETURN>. The Fourier display for SUM_3 will follow automatically. These analysis displays are shown in Fig. 5-9. Next prompt: Enter the letter N (to indicate that no further data inputs are to be read at this time). Press <RETURN>. Next prompt and printout: Enter the letter Y (to call for a volt-ampere and power analysis). A tabular display, followed by the summation, lists the required solutions for total system rms voltage, rms current, and power factor. The power and volt-amperes are for one phase; hence these values require multiplication by 3. These analysis results are shown in Fig. 5-10.

The significant results of the analysis are as follows:

```
---Numerical Integration Evaluation Of Fourier Coefficients---
              Number Of Ordinates = 161
              D-C Term, A(0) = -.8291667
                  -- Data Ordinates --
Point #  Ordinate   Point #  Ordinate   Point #  Ordinate   Point #  Ordinate   Point #  Ordinate
-------  --------   -------  --------   -------  --------   -------  --------   -------  --------
   0    -7055.000      1    -7058.000      2    -7008.000      3    -6977.000      4    -6943.000
   5    -6901.000      6    -6849.000      7    -6787.000      8    -6715.000      9    -6633.000
  10    -6541.000     11    -6439.000     12    -6343.000     13    -5791.000     14    -5820.000
  15    -5760.000     16    -5651.000     17    -5513.000     18    -5357.000     19    -5186.000
  20    -5006.000     21    -4816.000     22    -4618.000     23    -4413.000     24    -4201.000
  25    -3983.000     26    -3758.000     27    -3528.000     28    -3320.000     29    -3139.000
  30    -2953.000     31    -2764.000     32    -2572.000     33    -2377.000     34    -2178.000
  35    -1976.000     36    -1772.000     37    -1564.000     38    -1354.000     39    -1172.000
  40     -279.000     41       95.000     42      423.000     43      727.000     44     1017.000
  45     1297.000     46     1571.000     47     1840.000     48     2106.000     49     2367.000
  50     2625.000     51     2878.000     52     3127.000     53     3370.000     54     3609.000
  55     3748.000     56     3910.000     57     4065.000     58     4216.000     59     4360.000
  60     4498.000     61     4629.000     62     4754.000     63     4871.000     64     4980.000
  65     5082.000     66     5724.000     67     5955.000     68     6145.000     69     6301.000
  70     6432.000     71     6546.000     72     6647.000     73     6735.000     74     6813.000
  75     6880.000     76     6936.000     77     6982.000     78     7017.000     79     7042.000
  80     7055.000     81     7058.000     82     7007.000     83     6976.000     84     6942.000
  85     6900.000     86     6848.000     87     6786.000     88     6714.000     89     6632.000
  90     6540.000     91     6438.000     92     6341.000     93     5793.000     94     5821.000
  95     5759.000     96     5650.000     97     5511.000     98     5355.000     99     5184.000
 100     5003.000    101     4813.000    102     4615.000    103     4410.000    104     4198.000
 105     3980.000    106     3755.000    107     3525.000    108     3319.000    109     3136.000
 110     2950.000    111     2762.000    112     2570.000    113     2374.000    114     2176.000
 115     1974.000    116     1769.000    117     1561.000    118     1352.000    119     1169.000
 120      282.000    121      -94.000    122     -425.000    123     -730.000    124    -1019.000
 125    -1300.000    126    -1574.000    127    -1844.000    128    -2109.000    129    -2371.000
 130    -2628.000    131    -2881.000    132    -3130.000    133    -3374.000    134    -3612.000
 135    -3750.000    136    -3912.000    137    -4067.000    138    -4218.000    139    -4362.000
 140    -4500.000    141    -4631.000    142    -4755.000    143    -4872.000    144    -4981.000
 145    -5083.000    146    -5723.000    147    -5955.000    148    -6145.000    149    -6301.000
 150    -6433.000    151    -6547.000    152    -6648.000    153    -6736.000    154    -6814.000
 155    -6881.000    156    -6937.000    157    -6982.000    158    -7017.000    159    -7042.000
 160    -7055.000

HARMONIC      COS TERM      SIN TERM      MAGNITUDE      PHASE       PERCENT
--------      --------      --------      ---------      -----       -------
    1         -6936.23       -284.10       6942.04      -177.65      100.00
    2             0.60         -0.02          0.60        -2.13        0.01
    3            -1.86         34.22         34.27        93.10        0.49
    4             0.18         -0.08          0.20       -23.81        0.00
    5          -208.75        -37.64        212.11      -169.78        3.06
    6            -0.33          0.11          0.35       161.00        0.01
    7            82.66        114.14        140.93        54.09        2.03
    8             0.11         -0.06          0.12       -29.85        0.00
    9             1.31        -31.31         31.34       -87.60        0.45
   10             0.03         -0.11          0.12       -73.09        0.00
   11            83.31        -39.87         92.36       -25.57        1.33
   12             0.30          0.10          0.32        17.78        0.00
   13           -54.99         -6.28         55.35      -173.48        0.80
   14             0.01         -0.06          0.06       -81.91        0.00
   15            -0.76         26.42         26.43        91.65        0.38
   16             0.06          0.03          0.07        25.21        0.00
   17           -49.59         19.11         53.14       158.92        0.77
   18            -0.14         -0.09          0.17      -148.48        0.00
   19            27.70          8.15         28.88        16.40        0.42
   20             0.08          0.01          0.08         7.68        0.00
   21            -3.17        -24.55         24.75       -97.35        0.36
   22            -0.01         -0.13          0.13       -94.46        0.00
   23            32.07        -24.09         40.12       -36.91        0.58
   24             0.13          0.07          0.15        26.51        0.00
   25           -26.43          6.48         27.21       166.23        0.39

Fundamental Frequency = 62.5 Hertz
```

Figure 5-8 Fourier coefficients for voltage V_3 using Appendix 7.

Display of Figure 5-8 This display shows clearly that the power electronics system has introduced significant distortion on the source voltage waveform—fifth harmonic at 3.06%, seventh harmonic at 2.03% and so on.

Sec. 5-3   DC Power Supply System Analyses

```
---Numerical Integration Evaluation Of Fourier Coefficients---
              Number Of Ordinates =  161
              D-C Term, A(0) = .1437519
                      -- Data Ordinates --
Point #  Ordinate  Point #  Ordinate  Point #  Ordinate  Point #  Ordinate  Point #  Ordinate
-------  --------  -------  --------  -------  --------  -------  --------  -------  --------
   0     -36.100     1     -36.200     2     -36.200     3     -36.400     4     -36.600
   5     -36.700     6     -36.800     7     -36.800     8     -36.900     9     -36.800
  10     -36.800    11     -36.700    12     -36.500    13     -37.200    14     -38.000
  15     -38.500    16     -38.700    17     -38.700    18     -38.700    19     -38.600
  20     -38.500    21     -38.400    22     -38.200    23     -37.900    24     -37.600
  25     -37.300    26     -36.900    27     -36.500    28     -36.000    29     -35.300
  30     -34.100    31     -32.600    32     -30.700    33     -28.400    34     -25.800
  35     -22.800    36     -19.400    37     -15.600    38     -11.500    39      -8.800
  40      -7.400    41      -6.800    42      -6.200    43      -5.700    44      -5.200
  45      -4.700    46      -4.200    47      -3.700    48      -3.200    49      -2.700
  50      -2.100    51      -1.600    52      -1.100    53      -0.600    55       0.600
  56       1.600    57       3.000    58       4.800    59       6.900    60       9.400
  61      12.200    62      15.400    63      18.800    64      22.600    65      26.700
  66      30.100    67      31.300    68      32.200    69      32.900    70      33.400
  71      33.900    72      34.300    73      34.700    74      35.100    75      35.400
  76      35.700    77      35.900    78      36.100    79      36.300    80      36.400
  81      36.500    82      36.600    83      36.700    84      36.900    85      37.000
  86      37.100    87      37.200    88      37.200    89      37.200    90      37.100
  91      37.000    92      36.800    93      37.600    94      38.400    95      38.800
  96      39.000    97      39.000    98      39.000    99      39.000   100      38.800
 101      38.700   102      38.500   103      38.200   104      37.900   105      37.600
 106      37.200   107      36.800   108      36.300   109      35.600   110      34.400
 111      32.900   112      31.000   113      28.700   114      26.100   115      23.000
 116      19.600   117      15.900   118      11.700   119       9.100   120       7.700
 121       7.100   122       6.500   123       6.000   124       5.500   125       5.000
 126       4.500   127       4.000   128       3.500   129       2.900   130       2.400
 131       1.900   132       1.400   133       0.800   134       0.300   135      -0.300
 136      -1.300   137      -2.700   138      -4.500   139      -6.700   140      -9.100
 141     -12.000   142     -15.100   143     -18.600   144     -22.400   145     -26.500
 146     -29.900   147     -31.000   148     -31.900   149     -32.600   150     -33.200
 151     -33.600   152     -34.100   153     -34.500   154     -34.800   155     -35.100
 156     -35.400   157     -35.700   158     -35.900   159     -36.000   160     -36.100


     HARMONIC       COS TERM      SIN TERM     MAGNITUDE       PHASE       PERCENT
     --------       --------      --------     ---------       -----       -------
         1           -37.04        -16.06        40.37         -156.56      100.00
         2             0.00          0.02         0.02           79.15        0.05
         3            -0.00          0.07         0.07           91.47        0.18
         4             0.00          0.01         0.01           86.79        0.03
         5            -0.48          5.15         5.17           95.37       12.81
         6            -0.00         -0.00         0.00         -142.49        0.00
         7             1.83         -1.60         2.43          -41.18        6.02
         8            -0.00          0.00         0.00          141.32        0.01
         9             0.00         -0.05         0.05          -90.00        0.14
        10            -0.00         -0.00         0.00         -173.82        0.01
        11            -0.80         -0.88         1.18         -132.15        2.93
        12             0.00          0.01         0.01           78.74        0.02
        13             0.09          0.58         0.59           81.43        1.46
        14             0.00         -0.00         0.00          -48.06        0.01
        15             0.01          0.04         0.05           77.75        0.11
        16             0.00         -0.00         0.00          -27.23        0.01
        17             0.28          0.24         0.37           40.27        0.91
        18            -0.00          0.00         0.00          171.69        0.01
        19            -0.04         -0.15         0.16         -103.49        0.39
        20             0.00          0.00         0.00           51.47        0.00
        21            -0.01         -0.03         0.03         -113.74        0.08
        22            -0.00          0.00         0.00          116.16        0.01
        23            -0.21         -0.09         0.23         -156.18        0.57
        24             0.00          0.00         0.01           40.99        0.01
        25             0.08          0.10         0.13           49.49        0.32

     Fundamental Frequency = 62.5 Hertz
```

Figure 5-9 Fourier coefficients for SUM$_3$ using Appendix 7.

Display of Figure 5-9 This display shows the power electronics system has even more pronounced distortion in the delta leg summation currents—fifth harmonic at 12.81%, seventh harmonic at 6.02%, and so on.

```
Volt-Ampere & Power Analysis
-----------------------------

Harmonic  Voltage  V. Angle  Amperes  A. Angle   Power
--------  -------  --------  -------  --------   -----
   1      6942.04  -177.65    40.37   -156.56  130748.30
   2         0.60    -2.13     0.02     79.15       0.00
   3        34.27    93.10     0.07     91.47       1.25
   4         0.20   -23.81     0.01     86.79      -0.00
   5       212.11  -169.78     5.17     95.37     -46.38
   6         0.35   161.00     0.00   -142.49       0.00
   7       140.93    54.09     2.43    -41.18     -15.72
   8         0.12   -29.85     0.00    141.32      -0.00
   9        31.34   -87.60     0.05    -90.00       0.85
  10         0.12   -73.09     0.00   -173.82      -0.00
  11        92.36   -25.57     1.18   -132.15     -15.61
  12         0.32    17.78     0.01     78.74       0.00
  13        55.35  -173.48     0.59     81.43      -4.24
  14         0.06   -81.91     0.00    -48.06       0.00
  15        26.43    91.65     0.05     77.75       0.59
  16         0.07    25.21     0.00    -27.23       0.00
  17        53.14   158.92     0.37     40.27      -4.66
  18         0.17  -148.48     0.00    171.69       0.00
  19        28.88    16.40     0.16   -103.49      -1.12
  20         0.08     7.68     0.00     51.47       0.00
  21        24.75   -97.35     0.03   -113.74       0.39
  22         0.13   -94.46     0.00    116.16      -0.00
  23        40.12   -36.91     0.23   -156.18      -2.24
  24         0.15    26.51     0.01     40.99       0.00
  25        27.21   166.23     0.13     49.49      -0.79

RMS Voltage  = 4913.14
RMS Current  = 28.85101
Volt Amperes = 141749.1
       Power = 130660.5
Power Factor = .9217736
```

Figure 5-10 Single-phase volt-amperes/power/power factor using Appendix 7.

Display of Figure 5-10 This display summarizes our objective: determination of the rms line-to-line voltage, rms delta summation line current, system volt-amperes, system power, and system power factor. Without a transient analysis/transform approach, none of this quantitative information would have been available to us (i.e., other than by physically constructing the system and performing electrical measurements to determine the values).

Summary

This Fourier approach provides a powerful tool that allows us to assess/optimize the performance of a power electronics system *during its design phase*. Application of this system performance capability insight results in substantially minimized financial/delivery schedule problems that might otherwise be encountered during subsequent phases of the project.

Example 5-1

To demonstrate further the capabilities of this type of analysis, the transient values of delta primary current in one leg of the power supply transformer have been listed in DATA lines 2500–2610. How would you determine the rms current, volt-amperes, power, and power factor for this phase of the power supply portion of the system?

Solution: First, enter the command DELETE 2300–2430 to remove existing values of SUM_3, and second, repeat the analysis in the same manner as before. (*Note:* Do not terminate the run.) The results are rms current = 19.81 A, volt-amperes = 97,316 kVA, power = 91,183 W, and power factor = 93.70%.

Example 5-2

The value of the prime line currents have been listed as DATA statements 2700–2920. How would you determine the values of harmonics 5, 7, 11, 13, and 17?

Solution: Simply continue the run as was done in the previous analysis. The harmonic percentages are 12.98, 6.09, 3.06, 1.57, and 1.05, respectively.

Chapter 6

RF Power Generating Systems

The objective of this chapter is to develop a frame of reference between the preceding and subsequent chapters of this book. Some representative RF power-generating systems will be used to provide this frame of reference. The preceding chapters have been concerned with system areas external to the power electronics system proper. For example, typical electronic system input interface interactions with their public utility energy sources have been analyzed. In addition, typical electronic system functional interactions with their output interface's external load have been analyzed (e.g., at the end of electronic system's output transmission line). Finally, some elements of the system's frequency spectrum have been explored. Now we need to define the nature/requirements of the power electronics system itself. In the next four sections of this chapter we define power electronics in terms of:

1. Definitive documents and modulating waveforms
2. System functional block diagrams
3. System power amplification devices
4. Practical system applications

Even though the frame of reference to be used here concerns RF power-generating systems, try to visualize/modify the concepts in terms of *your own* power electronics equipment/interactions. You will find that the principles outlined in the first three sections are fully applicable to your power electronics application as well.

6-1 DEFINITIVE DOCUMENTS AND MODULATING WAVEFORMS

A typical power electronics system may be made up of a number of intricate and highly interrelated parts. How can an equipment designer deal effectively with this type of situation? One proven approach is, first, to define clearly the system's technical objectives by means of *definitive documents*; second, perform initial analyses to investigate the *feasibility* of the objectives; third, if the objectives are feasible, *perform refined analyses*, as the project progresses, to develop fully the initial findings into a design; and fourth, *fabricate* the power electronics hardware from elements that meet both the physical and the performance capabilities defined by the system documents.

Definitive Documents

It is fundamental that new or revised systems have clear definitions of their requirements. Two definitive documents are usually involved. The first is a *specification*—it defines the system's technical capability requirements. The second document is a *statement of work*—it defines the system's hardware requirements.

Sample specification. To illustrate a specification, consider the requirements listed in the outline of Fig. 6-1. It shows the two essential definitive characteristics of any RF power generator: the generator's RF frequency of operation and its RF power output.

SPECIFICATION

1. Operating Frequency(s)   3000-3500 MHz

2. Peak power output   350 kW

3. Pulse duration   5 uS

4. Pulse repetition frequency   1000 Hz

•

•

9. Pulse to pulse stability   0.5 Deg

•

Figure 6-1 Partial outline of a specification.

This generator must operate at any carrier frequency between 3000 and 3500 MHz with a minimum RF output of 350 kW peak.

Following this, the specification establishes the system's operating mode (i.e., is the generator part of a communication system, part of a radar system, etc.). The mode is practically always defined in terms of the *modulating waveforms*—in this case a radar's train of 5-μs-duration RF pulses repeated at a 1000-Hz pulse repetition rate.

Numerous additional requirements follow. Ultimately, the specification defines the *performance capabilities* of the system. The radar system of the example requires that electrical waveforms of adjacent RF pulses be identical within 0.5 electrical degree.

In all but the simplest cases, a system's performance capabilities require careful investigation. A study, conducted by system analysts, typically determines if an RF power generator system capable of meeting the performance requirement (and all other listed technical requirements) can in fact be devised.

Sample statement of work. Next consider the statement of work outline shown in Fig. 6-2. It sets fourth essential requirements involving the system's hardware design and construction. For example, a statement of work defines the major *input and output interfaces*. At its input interface, the system must operate from a public utility source that delivers 480-V three-phase 60-Hz power. At its output interface, it must deliver its 3000-MHz pulse train from a type WR 284 RF waveguide transmission line flange.

Physical definitions are essential parts of a statement of work. In this case overall dimensional limits of 2.29 m (90 in.) high by 1.52 m (60 in.) wide by 0.91 m (36 in.) deep define the radar system's enclosure. In addition, acceptable types of components and manufacturing techniques are defined. All of the radar's large power transformers are

STATEMENT OF WORK

1. Prime power — 480/277 V, 3 Ph., 60 Hz

2. RF output termination — WR 284 waveguide flange

3. Maximum encloseure size — 2.29 m H., 1.52 m W., .91 m D. (90 in H., 60 in W., 36 in D.)

4. Transformers over 500 V-A — Per ANSI C57.12.1290

•

7. Testing — Demonstrate RF spec. compliance

•

9. Delivery requirements — Date

Figure 6-2 Partial outline of a statement of work.

to be tested in accordance with the American National Standards Institute's Specification ANSI C57.12.90.

Ultimately, the document will address the *performance tests* that the equipment manufacturer must conduct before the unit will be accepted. Finally, the statement of work will clearly define the *delivery schedule* for the RF generator.

Modulating Waveforms

Consider an RF voltage wave phasor that is rotating in the conventional counterclockwise direction as a function of time. Its amplitude projection on the x axis would be

$$V = A \sin \phi \tag{6-1}$$

where ϕ is the accumulated phase angle of the voltage wave. We can also write

$$\omega = \frac{d\phi}{dt_1} \tag{6-2}$$

where ω is the radian frequency of the wave. By integration it follows that

$$\phi = \Phi + \int \omega \, dt_1 \tag{6-3}$$

where Φ is an initial phase angle. If ω is constant, we have, from Eq. (6-1),

$$V = A \sin(\omega t_1 + \Phi) \tag{6-4}$$

It represents a sine-wave voltage of constant amplitude and frequency.

Modulation will occur if any one of three independent magnitudes (amplitude A, phase Φ, or frequency ω) is subjected to a low-frequency *periodical change*.

Amplitude modulation. In this mode we can define the AM periodical change envelope as

$$A = A_0[1 + k_a \sin(\omega_m t)] \tag{6-5}$$

where A_0 is the average value of the envelope, k_a the modulation factor, and $A_0 k_a$ the amplitude of a superimposed low-frequency sine-wave modulating voltage of radian frequency ω_m. Then Eq. (6-4) becomes

$$V = A_0[1 + k_a \sin(\omega_m t)] \sin(\omega t_1 + \Phi) \tag{6-6}$$

but ω and Φ are constants, and since $(\omega t_1 + \Phi_0) = \omega_0 t$, we can write the amplitude response equation as

$$V = A_0[\sin(\omega_0 t) + k_a \sin(\omega_0 t) \sin(\omega_m t)] \tag{6-7}$$

We can trigonometrically expand the equation's second term and express the frequency spectrum equation as follows:

$$V = A_0[\sin(\omega_0 t) + 0.5 k_a \cos(\omega_0 - \omega_m)t - 0.5 k_a \cos(\omega_0 + \omega_m)t] \tag{6-8}$$

Example 6-1

What RF spectrum results if an ideal RF amplifier is amplitude modulated by a 1-V 0.5 duty rectangular pulse?

Solution In Section 5-1 we define the rectangular pulse coefficients as $0.5 + 0.636 \cos(2\pi \cdot \text{PRR} \cdot t) - 0.212 \cos(6\pi \cdot \text{PRR} \cdot t) + 0.127 \cos(10\pi \cdot \text{PRR} \cdot t) \cdots$. If it modulates an ideal amplifier, the RF spectrum is

$$V = 0.5 \sin(\omega_0 t) + 0.318 \sin[(\omega_0 + 2\pi \cdot \text{PRR})t] + 0.318 \sin[(\omega_0 - 2\pi \cdot \text{PRR})t]$$
$$- 0.106 \sin[(\omega_0 + 6\pi \cdot \text{PRR})t] - 0.106 \sin[(\omega_0 - 6\pi \cdot \text{PRR})t] \cdots$$

Phase modulation. A and ω are now constants. The value of Φ changes periodically as follows:

$$\Phi = \Phi_0[1 + k_p \sin(\omega_m t)] \quad (6\text{-}9)$$

If Eq. (6-9) is substituted into Eq. (6-3) and the result is substituted into Eq. (6-1), we have the following PM amplitude response equation:

$$V = A_0 \sin[\omega_0 t + m_p \sin(\omega_m t)] \quad (6\text{-}10)$$

where $m_p = k_p \Phi_0$, a constant referred to as the modulation index. By trigonometric expansion we obtain the following frequency spectrum equation:

$$V = A_0 \{\sin(\omega_0 t) \cos[m_p \sin(\omega_m t)] + \cos(\omega_0 t) \sin[m_p \sin(\omega_m t)]\} \quad (6\text{-}11)$$

Through the use of Bessel functions of the first kind and nth order, and the use of an additional trigonometric expansion, we obtain the following practical spectrum equation:

$$\begin{aligned} V = &\; A_0(J_0(m_p) \sin(\omega_0 t) \\ &+ J_1(m_p)[\sin(\omega_0 + \omega_m)t - \sin(\omega_0 - \omega_m)t] \\ &- J_2(m_p)[\sin(\omega_0 + 2\omega_m)t - \sin(\omega_0 - 2\omega_m)t] \\ &+ J_3(m_p)[\sin(\omega_0 + 3\omega_m)t - \sin(\omega_0 - 3\omega_m)t] - \cdots \end{aligned} \quad (6\text{-}12)$$

If modulation index m_p has a value greater than perhaps 0.25, there will be a significant number of additional sets of sidebands generated—even though only a single low modulating frequency, $f_m = \omega_m/2\pi$, is present.

Example 6-2

In spectrum analyses, how can you avoid the inconvenience of looking up Bessel functions in tables?

Solution Simply load Appendix 9 and issue a <RUN> command. Through the use of this program, determine the relative amplitudes of the carrier and up to the ninth order sidebands for modulation indices of 0.25, 1, 4, and 10.

Figure 6-3 shows the solutions.

Frequency modulation. A and Φ are now constants. The value of ω changes periodically as

$$\omega = \omega_0[(1 + k_f \cos(\omega_m t_1)] \quad (6\text{-}13)$$

If we substitute this into Eq. (6-3), integrate, set the constant of integration equal to 0, and then substitute into Eq. (6-1), we obtain the following FM amplitude response:

$$V = A_0 \sin[\omega_0 t + m_f \sin(\omega_m t)] \quad (6\text{-}14)$$

```
                         Bessel Functions Of First Kind

J 0 ( 0.25) = 0.9844;  J 1 ( 0.25) = 0.1240;  J 2 ( 0.25) = 0.0078;  J 3 ( 0.25) = 0.0003;  J 4 ( 0.25) = 0.0000;
J 5 ( 0.25) = 0.0000;  J 6 ( 0.25) = 0.0000;  J 7 ( 0.25) = 0.0000;  J 8 ( 0.25) = 0.0000;  J 9 ( 0.25) = 0.0000;
Figure 6-3a Modulation Index = .25
------------------------------------------------------------------------------------------------------------

                         Bessel Functions Of First Kind

J 0 ( 1.00) = 0.7652;  J 1 ( 1.00) = 0.4401;  J 2 ( 1.00) = 0.1149;  J 3 ( 1.00) = 0.0196;  J 4 ( 1.00) = 0.0025;
J 5 ( 1.00) = 0.0002;  J 6 ( 1.00) = 0.0000;  J 7 ( 1.00) = 0.0000;  J 8 ( 1.00) = 0.0000;  J 9 ( 1.00) = 0.0000;
Figure 6-3b Modulation Index = 1.0
------------------------------------------------------------------------------------------------------------

                         Bessel Functions Of First Kind

J 0 ( 4.00) = -.3971;  J 1 ( 4.00) = -.0660;  J 2 ( 4.00) = 0.3641;  J 3 ( 4.00) = 0.4302;  J 4 ( 4.00) = 0.2811;
J 5 ( 4.00) = 0.1321;  J 6 ( 4.00) = 0.0491;  J 7 ( 4.00) = 0.0152;  J 8 ( 4.00) = 0.0040;  J 9 ( 4.00) = 0.0009;
Figure 6-3c Modulation Index = 4.0
------------------------------------------------------------------------------------------------------------

                         Bessel Functions Of First Kind

J 0 (10.00) = -.2454;  J 1 (10.00) = 0.0436;  J 2 (10.00) = 0.2547;  J 3 (10.00) = 0.0584;  J 4 (10.00) = -.2196;
J 5 (10.00) = -.2341;  J 6 (10.00) = -.0145;  J 7 (10.00) = 0.2167;  J 8 (10.00) = 0.3179;  J 9 (10.00) = 0.2919;
Figure 6-3d Modulation Index = 10.0
------------------------------------------------------------------------------------------------------------
```

Figure 6-3 Comparison of PM/FM sideband amplitudes versus modulation index.

where the FM modulation index $m_f = k_f \omega_0/\omega_m$.

Note the similarity between FM amplitude in Eq. (6-14) and the PM amplitude in Eq. (6-10). The equations are identical except for the modulation index subscripts. Similarly, an FM spectrum equation can be written simply by replacing all m_p terms in Eq. (6-12) with m_f terms.

6-2 SYSTEM FUNCTIONAL BLOCK DIAGRAMS

Based on the groundwork set forth in Section 6-1, assume that we now possess a specification, a statement of work, and a definition of the controlling waveforms for an RF power generator system that is to be designed. Typically, our next course of action is to define the system *functionally*. With this in mind, the major objectives in this section are, first, to define a representative RF power-generating system in terms of its functional elements, and second, to define typical performance limitations associated with the functional elements. Our frame of reference is shown in the block diagram of Fig. 6-4. From a functional point of view, blocks 1 through 4 define the system's RF amplification path. Blocks 5–7 define the system's power and control functions.

A broad-brush pass through the RF path shows that a low-level RF input interface signal is delivered to block 1. The sequential RF power amplification capabilities of blocks 1 and 2 function to deliver the system's specified RF power output level at the RF output interface of block 2 (note that systems may use an alternate number of blocks to perform this function). The RF transmission line of block 3 functions to transport this high-level output to an RF radiator/load. The fourth block typically functions to terminate the line and radiate the RF energy from the line into space.

In the lower path the input to block 5 defines the prime power source input interface. Block 5, in turn, provides the the power distribution functions for the system. Block 6 functions to convert the ac energy into the *dc electrode energy sources* required to actuate

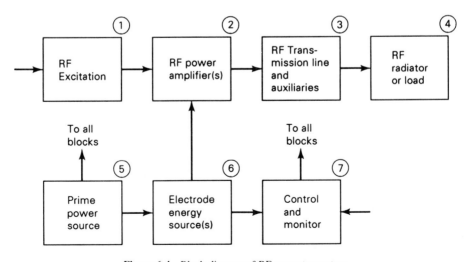

Figure 6-4 Block diagram of RF generator system.

the RF power amplifiers. Block 7 functions to provide all essential control and monitor functions for the RF power amplifier system.

In the following, the interfaces that exist between the seven major blocks will be reviewed briefly. Interface *interactions* that affect the system's performance capabilities will be noted. The prime areas of concern are, first, electrode energy interactions wherein the RF amplifier stages experience electrode voltage/current variations that adversely affect system performance, and second, RF transmission line interactions wherein the lines experience load impedance mismatches that affect system performance adversely.

Blocks 1 and 2: RF Amplifier Functions

Each RF stage has one primary function: power amplification of its input signal. Each stage has three interface areas: first, its RF input path; second, its RF output path; and third, its electrode energy path(s). If the transmission lines associated with either the input or output paths are not matched, the line will have a VSWR, and the input impedance of the line will be mismatched. As a result, a driving source interaction occurs, and if the driver's performance capabilities are seriously affected, vswr corrective measures similar to those noted in Chapter 4 (e.g., stub matching, circulators, etc.) will be required.

All RF amplifiers have phase and amplitude sensitivity definitions that express their performance capabilities as a function of applied electrode voltage/current stability. These values, in turn, establish one or more interaction area(s) that require definition at the amplifier's electrode energy sources.

Block 3: Output Transmission Line/Auxiliaries Functions

The output transmission line functions to transport power from the RF amplifier's output port to the RF radiator and load area. All significant interactions in this area are a function of line matching. Since the system's RF output load is typically mismatched, the system's performance capabilities may be seriously affected unless VSWR corrective measures (stubs, circulators, etc.) are used. If circulators are not available in the operating frequency range, the final amplifier stage must be designed to accommodate the adverse effects of load VSWR.

Block 4: RF Radiator/Load Functions

This area functions either to radiate the system's power into free space or to provide a dummy test load for system adjustment/evaluation. The final amplifier's output path is normally connected to a three-port RF switch. One output port goes to the radiator and the other goes to a matched dummy load.

Block 5: Prime Power Source Functions

The prime power source for most RF power generator systems supply 50-, 60-, or 400-Hz energy. In large systems 50- or 60-Hz three-phase public utility sources are typically used. A power control and distribution center functions in a manner similar to the centers/switchgear analyzed in Chapters 2 and 3. Line resistance and inductances function to degrade the system's performance capabilities. As noted in Chapters 2 and 3, all power sources introduce an appreciable degree of inductance. These values form a part of all overall RF power generator system analyses.

Block 6: Electrode Energy Source Functions

All RF amplifiers require at least one dc energy source of either a continuous or a pulsed nature. The function of an electrode energy source is to convert ac source energy from a prime power system into the specified form of dc source energy for an RF amplifier. It is essential that analyses be conducted to investigate the often critical interactions that exist between electrode energy source amplitude performance capabilities and RF phase and amplitude performance capabilities in the associated RF amplifier. These areas are pursued in Chapters 8, 9, 10, and 11.

Block 7: Monitor and Control Functions

This equipment area has two prime functions. First, it initiates system operating functions (e.g., turn on, electrode voltage-level adjustment, RF frequency changing, etc.), and second, it functions to monitor system performance (e.g., RF power output measured, system waveforms checked, etc.). While functionally important, this block has no direct impact on system performance capabilities.

6-3 SYSTEM POWER AMPLIFICATION DEVICES

The RF power amplifier functions established in the preceding section provide a logical starting point for any RF power generator system analysis. The objective of this section is to provide background on representative RF power amplifier assemblies.

Elementary Considerations

A microwave amplifier tube is normally produced as a complete *RF package*. The purchaser is supplied with an RF amplifier assembly—complete from its RF input port connector to its RF output port connector. On the other hand, a transistor or a triode/tetrode tube is normally supplied as a *device*. The purchaser is then responsible for designing and fabricating the required RF amplifier assembly(s). In the following we make frequent reference to types of RF amplifiers in terms of a package approach or a device approach.

Due to the wide and diverse range of RF amplifiers on the market, it is *inadvisable* to rely on representative amplifier information for equipment design purposes. A qualitative chart showing RF amplifier properties will be presented later—for use as a general guide. Practical design information on an amplifier's capabilities should always be obtained from the amplifier's manufacturer. The following list indicates equipment design areas where technical information is usually required.

RF frequency range. With packaged approaches, this may be expressed as either a bandpass range or a tuning range of operation. If a bandpass range is specified, the package manufacturer should supply performance response data that first, are within the system's specified RF power output region; second, extend from below to above the system's specified bandpass range; and third, are taken with constant RF drive applied. If a tuning range is specified, the manufacturer should supply data on the mechanical characteristics and life expectancy of the tuning mechanisms.

With device approaches, the device manufacturer should supply all electrical, mechanical, and thermal interface information required for the RF amplifier design. Bandpass characteristics and/or tuning range capabilities are the responsibility of the purchaser.

RF power output. For pulse operation with packaged approaches, the required values of peak power and duty factor should be specified by the user. Peak power is defined as the average power during a pulse. Duty factor is defined as product of pulse duration in seconds and pulse repetition rate in hertz (the maximum operating value of pulse duration for the RF amplifier should be specified by the user). The manufacturer should supply all required power-related operating parameters (electrode voltages and currents, electrode heat removal requirement/procedures, etc.).

For pulsed operation with device approaches, the device manufacturer must supply the same information as noted under frequency range above. The purchaser is responsible for establishing the peak power, average power, and all other operating conditions. If the RF amplifier's data define no limit on pulse duration, continuous-wave (CW) RF amplifier operating capabilities are implied. However, the package/device manufacturer should be consulted to confirm this assumption.

RF pushing. *Pushing* defines phase and RF amplitude variations in an RF amplifier's output as a function of electrode voltage and/or current variations from its dc energy sources. The amplifier manufacturer should define the pushing values associated with all electrodes of packaged RF amplifier approaches. Frequently, only RF phase pushing is given consideration because its effect on system performance is significantly greater than that of amplitude pushing. For RF device approaches, the pushing values are defined based on the purchaser's RF amplifier tests.

RF pulling. *Pulling* is defined as RF phase and amplitude variations in an RF amplifier's output as a function of input and output port VSWR variations. The manufacturer should clearly define the maximum allowable VSWR levels for packaged RF amplifier approaches. For RF device approaches, the customer must establish an RF amplifier design that is compatible with the RF device's ratings.

RF input/output linearity. Most packaged RF amplifiers do not have a linear RF drive voltage versus RF output voltage characteristic. In addition, they do not have a linear RF drive phase versus RF output phase characteristic. If RF phase and amplitude linearity are important considerations, their capabilities should be defined by the package manufacturer.

For RF device approaches, definitions based on tests similar to those of the RF pushing case are required.

RF harmonics. All RF amplifiers generate power output at harmonics of their fundamental frequency. When using packaged amplifiers contact the manufacturer for a definition of the harmonic levels, and establish the amount of RF filtering hardware that must be added to meet system specifications. In RF device approaches, harmonic attenuation is a functional part of the purchaser's design process.

Crossed-Field RF Power Tubes

A magnetron consists of a thermionic emission diode composed of a cathode and a special anode structure. Both are mounted within a vacuum. In addition, an externally generated dc magnetic field forms a part of the overall structure. The field lines pass through the vacuum parallel to the axis of the cathode.

The anode of a magnetron is physically formed into an RF structure that will interact with the cathode emission when the proper-level electrode energy source is connected across the diode. The magnetron will then oscillate at an RF frequency that is a function of the anode structure's physical dimensions. The left cross section in Fig. 6-5 shows an early version magnetron vacuum structure. A hot wire heater usually passes through the center of the cathode circle, and thermionic emission then takes place from the cathode's surface.

The anode consists of two parts, the outer metallic cylinder that forms the vacuum vessel, and a series of metallic *vanes* that are attached to the inner wall of the cylinder. These vanes form microwave cavities that establish the frequency of oscillation. A circular ring, known as a *strap*, connects to alternate vanes. It helps stabilize the frequency of oscillation. A coupling loop, shown at the bottom, provides the means for delivering the RF generator's output to an RF transmission line and radiator. Some general technical capabilities of these packaged RF generators are shown in Fig. 6-6.

A number of versions of crossed-field amplifiers (CFAs) have been developed. The right cross section of Fig. 6-5 shows an early version CFA structure. It has a strong resemblance to the magnetron's structure shown at its left. The cathode and anode structure

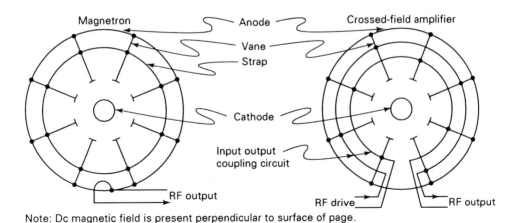

Note: Dc magnetic field is present perpendicular to surface of page.

Figure 6-5 Simplified cross sections of magnetron and cross-field amplifier.

	Frequency range (GHz)	Bandwidth	Power	Φ Pushing	Φ Pulling	Linearity of amplifier
Magnetron oscillator	1–20	NA	M–H	M–H	M–H	NA
CFA	1–10	M–H	M–H	L	M–H	NA
Klystron	0.4–25	L–M	Highest	M–H	M–H	L–M
TWT	1–20	Highest	M–H	M–H	M–H	L–M
Tetrode	HF–1	L–H	M–H	L	L	M–H
Transistor	HF–1	L–H	L (per unit)	L	L	M–H

L,low; M,medium; H,high; HF,3–30 MHz; NA,not applicable.

Figure 6-6 Comparison of RF power amplifier capabilities.

are similar. The fundamental difference lies in the input and output coupling circuits that are closely associated with the vane cavities. Some general technical capabilities of packaged CFAs are shown in Fig. 6-6.

Linear Beam Tubes

Klystron amplifiers and *traveling-wave tube* (TWT) amplifiers are the two major forms of linear beam tubes. Klystrons are noted for their RF high power-generation capabilities in the microwave region. TWTs are noted for their broad bandwidth capabilities. Below we present general information on both.

The cross-sectional views shown in Fig. 6-7 illustrate the main elements contained within the vacuum envelopes. Both have thermionic cathodes with circular emitting surfaces of a concave nature. Electrically, the cathode is placed at a high negative potential relative to the grounded interaction structure of the tube.

If the linear beam tube has a control electrode, beam voltage is provided by a high-voltage dc power supply, and the control electrode is pulsed on or off by a low-amplitude pulse applied between the control electrode and the cathode. If a control electrode is not present, the total beam voltage must be pulsed on and off between the cathode and the interaction structure.

The beam voltage between the cathode and the interaction structure causes the electrons emitted from the cathode to be formed into a beam of small diameter. Figure 6-7 shows how the beam converges as it approaches the interaction structure.

The beam must retain its narrow cross section as it passes down the interaction structure. Since the beam attempts to diverge when it enters the structure, a dc magnetic field is generated (with magnetic lines parallel to the beam) to produce beam focusing. Both electromagnetic and permanent magnetic fields are used. The focusing magnetic

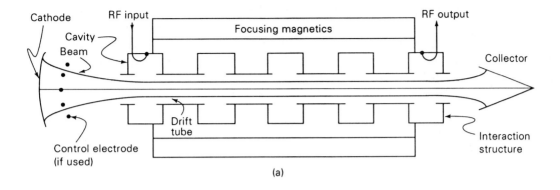

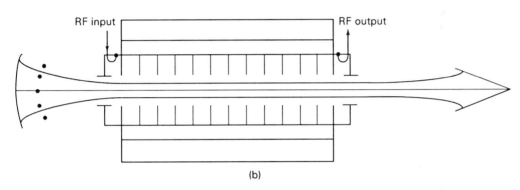

Figure 6-7 Longitudinal cross sections of representative linear beam amplifiers: (a) klystron; (b) traveling-wave tube.

cylinder's cross section can be seen in the figures (i.e., outer cylinder of the cross sections).

When the beam leaves the right-hand end of the structure its cross section increases again. The remaining beam energy is dissipated in the collector. In some tubes the collector is connected to ground, but in higher-efficiency tubes it is connected to a potential that is depressed relative to ground.

RF amplification is accomplished in the tube's *interaction structure*. In klystrons the interaction structure consists of three or more resonant cavities, centered along the beam, and connected together by means of *drift tube* sections. As the beam passes into the cavity closest to the cathode it interacts with the RF drive injected into the cavity. Electron bunching effects occur, and as the beam exits the cavity it exhibits a *bunched* RF pattern. In subsequent cavities bunching of increased magnitude occurs and RF gain results. In the last cavity a high-power replica of the RF drive signal is present. Power is then coupled from the tube's output port and delivered to the RF generator system's output transmission line. Similar bunching action takes place in TWT amplifiers, but TWT beam interaction structures are finely divided, as indicated in its cross-sectional in Fig. 6-7. Representative klystron amplifier and TWT technical capabilities are shown in Fig. 6-6.

Sec. 6-3   System Power Amplification Devices

Triode and Tetrode Amplifiers

The triode vacuum tube is the antecedent of all RF power generators. Triodes were the standard amplifier for high-power broadcast transmitters for many decades. When the need for pulsed radar transmitters arose during the World War II period, triodes were the first and only choice. Their capabilities were limited to the lowest radar frequency bands. As new and more powerful magnetrons and other packaged approaches were developed, triodes were gradually eliminated from the radar scene.

Tetrodes also have rather severe limitations as the RF frequency of operation rises above several hundred megahertz. As a result, there are few tetrodes used in current *microwave* radar systems. They have been replaced either by the packaged RF amplifiers or by solid-state RF amplifier systems. However, at lower RF frequencies, for example in the RF power generator role for the OTH-B radar system (see Section 6-4), tetrodes have proven to be an excellent choice.

Tetrodes are RF devices—the purchaser is responsible for the design of the RF generating system of which they are a part. Figure 6-8 summarizes several background areas associated with tetrodes. Figure 6-8a shows a cross section of a conventional tetrode's vacuum envelope. All amplifiers dissipate energy at their high-level electrode, and in high-power tetrodes the dissipation is appreciable. Typically, the anode is enclosed in a cooling jacket. A liquid, often water, is circulated through the jacket to remove the anode's dissipation.

The tube's screen grid is typically a spiral cylinder inside the anode structure, and the control grid is a second spiral structure inside the screen structure. A cathode/heater structure at the center of the tube furnishes the necessary levels of thermionic emission.

Figure 6-8b is an equivalent RF output current generator, where I_o is a linear function of input voltage V_I. The circuit is often used in RF power generator response analysis, but it gives no indication of the status of the amplifier's energy sources.

Figure 6-8c shows an idealized (linear) set of anode voltage/grid voltage characteristic curves for a tetrode with a 150-kW anode dissipation capability. To illustrate their use, consider an HF power amplifier example where first, the tetrode is operated as a class AB-1 RF amplifier; second, the dc anode voltage $V_a = 10$ kV; third, the dc screen voltage $V_a = 1.5$ kV (anode voltage excursions should never go below screen voltage); fourth, the dc grid bias $V_g = -200$ V, the anode current cutoff bias value; and fifth, a matched load line (215.5 Ω) has been constructed between the quiescent point (junction of V_g and V_a) and the maximum RF power output excursion point (anode voltage = 1.5 kV and anode current = 80 A).

Figure 6.8d shows a simplified circuit diagram for an RF power generator system with an operating frequency in the 5-MHz region. Its elements are as follows: first, RF excitation, labeled "RF Gen," provides the system drive signal at frequency f_0; second, capacitance C_i and inductance L_i form a parallel input circuit (resonant at f_o); third, the RF output circuit, also resonant at f_o, consists of capacitance C_o and inductance L_o; and fourth, the RF output load consists of a resistance equal to the matched load value.

If the voltage produced by "RF Gen" is a 200-V peak sine wave, a half sine wave of anode current, whose peak value is $I_{apk} = 80$ A, will be conducted through the tetrode. The values of L_o and C_o will interchange their stored energy at resonant frequency f_o. To

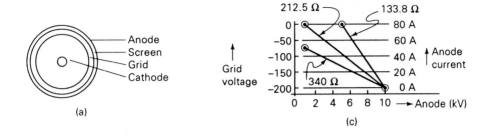

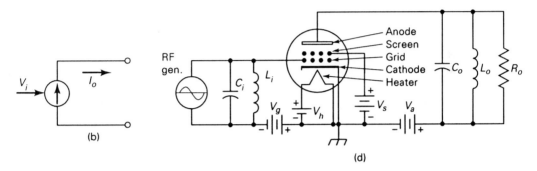

Figure 6-8 Configuration, symbology, and characteristics of an ideal tetrode: (a) cross section; (b) simplified equivalent circuit; (c) ideal tetrode characteristic; (d) symbology and simplified circuit.

a close approximation, if the output $L_o C_o R_o$ circuit has a Q value greater than 10, the voltage and current waves of R_O will be sine waves.

Based on the rationale above, the amplifier's performance can approximated by using the following eight-step analysis:

Step 1. The peak value of f_o current in R_o is

$$I_{pk} = \frac{80}{2} = 40 \text{ A} \qquad I_{orms} = \frac{40}{2^{1/2}} = 28.28 \text{ A}$$

Step 2. The anode voltage's sinusoidal f_o excursion goes from 10 kV, down to 1.5 kV, back up to 10 kV, then up to 18.5 kV and back to 10 kV. Its peak value is

$$V_{opk} = 10 - 1.5 = 8.5 \text{ kV} \qquad V_{orms} = \frac{8.5}{2^{1/2}} = 6.010 \text{ kV}$$

Step 3. The value of load resistance is

$$R_o = \frac{V_{orms}}{I_{orms}} = \frac{6010}{28.28} = 212.5 \text{ }\Omega$$

Sec. 6-3 System Power Amplification Devices

Step 4. The amplifier's output power is

$$P_o = V_{orms} I_{orms} = 6.010 \times 28.28 = 170 \text{ kW}$$

Step 5. The average power supply current is

$$I_{aav} = \frac{I_{pk}}{\pi} = \frac{80}{\pi} = 25.46 \text{ A}$$

Step 6. The dc power input to the anode is

$$P_{ain} = V_d I_{aav} = 10 \times 25.46 = 254.648 \text{ kW}$$

Step 7. The amplifier's anode conversion efficiency is

$$E = \frac{P_o}{P_{ain}} = \frac{170}{254.648} = 0.6675 \text{ or } 66.75\%$$

Step 8. The anode dissipation is

$$P_{adu} = P_{ain} - P_o = 254.648 - 170 = 84.688 \text{ kW}$$

Example 6-3

Assume that the 5-MHz amplifier analyzed above is used in an unattended transmitter application. The 133.8- and 340-Ω load lines shown in Fig. 6-16c represent the maximum and minimum resistive load equivalents on the system if a 1.6:1 VSWR were present. If the system is optimized for 212.5-Ω operation, but must also function in its unattended mode on the 1.6:1 VSWR load lines, how does the tetrode's RF output and anode dissipation compare with the matched load case?

A comparison using the eight-step approximate analysis shows the following:

Load (Ω)	RF output (kW)	Anode dissipation (kW)
133.8	106	148
212.5 (matched load)	170	85
340	106	53

What do these results indicate from a system design point of view?

Solution Two approaches are possible, depending on the system's operating frequency. *First approach:* If the amplifier must instantaneously provide operation at any point on a 1.6:1 VSWR circle (i.e., without recourse to tuning and matching procedures), its inherent design must be compatible with the wide range of operating conditions shown (i.e., especially with the worst-case conditions of the 133.8-Ω load line). *Second approach:* If an ideal circulator could be placed in the output transmission line, only the power reflected from the load would be lost in the circulator's dummy load (i.e., about 5.3% of the matched load power or 9 kW under 1.6:1 VSWR conditions). The effective RF output would then be 170 kW under matched conditions and 161 kW under 1.6:1 VSWR conditions. Unfortunately, high-powered circulators are available for use only in the microwave region, typically above about 1000 MHz. Therefore, the amplifier design must be compatible with the first approach.

Bipolar and Field-Effect Transistor Amplifiers

Why were the preceding three sections devoted to vacuum tubes in this era of solid-state devices? Because many high-powered RF systems still require tubes rather than transistors to generate and amplify signals at high frequencies [6]. Nevertheless, transistors have two outstanding advantages over vacuum tubes: (1) they are physically much smaller, and (2) they require no energy source to supply thermionic emission.

From an RF power generation point of view, the small physical size of a single transistor significantly limits its RF power output capabilities, and as the frequency of operation increases, the unit's RF power generation capabilities decrease. In the following we review: some techniques that have been used to improve the high-frequency performance capabilities of transistors and some techniques that can be used to further increase the RF power output capabilities of fundamental solid-state amplifier packages. Both *bipolar* transistors (NPN units) and *field-effect* transistors (N-channel MOSFETs) are considered.

Bipolar RF power amplifiers. In NPN transistors two junctions are formed by fabricating a very thin layer of P-type material between two layers of N-type material. Such an oversimplified NPN junction, along with several external circuit elements, can be connected into a solid-state RF power generator arrangement as shown in Fig. 6-9a. As an RF sine wave from V_S proceeds through its positive half-cycle, the collector current through R_O will rise and fall essentially in proportion to the current that V_S delivers to the base. On the negative half-cycle of the V_S wave, essentially no collector current will occur since the junction will be reverse biased. Hence in this example we have a class B amplifier system.

Class B arrangements deliver higher circuit efficiency than class A (ideally, class A is capable of 50% efficiency and class B is capable of about 79%). Class B amplifiers frequently use the push-pull circuit shown in Fig. 6-9b, where the alternate half-cycles of current through R_O are developed by a second transistor. The unfiltered output waveform is more nearly sinusoidal than with the single-ended class B amplifier. Appropriate coupling transformers or transmission line elements are necessary to properly implement a push-pull class B arrangement.

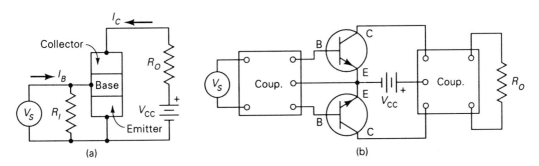

Figure 6-9 Two simplified NPN transistor amplifer cicuit configurations.

The capability of producing relatively high levels of RF power output from a single transistor operating in the HF/UHF region is difficult, if not impossible. On the other hand, it is a relatively simple task to produce low-power-level NPN transistors capable of delivering significant amplification beyond the UHF region. This capability has been capitalized on by means of unique fabrication techniques. High-powered transistors effectively parallel a large number of these low-level transistors within a single package. The interdigitated structure, shown in Fig. 6-10, is an example of how a number of small transistors with excellent high-frequency capabilities can be connected in parallel.

One of the problems involved in parallel operation of NPN units is to make them share the load equally. The base connection is the horizontal connection strip across the top of the Fig. 6-10, and the dark P-type base layers are attached to this strip. The emitters can be identified as the vertical N-type layers that emerge downward from the areas identified as resistors. The collectors are the N-type layers that emerge upward from the bottom common connection strip.

The emitter resistors define a give-and-take area. They force the NPN units to share the load more equally, but of course, they introduce additional loss into the overall unit. Still another give-and-take area lies in the size/density of the overall configuration. This consideration affects the power gain that can be achieved from the package.

Our objective here is simply to indicate that these and many other techniques have been studied, optimized, and applied by transistor designers. As a result, effective NPN transistors are now available for use in relatively high-powered RF generator systems.

MOSFET RF power amplifiers. Let us consider the elements that form a typical Motorola MOSFET structure. The initial building block is a substrate of high resistivity P-type material. Next, two low-resistivity N-type regions are diffused into the substrate (i.e., the regions are separated by a P substrate channel region). The two N regions become the MOSFET's drain and source, and the intermediate P channel becomes

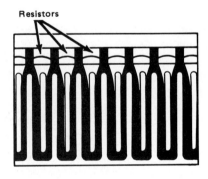

Enlargement of this "interdigitated" geometry shows emitter resistors that have been added to balance the current throughout the chip.

Figure 6-10 Interdiditated RF arrangement. ("Motorola RF Data Manual," 1986; Courtesy of Motorola Inc.)

its gate. The three regions are then covered by two insulating layers, one of insulating oxide and one of nitride. The gate and source connections are made to two metal plates that are attached to the insulating layers. The plates, positioned above the drain and source, are connected to the elements via holes in the layers. The gate connection is made to a third metal plate on the insulating layers. It is positioned above the P channel between the drain and source.

One significant difference between MOSFET and NPN transistors is that there is no physical connection between the gate terminal and the other electrodes. The coupling is strictly by means of the capacity of the gate's metal plate and the channel region. From a dc point of view the resistance between the gate and the source is effectively infinite, whereas in an NPN transistor the base-to-emitter dc resistance may be several hundred ohms or less. From an RF point of view the capacitances that exist between the electrodes are substantial, and as a result, the RF input *impedance* of a MOSFET is also fairly low.

From a transistor operating point of view, the three-element MOSFET can be applied in substantially the same manner as an NPN transistor. The diagrams and discussion presented on NPN transistors are similar with one exception. MOSFETs typically require the gate to be biased to a positive dc level of several volts in order to raise the device to its threshold of operation. The gate's control voltage region then encompasses several additional volts of positive excursion. In contrast, with an NPN transistor, the total control voltage excursion is perhaps a few tenths of a volt.

Packages for MOSFET assemblies are similar to those of NPN transistors. A relative comparison of transistors with other RF generators is given in Fig. 6-6.

RF amplifier configurations. Like triodes and tetrodes, transistors are devices. The purchaser is responsible for the design of the RF generator system of which they are a part. The RF output capability of an individual transistor is clearly much lower than required for many RF power generator system applications. A typical method used to overcome this situation can be illustrated with the following exercise.

Example 6-4

Consider an HF-region requirement for a CW power output of 17.5 kW into a matched load. One system approach is as follows. First, use HF transistors, rated to have 600 W RF output, for the basic building blocks. Second, use pairs of these transistors in the push-pull amplifier package configuration shown in the upper part of Fig. 6-11. Under ideal conditions this package will deliver 1200 W, and if 15 packages are effectively paralleled, the objective power level will be achieved. Third, assume that 16 of the packages are used to account for component losses. Fourth, use 50-Ω broadband ferrite transformers. It is common practice to design RF amplifier packages with input and output impedances of 50-Ω. This makes the packages compatible with conventional 50-Ω RF transmission lines. The transformers provide matching between the transistor's impedance levels and the 50-Ω terminating impedance levels. Fifth, use a corporate RF combiner approach to effectively parallel the RF outputs of the 16 RF amplifier packages as shown in the center of Fig. 6-13. First, eight pairs of packages are combined, next four 2400-W summations are combined, then two 4800-W summations are combined, and finally, two 9600-W summations are combined at the top of the corporate structure. Sixth, use 3-dB hybrids (i.e., the circles labeled H are four-port 3-dB hybrids). Two 50-Ω input ports accept the 1200-W RF levels of two RF amplifier packages

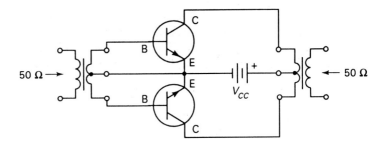

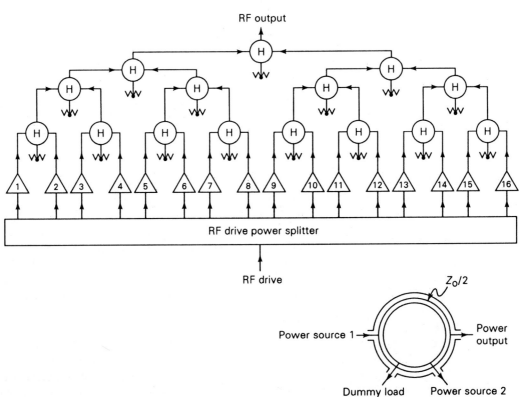

Figure 6-11 Some RF transistor power-combining arrangements.

and combine them to 2400 W at a third port. The fourth port connects to a dummy load. If the output of one 1200-W amplifier should fail, the remaining 1200-W amplifier will deliver half of its output to the normal output port and half of its output to the dummy load. This is a significant system reliability feature since the overall RF power generator system maintains it prime function—but at a somewhat reduced RF power output level.

For HF, 3-dB hybrids often use ferrite assemblies (i.e., similar to the ferrite transformers in the last paragraph). A VHF and UHF approach, shown in Fig. 6-11, uses a ring of RF line with one-quarter-wavelength port-to-port spacing in its lower portion and three-quarters-wavelength spacing in its upper portion. At low to moderate power levels, hybrids of this type are readily fabricated using microstrip. Higher power levels use coaxial/waveguide line structures.

In several respects transistor power amplifier systems are comparable to tetrode power amplifier systems. For example, first, transistor characteristic curves have the same approximate shape as tetrode characteristic curves, albeit the voltage and current axis scales are vastly different. Second, if a VSWR analysis similar to the one conducted for the tetrode amplifier in Example 6-2 were conducted for the high-powered transistor amplifier above, the system design considerations/results would be substantially the same in both cases (i.e., other than for the difference in the operating power levels involved).

6-4 PRACTICAL SYSTEM APPLICATIONS

As a means of providing background on some future topics of discussion, in the following sections we present brief reviews of three RF systems. Since AM, FM, and television RF power generator systems have been given extensive coverage in other books, the arrangements we review are associated with radar RF power generator system techniques/principles of relatively recent origin.

MTI Pulse Doppler Radar

Radar may be defined as the art of detecting by means of radio echos the presence of objects, determining their direction and range, recognizing their character, and employing the data thus obtained in the performance of military, naval, or civilian activities.

Figure 6-12 shows a block diagram of perhaps the most fundamental form of radar system. Typical functions and operations proceed as follows. First, the transmitter block starts a chain of events when it functions to generate a train of rectangular RF pulses and deliver them to antenna 1. Second, the radiated beam from antenna 1 directs the RF energy into a volume in space where an object is assumed to exist. Third, if there is an object present, the RF energy illuminates the target in space. Fourth, a miniscule portion of this energy (the echo) is reflected from the target and returned through space to the radar site. Fifth, a portion of the reflection is picked up by antenna 2 and then processed through a receiver. Sixth, the detected output from the receiver is displayed on an indicator that shows the target's range (based on electronic measurement of the period of time that elapses between a pulse's transmission until its echo's indication). Seventh, the azimuth of the target is determined by physically positioning the pointing direction of antenna 2's narrow beam. The beam is directed to a position in space that causes the echo on the indicator to have its maximum response. Eighth, when the radar operator observes the indicator over a period of time, he or she can establish if the target's range is changing (i.e., if it is in motion). Ninth, in summary, the system has allowed the operator to locate the position of a previously unknown moving or fixed object in space. The last objective of our initial definition—regarding future activities—is of course a function of the radar system's operating personnel.

It was evident to early era radar personnel that a more sophisticated moving-target

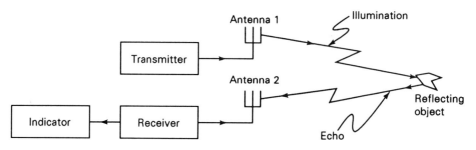

Figure 6-12 Elementary system for radio echo detection.

technique was required. Current *MTI* (moving-target indicator) radar systems are based on the *Doppler* principle. When the illumination from an amplitude-pulsed radar encounters a moving target, the target's change in range from pulse to pulse causes its reflected energy to be shifted in frequency, by an incremental amount, from pulse to pulse. This incremental shift (in either a positive or a negative direction) is known as the Doppler shift. The Doppler frequency shift is

$$F_D = kFV \qquad (6\text{-}15)$$

where F is the radar's frequency in MHz, and V is the radial velocity of the target, and $k = 0.00186$ for V in km/h (0.00299 for V in mi/h).

In conventional radar systems there is no significant difference between a fixed target and a moving target on the radar's indicator except for the relatively slow movement of the moving target on the display. Consider the situation around an airport, where the vast number of fixed targets (buildings, towers, fixed aircraft on the ground, etc.) makes a conventional radar virtually useless—the fixed targets mask most of the moving targets. However, if a system were devised that responded only to targets that have a Doppler shift, the entire situation would be reversed, and ideally, only moving targets will appear on the airport operator's radar indicator.

Most MTI radar implementation techniques are based on measurement of the phase difference between the illuminating signal and the illuminating signal plus or minus the Doppler shift. One time-proven measurement approach uses a delay line canceler. Figure 6-13 shows a block diagram of a simple MTI radar system with a delay-line canceler.

Early MTI systems often had their capabilities limited by an inherent lack of stability in the power oscillators used to generate their radiated RF pulses. More recent systems achieve an order-of-magnitude better stability by using an RF power amplifier driven from a stable low-level RF source, as shown in Fig. 6-13. The RF amplifier is driven by a mixer signal that combines the stable local oscillator output F_L plus the stable coherent oscillator output F_C.

When a radiated RF pulse is required, the pulse modulator is triggered. It provides a dc electrode energy pulse to the RF amplifier, and a high-powered RF output pulse is delivered to the transmit/receive switch. This electronic switch functions to automatically deliver the high-powered pulse to the common antenna system during the transmit period and to deliver the reflected echoes to the receiver during later portions of each interpulse period.

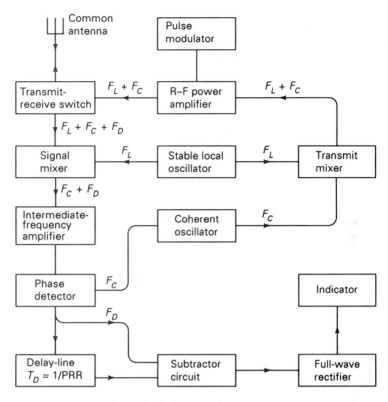

Figure 6-13 MTI pulse Dopler radar with delay-line canceler.

The superheterodyne receiver mixes the moving target echos ($F_L + F_C + F_D$) with a stable local oscillator output (F_D) to generate the intermediate frequency (IF) of the receiver ($F_C + F_D$), typically around 30 MHz. This signal is amplified by the IF amplifier.

The coherent oscillator is a stable oscillator operating at F_C, the receiver's IF. The phase detector compares the output of the IF amplifier ($F_C + F_D$) and the coherent output (F_C). The resultant phase detector output is a bipolar video Doppler signal, F_D (its polarity depends on how much phase shift occurs on successive pulses relative to the unshifted coherent oscillator phase).

Early MTI radars used A-scope indicators (where targets were displayed as peaks along a horizontal trace and range was defined by the distance from left to right along the trace). When the bipolar video was placed on these indicators, moving targets were immediately evident by their "butterfly" effect caused by the bipolar video, while fixed targets were stable peaks.

The purpose of the three blocks along the bottom of the diagram is to cancel out the fixed targets. Note that the output of the phase detector is routed two ways. The first path goes to the delay-line input (the line has a time delay exactly equal to 1/PRR seconds). The delay-line output goes to one of the subtractor inputs. The second phase detector path goes to the other subtractor input. The subtractor allows us to compare the latest Doppler return with the previous Doppler return. Therefore, if a target is fixed, the two signals will be equal and there will be no output from the subtractor, but if the target is moving, the subtractor output will be bipolar video. Since there is no further need for bipolar video, the output is rectified by a full-wave rectifier, and with a perfectly stable MTI radar system, only moving targets will appear on the radar's indicator.

The need for high-performance MTI radar systems is as great today as at any time in the history of radar. RF power generators possessing unusually high degrees of stability are essential for a high-performance MTI radar system. We will examine several ways of improving the MTI capabilities of RF power generators in Chapters 10 and 11.

Phased Array Radar Systems

Surveillance information is supplied by search radar systems. They function to provide 360° azimuth coverage of the volume surrounding military sites, airports, and so on. Their antenna systems are kept in a state of continual rotation to provide updated information. Typically, a PPI indicator displays the echos on a polar plot display of the surrounding area.

If a simultaneous radar tracking function were required, a second radar system would be required so that its antenna could provide the necessary tracking beam; if a third simultaneous radar task were required, a third radar system would be required; and so on.

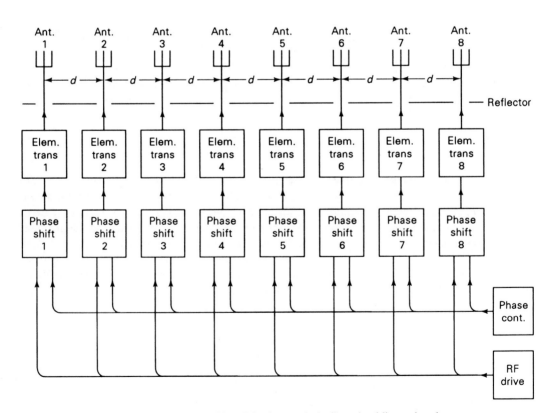

Figure 6-14 Simplified eight-element single-dimensional linear phased array.

The relatively slow response capabilities of mechanically driven antenna systems is one of the major factors that contributed to *phased array* radar system development. Phased arrays can provide virtually instantaneous antenna beam directivity. As a result, a single radar system can execute a number of simultaneous functions.

Consider the simplified phased array/transmitting system shown in Fig. 6-14. The system rationale is as follows: Eight RF radiating elements are arranged to form an array. Their center-to-center spacing is d. A vertical reflecting screen is placed behind the elements to minimize radiation/reception from the rear. Elemental RF power amplifiers deliver energy to their respective elements from behind the screen. RF excitation is delivered to each elemental amplifier via its associated RF phase shifter. A common RF exciter can serve all amplifiers.

Electronically controlled phase shifters are the heart of the system. The computer-actuated phase control block issues commands to the phase shifters. It can direct the antenna's beam to alternate locations within microseconds. The computer must also keep track of all antenna targets and steering functions for the multifunction radar system.

Alternative RF generator approaches can be employed. One common approach is to use a smaller quantity of higher-powered RF amplifiers. Each of these RF amplifiers then delivers its output to a power divider system. Each of the reduced RF output ports of the divider is then passed through an electronic phase shifter. Many further variations of this technique exist as well.

System complexity/cost—All phased array systems are significantly more complicated and more expensive than their single function counterparts. A number of alternative system configurations are usually reviewed (from cost, complexity, performance, etc., points of view) before an optimum system is established.

The simple array shown in Fig. 6-14 produces an antenna beam of limited capabilities. If distance d between elements is one-half wavelength at the RF operating frequency, the approximate half-power bandwidth in the horizontal plane is

$$\text{BW} = \frac{102}{N} \tag{6-16}$$

degrees, where N is the number of elements (BW = 102/8 = 12.75°). The beam in the vertical plane is wide and fan shaped. If we wanted to construct a pencil beam, a two-dimensional array is required. For example, if seven similar rows of equipment were stacked above the present row, the resulting 64-element array would have half-power beam widths of around 12.75° in both vertical and horizontal planes.

Over-the-Horizon Backscatter (OTH-B) Radar Systems

The long-range detection capabilities of OTH-B radar systems are currently receiving significant levels of attention. To understand the reason, consider the following approximate horizon range equation:

$$d = k(2h)^{1/2} \tag{6-17}$$

where d is the distance to the radar horizon, h the height of the target, and $k = 0.888$ for d in kilometers and h in meters ($k = 1$ for d in miles and h in feet).

Assume that we need continuously to detect aircraft flying at 3048 m (10,000 ft). The equation indicates that the radar horizon, and hence the maximum range of a microwave radar, will be approximately 226 km (141 mi). OTH-B radar detection ranges can be 10 or more times greater.

High frequencies (HF), in the range 3 to 30 MHz, have been the medium of long-distance radio transmissions for many years. The long-distance propagation capabilities in this frequency range are the result of wave reflection from the ionosphere's layers that exist over the surface of the earth. The ionosphere's properties are subject to wide variations as a function of time of day, month, and year. OTH-B radars systems operate in the HF region.

The essential principles of OTH-B system operation are generally the same as those of microwave systems, but there are a few significant differences. Radar illumination propagates out in conventional fashion, but ultimately it is reflected from the ionosphere and is returned toward earth. If it then encounters various elevated and surface objects, in conventional fashion, miniscule echos are reflected and returned via the ionosphere to the general vicinity of the radar system.

From a difference point of view, recent OTH-B systems typically operate in a CW mode. These systems are bistatic: that is, the transmit and receive sites are at separate locations, perhaps 100 mi or so apart. Figure 6-15 shows a very simplified OTH-B arrangement based on these principles. Insofar as the CW mode is concerned, frequency-modulated pulses (i.e., chirped pulses) are used rather than conventional pulses repeated at a PRF. At the end of one chirp the next chirp starts almost immediately. Hence there is no PRF, but rather, there is a *sweep repetition rate*. It is by this means that updated echo information is delivered to the system. Some OTH-B areas of interest are discussed below.

Frequency coverage.   To accommodate the variations in the ionosphere, systems typically operate over a range of frequencies from approximately 5 to 30 MHz.

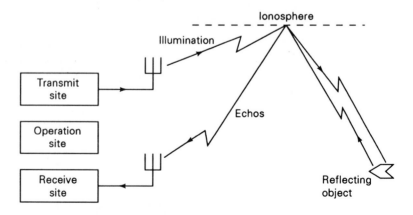

Figure 6-15   Principles of over-the—horizon backscatter radar.

Antenna systems. The radar uses two groups of phased array antennas, one group for illumination of targets and the other group for reception of echos. The antennas can be steered over angular segments of around 60°. The baseline length of an antenna system is typically 1 km (0.62 mi).

Radiated power. Twelve RF power amplifiers, each capable of around 100 kW, are used to drive the transmit antenna system. The RF generators have broadband characteristics, and in addition, they operate over several switched bands. These bands are selected in such a way that, first, they accommodate the antenna system bandwidth capabilities, and second, they accommodate the extremely high RF harmonic attenuation specifications placed on the system.

VSWR. The transmit antennas exhibit VSWR values of up to 1.6:1. The effect this level of VSWR produces on the bandpass RF power amplifier's characteristics will be analyzed later in an example.

Increased angular coverage. Typically, more than 60° of radar coverage is required. This is provided by constructing one or more identical systems, each perhaps a 1.6 km (1 mi) away from one another, with their antenna systems oriented to cover adjacent 60° sector(s).

Operations center. A remotely located operations site functions as the control center for the transmit and receive sites. An array of communications equipment links the three sites and ensures that coordinated system operation occurs.

Example 6-5

OTH-B system tetrode power amplifiers utilize several broadband output circuits. The bandpass circuits perform three major functions: first, provide a reasonably flat response curve across each band when the system is terminated in a matched load; second, provide a transformation ratio between the 50-Ω output transmission line impedance and the optimum output impedance of the power amplifier (approximately 200 to 250 Ω); and third, provide a high degree of harmonic suppression.

The capabilities of one of the 5- to 6.5-MHz bandpass circuit can be assessed by using Appendix 2. The tetrode can by simulated by an FET model shunted with a resistor. The top section of Fig. 6-16 lists the circuit elements and nodes for a representative bandpass circuit.

1. Under matched load conditions, what is the flatness of the response across the operating portion of the band?
   Approximately ± 0.22 dB (see the first run in Fig. 6-16).
2. What is the second harmonic attenuation?
   At least 63 dB (see the second run in Fig. 6-16).
3. What is the average input impedance of the bandpass circuit across the active band?
   Approximately 235 Ω (see the third run in Fig. 6-16).

```
The Input Information For This Analysis Is As Follows:
         1. Number of network nodes = 9
         2. FET = 1 , 0 , 2 , 1
         3. C1 = 2 , 0 , .00043
         4. L1 = 2 , 3 , 1.05E-06
         5. L2 = 3 , 0 , 7.56E-07
         6. C2 = 3 , 0 , .0000057
         7. L3 = 3 , 8 , 2.97E-06
         8. C4 = 8 , 0 , .0000101
         9. L4 = 8 , 4 , 2.97E-06
        10. C5 = 4 , 0 , .0000057
        11. L5 = 5 , 0 , 3.26E-07
        12. C6 = 4 , 5 , .000114
        13. C7 = 5 , 0 , .00239
        14. C8 = 5 , 6 , .00013
        15. C9 = 6 , 0 , .0000057
        16. L6 = 6 , 9 , 2.82E-06
        17. C10 = 9 , 0 , 9.62E-06
        18. L7 = 9 , 7 , 2.82E-06
        19. L8 = 7 , 0 , 5.940001E-07
        20. C11 = 7 , 0 , .001317
        21. R0 = 7 , 0 , 50
        22. RB = 2 , 0 , 1600
        23. End of list
```

--- This Is A Voltage Analysis ---
Input and output nodes are 1 , 7

FREQ.	VOLTS	DB	PHASE	DELAY
4.50E+06	14.15	23.01	-168.0	
4.75E+06	93.10	39.38	94.7	
5.00E+06	95.98	39.64	-9.1	
5.25E+06	97.97	39.82	-80.2	
5.50E+06	98.45	39.86	-137.7	
5.75E+06	99.12	39.92	166.0	
6.00E+06	94.26	39.49	115.4	
6.25E+06	99.14	39.93	64.8	
6.50E+06	99.15	39.93	5.0	
6.75E+06	99.44	39.95	-64.6	
7.00E+06	42.08	32.48	-160.1	

FREQ.	VOLTS	DB	PHASE	DELAY
9.00E+06	0.22	-13.05	119.7	
9.50E+06	0.12	-18.54	116.0	
1.00E+07	0.07	-23.23	113.2	
1.05E+07	0.04	-27.31	111.0	
1.10E+07	0.03	-30.92	109.3	
1.15E+07	0.02	-34.16	107.8	
1.20E+07	0.01	-37.09	106.6	
1.25E+07	0.01	-39.76	105.5	
1.30E+07	0.01	-42.22	104.6	
1.35E+07	0.01	-44.49	103.8	
1.40E+07	0.00	-46.59	103.1	

Input and output nodes are 1 , 2
The input or transfer impedance is:

FREQ.	REAL	IMAG.	AMPL.	ANGLE
4.50E+06	27.761	192.980	194.966	81.81
4.75E+06	243.626	230.385	335.307	43.40
5.00E+06	223.413	112.920	250.328	26.81
5.25E+06	225.315	50.912	230.996	12.73
5.50E+06	226.524	31.345	228.682	7.88
5.75E+06	229.699	-18.675	230.457	355.35
6.00E+06	204.555	-33.408	207.266	350.72
6.25E+06	231.040	-41.642	234.763	349.78
6.50E+06	239.701	-107.264	262.607	335.89
6.75E+06	257.049	-169.630	307.975	326.58
7.00E+06	85.237	-269.168	282.342	287.57

Figure 6-16 Matched load/second harmonic/input impedance for bandpass amplifier.

4. What variation in power output occurs when the VSWR is 1.6:1? (*Note:* This variation can be approximated by terminating the output of the filter with alternate values of 31.25 and 80 Ω—the two pure resistive terminations that occur on a 1.6:1 VSWR circle—and repeating the computer runs.)

The variation is on the order of $\pm$ 1.5 dB.

Chapter 7

Transient Analysis of Power Electronics Systems

The objective of the preceding six chapters was to review some techniques for performing *steady-state* analyses on power electronics systems. The objective of the remainder of the book is to develop a set of techniques for performing *transient* analyses on typical power electronics systems. In Chapter 8 we establish the highly transient-related behavior of dc power supply energy sources (including energy sources that operate from low-frequency, high-frequency, and/or dc prime power sources). In Chapters 9 and 10 the transient-related behavior of high-level energy storage and switching techniques is established (including both low-level and high-level pulser circuits). In Chapter 11 we analyze the specialized transient performance capabilities common to typical RF power amplifier systems. However, our objective in this chapter is to develop a transient problem solution technique—a procedure that we can use to obtain solutions to the problems in all subsequent chapters.

7-1 TRANSIENT ANALYSIS APPROACHES

The solutions for transient analyses practically always require the solution of sets of differential equations. A differential equation may be defined as an equation that includes derivatives and differentials in some of its terms. For example,

$$\frac{dy}{dx} = 2x \tag{7-1}$$

is a simple differential equation. The solution (integral) of a differential equation is the relation between the variables involved by which the equation is satisfied. For Eq. (7-1) by straightforward integration we obtain

$$y = x^2 + C \tag{7-2}$$

which is the solution. Unfortunately, most of the equations we will encounter cannot be disposed of with such ease.

There is no specific equation solution method that is best for all situations [7]. Three methods commonly used for the solution of differential equations are described below.

Classical Method

This approach is based on the use of a number of techniques, developed over the centuries, that have been found effective in solving certain forms of differential equations. In general, a good background in mathematics is helpful when using this method.

Laplace Transform Method

This approach uses the technique of transforming the equations to forms that have lower levels of complexity. The transformed equations can then usually be solved by using less sophistocated mathematics.

Approximation Method

This approach employs numerical integration for approximating the desired solutions. Differential equation solutions can be obtained simply by using repetitive computer analysis loop procedures. In future chapters we use the approximation method for all of our analyses. In the next two sections we review the network diagram element equations used and define the principles of the analysis program. Following this, we demonstrate the program by solving differential equations associated with a number of elementary power electronic system problems.

7-2 NETWORK BUILDING BLOCK DEFINITIONS

The following six building blocks will allow us to prepare almost any circuit diagram we need:

1. Voltage and current sources
2. Resistances
3. Capacitances
4. Inductances

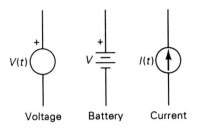

Figure 7-1 Active network element summary.

5. Mutual inductances
6. Time-variant elements

Most of our analyses are concerned with network problems rather than field problems. In network problems the various parameters (resistance, inductance, etc.) are assumed to be lumped, whereas in field problems the parameters are distributed (e.g., as they were in the long-line analyses of Chapter 4).

Active Network Elements

Electrical networks are made up of active and passive elements. *Active elements* are power or energy sources. *Passive elements* are resistors, capacitors, and inductors. The choice of a voltage source or a current source is governed primarily by the type of analysis. We use the type of source that leads to the simplest set of analysis equations.

Figure 7-1 illustrates the three sources we will use. The polarity signs on a voltage source define the source at the instant the voltage $V(t)$ has a positive value. The direction of the arrow on a current source defines the direction of flow of conventional current when $I(t)$ has a positive value. On a network diagram the voltage definition allows us consistently to define the polarity of voltages as we proceed around a loop. The current definition allows us to consistently sum up the currents that exist at a given circuit junction (node).

Resistance

Resistance is the circuit element that dissipates electrical power in the form of heat. Figure 7-2 defines resistance in its network diagram format. The voltage and current, both expressed as a function of time, are

$$V_R(t) = RI_R(t) \tag{7-3}$$

$$I_R(t) = \frac{V_R(t)}{R} \tag{7-4}$$

There is a voltage drop across a resistance when conventional current has the direction indicated by the arrowhead.

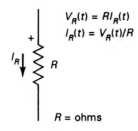

Figure 7-2 Resistance element summary.

Capacitance

Capacitance is the passive circuit element in which electrostatic energy is stored. The charge on a capacitor is defined as the integral of $I_c dt$, and the voltage drop in the direction of positive conventional current is

$$V_c(t) = \frac{1}{C}\int I_c(t)\, dt \tag{7-5}$$

For our analyses we need to include any initial voltage that exists across the capacitor. Then we have

$$V_c(t) = \frac{1}{C}\int_0^t I_c(t)\, dt + V_c(0) \tag{7-6}$$

where $V_c(0)$ is the voltage across the capacitor due to the initial charge at $t = 0$. Equation (7-6) can be solved by differentiating it,

$$I_c(t) = C\frac{dV_c(t)}{dt} \tag{7-7}$$

We will use Eq. (7-7) extensively.

Figure 7-3 defines capacitance in its network diagram format. If the initial voltage is as shown by the figure's polarity marks, the sign of the voltage term in Eq. (7-6) is positive, and if the polarity is the reverse of that shown, the sign of $V_c(0)$ is negative. The markings indicate the polarity of the voltage drop term, $(1/C)\int I_c(t)\, dt$, when current has a positive value in the direction of the arrowhead.

Inductance

Inductance is the passive element in which magnetic energy is stored. The voltage drop in the direction of positive current—when the current has a positive derivative—is as follows:

$$V_L(t) = L\frac{dI_L(t)}{dt} \tag{7-8}$$

Sec. 7-2   Network Building Block Definitions

When the derivative is negative a voltage rise rather than a voltage drop will occur. We will use this equation extensively. If Eq. (7-8) is integrated, we obtain

$$I_L(t) = \frac{1}{L}\int_0^t V(t)\, dt + I_L(0) \tag{7-9}$$

where coupling with other inductances is zero. The constant $I_L(0)$ defines the initial current through the inductance at $t = 0$. This constant must be assigned the proper sign in accordance with the assumed direction of positive current in the circuit. Figure 7-4 summarizes these considerations.

Mutual Inductance

Mutual inductance is derived from magnetic coupling between inductances; hence it is not literally considered a network element. Mutual inductance is represented as shown in Fig. 7-5. The total energy stored can be segregated as, first, energy due to the inductance of circuit 1; second, energy due to the inductance of circuit 2; and third, energy due to the mutual inductance between circuit 1 and circuit 2.

If current $I_1(t)$ is varying, the open-circuit voltage induced in circuit 2 is

$$V_2(t) = M\frac{dI_1(t)}{dt} \tag{7-10}$$

and if current $I_2(t)$ is varying the open-circuit voltage induced in circuit 1 is

$$V_1(t) = M\frac{dI_2(t)}{dt} \tag{7-11}$$

The sign of M is always positive, but the polarity of the voltage drops depend on the winding sense of the coils. The polarity of these voltages must be referred to the assumed direction of current on the network diagram, as follows:

1. If a positive rate of change of current in circuit 1 induces a voltage drop in circuit 2 in the arrow direction of current, the sign of the voltage term is plus. If a voltage rise is produced, the sign is minus.
2. If a positive rate of change of current in circuit 2 induces a voltage drop in circuit 1 in the arrow direction of current, the sign of the voltage term is plus. If a voltage rise is produced, the sign is minus.

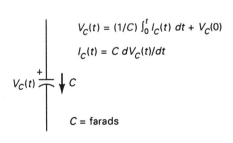

Figure 7-3 Capacitance element summary.

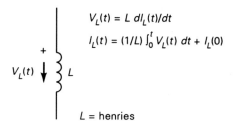

Figure 7-4 Inductance element summary.

Note that in the two statements above a pairing of signs always occurs—either both signs are plus or both are minus.

While ideal transformers can be used in simplified analyses, in our high-frequency and pulse analyses the winding and leakage inductances will be included. These transformers will be based on the fundamental steady-state principles developed in Section 2-2.

Time-Variant Elements

Most of the equations in our analyses have constant coefficients. In some cases we may want to vary a coefficient to define two operating extremes, for example open-circuit and short-circuit conditions. These two case solutions can easily be accommodated by directing the program with alternate conditional statements.

On some occasions we will want a coefficient to assume a time-variant value as the program progresses. Microwave power tubes (and most other active RF elements) introduce this requirement. The numerical integration program we will be using can accommodate time-variant elements.

Linear beam tubes have the following beam current versus beam voltage characteristic,

$$I = kV^{1.5} \tag{7-12}$$

where I is the beam current, V the beam voltage, and k the *perveance* factor associated with an amplifier's physical structure. When the tube is connected to a pulser, it represents a time-variant load resistance as the beam voltage rises from zero to full operating potential. We can analyze this condition by introducing Eq. (7-12) within the equation set defining a system of this type.

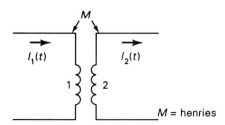

Figure 7-5 Mutual inductance summary.

Sec. 7-2   Network Building Block Definitions

7-3 NUMERICAL INTEGRATION PROGRAM

The mathematical approach we use is referred to as either numerical approximation or numerical integration. In its simplest form this approach assumes that for a given differential equation, and for a given point in time, we know the equation's solution and its rate of change as a function of time. Then for any future point in time we predict that the solution of the equation will simply be the present value plus the rate of change times the difference in the two values of time. A reduction in the time-step size typically improves the solution accuracy. This can be itemized as follows: first, we choose an integrating interval; second, we then compute the derivative using this interval; third, we say that the new value of the variable is the old value plus the derivative times the integrating interval; and fourth, we repeat the process.

Assume that we want to calculate the value of a current that varies as a function of time. We know that its present value is 100 A and that it is changing at rate of 1000 A/s. We can write

$$I(t) = 100 + 1000t \tag{7-13}$$

Suppose that we let our integrating interval be 0.001 s. The solution is then $I(t)$ = 101 A—perhaps a very reasonable solution. However, if we would have instead let the integrating interval equal 1 s, that solution, $I(t)$ = 1100 A, is very questionable. This is because the derivative could go through a very large change in this time period (except for the special case where the I versus t function is a straight line). Typically, a higher-order *algorithm* is used instead of the simple approach used in this example. This usually improves the analysis program's capabilities.

Analysis Algorithms

It is essential that a numerical integration program have a step-by-step problem solution procedure—its algorithm. We will consider only three of the numerous procedures that have been set forth.

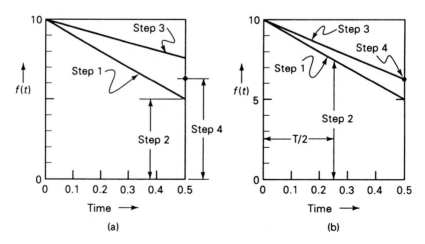

Figure 7-6 Second-order algorithms: (a) Euler–Cauchy; (b) Runge–Kutta.

First-order algorithm. Consider the following differential equation:

$$\frac{dy}{dx} = -y \quad \text{where } y = 10 \text{ at } t = 0 \qquad (7\text{-}14)$$

A first-order algorithm utilizes the simple approach we used to solve our demonstration problem in Eq. (7-13). Suppose that we want to obtain the value of y at $t = 0.5$ s. We could write

$$y(t) = y(t_o) + (t - t_o)\frac{dy}{dt(t_o)}$$
$$= 10 + (0.5 - 0)(-10)$$
$$= 5$$

The differential equation has an exact solution:

$$y = 10e^{-t} \qquad (7\text{-}15)$$

The exact solution gives a value of $y = 6.06530695$. We can see that the first-order algorithm's value of 5 is in poor agreement for a time increment of 0.5. An improved approach is required.

Second-order Euler–Cauchy algorithm. The approach used by this algorithm is shown in Fig. 7-6. The rationale stated in words is as follows:

1. Compute the derivative at $t = 0$ (as in the first-order approach).
2. Estimate the value of the variable at time $= T$ by using the derivative above (as in the first-order approach).
3. Compute a new derivative by using the estimated value of step 2.
4. Take the average of the two derivatives.
5. Use the derivative of step 4 to compute the variable at $t = T$.

Now if we solve the problem using this algorithm:

1. Initial derivative $= -10$
2. Variable estimate $= 5$
3. New derivative $= -5$
4. Average derivative $= -7.5$
5. $y = 10 + 0.5(-7.5) = 6.25$

Thus we see that the second-order Euler–Cauchy gives a much closer approach to the correct value despite the quite long integrating interval used.

Second-order Runge–Kutta algorithm. This algorithm is also illustrated in Fig. 7-6. Its rationale is as follows:

1. Compute the derivative at $t = 0$.
2. Estimate the variable at $t = T/2$ using the derivative from step 1.

3. Compute the new derivative using the estimated value from step 2.
4. Use the new derivative from step 3 to compute the variable at $t = T$.

For our example we then obtain:

1. Initial derivative $= -10$
2. Variable estimate $= 10 + 0.25(-10) = 7.5$
3. Variable estimate $= dy/dt = -y = -7.5$
4. Variable at $t = 0.5$ is $y = 10 + 0.5(-7.5) = 6.25$

For this particular problem, since the derivative is a linear function of time, both algorithms give the same answer. They do not agree for all cases.

All algorithms are subject to *numerical instability* if the integrating interval is made too large. This situation becomes almost immediately apparent since the computer will signify that an arithmetic overflow condition has taken place.

Higher-order algorithms are sometimes used where more precise solutions are required. These algorithms are less susceptible to numerical instability. The program we use for our analyses employs the second-order Euler–Cauchy because of its simplicity and generally acceptable performance.

Program Outline and Elementary Solutions

This program is an effective means of obtaining solutions to most transient system problems. It was introduced at General Electric by W. G. Wright[8] for use on the company's time-share computer system in the 1970s and has been widely used both within and outside the company. The version we use adapts the same principles to personal computer use.

Appendix 12 lists the program in its fundamental form. Its outline is as follows. Lines 12–18 dimension the four arrays used for storing analysis values. Lines 32–36 establish the three time periods involved in the analysis. T_1 sets the time step used for integration. T_2 sets the printout step size of the solutions. T_3 sets the upper time limit of the analysis.

Lines 80 and 82 head up the solution printout with the integration step T_1. Lines 84 and 86 solve the system equations for the initial analysis step. Subroutine 1000 contains the equations for the system, and subroutine 2000 processes the printout for each T_2 increment. Lines 100–300 are the heart of the program. Each step of the analysis uses the second-order Euler–Cauchy algorithm that is defined between lines 100 and 200 of the listing.

Lines 1000–1999 list the equations we have to process to obtain a solution. You will note that only the single equation we want to solve in the first demonstration is listed. It is a traditional battery, inductance, and resistance circuit. Since only one equation is used, we enter $N = 1$ in line 20 to advise the program of this fact.

Lines 50–70 define the network elements shown in Fig. 7-9b: resistor $R = 1\,\Omega$, inductance $L = 1$ H, and battery $V = 1$ V. Lines 2000–2999 are the printout statements. Line 2005 sets up the column headings for the printout, and 2010 prints out the solution results.

To illustrate the program, load Appendix 12 and issue a <RUN> command. The printout you will obtain is shown in Fig. 7-7. It shows how the current builds up in the

Integration Step = .1 Seconds

Transient Numerical Integration Program

Seconds	Current
0	0
.1	.095
.2	.181
.3	.259
.4	.329
.5	.393
.6	.451
.7	.5030001
.8	.55
.9	.593
1	.631
1.1	.666
1.2	.698
1.3	.727
1.4	.753
1.5	.776
1.6	.798
1.7	.817
1.8	.834
1.9	.85
2	.864
2.1	.877
2.2	.889
2.3	.899
2.4	.909
2.5	.918
2.6	.925
2.7	.932
2.8	.939
2.9	.945
3	.95

Figure 7-7 Appendix 12 *LR* circuit analysis.

circuit as a function of time after the switch is closed at $t = 0$. This is a traditional electrical transient exercise with well-publicized results; hence we can see that the program is functioning as it should and that the solution is correct. Next let us define how we set up the system equations.

In a classical solution of the problem we could write the following loop equation:

$$iR + L\frac{di}{dt} - V = 0 \qquad (7\text{-}16)$$

where i is the transient current in the circuit. In this program a variable is defined as $X(\cdot)$, or $D(\cdot)$, where the parentheses contain an integer that allows us to identify different variables. The $X(\cdot)$ terms are typically either voltages or currents. The $D(\cdot)$ terms are derivatives of $X(\cdot)$ terms that have the same integer within the parentheses. That is essentially the complete program rationale. For the example of Fig. 7-7 we can write

$$X(1)\,R + L\,D(1) - V = 0$$

Sec. 7-3 Numerical Integration Program

Integration Step = .1 Seconds

Transient Numerical Integration Program

Seconds	Current
0	1
.1	.9
.2	.815
.3	.737
.4	.667
.5	.604
.6	.546
.7	.494
.8	.447
.9	.405
1	.367
1.1	.332
1.2	.3
1.3	.272
1.4	.246
1.5	.222
1.6	.201
1.7	.182
1.8	.165
1.9	.149
2	.135
2.1	.122
2.2	.111
2.3	.1
2.4	.091
2.5	.082
2.6	.074
2.7	.067
2.8	.061
2.9	.055
3	.05

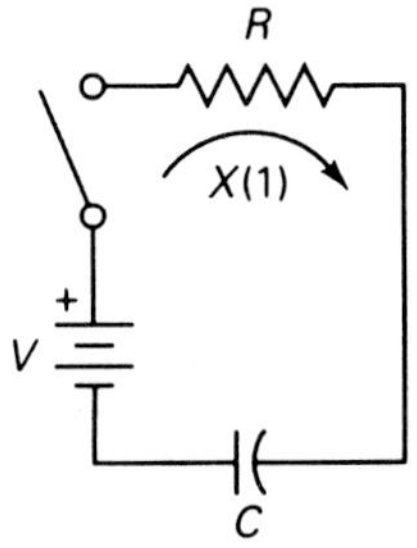

Figure 7-8 Appendix 12 *RC* circuit analysis.

With a term rearrangement and a shift to program format, we have

$$D(1) = (V - X(1)\ R))/L \qquad (7\text{-}17)$$

All of our analysis equations will be written in a mathematically acceptable form. They can be converted to BASIC simply by entering multipliplication symbols (i.e., *) and modifying exponentials [e.g., $X(2)^2$ to $X(2)^\wedge 2$]. Lower-level brackets will be used throughout, as in BASIC.

How can we handle integrals in a program that deals only with derivatives? The next example demonstrates the approach—we change the inductance of the last example to a capacitor as shown in Fig. 7-8. We know from Eq. (7-5) that the voltage across a capacitor is $1/C \int I(t)\ dt$. In addition, we know from Eq. (7-7) that the current in a capacitor is $C[dV(t)/dt]$. When we write our loop equation we define the voltage across the capacitor as $X(1)$. To illustrate, first we can write

Integration Step = .1 Seconds

Transient Numerical Integration Program

Seconds	Current
0	0
.1	.05
.2	.188
.3	.391
.4	.634
.5	.889
.6	1.13
.7	1.335
.8	1.487
.9	1.577
1	1.602
1.1	1.565
1.2	1.478
1.3	1.352
1.4	1.204
1.5	1.05
1.6	.907
1.7	.787
1.8	.699
1.9	.649
2	.638
2.1	.664
2.2	.72
2.3	.797
2.4	.887
2.5	.979
2.6	1.064
2.7	1.135
2.8	1.185
2.9	1.213
3	1.217

Figure 7-9 Appendix 12 *RLC* circuit analysis.

$$I = C\, D(1) \qquad (7\text{-}18)$$

Then the voltage loop equation becomes

$$D(1) = (V - X(1))/(C\, R) \qquad (7\text{-}19)$$

Next, the following modifications are inserted into the program: first, in line 60 change L to C; second, in line 1010 enter Eq. (7-19); and third, in line 2010 change X(1) to D(1) C. Now enter a <RUN> command. The transient current shown in Fig. 7-8 tabulation shows that capacitor current decays in traditional fashion after the switch is closed at $t = 0$.

One more introductory example will help round out the principles involved in preparing program equations. Consider the diagram of Fig. 7-9—an *LRC* circuit. Assume that we want to observe the transient voltage across the capacitor. Using the previous line of reasoning we can write

Sec. 7-3 Numerical Integration Program   139

$$D(1) = X(2)/C \tag{7-20}$$

$$D(2) = (V - X(2) R - X(1))/L \tag{7-21}$$

Load Appendix 12 and enter the following modifications: first, change N in line 20 to 2 to accommodate two equations; second, add a new line: 65 C = .1; third, enter Eq. (7-20) as line 1010; and fourth, add line 1015 and enter Eq. (7-21).

Enter a <RUN> command. Figure 7-9 shows the tabular display that will appear on your screen. It shows a damped oscillation results from the circuit elements selected, and the capacitor steady-state voltage will have a value of 1 V.

7-4 SIMPLIFIED DC POWER SUPPLY ANALYSIS

In this section we analyze a frequently encountered transient situation. Figure 7-10 describes a simplified dc power supply circuit. The simplification concerns the source. Instead of an ac source, dc source V_{IN} is used. In multiple-phase rectifier systems the dc input does not depart greatly from a constant dc level, so in this particular analysis the simplification in not significant.

The problem to be investigated concerns how much the dc output voltage fluctuates when the load on the supply is removed instantaneously. Assume that the 166.7-Ω value of R_1 simulates the normal load presented by an RF amplifier tube. The full-load output values are 10,000 V, 60 A, and 600 kW.

Assume that the system requires a high degree of ripple filtering. Therefore, the LC filter uses a 2-H inductance and a 270-μF capacitor. With our network diagram established, the next step is to set up the system equations.

Two new features are involved in this analysis: first, initial conditions must be set up for L and C and second, an equivalent diode circuit must be added to simulate the rectifier portion of the power supply. The approach to writing the equations is similar to that for the LRC circuit of our last analysis—write a mesh equation for the source, R, L, and C circuit, and then write a current summation at the L, C, and R_1 node. As a first step, convert the descriptive currents and voltages from the diagram into $X(\cdot)$ and $D(\cdot)$ format as follows:

$$\begin{aligned} X(1) &= \text{IIN} \\ D(1) &= dX(1)/dt \\ X(2) &= \text{VO} \\ D(2) &= dX(2)/dt \end{aligned} \tag{7-22}$$

We can now write the two program equations as follows:

$$D(1) = -(RL/L) X(1) - (1/L) X(2) + VI/L \tag{7-23}$$

$$D(2) = (1/C) X(1) - (1/(R1\ C)) X(2) \tag{7-24}$$

Appendix 13 lists the program. These two equations are shown on lines 1005 and 1010. Other significant changes made in the previous program are as follows. First, the time requirement of lines 32, 34, and 36 are modified to fit this analysis. Second, lines

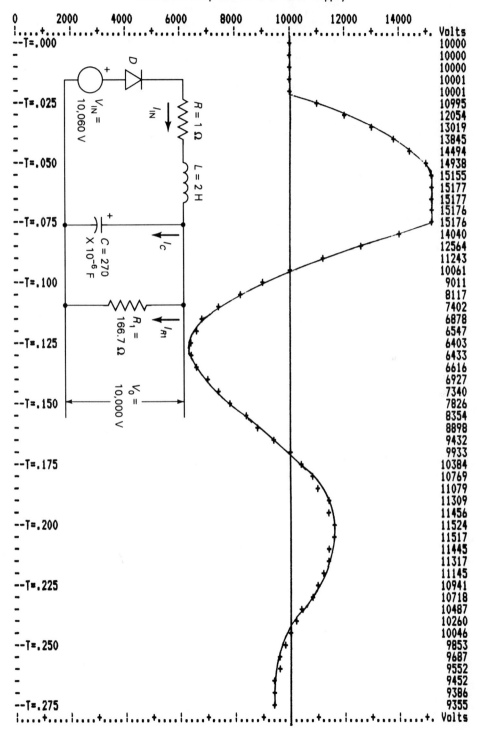

Figure 7-10 Appendix 13 analysis of simplified dc power supply.

50 and 60 introduce the initial conditions for L and C. Third, lines 70–78 define the circuit elements. Fourth, line 112 causes the integration routine to simulate full-load operation during the first 20 ms of the run. R_1 then becomes 1 MΩ. Fifth, after 75 ms of operation with this load, line 114 returns R_1 to its full-load value of 166.7 Ω. Sixth, lines 152 and 192 were added to simulate the effect of rectifier action in the power supply (current can not flow in the reverse direction in L). Seventh, lines 11 and 3540 allow you to select either a tubular or a plotted output response. Eighth, the program segment from lines 2002–3210 has been modified to provide tabular/plot capabilities.

Load Appendix 13 and issue a <RUN> command. Turn on your printer, and then respond to the prompt with the letter P. A response similar to Fig. 7-10 will start to develop on your printer. Two significant result of the analysis are:

1. When the load open circuits at 20 ms a rapid rise in dc output voltage occurs. The rise is in excess of 50% during the next 35 ms. At this point in time, L has discharged all of its energy into C, and since the rectifier diodes prevent current from reversing, the voltage holds at about 15,000 V until $T = 75$ ms.
2. At this point in time R_1 becomes 166.7 Ω again. A severe damped oscillation ensues. In an RF power generator application, the resulting excursion of around 9 kV would probably introduce degraded RF amplifier performance.

This effect can occur in all power supplies, regardless of their voltage operating level. It is a worthwhile precaution to investigate any new design with a simple analysis such as this.

This particular analysis is based on an actual case where no prior investigation had been performed. As a result, it was necessary to operate the system at a substantially reduced power level until equipment modifications were introduced to correct the problem—at a much later point in time.

7-5 OPERATIONAL AMPLIFIER TRANSIENT ANALYSIS

Transient response is often an important consideration when operational amplifiers are used as pulse amplifiers or in similar transient waveform roles. Figure 7-11 shows the circuit arrangement we will analyze. It is the performance evaluation circuit recommended by a number of operational amplifier manufacturers. It checks pulse response in the unity-gain mode with an external 100-pF capacitor across the amplifier's output terminals.

The following rationale is behind the equation development for the network diagram: If we sum up the currents at node A we get

$$D(1) = ((VI - X(1) - X(4))/R1 - X(3)/C1 \qquad (7\text{-}25)$$

A fundamental circuit relationship leads to

$$D(2) = X(3)/C2 \qquad (7\text{-}26)$$

If we write the mesh B equation, we obtain

$$D(3) = (X(1) - X(2) - R2\,X(3))/L3 \qquad (7\text{-}27)$$

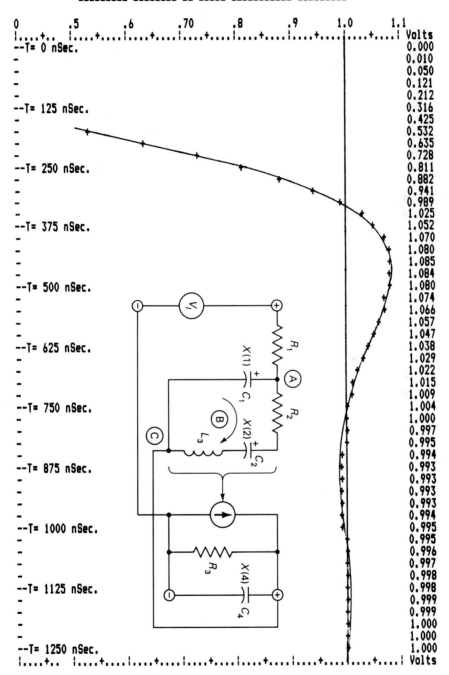

Figure 7-11 Appendix 14 analysis of operational amplifier.

Sec. 7-5 Operational Amplifier Transient Analysis

Finally, if we sum up the currents at node C, we have

$$D(4) = (GM\ (L3\ D(3) + X(2)) + X(3) + C1\ D(1) - D(4)/R3)/C4 \quad (7\text{-}28)$$

You may note an apparent mathematical error in Eq. (7-28). There are derivatives on both sides of the equation, but this is a valid arrangement since we have solved for $D(1)$ and $D(3)$ before we entered them on the right side of Eq. (7-28). The equations should be entered into the program in the order shown to obtain the highest solution accuracy.

Appendix 14 lists the program. The modifications made are as follows. First, line 30 has been modified to handle four equations. Second, lines 32–36 have had their time scales modified to suit the requirements of this analysis. Third, lines 112, 114, 152, and 192, used in the preceding problem, have been removed. Fourth, Eqs. (7-25)–(7-28) have been entered as lines 1005–1020. Fifth, lines 3010–3030 have been modified to handle the values of this analysis.

Load Appendix 14 and issue a <RUN> command followed by a P command. A plot similar to Fig. 7-11 should develop slowly on your printer. To avoid numerical instability it was necessary to use 0.25-ns T_1 steps. Hence 100 integrations are required for each output plot ordinate, and the program runs slowly.

Data sheets for 741 operational amplifiers list a rise time of 300 ns and an overshoot of 5% as the typical result this test should produce. Our model/analysis correlates closely.

7-6 BANDPASS AMPLIFIER TRANSIENT RESPONSE

This analysis establishes the transient response of a bandpass amplifier (i.e., the amplifier analyzed in Example 6-4). Several simplifications were made in the network diagram, shown in Fig. 7-12, in order to hold down the number of equations required. These simplifications have only a minor effect on the analysis results.

The diagram may seem overwhelming at first glance, yet the equation development is straightforward. Sum up the currents at node a:

$$D(1) = (SIN\ (W\ T) - X(1)/R1 - X(2))/C1 \quad (7\text{-}29)$$

Sum the voltages around meshes $X(2)$ and $X(3)$. Then simplify as follows:

$$D(2) = ((1 + L3/L2)\ X(1) - X(4) - X(5))/(L1 + L1\ L3/L2 + L3) \quad (7\text{-}30)$$

$$D(3) = ((L1 + L2)\ D(2) - X(1))/L2 \quad (7\text{-}31)$$

Write the following circuit relationship:

$$D(4) = X(3)/C2 \quad (7\text{-}32)$$

Sum up the currents at node b:

$$D(5) = (X(3) - X(6) - X(7))/C3 \quad (7\text{-}33)$$

Write the following circuit relationship:

$$D(6) = X(5)/L4 \quad (7\text{-}34)$$

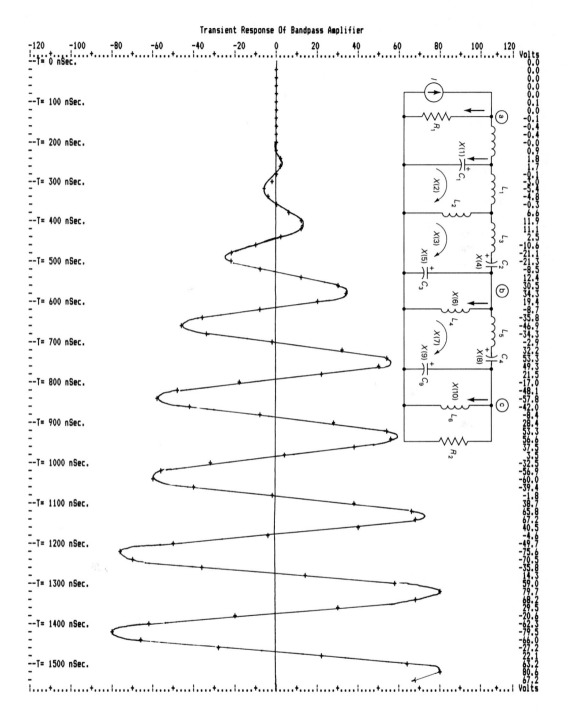

Figure 7-12 Appendix 15 bandpass amplifier transient response.

Sec. 7-6 Bandpass Amplifier Transient Response 145

Sum up the voltage around mesh $X(7)$:

$$D(7) = (X(5) - X(8) - X(9))/L5 \qquad (7\text{-}35)$$

Write the following circuit relationship:

$$D(8) = X(7)/C4 \qquad (7\text{-}36)$$

Sum up the currents at node c:

$$D(9) = (X(7) - X(9)/R2 - X(10))/C5 \qquad (7\text{-}37)$$

Write the following circuit relationship:

$$D(10) = X(9)/L6 \qquad (7\text{-}38)$$

Appendix 15 lists the program. The modifications are as follows. First, line 30 was changed to define a 10-equation solution. Second, lines 32–36 were changed to fit this analysis. Third, lines 50–74 list the circuit elements. Fourth, line 76 defines ω, the radians/second operating frequency [i.e., this term in conjunction with the Eq. (7-29) establishes a constant current (1 A peak) generator as the source]. Fifth, lines 1005–1050 enter Eqs. (7-29)–(7-38). Sixth, lines 3007–3220 were changed to handle the new analysis scale values.

Load Appendix 15, followed by a <RUN> command, followed by a P. A plot similar to Fig. 7-12 will develop on your printer. Three items of interest follow:

1. There is a time delay of about 200 ns before anything appears at the output port of the filter—the time delay of the network.
2. The output builds up in traditional damped oscillation style.
3. If allowed to run to steady state, it will be found that the output agrees with the steady-state analysis of Example 6-4 to a close degree. A cross-check such as this is valuable in any analysis.

Chapter 8

DC Power Sources and Inverters

In this chapter we analyze two techniques: ac-to-dc power supply systems (ac power *conversion* systems) and dc-to-ac power supply systems (dc power *inversion* systems). The building blocks for these systems are inductances, capacitances, resistances, and active switching elements—inclusive. Power switching functions accomplish the fundamental processes of either conversion or inversion. Switching functions, in turn, are synonymous with transient functions. As a result, quantitative evaluations of the capabilities of converter/inverter systems require transient analyses. In this chapter we evaluate a wide range of converters and inverter systems (1) in terms of their inherent voltage and current transients, and (2) in terms of their input and output interface capabilities.

General Considerations

Practically any electrical text that addresses electronic system design contains a table that lists the ideal steady-state characteristics/performance that can be expected from a wide range of dc power supply configurations. These tables are useful as a frame of reference, but they usually do not address the following three significant areas: first, the effects of rectifier transformer leakage inductance; second, the effects of *commutation* (the transients that occur when current transfers from one rectifier to the next, as a function of time); and third, the effects of less-than-ideal elements such as inductors, capacitors, transistors, and rectifiers.

TABLE 8-1 STEADY-STATE CHARACTERISTICS OF DC POWER SUPPLIES

	One-phase full-wave center tap	One-phase full-wave bridge	Three-phase full-wave bridge
Secondary Circuit Characteristics			
Current waveform	(each half)		(one phase)
Rms current	$I_{dc}/\sqrt{2}$	I_{dc}	$I_{dc}\sqrt{2/3}$
Rms voltage	$\pi V_{dc}/(2\sqrt{2})$ (each half)	$\pi V_{dc}/(2\sqrt{2})$	$\pi V_{dc}/(3\sqrt{2}\sqrt{3})$ (each phase)
Volt-amperes	$2V_{rms}I_{rms}$	$V_{rms}I_{rms}$	$3V_{rms}I_{rms}$
Utility factor	$\pi/2$	$(2\sqrt{2})/\pi$	$3/\pi$
Primary Circuit Characteristics: n = primary turns/secondary turns			
Current waveform			
Rms current	I_{dc}/n	I_{dc}/n	$I_{dc}\sqrt{2/3}/n$
Rms voltage	$nV_{dc}\pi/(2\sqrt{2})$	$nV_{dc}\pi/(2\sqrt{2})$	$nV_{dc}\pi/(3\sqrt{2}\sqrt{3})$
Volt-amperes	$V_{rms}I_{rms}$	$V_{rms}I_{rms}$	$V_{rms}I_{rms}$
Utility factor	$(2\sqrt{2})/\pi$	$(2\sqrt{2})/\pi$	$3/\pi$
Primary Source Circuit Characteristics			
Current waveform	Same as primary	Same as primary	
Rms current	Same as primary	Same as primary	$\sqrt{2}\,I_{dc}/n$
Rms voltage	Same as primary	Same as primary	Same as primary
Power Supply DC Voltage Regulation Characteristics			
% Regulation	$0.35355(\% I_X)$	$0.70711(\% I_X)$	$0.50000(\% I_X)$

Transformer leakage inductance in conjunction with rectifier commutation can significantly effect the dc voltage regulation of a supply. To illustrate, in a three-phase bridge power supply with an inductance input ripple filter, if the rectifier transformer has a leakage reactance of 10%, a no load-to-full load voltage regulation of 5% will be present at the dc output interface of the system. Most tables of ideal power supply characteristics do not indicate this regulation exists. In addition, the dc output voltage transient response of a power supply is not predictable from table listings. With the relatively simple computer programs we consider here, this characteristic can be established.

In our discussion of MTI radar systems in Section 6-4 we observed the importance of pulse-to-pulse phase similarity in the transmitter's RF output waveforms. Phase pushing in an RF amplifier is related to the stability of the output level of its associated dc power supply(s). In Chapter 11 the power supply programs developed here are used in conjunction with pulser programs developed in Chapters 9 and 10 to illustrate quantitatively how dc voltage stability is evaluated. This information is then used to assess the performance capabilities of an MTI system.

To establish our own point of reference for some of the power supply systems to be analyzed, Table 8-1 lists the *ideal* operating characteristics of three extensively used dc power supply arrangements. The characteristics shown are based on the use of inductive input dc ripple filters. In addition, the table includes steady-state power supply regulation factors for each system. We refer to this information as we proceed through our analyses.

8-1 SINGLE-PHASE CENTER-TAP FULL-WAVE DC POWER SUPPLY

One of the most frequently used dc power supply circuits is shown in its simplified schematic format in the left-hand diagram of Fig. 8-1. In the arrangement shown, a rectifier transformer with three windings (i.e., L_1, L_2, and L_3) is indicated. It is often considered a two-winding transformer with the secondary center tapped—the rationale from which the circuit terminology was derived.

The configuration shown also has a single-section LC ripple filter and a constant resistance load R. Two diodes provide the ac-to-dc conversion function. Some of this circuit's characteristics/capabilities are as follows. First, reasonably good utilization of the transformer's capabilities are obtained because an inductance input filter and full-wave rectification are used. Second, only two rectifiers are required; hence the losses due to their voltage drops are low—a significant advantage in low-voltage dc applications. Third, because of the inductance input filter arrangement, the rectifier peak currents are held down to reasonable levels.

One shortcoming of the circuit is the need for a three-winding transformer. Since each half of the secondary winding functions only half of the time, the transformer's utilization factor is not as high as it would be with a full-wave bridge configuration. Reference to Table 8-1 indicates that the secondary winding utility factor is about 0.637, appreciably below the ideal value of 1.0.

What is commutation, and why are we concerned with it? Commutation introduces transient voltages and currents into all practical converter/inverter circuits. These transients invariably deteriorate the system's performance capabilities.

Commutation occurs because it is impossible to construct a practical power transformer without introducing leakage inductance. Hence each side of the transformer's secondary winding will have a significant amount of leakage inductance associated with it. To visualize the effects of commutation, consider the case where the upper half of the secondary winding in Fig. 8-1 is executing its positive half-cycle of voltage relative to the center tap potential. As this half-cycle terminates and the wave passes through zero volts, the upper diode does not cease conducting current instantaneously. On the other

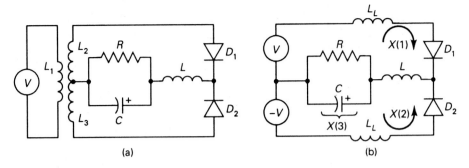

Figure 8-1 Single-phase center-tap dc power supply: (a) circuit diagram; (b) network diagram.

hand, the lower diode does begin to conduct when the voltage on the lower half of the transformer starts to execute its positive half-cycle of voltage and when its value exceeds the input voltage of the filter. Therefore, we encounter a situation where both diodes conduct for a period of time. They thereby create a virtual short circuit across the transformer's secondary winding.

Current decreases in the upper diode and increases in the lower diode during this short-circuit conduction interval (referred to as the *commutation period* of the rectifiers). The behavior of the currents in the two halves of the transformer are clearly the result of leakage inductance (i.e., because the current in any inductance cannot change instantaneously). The duration of the commutation period is a function of the amount of leakage inductance present in the circuit. Some quantitative considerations associated with commutation are expanded on in Section 8-2.

With regard to transformer properties, transformer designers must exercise a considerable amount of unique and specialized expertise. On the one hand, they must pursue a design that minimizes leakage inductance. On the other hand, they must pursue design and cost areas that optimize all other desirable transformer features. It is inevitable that a transformer design compromise results.

In addition to the power supply transformer's leakage inductance, any other inductance present in the prime power system will influence the power supply's performance capabilities. The quantitative considerations associated with this inductance will also be pointed out in Section 8-2.

Now let us consider the network diagram shown in the right half of Fig. 8-1. For simplification, two generators, V_1 and $-V_1$, replace the transformer and its three windings. The leakage inductance L_L associated with each half of the secondary winding, is connected in series with each generator. The remainder of the network diagram is essentially the same as the circuit diagram.

This is a practical circuit arrangement to use in our analyses. Mathematically, leakage inductance can be calculated in terms of the value of primary inductance and the value of coefficient of coupling (or the value mutual inductance) present in the transformer windings. The technique involved is shown in Eqs. (2-1)–(2-8). All parameters are in terms of measurable transformer properties. Transformer exciting current has not been considered in this case because its value is small enough to be ignored in typical low-frequency transformers.

System Equations and Analysis Program

In preparing a transient analysis program for this power supply, three operating periods must be taken into account: (1) the period when the upper rectifier alone is delivering a current to the filter/load circuit; (2) the period when the lower rectifier alone is delivering a current to the filter/load circuit; and (3) the period when commutation is taking place.

For the first period we can write the following differential equation around mesh 1:

$$(L + LL) D(1) - V + X(3) = 0$$

When entered into our transient analysis format this becomes

$$D(1) = (V - X(3))/(L + LL) \tag{8-1}$$

If we sum the currents at the junction of L, R, and C and arrange the result in transient analysis format, we have

$$D(3) = (X(1) - X(3)/R)/C \tag{8-2}$$

Equations (8-1) and (8-2) fully describe the first period.

In similar fashion we can write the equations for the second period as follows:

$$D(2) = (-V - X(3))/(L + LL) \tag{8-3}$$

$$D(3) = (X(2) - X(3)/R)/C \tag{8-4}$$

For the commutation period we can write the following equation around mesh 1:

$$(L + LL) D(1) + X(3) - V) = 0$$

Around the mesh containing the two generators and the two diodes we have

$$-V + LL\, D(1) - LL\, D(2) - V = 0$$

If these two equations are solved for $D(1)$, we obtain

$$D(1) = (V + 2L\,V/LL - X(3))/(2L + LL) \tag{8-5}$$

The other two equations that follow directly are

$$D(2) = D(1) - 2V/LL \tag{8-6}$$

$$D(3) = (X(1) + X(2) - X(3)/R)/C \tag{8-7}$$

Appendix 16 lists Eqs. (8-1)–(8-7) between lines 1125 and 1215 in the program's equation subroutine.

We will analyze the quantitative capabilities of the three alternate dc power supply system configurations listed in Table 8-1. A 200-kW dc supply whose nominal output is 10 kV will be analyzed for each arrangement. For this circuit several additional instructions are required to direct the program to the proper set of rectification equations and to further define the analysis input/output requirements.

In line 32 a very short integration time has been used for T_1 to ensure the accuracy of the run. This value will be increased in some later cases to speed up the program's printout rate. Lines 152, 154, 192, and 194 were inserted in the integration routine to simulate rectification by the two diodes. The rectifiers are assumed to be ideal. These lines state that if the forward current has a positive value it will encounter no impedance, but when current attempts to flow in the reverse direction through the ideal diodes, its impedance will be infinite.

Line 1005 in the equation subroutine defines the ac waveform as a function of time. Then lines 1010–1045 simply direct the program to the proper set of circuit equations, based on the status of $X(1)$ and $X(2)$. Line 30 lists the number of equations as three (this accounts for either the two normal cases or the commutation case).

For the first analysis system *snap-on* response will be simulated. In this case the initial voltage on C and the initial current through L are both zero. Snap-on is not a recommended mode of operation for power supplies, but it provides a useful yardstick for comparing the performance of alternate configurations. Lines 50–70 denote the initial conditions.

Line 71 defines the radian frequency of a power supply operating at 62.5 Hz. Line 72 is the peak value of voltage V and is obtained by multiplying the secondary rms voltage value by the square root of 2 (see Table 8-1, "One-phase full-wave center tap").

Under ideal conditions the supply should produce an average value of 10 kV across the load. However, the transformer has been assigned a value of 10% leakage reactance; hence the steady-state dc output will, of course, be lower. The value of leakage inductance is given on line 73. Lines 74 and 75 list the values of the filter inductor and filter capacitor used. They complete the list of circuit elements for the first analysis.

Analysis Results

If you now load Appendix 16 into your computer, issue a <RUN> command, and then enter the letter P to signify that you want a plot of the response, a printout similar to that shown in Fig. 8-2 will develop on your printer. In addition to the plot of output voltage, the numerical values of the three X functions are printed out in tabular fashion. Several things of interest show up on the display.

Snap-on. Even though the capacitor is shunted by its normal load resistance during the complete snap-on period, the output voltage overshoots its steady-state value by over 4 kV.

Rectifier currents. The peak current for the two rectifiers is approximately 23.5 A under steady-state conditions. The worst-case peak currents in the rectifiers, even for this severe case where no in-rush current limiting has been provided, is just slightly over twice normal.

System oscillations. The initial voltage overshoot on the waveform is part of a superimposed low-frequency oscillation. This is due to resonance taking place between filter inductor L and filter capacitor C. This oscillation damps out at a slow rate, and since its value is not of significant interest, the intermediate portion of the run has been deleted. The final portion of the selected analysis range shows that the waveform is headed for an average steady-state value of approximately 9.65 kV. Also, the ripple on the dc can be clearly detected after the amplitude of the low-frequency oscillation has attenuated somewhat.

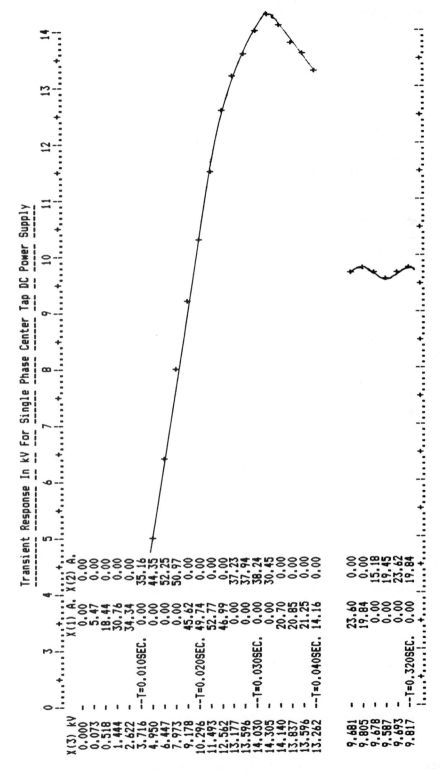

Figure 8-2 Appendix 16 analysis of one-phase center-tap dc supply.

Capacitor Input Filter Evaluation

Some designers consider inductor input filters to be rather archaic. The additional inductor is regarded as excess baggage that can be eliminated by the simple addition of filter capacitance. For some designs this may be the case; however, by introducing a few modifications to Appendix 16 we can quantitatively evaluate the case in hand.

To illustrate this comparison, assume that we effectively eliminate the filter inductor from the circuit we have just analyzed. For simplicity, just reduce the value of line 74 on your program to .001. Also, to speed things up a bit, change line 32 to .00005 and line 36 to .08. In addition, modify the scale divisions in line 3020 by a factor of 2. Finally, change the divisor in lines 3044 to 200. Then issue a <RUN> command to the program. A display similar to Fig. 8-3a should result, which shows the following.

Snap-on. The initial dc overshoot exceeds 22 kV (demonstrating that oscillatory conditions still exist in terms of the leakage inductance and filter capacity).

Overshoot rectifier currents. During the first half cycle the peak rectifier current, $X(1)$, exceeds 269 A.

Steady-state rectifier currents. Under nearly steady-state conditions, $X(1)$ and $X(2)$ have values of about 64 A.

Steady-state dc voltage. The steady-state voltage is much too high and its ripple is far in excess of that in the inductor input supply.

It is clear that for this capacitor input arrangement, modifications will have to be incorporated in the rectifier transformer, the capacitor, and the rectifiers in order to compare the system with the inductor input case. For example, to obtain essentially the same dc output voltage and ripple as the inductive filter case, the secondary rms voltage of the transformer should be changed to approximately 12,500 V in line 72, and the filter capacitor value should be increased to approximately 300 μF in line 75. Return lines 3020 and 3044 to their original values.

If you enter these revised conditions into the program and issue another <RUN> command, a plot similar to Fig. 8-3b should develop on your printer. It shows the following.

Snap-on. The initial overshoot of dc voltage has been eliminated because revised transient interactions now occur between the transformer's leakage inductance, the large value of filter capacitance, and load R.

Overshoot rectifier current. Because of the larger filter capacitor, a slower rate of rise in dc output voltage occurs. The peak rectifier current during the first rectification period now exceeds 370 A, as compared to about 53 A for the inductance input case.

Steady-state rectifier current. Steady-state peak currents for the capacitor filter case are about 50 A, and for the inductor filter about 23.5 A.

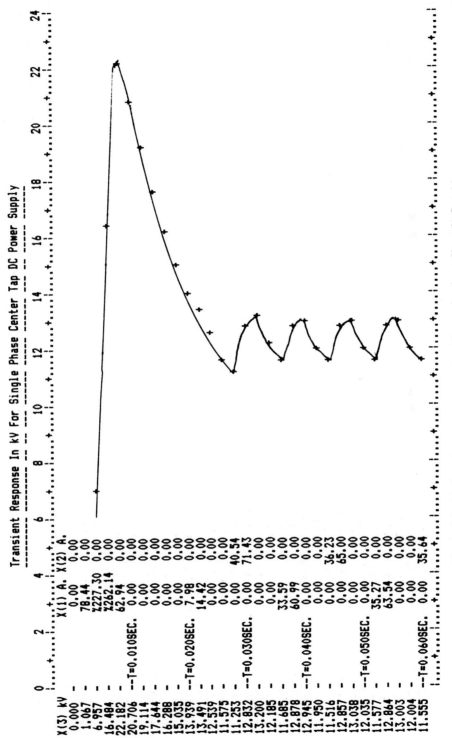

Figure 8-3 (a) One phase center-tap supply with capacitor input filter. (b) Part (a) revised.

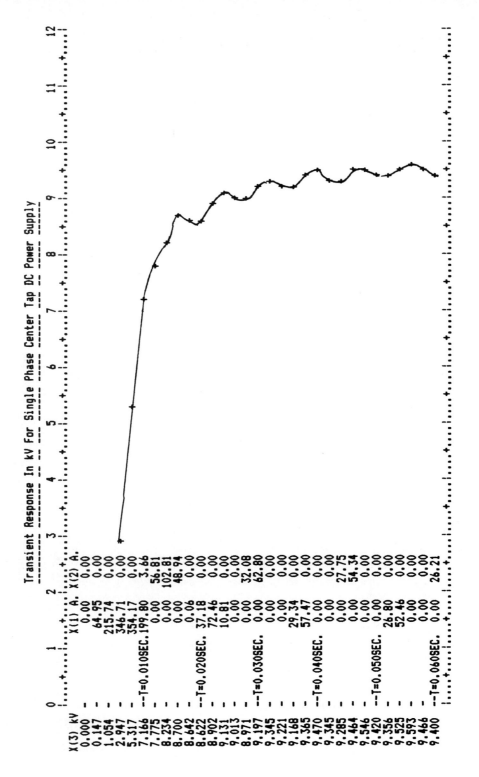

Rms currents. An elementary rms value analysis showed that the transformer current under steady-state conditions for the capacitor input filter was approximately 27.8 A. For the inductor input filter the rms current was approximately 19.7 A.

The use of Appendix 16 can be of significant value when evaluating design trade-off alternatives. For this example we have seen that a change from an inductor input to a capacitor input system would probably result in increased transformer size and cost due to the higher rms currents, probably increased cost in the rectifiers due to higher peak and rms currents, and significantly increased size and cost in the capacitors. The size and cost of the filter inductor would be eliminated.

Other performance areas should also be investigated if they are of importance in the design. For example, several computer runs with different values of load resistor R will show the comparative voltage regulation capabilities of the two approaches.

This program has been simplified to the greatest extent possible. If more comprehensive degrees of insight are required, additional elements can be added to the program equations. For example, representative values of resistance could be inserted into the transformer, rectifier, inductor, and capacitor circuit paths.

8-2 POWER SUPPLY COMMUTATION CONSIDERATIONS

In Section 8-1 we discussed commutation from a qualitative point of view. We now perform an analysis based on a steady-state quantitative point of view. With the use of Table 8-1, we can establish how commutation effects the steady-state voltage regulation of the power supply. The equation for the peak value of voltage for the transformer is

$$V_{pk} = \frac{\pi V_{dc}}{2} \tag{8-8}$$

The transformer's effective value of load, R_L, is its rms secondary voltage divided by the dc load current:

$$R_L = \frac{V_{rms}}{I_{dc}} = \frac{\pi V_{dc}}{2 \cdot 2^{1/2} I_{dc}} \tag{8-9}$$

In traditional transformer terminology, the voltage drop that occurs in a power transformer due to its leakage inductance L_L, or more correctly stated, due to its leakage reactance $X = 2\pi f L_L$, is referred to as the transformer's I_X. This is simply the voltage drop caused by the rated current flowing through the leakage reactance—when the unit is considered as a conventional power transformer. Hence for the case we have been considering, a transformer with 10% reactance, the I_X drop is 10% of the transformer's no-load output voltage and its reactance, X, is 10% of R_L.

Under commutation circumstances, we can then define the transformer's peak commutation current as

$$\begin{aligned} I_{pk} &= \frac{V_{pk}}{X} = \frac{\pi V_{dc}/2}{k\pi V_{dc}/(2 \cdot 2^{1/2} I_{dc})} \\ &= \frac{2^{1/2} I_{dc}}{k} \end{aligned} \tag{8-10}$$

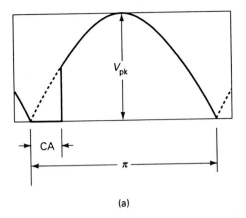

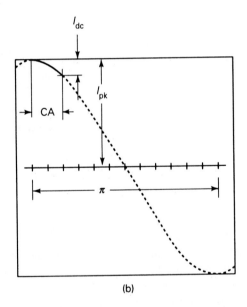

Millisec.	X(1)	X(2)
304.000	0.00000	19.83000
304.025	0.00000	19.71477
304.050	0.00000	19.59774
304.075	0.00000	19.46831
304.100	0.03631	19.31376
304.125	0.09988	19.13196
304.150	0.19068	18.92293
304.175	0.30870	18.68668
304.200	0.45394	18.42322
304.225	0.62635	18.13258
304.250	0.82594	17.81479
304.275	1.05266	17.46985
304.300	1.30650	17.09781
304.325	1.58742	16.69869
304.350	1.89540	16.27253
304.375	2.23039	15.81936
304.400	2.59235	15.33923
304.425	2.98126	14.83216
304.450	3.39706	14.29821
304.475	3.83971	13.73742
304.500	4.30917	13.14984
304.525	4.80537	12.53552
304.550	5.32828	11.89451
304.575	5.87782	11.22688
304.600	6.45395	10.53268
304.625	7.05661	9.81197
304.650	7.68572	9.06482
304.675	8.34123	8.29129
304.700	9.02306	7.49145
304.725	9.73115	6.66538
304.750	10.46542	5.81314
304.775	11.22579	4.93482
304.800	12.01219	4.03050
304.825	12.82453	3.10025
304.850	13.66273	2.14416
304.875	14.52671	1.16232
304.900	15.41638	0.15482
304.925	15.84325	0.00000
304.950	15.79535	0.00000
304.975	15.74915	0.00000
305.000	15.70465	0.00000

Figure 8-4 Waveforms of rectifier commutation: (a) voltage waveform at output of rectifier; (b) commutation current waveform.

where k is the decimal equivalent of % I_X.

The commutation angle CA can be defined from Fig. 8-4 as follows:

$$\text{CA} = \cos^{-1} \frac{(2^{1/2} I_{dc}/k) - I_{dc}}{2^{1/2} I_{dc}/k}$$
$$= \cos^{-1}(1 - \frac{k}{2^{1/2}}) \qquad (8\text{-}11)$$

158   DC Power Sources and Inverters   Chap. 8

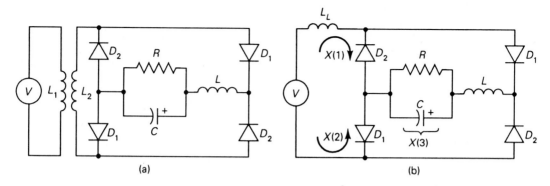

Figure 8-5 Single-phase bridge dc power supply: (a) circuit diagram; (b) network diagram.

The desired dc output voltage can then be determined by integrating the notched waveform of Fig. 8.4a as follows:

$$V'_{dc} = \frac{1}{\pi} \frac{\pi V_{dc}}{2} \int_{CA}^{\pi} \sin(t) = \frac{V_{dc}}{2} \left| -\cos(t) \right|_{CA}^{\pi}$$

$$= \frac{V_{dc}}{2[1 + \cos(CA)]} = \frac{V_{dc}}{2[1 + (1 - k/2^{1/2})]} \quad (8\text{-}12)$$

$$V'_{dc} = V_{dc}(1 - 0.35355k)$$

Note that the numerical constant value in Eq. (8-12) is the same as that given for voltage regulation in Table 8-1 (for a single-phase full-wave center-tap power supply). Hence Eq. (8-12) permits us to accurately specify transformers for this type of power supply. The coefficient will, of course, take on different values for other power supply configurations. Alternative system coefficient values can readily be determined by using the foregoing principles.

Finally, let us consider the impedance introduced by the power source. When we evaluated transformer regulation in the equations above we assumed that all of the system reactance was associated with the transformer. Actually, the reactance we are concerned with is the total system reactance. For example, assume that each of the generators we used to simplify our analysis in Fig. 8-1 also had 10% I_X. Then we would be involved with a power supply that has 20% reactance, and it would be essential that the transformer used in the power supply have additional capabilities to provide the required dc output voltage.

8-3 SINGLE-PHASE FULL-WAVE BRIDGE DC POWER SUPPLY

The left portion of Fig. 8-5 shows a single-phase bridge circuit configuration. Table 8-1 sets forth the characteristics that can be expected from the ideal form of this arrangement. Several advantages/disadvantages exist relative to the center-tap power supply arrangement. First, only one secondary winding is required; hence the arrangement provides considerably better transformer utilization of copper. It can be seen that the utilization factor for this arrangement is now approximately 0.9 as compared with 0.637 for the center-tap circuit. Second, the transformer design is considerably simpler since it has one less winding and one less lead and terminal connection. Third, on the other hand, two sets of rectifiers are required; hence in practical systems, additional voltage drops and power losses are present. To compare the performance capabilities of the two circuits quantitatively, an analysis similar to that of Section 8-1 must be conducted.

System Equations and Analysis Program

The equivalent network diagram is shown in the right-hand portion of Fig. 8-5. Equations can be set up for the three circumstances of operation: normal conduction due to positive voltage from the transformer, normal conduction due to negative voltage from the transformer, and commutation conduction when all four diodes are involved. For the case where only current $X(1)$ is active, we can write the following mesh equation:

$$(L + LL) D(1) + X(3) - V = 0$$

In our program format this becomes

$$D(1) = (V - X(3))/(L + LL) \qquad (8\text{-}13)$$

Similarly,

$$D(2) = (-V - X(3))/(L + LL) \qquad (8\text{-}14)$$

The current summation equation becomes

$$D(3) = (X(1) - X(3)/R)/C \qquad (8\text{-}15)$$

For the case where current $(X(2)$ is the only active path, Eqs. (8-14) and (8-15) are the same, but the current summation equation is

$$D(3) = (X(2) - X(3)/R)/C \qquad (8\text{-}16)$$

When commutation occurs we can write for the $X(1)$ mesh,

$$LL\, D(1) + L\, D(1) + X(3) - V - LL\, D(2) + L\, D(2) = 0$$

For the $X(2)$ mesh we have

$$-LL\, D(1) + L\, D(1) + X(3) + V + LL\, D(2) + L\, D(2) = 0$$

When these two equations are solved and put in program form the results are

$$D(1) = V/2LL - X(3)/2\,L \qquad (8\text{-}17)$$

$$D(2) = -V/2LL - X(3)/2\,L \qquad (8\text{-}18)$$

The current summation equation is

$$D(3) = (X(1) + X(2) - X(3)/R)/C \qquad (8\text{-}19)$$

Appendix 17 lists the program for the single-phase full-wave bridge power supply. The circuit elements have the same values as in the center-tap supply arrangement in lines 71–76. In the equation subroutine, appropriate modifications have been made to lines

1010–1045 to direct the program to the proper analysis equations. The analysis equations derived above are listed on lines 1105–1200. To perform a snap-on analysis, the initial conditions for $X(1)$, $X(2)$, and $X(3)$ listed on lines 50–70 are set at 0.

Analysis Results

Now if you load Appendix 17 into your computer, issue a <RUN> command, and then follow this with a P command, a plot similar to Fig. 8-6 should develop on your printer. At first glance it may appear to be the same as the center-tap case, but there are several areas that differ.

Snap-on. The initial overshoot of voltage is somewhat lower with the single-phase center-tap LC filtered system.

Steady-state output voltage. The steady-state value at which the run concludes is now at only about 93% of ideal output voltage, as compared to about 96.5% for the center-tap power supply case. If we refer back to Eq. (8-11), the reason for these changes can be established. In the center-tap power supply case, when we set up the equation for the commutating angle it was only necessary for the rectifier current to decrease from its peak value by an amount equal to I_{dc} for the initially conducting rectifier to assume its current blocking mode. In the bridge case, since we have only one winding, it is necessary for the current not only to drop to zero, but it is also necessary that it reverse through the transformer winding to provide conduction for the second rectifier. Hence in Eq. (8-11) for the bridge, $2I_{dc}$ should be used in the numerator instead of I_{dc}. This will substantially increase the commutating angle, and Eq. (8-12) becomes

$$V'_{dc} = V_{dc}(1 - 0.70711\ k) \qquad (8\text{-}20)$$

This is also responsible for the somewhat lower values of dc output voltage for the bridge power supply circuit.

We now have two power supply analysis programs at our disposal for determining the most effective course of action in a system design. It is likely that neither of the two would be used for a system operating at the 200-kW power level of our example. The three-phase full-wave system described in the next section would probably be a more advisable selection—if an appropriate three-phase prime power source were available for its operation.

8-4 THREE-PHASE FULL-WAVE BRIDGE DC POWER SUPPLY

Three-phase power supplies offer several significant advantages over the single-phase configurations we have reviewed. First, for a given set of dc filter elements, the ripple reduction will be about 100 times greater. Second, the prime power source current waveform more closely approaches sinusoidal. Third, the transformer utilization factor is higher. Fourth, the power factor is higher (i.e., power factor is defined as the ratio of the total active power in watts to the total volt-amperes).

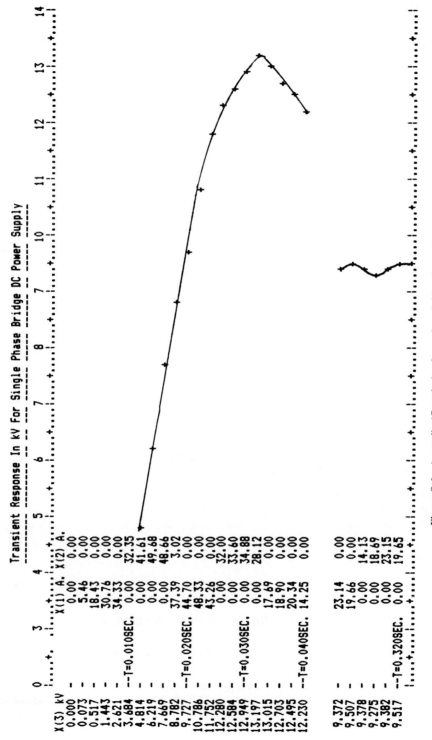

Figure 8-6 Appendix 17 analysis of one phase bridge dc supply.

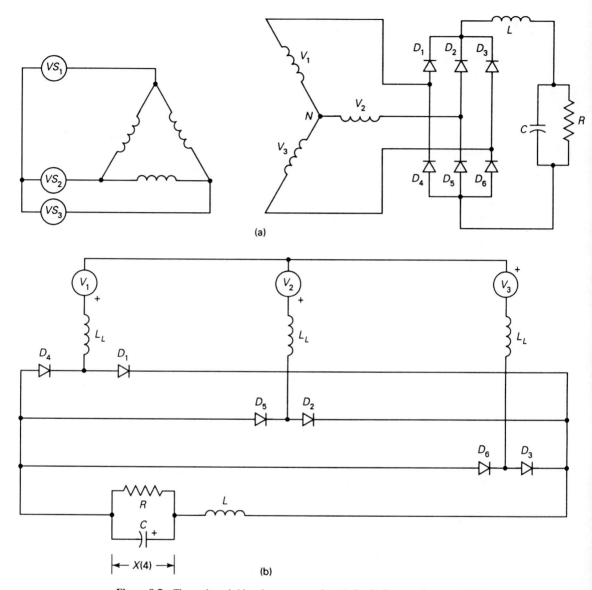

Figure 8-7 Three-phase bridge dc power supply: (a) circuit diagram; (b) network diagram.

The circuit diagram of Fig. 8-7 indicates a typical arrangement for a three-phase full-wave dc power supply. Prime power source generators VS_1, VS_2, and VS_3 produce three sinusoidal voltage waveforms of equal amplitude. The voltages are separated in phase by 120°. These are applied to either a three-phase transformer, or three single-phase transformers whose primary windings are delta connected. The secondary windings are wye connected and the neutral of the wye is a floating connection. The wye output voltages (V_1, V_2, and V_3) are converted to dc by means of six rectifiers (D_1–D_6). The ripple filter and load connections are identical to the single-phase circuits.

Sec. 8-4   Three–Phase Full–Wave Bridge DC Power Supply

Just as in the single-phase bridge circuit, at least two rectifier diodes are active during the process of rectification. When commutation occurs, three diodes are involved. Because of the multiple paths and the 120° phase relationships encountered, rather careful bookkeeping is required or possible confusion may result when performing an analysis.

A colloquial, yet very definitive description of the waveforms generated by this circuit is: "They are just like you would expect from a three-spoke wagonwheel rolling along the ground—except that the wheel has no rim." Figure 8-8 is intended to depict this description. The numbers 1, 2, and 3 are intended to represent the peak values of the voltages V_1, V_2, and V_3 of the circuit of Fig. 8-7.

In Fig. 8-8 the relative secondary voltage positions are shown at every 30° increment of the prime power cycle. For the ideal conditions shown (no leakage inductance), each pair of rectifiers will conduct for 120°; however, because the supply has a full-wave configuration, the fundamental ripple frequency is six times the power line frequency.

Consider the voltage pattern shown at 0°. At this instant, V_1 and V_2 have attained positions where they are equal. Hence whereas current was previously flowing out of phase 2 and back into phase 3 of the transformer, it will now commutate instantaneously and flow out of phase 1 and into phase 3. The dashed line labeled C is intended to denote that commutation ideally takes place at this angle. Clearly, an actual transformer with leakage inductance would introduce a commutation time-delay period during which the current from V_2 would decay to zero and the current from V_1 would rise to the full dc value.

Now proceed to the 30° diagram. It can be seen that the voltage delivered to the rectifier diodes reaches its maximum value during this conduction period. It then returns to a set of amplitudes comparable to those it had initially, when we reach the 60° diagram. At 60° the second commutation period is encountered. For the ideal case, the current would then instantaneously start flowing out of V_1 and back into V_2 on the transformer. As we proceed through the remaining diagrams we encounter four additional commutation cycles of a similar nature.

System Equations and Analysis Program

With Fig. 8-8 to help keep our bookkeeping straight, we can now return to Fig. 8-7 and start writing the system equations. Let us consider the 0 to 60° period described above. As in the single-phase analyses, each secondary leg of the transformer is simulated by a generator, and each has a leakage inductance L_L connected in series, as indicated on the network diagram.

Two meshes are involved. The first mesh path is from V_2, through L_L, through D_2, through the load circuit, through D_6, through L_L, through V_3, and back to V_2. The second mesh path is from $V1$, through L_L, through D_1, through the load circuit, through D_6, through L_L, through V_3, and back to V_1.

To hold down the number of variables, the currents from the generators are always referred to as $X(1)$, $X(2)$, and $X(3)$. The voltage across the load is referred to as $X(4)$. There are, of course, six diodes; however, we can obtain the diode current values from the four variables noted above.

For the first mesh we can write

$$(L + 2LL)\,D(1) + (L + LL)\,D(2) + X(4) - V1 + V3 = 0$$

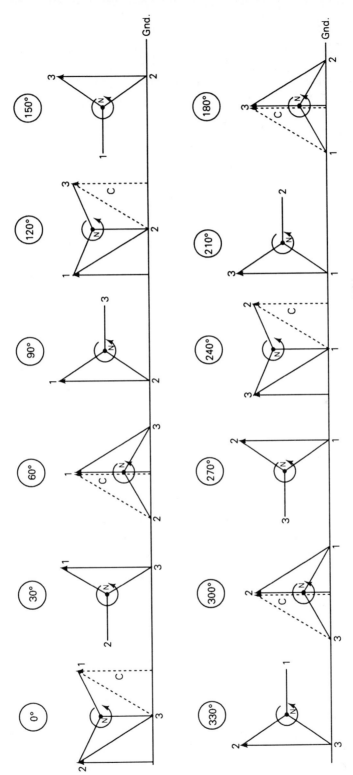

Figure 8-8 Three-phase secondary voltage amplitudes at 30° increments.

For the second mesh we have

$$(L + LL) D(1) + (L + 2LL) D(2) + X(4) - V2 + V3 = 0$$

If we solve these two equations and put the result in program format, we obtain

$$D(1) = ((V3 - V2 + X(4))K3 - (V3 - V1 + X(4))K4)/(K4^2 - K3^2) \quad (8\text{-}21)$$

$$D(2) = ((V3 - V2 + X(4))K4 - (V3 - V1 + X(4))K3)/(K3^2 - K4^2) \quad (8\text{-}22)$$

From the above it can be seen that

$$K3 = L + LL \quad K4 = L + 2LL \quad (8\text{-}23)$$

$$D(3) = D(1) + D(2) \quad (8\text{-}24)$$

A current summation at the junction of L and R gives

$$D(4) = (X(1) + X(2) - X(4)/R)/C \quad (8\text{-}25)$$

These are the program equations required to solve for the commutating portion of the first conduction period of the power supply. The equations for the normal conduction portion can be written directly as

$$D(1) = (V1 - V3 - X(4))/K4 \quad (8\text{-}26)$$

$$D(4) = (X(1) - X(4)/R)/C \quad (8\text{-}27)$$

The network equations for the other five conduction periods have an identical format; however, the D variables involved the voltages V_1, V_2, V_3 used in these equations, and the currents $X(1)$, $X(2)$, $X(3)$ of the summation equation assume different arrangements and signs in other periods. A complete listing of all equations along with the paths involved for each of the conduction periods are shown in Appendix 18 on lines 1010–1124.

Because of the greater number of variables, Appendix 18 is somewhat more involved than the single-phase power supply programs. In quick summation: Line 30 lists the number of equations as four. Lines 53–66 list the system elements. The new generator voltages were determined from the last column of Table 8-1. The *LC* filter elements are the same as for the single-phase analyses. Lines 400–409 form a new subroutine that directs the program to the proper set of equations as a function of time. Lines 2013–2025 perform the operations required to derive the six diode currents from the three transformer current solutions of the program.

Analysis Results

With this background established, if you load Appendix 18 into your computer, enter a <RUN> command, and follow this with a P command, a snap-on response plot similar to that shown in Fig. 8-9 should develop on your printer. The general characteristics of the plot are similar to the single-phase runs except that:

1. Dc voltage ripple at the output of the supply is almost invisible.
2. Dc output voltage is now settling in at about 9500 V.

If a commutation analysis similar to that of Section 8-2 is performed, the following equation for dc output voltage will be obtained:

$$V'_{dc} = V_{dc}(1 - 0.50000k) \qquad (8\text{-}28)$$

Program Revision in Appendix 18C

The capabilities programs of Appendices 16, 17, and 18 can be improved significantly with two additions: (1) a dc transient suppressor, and (2) a more accurate commutation routine.

First program revision. A series RC circuit connected across the dc input of the LC ripple filter can be used to effectively limit transients. From a computer analysis point of view, the variable nature of the point in time at which commutation occurs necessitates a very short integration step T_1 if numerical stability is to be maintained. When an RC snubber is added, we can typically increase T_1.

Adding the suppressor necessitates revision of our basic sets of analysis equations. You can perform the revisions in a straightforward manner by employing the same principles we used earlier in this section. If lines 1010–1024 of Appendix 18 are compared to similar line numbers in Appendix 18C, the nature of the revisions can be observed. Each of the six rectification periods now uses six equations instead of the previous four equations. Nevertheless, program execution is more rapid due to the fact that a larger value of T_1 can now be used.

Second program revision. In Appendices 16, 17, and 18 rectifier commutation is processed by program lines 152 and 192—the program recognizes commutation simply by means of a reversal of rectifier current. One way the program can be speeded up is to monitor the rectifier, which has decreasing current during commutation, determine the time at which zero current interception occurs, and repeat the T_1 time period. Then execute a two-step process during the repeated period. Initially, use a revised T_1 time (from the start of this period to the interception time) and recalculate all active currents. Then continue using the remaining portion of the original T_1 time period and again recalculate all active current values.

To accomplish this, in Appendix 18C we first, detect zero interception by means of the conditional statements shown in lines 202 and 204. Second: Reset the time scale to $T = T - T_1$. Calculate T_C, the zero-current interception point by linear interpolation of the current at T. Calculate all $X(\cdot)$ values using the equivalent of line 150 but using T_C instead of T_1. Calculate the values of V_1, V_2, and V_3 at T_C. These steps are shown in line 206 of the program. Third, based on the value of T_S, go to the system derivative calculation subroutine required. This is shown in line 208 of the program. Fourth, let $T_X = T_P - T_C$, the remainder of the time in the overall integration period (note that T_P has been set equal to T_1 in step 32). Then calculate all $X(\cdot)$ that exist at the end of the step using an equivalent of line 190 except with T_X instead of T_1. This then provides the corrected commutation values. The program then returns to normal operation (after several housekeeping procedures). The steps above are shown in line 210 of the program. Fifth,

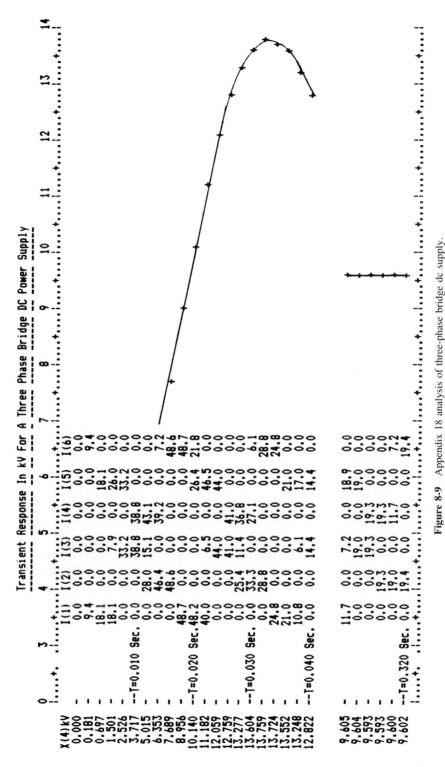

Figure 8-9 Appendix 18 analysis of three-phase bridge dc supply.

delete lines 152 and 192. Modify line 165 with V_1, V_2, and V_3 calculations. Add selection statement 170.

We use Appendix 18C for analyses in Chapter 10. The principles were pointed out at this time because they are also used for a program in the next section.

8-5 TWELVE-PULSE BRIDGE DC POWER SUPPLY ANALYSIS

The preceding power supply analyses clearly illustrate that the input and output interface performance capabilities of a power supply improve as a function of the number of rectification periods present during each cycle of the prime power wave. Typically, this improved performance is achieved at the expense of increased circuit complexity. However, the increased circuit complexity is often essential in order to achieve the technical performance levels defined in the equipment's overall system specification.

A 12-pulse system is simply one that uses 12 rectification periods per cycle, in contrast with the maximum of six that we have used up to this point. What major performance capabilities can be improved by going from a 6-pulse bridge to a 12-pulse bridge? Three fundamental areas of improvement—for ideal 6- and 12-pulse systems—are illustrated in Table 8-2. The improvements are discussed below.

Prime Line Current Interface Harmonics

A tabulation of 6- and 12-pulse line current harmonics, derived through the use of Eqs. (6-5) and (6-6), are shown in Table 8-2. A comparison shows that 12-pulse systems significantly attenuate most harmonics of the fundamental line frequency. As a result, the line current waveform of the 12-pulse system approaches that of a sine wave to a much closer degree than does the waveform of a 6-pulse system.

TABLE 8-2 COMPARISON OF IDEAL 6- AND 12-PULSE SYSTEMS

Fourier series Coefficient	6-Pulse system	12-Pulse system
1	0.95493	1.02349
5	0.19099	0.00000
7	−0.13642	0.00000
11	−0.08681	−0.09304
13	0.07346	0.07873
17	0.05617	0.00000
19	−0.05026	0.00000
23	−0.04152	−0.04450
25	0.03820	0.04094

Power Factor
1. 6-Pulse power factor = 0.95493
2. 12-Pulse power factor = 0.98862

Peak Ripple Levels for Average DC Level of 1.0
1. 6-Pulse maximum = 1.0472; 6-pulse minimum = 0.90690
2. 12-Pulse maximum = 1.0115; 12-pulse minimum = 0.97705

Input Interface Power Factor

Table 8-2 indicates that the prime line power factor is improved from approximately 95.5% to 98.9% when changing from a 6- to a 12-pulse system.

Output Interface dc Ripple

A comparison of peak-to-peak dc ripple (delivered by the full-wave rectifiers to the ripple filter) is shown in Table 8-2. It indicates that a reduction from approximately 14.03% to 3.447% is achieved by going from a 6-pulse to a 12-pulse system. In addition, the 12-pulse ripple is at twice the frequency of the 6-pulse ripple. Hence for a specified figure of filtered dc ripple output, the amount of ripple filtering required in the 12-pulse system will be vastly reduced from that required for a 6-pulse system.

Network Diagram and Delta–Wye Supply Equations

One common circuit arrangement for a 12-pulse converter is shown in Fig. 8-10. The rationale of this circuit is as follows:

Delta-to-wye power supply. A set of delta-connected three-phase transformer primary windings supply energy to a set of wye-connected windings denoted secondary 1. The upper set of bridge rectifiers provide 6-pulse rectification via the currents indicated as I_1–I_6. This arrangement is similar to the system analyzed in Section 8-4.

Delta-to-delta power supply. The set of delta-connected primary windings also supplies prime line energy to secondary 2. The secondary delta winding terminals then connect to the lower set of bridge rectifiers, and 6-pulse rectification is provided via the currents indicated as I_7–I_{12}.

Thirty-degree phase shift. Three-phase transformer phase relationships automatically establish a 30° phase shift between the ac voltages delivered by appropriately selected sets of wye and delta secondary terminals. The rectified output of the lower bridge rectifier then effectively leads the rectified output of the upper bridge rectifier by 30°. Hence if the upper and lower sets of rectifiers are connected in series, as shown, a 12-pulse rectifier system will result.

Transformer characteristics. The highest level of system performance capabilities is achieved if the 12-pulse rectifier produces identical average dc output voltages from each set of rectifiers. This will be achieved if, for example, the ratio of each delta primary winding to each wye secondary winding leg has a ratio of 1:1 and the ratio from each delta primary winding to each delta secondary winding has a ratio of $1:3^{1/2}$.

In a practical configuration, the effective leakage inductances of the secondary wye and delta windings (denoted L_L and L_{LD}) should also be equal. This ensures that the dc output voltage regulation of the upper and lower sets of rectifiers are equal to one another.

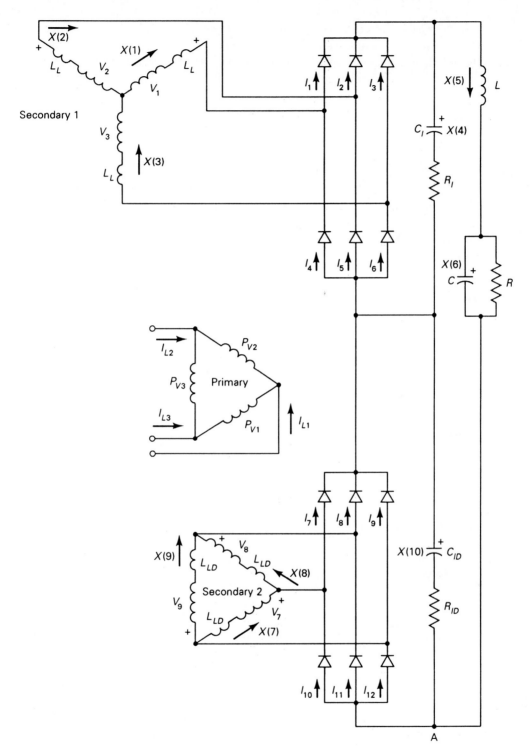

Figure 8-10 Network diagram for a 12-pulse bridge dc power supply.

Dc output circuit. The rectified output of both the upper and lower rectifier banks are initially terminated in RC transient circuits consisting of R_I, C_I, R_{ID}, and C_{ID}. The dc output ripple filter circuit is identical to the arrangements used previously, except in this case the input voltage is the sum of the voltages across series-connected elements R_I, C_I, R_{ID}, and C_{ID}.

Delta-to-Delta Power Supply Rationale

In what respects are 6-pulse systems with delta-to-wye transformers and 6-pulse systems with delta-to-delta transformers similar, and in what respects are they different? The two systems are similar in that as line-to-line voltages run through a 360° cycle, both will encounter similar sets of rectification periods.

The two systems differ, first, in regard to the process of commutation. With a wye secondary, open-circuit winding commutation occurs. With a delta secondary, this cannot occur, simply because all three windings are connected in series. Second, the systems differ in regard to transformer winding currents. Typically, with a delta secondary we encounter a situation where one winding has an instantaneous voltage that is higher than that of the other two windings. Electrically, it is connected in parallel with two series-connected windings. The sum of the instantaneous voltages of the series windings must (in accordance with Kirchhoff's voltage law) be equal to that of the higher-voltage winding. Hence in an ideal system (one that has no leakage inductance in the transformer) this results in a situation where the higher voltage winding delivers two-thirds of the rectified current and the two series windings deliver one-third of the rectified current.

In an ideal delta secondary system, commutation occurs at the instant when the terminal voltages of two of the delta windings become equal. At an infinitesimal point in time later, a new winding has assumed two-thirds of the rectified load current and the other two windings assume one-third of the rectified load current.

So much for ideal system background. Before we wrote analysis equations for the practical delta-to-wye system in Section 8-4 it was necessary to develop a rationale that defined how the circuit functioned—in order to provide the background for the system equations. A similar approach will be used here to establish a line of reasoning for the delta-to-delta rectifier. It proceeds as follows:

Step 1. Refer to Figure 8-10 and consider the point in time where the currents in secondary 2 are flowing as shown. Assume that current $X(7)$ is equal to current $X(8)$. Then diode path current I_8 is the sum of $X(8)$ and $X(9)$, and the current flow through diode path I_7 is zero. The return path for the circuit is via diode path I_{12}. Hence this point in time defines an event during the normal conduction portion of a rectification period.

Step 2. Following the passage of a short additional period of time, a situation develops where the terminal voltages across the winding 7 and winding 9 become equal to one another (note that these terminal voltages also include the voltage drops across leakage reactances L_{LD}). At this instant, the current in winding 7 instantaneously transfers and flows though diode path I_7, and the current through winding 9 continues to flow through diode path I_8. Since both ends of winding 8 are then at the same potential, it is effectively short circuited, and the current through winding 8 starts to reverse, because V_8 is in the process of reversing its polarity. This instant defines the start of commutation.

Step 3. Following the passage of a short additional period of time, current $X(8)$ in the shorted winding reverses and becomes equal to $X(9)$. Current then instantaneously ceases to flow through diode path I_8, and with current $X(8)$ remaining equal to $X(9)$, the combined currents of this series path and current $X(7)$ flow through diode path I_7. This instant defines the conclusion of commutation. The second portion of this rectification period proceeds with this circuit arrangement in a manner similar to that described in step 1.

Step 4. The next rectification period begins when the voltage across leg 8 becomes equal to the voltage across leg 7. A sequence of events similar to steps 2 and 3 are then repeated for the new paths.

System Equations and Analysis Program

Referring to Fig. 8-10, a set of system equations can now be developed by first defining the equation for current in series path R_{ID} and C_{ID} from a Kirchhoff current summation at node A,

$$X(7) + X(9) = CID\ D(10) + X(5)$$

which becomes, in derivative form,

$$D(10) = (X(7) + X(9) - X(5))/CID \qquad (8\text{-}29)$$

Next the voltages around the system's dc output mesh can be defined as

$$X(4) + RI\ CI\ D(4) + X(10) + RID\ CID\ D(10) = L\ D(5) + X(6)$$

which becomes, in derivative form,

$$D(5) = (X(4) + RI\ CI\ D(4) + X(10) + RID\ CID\ D(10) - X(6))/L \qquad (8\text{-}30)$$

As in previous analyses,

$$D(6) = (X(5) - X(6)/R)/C \qquad (8\text{-}31)$$

Based on the rationale outlined above, during the commutation period, while winding 8 is short circuited,

$$D(8) = V8/LLD \qquad (8\text{-}32)$$

Two mesh equations complete the set. First,

$$LLD\ D(9) + V9 = RID\ CID\ D(10) + X(10)$$

This becomes

$$D(9) = (-V9 - RID\ CID\ D(10) - X(10))/LLD \qquad (8\text{-}33)$$

In similar fashion the second mesh yields

$$D(7) = (V7 - RID\ CID\ D(10) - X(10))/LLD \qquad (8\text{-}34)$$

Equations (8-33)–(8-35) are accessed by the analysis program during commutation— when $X(9) => X(8)$.

When commutation is complete, $X(9) = X(8)$. Equation (8-34), defining $D(7)$, remains unchanged during this period, however two new equations are required. Since $X(8) => X(9)$ the first equation is simply

$$D(8) = D(9) \tag{8-35}$$

A voltage mesh including series windings 8 and 9 can be written as follows:

$$LLD\ D(8) + V(8) + LLD\ D(9) + D(9) + RID\ CID\ D(10) + X(10) = 0$$

This simplifies to

$$D(8) = (-V8 - V9 - RID\ CID\ D(10) - X(10))/(2\ LLD) \tag{8-36}$$

Equations (8-29) through (8-36) describe the delta-to-delta system over one of its six rectification periods. Through the use of the equation techniques noted previously in connection with Appendix 18C, the delta-to-wye equations can be written.

Program Structure

The computer program listed as Appendix 18G has been developed based on the rationale and equations above. The program structure and some of the simplifications used are as follows:

Lines 10–310. These lines define the initial program parameters in a manner similar to previous systems.

Lines 320–930. These lines define the structure of the delta-to-wye rectifier system.

Lines 660 and 930. In this area the program has been compressed by eliminating the six fundamental rectification segments used previously. Instead, lines 750–800 define the segment variables. This allows lines 830–930 to be used in performing calculations in any segment. While a simpler program results, the condensed format was not used in previous programs because the procedure is more difficult to comprehend and debug (e.g., when adapting the program to new and different problem solutions).

Transfer from delta–wye to delta–delta. After a time analysis value passes through the delta-to-wye portion of the program, line 540 directs it to line 1930, and the delta-to-delta circuits are calculated.

Delta–delta analysis. The fundamental delta-to-delta analysis parameters are developed in a manner similar to those in the delta-to-wye system. They are listed between lines 2120 and 2410. Lines 2290–2400 utilize Eqs. (8-29)–(8-36) developed above.

Analysis Results

With this background established, it can be seen from the program that the delta-to-wye system elements used are similar to the 200-kW 10-kV region converters we have analyzed previously. The delta-to-delta portion approximately doubles the dc output voltage—hence we are dealing with a converter with a dc output of approximately 400 kW at 20 kV. Several computer printouts will illustrate some of this system's performance capabilities.

Snap-on analysis. To demonstrate the fundamental characteristics of the converter, load Appendix 18G, and change line 160 to T3 = .048#. In line 210, change to $X(6) = 0$ and $D(6) = 0$ (i.e., this starts the run with the dc filter capacitor fully discharged). Then turn on your printer, issue a <RUN> command, and respond to the prompts with any letter other than P. Highlights of the resulting tabular display are as follows. First: Wye winding rectifier currents $I(1)$–$I(6)$ commutate in normal fashion insofar as phase-to-phase current transfer is concerned. Second: Delta rectifier currents $I(7)$–$I(12)$ are similarly commutating properly. Third: Through reference to the time scale versus the start of commutation on each set of rectifiers, the delta and wye waves are properly superimposed (i.e., the delay time is typically 0.0014 s or approximately $\frac{1}{12}$ of a line waveform period). Fourth: By reference to transient suppression capacitor voltages $X(4)$ and $X(10)$, the two rectifiers are delivering similar voltages as a function of time. Fifth: From an operational point of view, filtered dc current $X(5)$ reaches a maximum value of 89.2 A (at $T = 0.0138$ s), dc output voltage $X(6)$ overshoots to a maximum value of 29,482.1 V (at $T = 0.0304$ s), typical primary line current I_{L1} reaches a peak value of 270.8 A (at $T = 0.0132$ s). Following the overshoot of voltage, the $X(5)$ and I_{L1} values decay to zero (at $T = 0.0376$ s). These two currents will build up after R discharges C into the rectifier operating zones (at $T = 0.046$ s).

A snap-on run such as this is a valuable debugging guide when initially evaluating a new program. In addition, it serves as a worst-case system operation evaluation tool.

Line current waveform. We noted that the line current waveform was improved when changing from a 6-pulse to a 12-pulse system. A representative steady-state waveform can be displayed if we change line 160 to T3 = .016# and we reinstate $X(6) = 19{,}234.89$ and $D(6) = -688.9628$ in line 210. Follow this with a <RUN> command. Then respond to successive prompts with the letters Y, P, and L.

A printout similar to Figure 8-11 will result. It can be seen that, first, the 12 commutating steps per cycle show up as 12 steps of moderate distortion superimposed on a sine wave; second, the two halves of the cycle are symmetrical; and third, if a more exact assessment of harmonic levels is required, an analysis similar to that given in Section 5-3 could be performed by using the tabular values from the plot and inserting them into Appendix 7.

Dc input/output waveform analysis. An output interface analysis can be conducted simply by loading the original Appendix 18G program into your computer. The initial conditions listed in the program are the values that existed following an

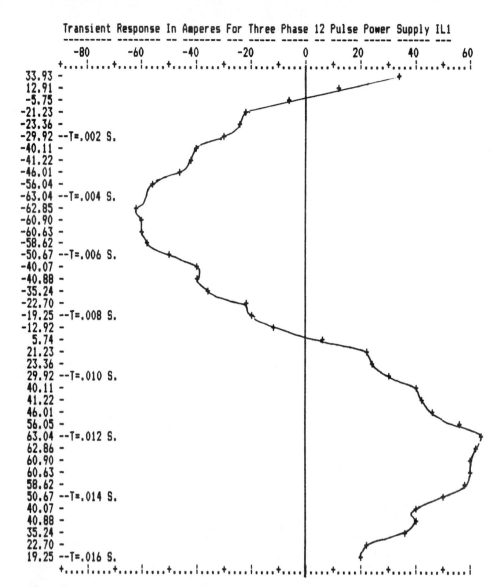

Figure 8-11 Apendix 18G analysis of 12-pulse system ac line current.

extended run that has produced substantially steady-state operating conditions within the system.

If you enter a <RUN> command, and follow this with successive prompt responses of Y, P, and V, output similar to Figure 8-12 and Figure 8-13 will develop on your printer. Figure 8-12 presents a combined tabular and plot display describing tabular values of output voltage $X(6)$, rectifier currents $I(1)$–$I(12)$, and a plot of the envelopes of the voltages across the series-connected wye transient suppressor R_I and C_I (denoted by the + plotting symbols, and across the series-connected delta transient suppressors RID and

Transient Response In kV For A Three Phase 12 Pulse DC Power Supply

Vdc Out	A./Rect.	I(1)	I(2)	I(3)	I(4)	I(5)	I(6)	I(7)	I(8)	I(9)	I(10)	I(11)	I(12)
19243.89-		0.0	19.2	0.0	0.0	0.0	19.2	0.0	21.9	0.0	0.0	0.0	7.2
19243.80-		1.1	17.8	0.0	0.0	0.0	18.9	0.0	22.1	0.0	0.0	0.0	22.1
19243.87-		4.8	14.2	0.0	0.0	0.0	18.9	0.0	21.8	0.0	0.0	0.0	21.8
19244.00-		10.8	8.1	0.0	0.0	0.0	18.9	0.0	21.1	0.0	0.0	0.0	21.1
19243.97-		19.0	0.0	0.0	0.0	0.0	19.0	0.0	20.3	0.0	0.0	0.0	20.3
19243.75-	T=.00125 S.	21.5	0.0	0.0	0.0	0.0	21.5	0.0	19.5	0.0	0.0	0.0	19.5
19243.69-		22.2	0.0	0.0	0.0	0.0	22.2	0.4	18.5	0.0	0.0	0.0	18.9
19243.82-		21.9	0.0	0.0	0.0	0.0	21.9	3.3	15.7	0.0	0.0	0.0	18.9
19244.06-		21.4	0.0	0.0	0.0	0.0	21.4	8.5	10.4	0.0	0.0	0.0	19.0
19244.22-		20.6	0.0	0.0	0.0	0.0	20.6	16.1	2.7	0.0	0.0	0.0	18.8
19244.08-	T=.00250 S.	19.8	0.0	0.0	0.0	0.0	19.8	21.0	0.0	0.0	0.0	0.0	21.0
19243.91-		19.0	0.0	0.0	0.0	0.0	19.0	22.1	0.0	0.0	0.0	0.0	22.1
19243.90-		18.9	0.0	0.0	0.0	2.0	16.9	22.1	0.0	0.0	0.0	0.0	22.1
19244.02-		18.9	0.0	0.0	0.0	6.5	12.4	21.6	0.0	0.0	0.0	0.0	21.6
19244.14-		18.9	0.0	0.0	0.0	13.3	5.6	20.9	0.0	0.0	0.0	0.0	20.9
19244.02-	T=.00375 S.	18.9	0.0	0.0	0.0	20.2	0.0	20.1	0.0	0.0	0.0	0.0	20.1
19243.81-		21.9	0.0	0.0	0.0	21.9	0.0	19.2	0.0	0.0	0.0	0.0	19.2
19243.81-		22.2	0.0	0.0	0.0	22.2	0.0	18.9	0.0	0.0	0.0	1.1	17.8
19243.97-		21.8	0.0	0.0	0.0	21.8	0.0	18.9	0.0	0.0	0.0	4.8	14.2
19244.20-		21.1	0.0	0.0	0.0	21.1	0.0	18.9	0.0	0.0	0.0	10.8	8.1
19244.27-	T=.00500 S.	20.4	0.0	0.0	0.0	20.4	0.0	19.2	0.0	0.0	0.0	18.7	0.0
19244.06-		19.5	0.0	0.0	0.0	19.5	0.0	21.6	0.0	0.0	0.0	21.6	0.0
19243.93-		18.5	0.0	0.4	0.0	18.9	0.0	22.2	0.0	0.0	0.0	22.2	0.0
19243.96-		15.6	0.0	3.2	0.0	18.9	0.0	21.9	0.0	0.0	0.0	21.9	0.0
19244.10-		10.4	0.0	8.5	0.0	18.9	0.0	21.4	0.0	0.0	0.0	21.4	0.0
19244.16-	T=.00625 S.	2.7	0.0	16.1	0.0	18.8	0.0	20.6	0.0	0.0	0.0	20.6	0.0
19243.96-		0.0	0.0	21.0	0.0	21.0	0.0	19.8	0.0	0.0	0.0	19.8	0.0
19243.82-		0.0	0.0	22.1	0.0	22.1	0.0	19.0	0.0	0.0	0.0	19.0	0.0
19243.89-		0.0	0.0	22.1	0.0	22.1	0.0	16.9	0.0	2.0	0.0	18.9	0.0
19244.10-		0.0	0.0	21.6	0.0	21.6	0.0	12.5	0.0	6.5	0.0	19.0	0.0
19244.31-	T=.00750 S.	0.0	0.0	20.9	0.0	20.9	0.0	5.6	0.0	13.3	0.0	18.9	0.0
19244.27-		0.0	0.0	20.1	0.0	20.1	0.0	0.0	0.0	20.2	0.0	20.2	0.0
19244.05-		0.0	0.0	19.2	0.0	19.2	0.0	0.0	0.0	21.9	0.0	21.9	0.0
19243.98-		0.0	0.0	18.9	1.1	17.8	0.0	0.0	0.0	22.1	0.0	22.1	0.0
19244.05-		0.0	0.0	18.9	4.8	14.2	0.0	0.0	0.0	21.8	0.0	21.8	0.0
19244.19-	T=.00875 S.	0.0	0.0	18.9	10.8	8.1	0.0	0.0	0.0	21.1	0.0	21.1	0.0
19244.17-		0.0	0.0	18.9	18.9	0.0	0.0	0.0	0.0	20.3	0.0	20.3	0.0
19243.95-		0.0	0.0	21.5	21.5	0.0	0.0	0.0	0.0	19.5	0.0	19.5	0.0
19243.90-		0.0	0.0	22.2	22.2	0.0	0.0	0.0	0.0	18.9	0.4	18.5	0.0
19244.04-		0.0	0.0	21.9	21.9	0.0	0.0	0.0	0.0	18.9	3.3	15.7	0.0
19244.28-	T=.01000 S.	0.0	0.0	21.4	21.4	0.0	0.0	0.0	0.0	19.0	8.5	10.4	0.0
19244.46-		0.0	0.0	20.6	20.6	0.0	0.0	0.0	0.0	18.8	16.1	2.7	0.0
19244.32-		0.0	0.0	19.8	19.8	0.0	0.0	0.0	0.0	21.0	21.0	0.0	0.0
19244.16-		0.0	0.0	19.0	19.0	0.0	0.0	0.0	0.0	22.1	22.1	0.0	0.0
19244.15-		0.0	0.2	16.9	18.9	0.0	0.0	0.0	0.0	22.1	22.1	0.0	0.0
19244.28-	T=.01125 S.	0.0	6.5	12.4	18.9	0.0	0.0	0.0	0.0	21.6	21.6	0.0	0.0
19244.41-		0.0	13.3	5.6	18.9	0.0	0.0	0.0	0.0	20.9	20.9	0.0	0.0
19244.29-		0.0	20.2	0.0	20.2	0.0	0.0	0.0	0.0	20.1	20.1	0.0	0.0
19244.09-		0.0	21.9	0.0	21.9	0.0	0.0	0.0	0.0	19.2	19.2	0.0	0.0
19244.09-		0.0	22.2	0.0	22.2	0.0	0.0	0.0	1.1	17.8	18.9	0.0	0.0
19244.26-	T=.01250 S.	0.0	21.8	0.0	21.8	0.0	0.0	0.0	4.8	14.2	18.9	0.0	0.0
19244.49-		0.0	21.1	0.0	21.1	0.0	0.0	0.0	10.8	8.1	18.9	0.0	0.0
19244.57-		0.0	20.4	0.0	20.4	0.0	0.0	0.0	18.7	0.0	19.2	0.0	0.0
19244.36-		0.0	19.5	0.0	19.5	0.0	0.0	0.0	21.6	0.0	21.6	0.0	0.0
19244.23-		0.0	18.9	0.0	18.5	0.0	0.4	0.0	22.2	0.0	22.2	0.0	0.0
19244.27-	T=.01375 S.	0.0	18.9	0.0	15.6	0.0	3.2	0.0	21.9	0.0	21.9	0.0	0.0
19244.41-		0.0	18.9	0.0	10.4	0.0	8.5	0.0	21.4	0.0	21.4	0.0	0.0
19244.48-		0.0	18.8	0.0	2.7	0.0	16.1	0.0	20.6	0.0	20.6	0.0	0.0
19244.27-		0.0	21.0	0.0	0.0	0.0	21.0	0.0	19.8	0.0	19.8	0.0	0.0
19244.14-		0.0	22.1	0.0	0.0	0.0	22.1	0.0	19.0	0.0	19.0	0.0	0.0
19244.21-	T=.01500 S.	0.0	22.1	0.0	0.0	0.0	22.1	0.0	18.9	0.0	16.9	0.0	2.0
19244.43-		0.0	21.6	0.0	0.0	0.0	21.6	0.0	19.0	0.0	12.5	0.0	6.5
19244.64-		0.0	20.9	0.0	0.0	0.0	20.9	0.0	18.9	0.0	5.6	0.0	13.3
19244.60-		0.0	20.1	0.0	0.0	0.0	20.1	0.0	20.2	0.0	0.0	0.0	20.2
19244.38-		0.0	19.2	0.0	0.0	0.0	19.2	0.0	21.9	0.0	0.0	0.0	21.9

Figure 8-12 Appendix 18G analysis of 12-pulse system rectified input voltage.

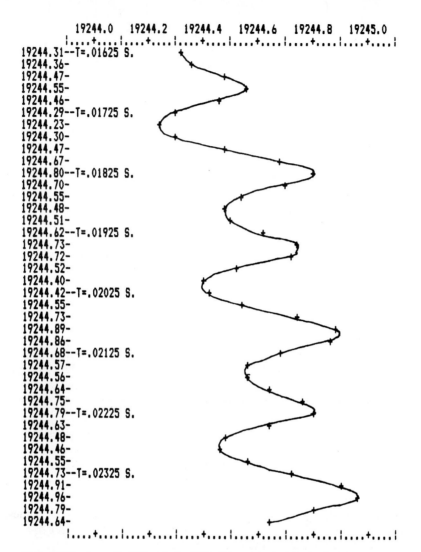

Figure 8-13 Appendix 18G analysis of 12-pulse system dc output voltage.

C_{ID} (denoted by the X plotting symbols). The plots show that (1) the two envelopes are similar; (2) the peak excursions are nearly identical, but they are not identical, due to the differences between the commutation procedures present in the two cases; and (3) the sum of these two sets of values is of course the rectified voltage applied to the input terminals of the system's ripple filter.

Figure 8-13 shows a continuation of the analysis with filtered dc output voltage $X(6)$ (tabulated and plotted) using an expanded dc voltage plotting scale. Because the level of dc ripple is so small, the program engages the computer's double precision mode to present the results accurately. The system has virtually reached steady-state conditions—peak-to-peak ripple of about 0.5 V is superimposed on the average dc output of 19,224 V.

8-6 REGULATORS AND SWITCHING DC POWER SUPPLIES

Low-Frequency Power Supply Regulators

The four types of low source frequency (i.e., typically 50 to 400 Hz) dc power supplies that we analyzed in Sections 8-1, 8-3, 8-4, and 8-5 are often referred to as "linear" power supplies. They function (1) to provide a voltage transformation that modifies the source's voltage to a level consistent with the dc output voltage required, (2) to provide the rectification essential for ac-to-dc conversion, and (3) to provide a degree of dc ripple filtering consistent with the intended application of the supply. In an ideal power supply these could be considered to be linear functions. As we know from our analyses, when the practical issues associated with commutation and dc loading changes are taken into account, the problem can no longer be considered linear. Rather, a determination of the dc voltage regulation characteristics of the system requires that a transient analysis, involving both power source and power supply parameters, must be performed.

If the dc voltage regulation of a conventional low-source frequency power supply is not adequate for the application under consideration, either some form of *regulator* must be added to the existing system or an alternative power supply design approach must be utilized. The course of action involves primarily the period of time within which the dc voltage correction must be accomplished. For example, in public utility prime power systems, voltage regulation is a continuous problem. Mechanically driven, automatic-controlled tap changers on substation power transformers proved the normal solution (these voltage corrections require fractions of seconds to several seconds to accomplish). Similar mechanical regulation techniques are often applied to electronic power supply systems when the mechanical response times are consistent with the application.

Often much shorter response times are required from the regulator system. Electronic regulators are then normally used. The traditional approach has been to use a linear regulator—effectively an electronically controlled resistor connected in series with the dc load on the power supply. Transient response times measured in microseconds can be accomplished with this approach. Alternatively, an electronically controlled shunt resistance arrangement is sometimes used to accomplish similar regulation capabilities. Both of these approaches have been in use for many decades. The power-regulating elements may be either vacuum tubes or solid-state devices.

High-Frequency Switching DC Power Supplies

There are two shortcomings associated with linear regulators: (1) if the dc load current and/or load voltage is very high, the regulating system tends to be quite costly; and (2) the power losses introduced into the overall dc power supply system are usually quite high.

In recent years, *switching power supplies* have seen widespread application in areas where the regulation capabilities of linear regulators are required. Switching power supplies take a number of forms, but in essence they consist of a high-frequency (i.e., typically in the range 10 to 50 kHz) electronic power generator whose output is rectified and filtered in a manner similar to the low-source-frequency units we considered previously. A second dc power source (that obtains its energy from a low-frequency prime power source) typically supplies energy to the high-frequency power generator. Hence these units are frequently referred to as *dc-to-dc converters*. Converter systems are characterized with considerably higher efficiency figures than systems using linear regulators. An elementary form of dc-to-dc converter will be analyzed in the next section.

8-7 FIFTY-KILOHERTZ DC-TO-DC CONVERTER POWER SUPPLY SYSTEM

The circuit diagram of Figure 8-14 illustrates the configuration we will analyze. This arrangement is sometimes referred to a *half-bridge forward converter*. The dc power source charges capacitors C_1 and C_2. These capacitors serve as the storage elements from which this power electronic system draws energy on alternate half-cycles of its waveform generation process. In the arrangement shown, two power-type field-effect transistors provide the switching function. Their gates are alternately turned on for 10-μs intervals, thereby causing the FETs to generate a 50-kHz waveform. This wave is connected to the primary of high-frequency transformer T. The secondary of the transformer is connected to a full-wave bridge rectifier, whose output is filtered by capacitor C, and dc energy is supplied to load resistor R.

The network diagram of Fig. 8-14 indicates our analysis approach. For purposes of analysis simplification, two batteries, both with a voltage V, are used to simulate the dc power supply and capacitors C_1 and C_2. The upper switch, FET_1, is simulated by the two diodes and switch designated as R_5. The rationale for this arrangement is that the diode and switch represent the gated on/off characteristics of the FET. For the gated-on case, the switch is closed, and if the circuit current is in the direction indicated by the diode, its resistance, R_5, has a value of 1 Ω.

When the FET's gate is not turned on and the current is in the same forward direction, the switch looks like a leakage resistance, R_5, of 1 MΩ. However, if current should attempt to flow in the reverse direction through the FET, the second diode, shown to the right of the diode/switch pair, will pass current, and it will exhibit a resistance, R_5, of 1 Ω.

The resistance values noted above are representative of those one might encounter with SIT power FETs capable of handling the voltage and currents involved in the analysis we will perform. This example will not consider the interelectrode capacitances associated with the FETs. FET_2 is assumed to have identical characteristics to those of FET_1, and its resistance values are designated as R_4.

In high-frequency circuits, it is usually advisable to document a more complete transformer description than we did in our low-frequency analyses. For example, in the network diagram the primary inductance, the secondary inductance, and the mutual inductance between them is indicated. The transformer secondary connects to a conventional bridge rectifier and capacitor input filter. The only new feature encountered in this area is the greatly reduced amount of capacity required to provide a degree of ripple filtering comparable to that obtained with the low-frequency supplies.

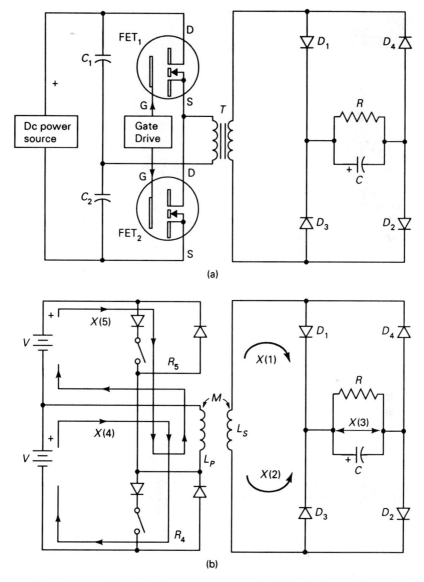

Figure 8-14 Half-bridge dc–dc converter power supply: (a) circuit diagram; (b) network diagram.

The network diagram indicates that on the secondary side of the transformer we are dealing with three variables familiar to us from our low-frequency analyses: the two rectifier currents, $X(1)$ and $X(2)$, and the capacitor voltage, $X(3)$. On the power generator side of the transformer we are dealing with two unfamiliar currents, $X(4)$, the current passing through the primary due to R_4, and $X(5)$, the current passing through the primary in the opposite direction due to R_5.

Sec. 8-7 Fifty–Kilohertz DC-to-DC Converter Power Supply System

System Equations and Analysis Programs

One way of setting up the system equations is to segregate each operating cycle time period of the supply (i.e., 20 µs) as follows: a first period, when R_4 is gated on and rectification is via path $X(2)$; a second period, when R_4 is gated on and rectification is via path $X(1)$; a third period, when R_5 is gated on and rectification is via path $X(1)$; and a fourth period, when R_5 is gated on and rectification is via path $X(2)$.

For the first period, which we will identify as SR = 1, we can write the following equation around the active paths on the primary side of the transformer:

$$LP\ D(4) + R4\ X(4) - V + M\ D(2) = 0$$

The fourth term of the equation defines the reflection, back to the primary, of the events occurring in the secondary circuit of the transformer. We can write the equation around the active path on the secondary side of the transformer as

$$M\ D(4) + X(3) + LS\ D(2) = 0$$

Now if we solve these two equations and arrange them in program format, we have

$$D(4) = (LS\ V - LS\ R4\ X(4) + M\ X(3))/(LP\ LS - M^2) \qquad (8\text{-}37)$$

$$D(2) = (-X(3) - M\ D(4))/LS \qquad (8\text{-}38)$$

A current summation at the junction of R and C at the load gives

$$D(3) = (X(2) + X(3)/R)/C \qquad (8\text{-}39)$$

If you refer to Appendix 19 these equations are listed on lines 1102–1106. In a format similar to the previous power supply appendices, the equation subroutine has directed the program to this set of equations by means of lines 1005–1020 since SR = 1. It should be noted that on line 1101 the integration interval T_1 has been set for 1 ns when using these equations. This has been done because of the high rate of change of current during this portion of the cycle. Some other bookkeeping equation statements, consistent with the FET mode of operation described earlier, have also been introduced in equations 1101 and 1102.

Now consider the second time interval, when SR = 2. We can write the primary and secondary mesh equations as follows:

$$LP\ D(4) + R4\ X(4) - V - M\ D(1) = 0$$
$$-M\ D(4) + X(3) + LS\ D(1) = 0$$

When solved and put in program format we have

$$D(4) = (LS\ V - LS\ R4\ X(4) - M\ X(3))/(LP\ LS - M^2) \qquad (8\text{-}40)$$

$$D(1) = (M\ D(4) - X(3))/LS \qquad (8\text{-}41)$$

A secondary current summation gives

$$D(3) = (X(1) - X(3)/R)/C \qquad (8\text{-}42)$$

These equations are listed on program lines 1112–1116, and the program is directed to these lines by program line 1010 because SR = 2. Program line 1111 has increased the

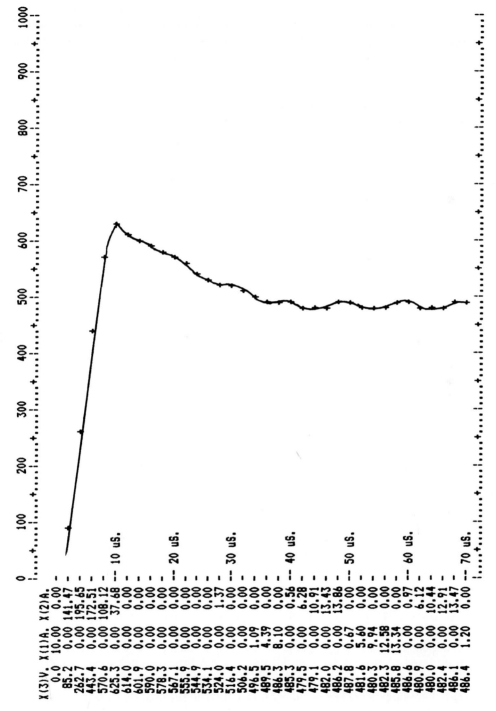

Figure 8-15 Appendix 19 analysis of single-phase 50-kHz dc–dc converter.

Sec. 8-7 Fifty–Kilohertz DC–to–DC Converter Power Supply System 183 integration interval T_1 by a factor of 100 since the currents are now changing much more slowly—hence the program execution is speeded up greatly during this period.

The equations during periods 3 and 4 can be developed in the same manner as outlined for SR = 1 and SR = 2. These equations are listed as program lines 1120–1126 and lines 1130–1136, respectively.

Now let us consider the program line inputs required to investigate a practical dc-to-dc converter system situation. Lines 111–118 set up the operating frequency of the power generator and define the SR periods. A 50-kHz frequency has been defined. Line 30 defines that we are using five equations. Lines 60–62 define the primary, secondary, and mutual inductance of a 50-kHz power transformer with a 1:1 turns ratio.

Line 64 sets the filter capacitor value at 1 μF, and line 66 sets the load resistance value at 50 Ω. Line 68 defines a source power supply voltage in the operating region of typical high-power-level FETs—500 V dc. Lines 50–58 define the initial conditions for a snap-on run.

Analysis Results

Now if you load Appendix 19 into your computer, issue a <RUN> command, and then enter a P command, your printer will produce a plot similar to that shown in Figure 8-15. Some general comments on the run are as follows.

Dc output voltage. The dc voltage has a steady-state value in the 480-V region. This would indicate fairly high operating efficiency based on our 500-V source level. An actual system would of course have additional loss elements not included in our analysis.

Dc ripple. The ripple frequency is now 100 kHz. As a result, a very small capacitor input filter (by low-frequency source standards) is all that is required for systems that operate with a constant dc load.

Snap-on currents. In actual circuits the currents associated with the initial overshoot of voltage could prove disastrous to the FETs—hence this mode of turning on the supply is not recommended. Snap-on computer runs provide design insight under worst-case operating conditions.

Dc regulator approaches. As noted in the introduction to this section, converters of this form are often used as regulated power supplies. The techniques used vary, but one common method is to use pulse width modulation of the gate drive signals. Based on the response characteristics of this run, it seems likely that this system would have a response time in the region of tens of microseconds. We investigate the pulse regulation capabilities of a more elaborate dc-to-dc regulator in Chapter 11.

8-8 CONTROLLED THREE-PHASE BRIDGE DC POWER SUPPLIES/INVERTERS

In more precise terms, ac-to-dc and dc-to-ac power supplies should be referred to as ac–dc converters, dc–ac inverters, or dc–dc converters. These more precise definitions assume significance whenever controlled switching elements are used.

In Section 8-7 we analyzed a single-phase 50-kHz dc–dc converter. It is possible, with proper sets of control commands to the switching elements, to utilize controlled converters either for ac–dc conversion or dc–ac inversion. Prime line frequency electronic converters frequently make use of these reversible power flow capabilities for industrial dc motor and generator drive systems. We examine some of the principles involved in this section.

Prime line conversion and inversion principles can be illustrated by the circuit arrangement shown in the industrial motor drive arrangement of Figure 8-16a. The circuit is fundamentally the same as the three-phase bridge circuits we have used previously except that each diode has been replaced by a thyristor [i.e., also referred to as a silicon-controlled rectifier (SCR)], and the load has been indicated as a dc motor.

The SCR switches assume substantially open-circuit characteristics between their anode and cathode connections until they are triggered. While the anode is positive with respect to the cathode, an appropriately timed voltage trigger is applied between the gate (denoted the stub extending from the cathode) and the cathode. The SCR then almost instantaneously assumes the substantially short-circuit properties of a forward-biased silicon diode. Therefore, if each of the active SCRs has its gate triggered at exactly the point in time at which its anode goes positive relative to its cathode, the circuit action will be identical to an uncontrolled ac–dc converter. However, if each of the triggers is delayed from its respective reference point by a period of time T_F, the dc output voltage of the power supply will be reduced.

In an ideal three-phase bridge ac–dc converter arrangement, under steady-state conditions, the level of the dc output voltage will be

$$V_O = \left(3 \cdot 2^{1/2} \frac{V_{LL}}{\pi}\right) \cos(360 T_F F) = 1.35 V_{LL} \cos \alpha \qquad (8\text{-}43)$$

where F is the frequency of the prime source in hertz, delay T_F is in seconds, V_{LL} is the rectifier's line-to-line rms voltage level, and α is the firing angle (i.e., the equivalent of $360 \times T_F \times F$ in degrees).

Motor under Steady-State Conditions

Note the armature circuit of the dc shunt motor with externally applied constant field excitation shown in Fig. 8-16a. It consists of a three-element series circuit: armature inductance L, armature resistance as R_A, and counter emf V (i.e., a series dc generator with a terminal voltage V). Voltage V is a function of both the amount of field excitation applied and the speed of the motor. When a shunt motor is loaded, its speed and the value

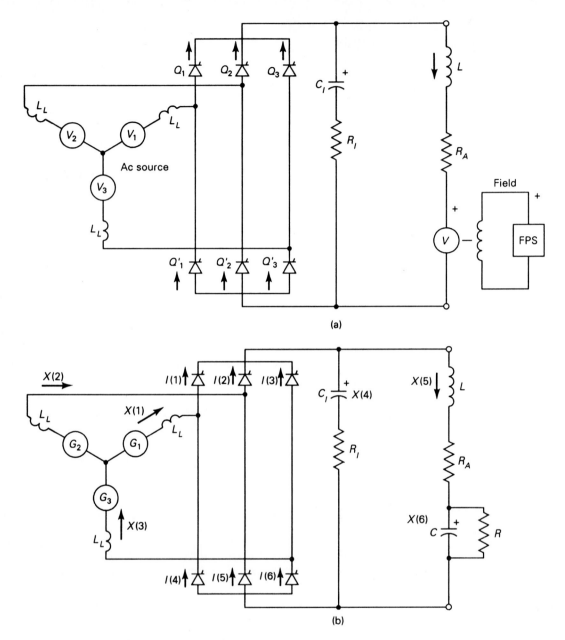

Figure 8-16 Controlled ac–dc converter for industrial dc shunt motor drive: (a) three-phase bridge ac–dc converter with dc shunt motor load; (b) network diagram for dc motor acceleration analysis.

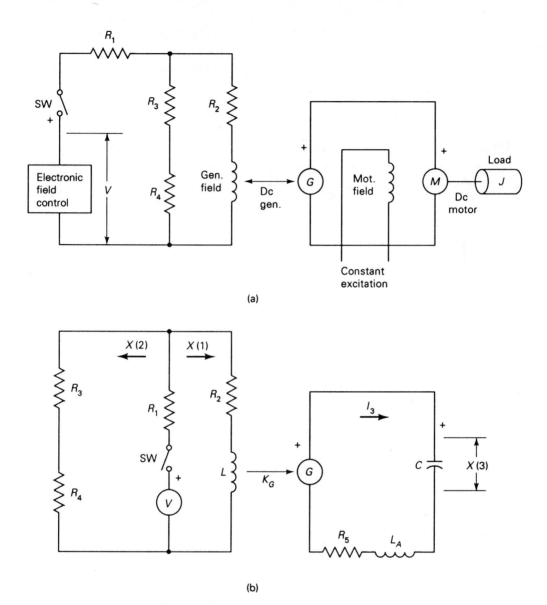

Figure 8-17 Industrial drive system with high-inertia loading: (a) system configuration with high-inertia motor loading; (b) network diagram of system.

of V decrease relative to their no-load operating values. If the motor has a constant value of applied terminal voltage, under steady-state conditions the armature current is

$$I_a = \frac{V_t - V}{R_A} \qquad (8\text{-}44)$$

where V_t is the value of terminal voltage.

Sec. 8-8   Controlled Three–Phase Bridge DC Power Supplies/Inverters

Motor under Acceleration or Deceleration Conditions

The relationships involved in rotational transient analyses require the development of electromechanical relationships. Consider the system shown in Figure 8-17a. Assume that the electronic field control is set to deliver a constant dc voltage output. When electronic switch SW is closed, the dc generator delivers energy to the dc motor that drives a high-inertia load. The load is accelerated from standstill to rated motor rpm at frequently repeated intervals. To draw a network diagram it is necessary to define quantities pertaining to the inertia load on the motor. From Newton's second law the motor torque is known to be proportional to the acceleration of the load. Based on torque T in N•m (ft-lb), the motor's angular velocity ω in rad/s, and the load's polar moment of inertia J in kg•m² (slug-ft²), the torque is

$$T = \frac{J d\omega}{dt}$$

If an ideal dc motor operates with constant field excitation, its torque will be proportional to its armature current I_3. A constant K_T, expressed in N•m/A (lb-ft/A), can be defined from the motor's nameplate values—its horsepower, speed, and current. This relationship is $T = K_T I_3$. In addition, the motor's generated voltage V (its counter emf) can be defined in terms of a constant of proportionality K_V (volt•seconds per radian), obtained from the motor's nameplate values of rated voltage and speed. This relationship is $V = K_V \omega$. When these expressions are combined with the expression from the first part of this exercise,

$$dV = \frac{K_T K_V}{J} I_3 dt$$

When this equation is integrated we obtain

$$V = \frac{K_T K_V}{J} \int_0^t I_3 dt$$

If this equation is compared to Eq. (7-5), we see they have the same mathematical structure. This indicates that during acceleration the motor can be replaced by a capacitor whose value is

$$C = \frac{J}{K_T K_V} \qquad (8\text{-}45)$$

Example 8-1

Based on the concepts above, the systems network diagram can be drawn as shown in Figure 8-17b. The system's parameters are as follows: first, $V = 250$ V (assume constant for this analysis); second, $R_1 = 25\ \Omega$; third, $R_2 = 50\ \Omega$ (field resistance); fourth, $R_3 = 90\ \Omega$; fifth, $R_4 = 10\ \Omega$; sixth, $R_5 = 0.05\ \Omega$ (sum of both armature resistances); seventh, $L = 50$ H (field inductance); eighth, $K_G = 90$ V/A (generator gain factor); ninth, $J = 3.25$ kg•m² (2.4 slug-ft²); tenth, $K_T = 0.81$ N•m/A (0.6 lb-ft/A); eleventh, $K_V = 1$ V•s/rad; and twelfth, $C = 4$ F.

Now let us make the following assumptions: (1) the drive motor friction losses are negligible; (2) the drive motor load torque is zero (i.e., inertia acceleration torque is the only

Seconds	Current	Seconds	Current
0	0	0	0
.05	307.599	.05	180.438
.1	525.939	.1	304.794
.15	676.626	.15	398.035
.2	775.933	.2	466.909
.25	836.444	.25	516.695
.3	867.878	.3	551.528
.35	877.727	.35	574.649
.4	871.754	.4	588.602
.45	854.381	.45	595.387
.5	828.991	.5	596.575
.55	798.157	.55	593.407
.6	763.832	.6	586.86
.65	727.485	.65	577.706
.7	690.212	.7	566.558
.75	652.826	.75	553.902
.8	615.915	.8	540.123
.85	579.901	.85	525.529
.9	545.079	.9	510.367
.95	511.641	.95	494.832
1	479.709	1	479.083
1.05	449.348	1.05	463.247
1.1	420.582	1.1	447.423
1.15	393.404	1.15	431.695
1.2	367.785	1.2	416.129
1.25	343.683	1.25	400.776
1.3	321.041	1.3	385.679
1.35	299.799	1.35	370.873
1.4	279.89	1.4	356.382
1.45	261.249	1.45	342.227
1.5	243.805	1.5	328.426

a. Original Circuit Motor Current   b. Revised Circuit Motor Current

Figure 8-18 Dc motor acceleration analysis using Appendix 11: (a) original circuit motor current; (b) revised circuit motor current.

load); (3) the generator has a linear field voltage versus output voltage characteristic; and (4) the combined armature inductance value of L_G is negligible.

We can write the system equations as follows:

$$X(2) = (V - R1\ X(1))/(R1 + R3 + R4)$$

$$D(1) = (V - (R1 + R2)\ X(1) - R1\ X(2))/L$$

$$D(3) = (KG\ X(1) - X(3)/(R5\ C)$$

Appendix 11 lists the program for this system. If the program is loaded and a <RUN> command is issued, the armature current response shown in Figure 8-18a will develop. The significant results are: (1) a peak current of 877.7 A is required from the generator to provide acceleration; and (2) the complete acceleration cycle is substantially complete within 2 s.

To minimize armature deterioration in any dc motor or generator, maximum armature current limits are established. Assume that this motor's specification limits the peak current to 600 A. An auxiliary analysis of the system has indicated that if an RC circuit is connected across R_3, the desired degree of limiting should occur. The value of $R_6 = 10\ \Omega$ and $C_1 = 0.03$ F. Call the current through this new mesh $X(4)$. Then we can write a new set of equations as follows:

$$X(2) = (V - R1\ X(1) + R3\ C1\ D(4))/(R1 + R2 + R3)$$

$$D(4) = (R3\ X(2) - X(4))/((R(3) + X(6))\ C1)$$

Here $D(1)$ and $D(2)$ are the same as before. When these revised equations are inserted in Appendix 11 and the program is repeated, the response of Figure 8-18b results. It meets the system's 600-A peak current requirement.

DC Generator Representation

A dc shunt generator with constant field excitation can also be represented as shown in Fig. 8-16a. The field, armature resistance, armature inductance, and counter emf (in this case it is the generated emf) rationale is similar to that of a dc motor; however, a generator requires mechanical drive system energy to provide its rotational operating speed.

When a generator replaces the motor on the diagram, the polarity connections to the armature's generated voltage is reversed. Reverse energy flow then occurs through the power electronics system, and we are dealing with an inverter system. In a typical industrial power system, the generator functions to pump back converted dc energy into the ac power circuit. This is the same fundamental principle illustrated in Section 8-7 for the single-phase 50-kHz dc–dc inverter, except that the inverter's ac output was then rectified to complete the dc–dc system function. In the industrial power case, the power electronics system delivers its converted energy into a three-phase power system and/or any other appropriate form of ac load.

8-9 ANALYSIS OF AN AC-TO-DC CONVERTER FOR A SHUNT MOTOR

Example 8-1 illustrated an acceleration analysis of a traditional industrial dc motor drive using a dc generator power source system. Many present-day dc motor drive systems use an ac–dc converter to replace the dc generator—typically, to reduce system costs. Figure 8-16b shows a representative network diagram for this type of system.

We will analyze a system similar to Example 8-1—a blooming/slabbing mill in a steel processing facility where it is necessary frequently to accelerate a set of high-inertia mill rolls. When they are up to speed, a red-hot metal billet or slab is fed between the rolls to reduce its thickness. The mill's rolls are then reversed, and on successive passes through the mill the slab is reduced to the required thickness.

For our transient analysis we modify the three-phase bridge power supply program developed in Appendix 18. The background for these modifications follows.

DC Motor Characteristics

The analysis values are (1) rated output = 100 hp; (2) rated armature current = 300 A; (3) armature resistance $R_A = 0.05 \, \Omega$; (4) armature circuit inductance $L = 1$ mH; and (5) counter emf at full load, $V = 250$ V.

AC Source Voltage

Based on Eq. (8-45), an ideal three-phase bridge supply requires a line-to-line rms voltage of 185.2 V when the phase-delay angle α has a value of zero degrees. In a practical power supply, allowance must also be made for the 15-V loss due to the motor's full-load current of 300 A flowing through armature resistance R_A of 0.05 Ω.

Assume that the transformer used to supply energy to the power supply has 8% reactance. Based on our previous analyses, the supply voltage should then be 4% higher than its ideal value to account for voltage regulation. Eight percent reactance, based on the dc supply full-load power, was also used to determine the value of L_L, the leakage inductance per phase. If these two effects are taken into account, the line-to-line voltage selected is approximately equal to (185 + 15)/0.96 or 208 V rms. This value has been used in the revised program to establish the value of V_P, the peak value of phase-to-neutral voltage.

Armature Current Regulator

Dc motors will have their armature elements (e.g., their brushes and commutator segments) severely damaged if excessive current passes through the armature circuit. In Example 8-1 the design approach for limiting armature current during acceleration was to limit the rate of rise of field current on the system generator. In this example, armature current will be limited by adjusting the delay time in firing the SCRs. The approach to be employed in this analysis is to use a very elementary two-step regulator. It functions as follows. First, when the current exceeds 350 A, each successive pass through the program will increase the firing delay of all SCRs by a small increment, until eventually an α value of 90° could be reached and the output of the supply could become zero volts. Second, when a system armature current decrease starts (i.e., the series power circuit contains transformer leakage reactance, motor armature inductance, and SCRs—all acting together to produce a long-time-constant power loop) the current will encounter a second control limit at 325 A. Then each pass through the program will cause the value of firing delay α to decrease by a small increment. It could eventually reach zero, thereby causing maximum power supply output voltage to occur. Third, the system voltage will fluctuate between these two extreme power supply output voltage levels, usually not reaching either, and thereby provide a coarse degree of regulation to hold the system within a safe range of commutating current levels.

Program Modifications

Several straightforward modifications of Appendix 18C based on the rationale above, are shown in Appendix 18D. The major changes are as follows. Program line 400 in conjunction with statements at the introduction of each of the six rectification periods of the power supply (i.e., on program lines 1011, 1031, 1051, 1071, 1091, and 1111) provide a set of conditional commands that select both the correct system operating period as a function of time and the correct SCR firing time delay, TCY. Program line 110 in conjunction with terms CP and TD in program line 13, establishes the value of TCY. Program lines 111–116 function to enable the two-step regulator process described above. *Friction loading*: The system is programmed to have virtually no friction loading during acceleration of the rolls (due to the value of R selected in line 64). After acceleration has been completed, line 118 introduces the full-load value of friction loading to simulate the entry of the billet between the rolls. *Resistance RA*: Several minor changes have been made to a number of the rectifier statements between lines 1010 and 1126 to introduce the term RA. It was not used in previous programs.

Analysis Results

With this background, load Appendix 18D into your computer, enter a <RUN> command, and follow this by a P command. A plot similar to Figure 8-19 will develop slowly on your computer. It shows the following: Along the bottom scale a tabular printout of the values of armature current $X(5)$ and counter emf $X(6)$. Each set shown represents the final values present at the conclusion of each cycle of the prime power waveform. The highest value of $X(5)$ flowing through the regulator appears to be 480 A, a value within the commutating capabilities of the motor. The plot indicates that the motor accelerates to the required counter emf value, $X(6)$, in approximately 0.72 s. A slight overshoot in $X(6)$ and motor speed occurs before the full value of friction loading is applied at 0.8 s. After load is applied, by the end of the run the motor has nearly settled in at its nominal full-load steady-state values of $X(6) = 250$ V and $X(5) = 300$ A.

Despite the fact that this is a very elementary problem, the general principles illustrated can be readily expanded to encompass considerably more complex system analyses.

8-10 ANALYSIS OF A DC-TO-AC INVERTER USING A SHUNT GENERATOR

Based on analysis background from Section 8-9, development of a program for processing reverse energy flow is a straightforward procedure. The inverter system we consider is shown in Figure 8-20. We use the same shunt machine for this example, but instead of using it as a shunt motor, it will be used as a shunt generator. Note in the figure that *a prime mover* (e.g., a diesel engine drive) has been added to rotate the shunt machine's shaft at the required speed. The shunt generator has external field excitation. The armature polarity connections to the shunt generator have been reversed. This allows the machine's generated energy to be pumped back through the same power supply system that we used in our previous analysis. The polarity marking on capacitor C_I have been reversed to agree with the new generator polarity. A set of three resistors, denoted R_L, have been added across the three power supply generator phases. These resistors can be used to simulate a load into which the shunt generator's energy can be pumped back.

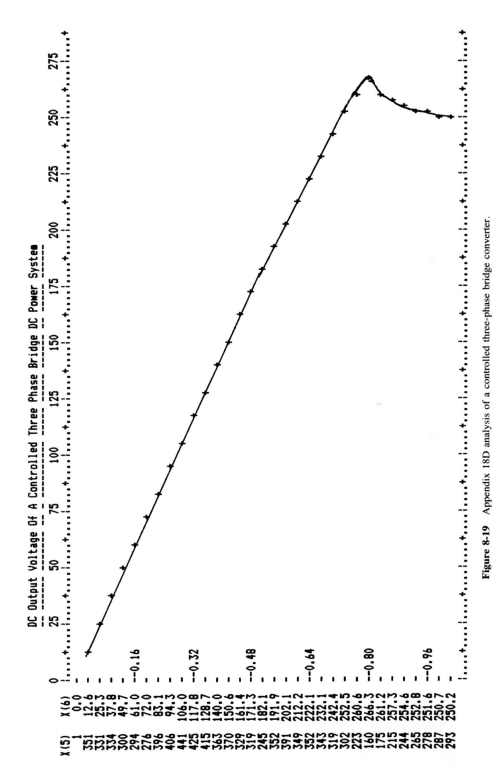

Figure 8-19 Appendix 18D analysis of a controlled three-phase bridge converter.

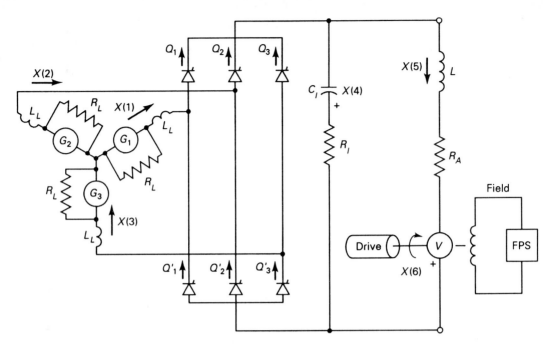

Figure 8-20 Controlled dc–ac inverter for industrial dc generator drive.

Program Modifications

A new program revision, Appendix 18E, simulates the system of Fig. 8-20. The major revisions are: Polarity signs (i.e., associated with terms in the equations between lines 1010 and 1122) have been revised to be consistent with the energy flow path. Delay time has the same significance as in the motor analysis: When $T_D = 0$ the power supply is at maximum voltage and when $T_D = 4$ ms the supply is at minimum voltage. The current regulator function has been removed. The generated voltage has been simulated as a constant value of 250 V simply by making C very large. The loading of resistor R has been effectively eliminated by assigning it a very high value. The values of R_L have been selected to dissipate the 75 kW—the same power as the motor delivered to its viscous friction load in the previous analysis. For the analysis, $T_D = 0$. The tabular printout format has been revised to include the three ac load currents and the three ac generator currents.

Analysis Interpretation

Figure 8-21 shows the result of a short analysis. Two conclusions are as follows. The system current flows in all paths appear to be in accordance with reversed energy flow in the system. Insight shortcomings—This system effectively has two power sources: stepped current waveforms generated by the power supply and dc transient current waveforms produced by the generator. It is difficult to gain a clear concept of the manner in which these system functions interact (i.e., the analyst's insight is very limited).

Sec.	TS	V1	V2	V3	I(1)	I(2)	I(3)	I(4)	I(5)	I(6)	X(6)	X(5)	Vin	X(1)	x(2)	X(3)	Del 1	Del 2	Del 3	IL1	IL2	IL3	I61	I62	I63
.00000	12	85	85	-170	0	300	0	0	0	300	-250	300	-250	0	300	-300	-300	600	-300	147	147	-295	147	-153	5
.00050	1	112	55	-166	46	249	0	0	0	295	-250	295	-254	46	249	-295	-203	544	-341	194	95	-289	148	-154	6
.00100	1	135	22	-157	185	98	0	0	0	284	-250	287	-242	185	98	-284	87	382	-469	234	38	-272	48	-60	11
.00150	2	152	-11	-141	296	0	0	0	0	296	-250	283	-290	296	0	-296	296	296	-592	264	-19	-245	-32	-19	51
.00200	2	164	-44	-120	298	0	0	0	0	298	-250	295	-283	298	0	-298	298	298	-596	285	-76	-208	-13	-76	90
.00250	2	169	-75	-94	297	0	0	0	0	297	-250	300	-266	297	0	-297	297	297	-594	294	-130	-164	-3	-130	133
.00300	3	168	-103	-65	293	0	0	0	21	272	-250	297	-254	293	-21	-272	314	251	-565	292	-179	-113	-1	-158	159
.00350	3	161	-128	-33	286	0	0	0	131	155	-250	290	-246	286	-131	-155	417	24	-441	279	-222	-57	-7	-91	98
.00400	4	147	-147	-0	290	0	0	0	290	0	-250	280	-277	290	-290	0	580	-290	-290	255	-255	0	-35	35	-0
.00450	4	128	-161	33	297	0	0	0	297	0	-250	292	-287	297	-297	0	594	-297	-297	222	-279	57	-75	18	57
.00500	4	103	-168	65	297	0	0	0	297	0	-250	299	-272	297	-297	0	595	-298	-297	179	-292	113	-118	5	113
.00550	5	75	-169	94	288	5	0	0	293	0	-250	298	-255	288	-293	5	582	-373	-283	130	-294	164	-158	-1	159
.00600	5	44	-164	120	205	84	0	0	289	0	-250	282	-249	205	-289	84	494	299	-121	76	-285	208	-128	4	124
.00650	5	11	-152	141	22	254	0	0	277	0	-250	287	-236	22	-277	254	299	-531	232	19	-264	245	-38	12	-9
.00700	6	-22	-135	157	0	296	0	0	296	0	-250	297	-290	0	-296	296	-0	-592	296	-38	-234	272	62	-24	-9
.00750	6	-55	-112	166	0	297	0	0	297	0	-250	299	-278	0	-297	297	-0	-595	297	-95	-194	289	103	-95	-8
.00800	6	-85	-85	170	0	294	0	0	294	0	-250	294	-259	0	-294	294	0	-588	294	-147	-147	295	147	-147	0
.00850	7	-112	-55	166	0	291	47	0	244	0	-250	291	-252	-47	-244	291	196	-534	338	-194	-95	289	147	149	-2
.00900	7	-135	-22	157	0	281	188	0	94	0	-250	286	-241	-188	-94	281	-94	-375	469	-234	-38	272	-46	55	-9
.00950	8	-152	11	141	0	295	295	0	0	0	-250	282	-290	-295	0	295	-295	-295	590	-264	19	245	31	19	-50
.01000	8	-164	44	120	0	297	297	0	0	0	-250	294	-283	-297	0	297	-297	-297	594	-285	76	208	12	76	-88
.01050	8	-169	75	94	0	296	296	0	0	0	-250	300	-266	-296	0	296	-296	-296	592	-294	130	164	2	130	-132
.01100	9	-168	103	65	0	271	292	21	0	0	-250	296	-254	-292	21	271	-313	-250	563	-292	179	113	-0	158	-158
.01150	9	-161	128	33	0	154	285	131	0	0	-250	289	-246	-285	131	154	-416	-23	439	-279	222	57	6	90	-97
.01200	10	-147	147	-0	0	0	289	289	0	0	-250	279	-278	-289	289	0	-578	289	289	-255	255	-57	34	-34	0
.01250	10	-128	161	-33	0	0	296	296	0	0	-250	291	-287	-296	296	0	-592	296	296	-222	279	-113	75	-17	-57
.01300	10	-103	168	-65	0	0	297	297	0	0	-250	299	-272	-297	297	0	-593	297	297	-179	292	-113	117	-5	-113
.01350	11	-75	169	-94	0	0	288	204	0	5	-250	298	-255	-288	293	-5	-580	297	283	-130	294	-164	157	1	-159
.01400	11	-44	164	-120	0	0	288	204	0	84	-250	292	-249	-204	288	-84	-492	372	120	-76	285	-208	128	-4	-124
.01450	11	-11	152	-141	0	0	276	22	0	254	-250	281	-236	-22	276	-254	-298	530	-232	-19	264	-245	3	-12	9
.01500	12	22	135	-157	0	0	296	0	0	296	-250	287	-290	0	296	-296	-296	591	-296	38	234	-272	38	-62	23
.01550	12	55	112	-166	0	0	297	0	0	297	-250	297	-278	0	297	-297	-297	593	-297	95	194	-289	95	-102	8

Cycle 1 Line Voltage RMS Values Are: V1 = 119.9961 V2 = 119.9961 V3 = 119.9959 VRMSavg = 119.9996
Cycle 1 Line Current RMS Values Are: X(7) = 230.8697 X(8) = 231.4302 X(9) = 232.2027 IRMSavg = 231.5009

| .01600 | 12 | 85 | 85 | -170 | 294 | 0 | 0 | 0 | 0 | 294 | -250 | 299 | -259 | 0 | 294 | -294 | -294 | 587 | -294 | 147 | 147 | -295 | 147 | -146 | -1 |

>>> Vdc = -250 <<<

Figure 8-21 Dc generator pump back in synchronous three-phase inverter system.

Inverter System Conservation-of-Energy Analysis

How can the preceding computer run be modified to provide practical insight concerning just how transient energy transformation systems of this type accomplish their end function? To provide this insight, a different form of quantitative analysis is required. Consider an analysis procedure involving *conservation of energy*.

The usual objective of a conservation-of-energy analysis is to bring order to an otherwise tangled maze of observational data. Reference to a physics textbook tells us that for all ordinary transformations of energy we should proceed (1) to add up all of the different sorts of energy that go into the transformation (e.g., add up the energy transformed in the ac generator/dc generator/inverter system); (2) to add up all of the energy produced (e.g., add up the energy produced in the external ac power system load); and (3) to observe if these two sums are precisely equal (i.e., if the sums are not equal, the analysis equations for the system should be investigated).

In Appendix 18E, there are 14 elements associated with the energy that is transformed and the energy that is produced. The energy associated with these elements will be calculated during each step of program execution if requested. You can demonstrate the conservation-of-energy feature of the program by (1) loading the program and entering a <RUN> command, and (2) entering the letter Y in response to the prompt concerning performance of an energy analysis. Note that system energy summations will then be printed out for each T_1 analysis interval that occurs coincident at an arbitrary point occurring at each 4 ms of program execution (i.e., alternate printout time spacings can be obtained by modifying the denominator of RIN1 in program line 34).

In the analysis shown in Figure 8-22, four energy analysis groupings are listed as follows: The first element energy grouping indicates that the energy stored or delivered by each of the inductive and capacitive elements. The final value sums up these energies and expresses the result as a power level (i.e., by dividing the result by T_1). All succeeding energy listings have also been converted to their power equivalents for the T_1 analysis period. The second grouping indicates the power lost in heat in resistors RA and RI along with a power summation. The third grouping indicates the power supplied or absorbed by each of the generator elements, a summation of these generator powers, and a summation of the three power grouping summations. The summation indicates the power delivered to (and dissipated by) the external ac power system resistors. It can be seen that the internal power summation of the first three groupings is substantially equal to the power dissipated in the external ac power system load, in comformance with the principles that define conservation of energy.

It is enlightening to compare the 7.45–7.5-ms transient analysis values with the 3.45–3.5-ms analysis. For example, the L and C energy levels can be seen to depart vastly from one another between the two analyses. A closer investigation indicates that the probable cause of this situation is that during the 3.45–3.5-ms period the inverter is in the process of commutation, while during the 7.45–7.5-ms period the commutation process has been completed.

Another area of observation shows that during the 3.45–3.5-ms period the sum of the generated power levels are higher than the ac load power. During the 7.45–7.5-ms period the generated power levels have a lower value than the ac load power.

Sec.	TS	V1	V2	V3	I(1)	I(2)	I(3)	I(4)	I(5)	I(6)	X(6)	X(5)	Vin	X(1)	X(2)	X(3)	Del 1	Del 2	Del 3	IL1	IL2	IL3	I61	I62	I63
.00000	12	85	85	-170	0	300	0	0	0	300	-250	300	-250	0	300	-300	-300	600	-300	147	147	-295	147	-153	5
.00050	1	112	55	-166	46	249	0	0	0	295	-250	295	-254	46	249	-295	-203	544	-341	194	95	-289	148	-154	6
.00100	1	135	22	-157	185	98	0	0	0	284	-250	287	-242	185	98	-284	87	382	-469	234	38	-272	48	-60	11
.00150	2	152	-11	-141	296	0	0	0	0	296	-250	283	-290	296	0	-296	296	296	-592	264	-19	-245	-32	-19	51
.00200	2	164	-44	-120	298	0	0	0	0	298	-250	295	-283	298	0	-298	298	298	-596	285	-76	-208	-13	-76	90
.00250	2	169	-75	-94	297	0	0	0	0	297	-250	300	-266	297	0	-297	297	297	-594	294	-130	-164	-3	-130	133
.00300	3	168	-103	-65	293	0	0	0	21	272	-250	297	-254	293	-21	-272	314	251	-565	292	-179	-113	-1	-158	159
.00350	3	161	-128	-33	286	0	0	0	131	155	-250	290	-246	286	-131	-155	417	24	-441	279	-222	-57	-7	-91	98

For period T = .00345 to T = .00350 Sec.

```
STORAGE ELEMENTS
  WLL1 =    -0.04
  WLL2 =     0.27
  WLL3 =    -0.39
  WCI  =    -0.05
  WL   =    -0.26
  PSUM =  -9299.09

DISSIPATIVE ELEMENTS
  PRI  =    30.84
  PRA  =  4210.21
  PDSUM=  4241.05

GENERATORS
  PDC  = 72544.84
  PAC1 = -1082.27
  PAC2 = 12152.16
  PAC3 = -3581.20
  PGSUM= 80033.53
  PSSUM+PDSUM+PGSUM =  74975.49

AC LOAD ELEMENTS
  PRL1 = 45121.97
  PRL2 = 27773.47
  PRL3 =  2099.59
  PSUM = 74995.03
```

Figure 8-22 Conservation-of-energy analysis for Appendix 18E inverter system.

```
.00400  4   147  -147   -0  290    0  290  -250  280 -277  290 -290    0  580 -290 -290  255 -255   -0  -35   35   -0
.00450  4   128  -161   33  297    0  297  -250  292 -287  297 -297    0  594 -297 -297  222 -279   57  -75   18   57
.00500  4   103  -168   65  297    0  297  -250  299 -272  297 -297    0  595 -297 -297  179 -292  113 -118    5  113
.00550  5    75  -169   94  288    0  293  -250  298 -255  288 -293    5  582 -298 -283  130 -294  164 -158   -1  159
.00600  5    44  -164  120  205    0  289  -250  292 -249  205 -289   84  494 -373 -121   76 -285  208 -128    4  124
.00650  5    11  -152  141   22    0  277  -250  282 -236   22 -277  254  299 -531  232   19 -264  245   -3   12   -9
.00700  6   -22  -135  157    0    0  296  -250  287 -290    0 -296  296  296 -592  296  -38 -234  272  -38   62  -24
.00750  6   -55  -112  166    0    0  297  -250  297 -278    0 -297  297  297 -595  297  -95 -194  289  -95  103   -8
```

For period T = .00745 to T = .00750 Sec.

```
STORAGE ELEMENTS
  WLL1 =     0.00
  WLL2 =     0.00
  WLL3 =     0.00
   WCI =     0.00
    WL =     0.21
  PSSUM =  4338.44

DISSIPATIVE ELEMENTS
   PRI =     0.44
   PRA =  4410.94
  PDSUM =  4411.38

GENERATORS
   PDC =  74254.10
  PAC1 =   4874.64
  PAC2 = -11407.40
  PAC3 =  -1478.47
  PGSUM =  66242.86
  PSSUM+PDSUM+PGSUM = 74992.68

AC LOAD ELEMENTS
  PRL1 =   4874.64
  PRL2 =  22223.33
  PRL3 =  47897.05
  PSUM =  74995.02
```

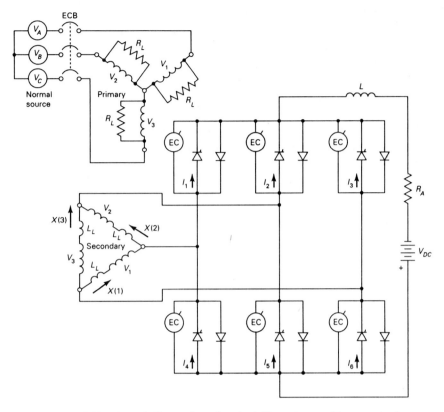

Figure 8-23 Network diagram for a six-pulse bridge uninterruptible power supply.

The two areas above are indicative of the insight that one can acquire concerning system functions, with the aid of a conservation of energy analysis. The same approach can, of course, be applied to any power electronics system analysis where a better understanding of system operation is required.

8-11 ANALYSIS OF AN UNINTERRUPTIBLE AC POWER SOURCE

Uninterruptible ac power sources (UPS) applications are widespread. Typically, a UPS functions to provide a means of preserving extensive computer memory files. For example, UPS sources provide the prime power for the memory file hardware of most commercial airlines. Catastrophic reservation network disruptions of a nationwide nature would occur if a major airline were to lose its central files due to a public utility power outage.

The dc-to-ac inverter techniques of Sections 8-9 and 8-10 describe a synchronous inverter approach. These techniques are similar to those of UPS approaches. The major difference is that when a power outage occurs in a synchronous system, its source of system control/timing no longer exists. With a UPS, the system control must function to accommodate operation with or without utility power present.

Circuit Configuration

How can we modify the dc-to-ac inverter circuit shown in Fig. 8-20 to implement an elementary form of UPS? One approach is as follows. First, add a turn-off commutating circuit across each thyristor. The system would then function to deliver turn-off control triggers in the same manner as it delivers turn-on control triggers to the thyristors. Typical technology in the turn-off area functions to discharge a capacitor (part of an energized *LC* circuit) by means of a small auxiliary thyristor. This discharge diverts the load current of the main thyristor and reverse biases it into a cutoff state. Second, add a shunt diode across each thyristor to permit reverse current conduction. These diodes provide energy discharge paths for stored energy (e.g., energy stored in power transformer leakage inductances) when the turn-off circuit functions. Third, add a prime line disconnect circuit to isolate the public utility lines from the UPS during the duration of the commercial power outage. Fourth, add appropriate control circuits that function to provide (a) the necessary turn-on and turn-off triggers for the thyristors, (b) the prime line disconnect function when an outage occurs, and (c) proper synchronization of the overall system so that the prime line disconnect circuit will have a precisely coordinated reclosure when commercial power again becomes available.

Network Diagram

The network diagram for a six-pulse bridge UPS is shown in Figure 8-23. It includes (1) six thyristor main switches, (2) six turn-off commutating circuits (EC) for control of the main switches, (3) six shunt diodes across the main switches, (4) a 1:1 ratio three-phase power transformer with wye-connected primary windings and delta-connected secondary windings, (5) a balanced three-phase resistance load connected across the primary windings, (6) an electronic switch (ECB) to isolate the normal utility source during a power outage, and (7) a battery dc source (V_{DC}), dc circuit inductance (L), and dc circuit resistance (R_A).

Triggering Sequence

A triggering sequence table for the thyristors and EC circuits—over the time period (CP) required to develop one cycle of UPS waveform—is shown in Table 8-3. The symbol TH_1 designates firing of the power transistor associated with current path I_1 on the diagram, symbol EC_1 designates firing of the turn-off circuit for TH_1, and so on. Assume that firing delay $T_F = 0$.

System Equation Rationale

Based on the system rationale above, let us proceed by first, assuming that ECB is open and then developing a set of elementary UPS equations for waveform time period $T = 0$ to $T = CP/6$. Note that in Fig. 8-23, the equivalent value of each primary load resistor R_L is reflected back to the secondary via a 1:1 transformer ratio. Hence from an analysis point of view, we will simply use resistance R_L connected in series with transformer leakage inductance L_L to replace each of the delta secondary circuits shown.

TABLE 8-3 TRIGGERING SEQUENCE FOR FIGURE 8-23

Time	TH_1	TH_2	TH_3	TH_4	TH_5	TH_6	EC_1	EC_2	EC_3	EC_4	EC_5	EC_6
0	×	×				×				×		
CP/6	×				×	×		×				
CP/3	×		×		×							×
CP/2			×	×	×		×					
2CP/3		×	×	×							×	
5CP/6		×		×		×			×			

Next, let us analyze the circuit with V_{DC} acting as the system power source. Three equations are involved. The first equation concerns current I_4 (i.e., according to Table 8-3, TH_4 must be cut off by triggering EC_4). When this path is open, current I_2, flowing in the L_L and R_L elements of delta leg 2, decays to zero through a short-circuit path consisting of TH_2 and the shunt diode across TH_1. The equation is simply

$$D(2) = -RL\ X(2)/LL$$

The second equation concerns initiation of current I_1. To simplify the analysis program, in the following equations let us assume that at the same instant EC_4 turns off TH_4 and another trigger turns on TH_1 as shown in Table 8-3 (i.e., practical circuits usually introduce a short TH_4 recovery time delay before TH_1 is turned on). The loop equation, via the dc source path, is

$$D(1) = (VDC - (RL + RA)\ X(1))/(L + LL)$$

The third equation, defining I_3, can be written in terms of the voltage loop around the paths involving delta legs 1 and 3 as follows:

$$D(3) = (RL\ (X(1) - X(3)) + LL\ D(1))/LL$$

The UPS must also possess ac voltage-level adjustment capabilities. One way we can accomplish this is by introducing a time delay T_F during the initial part of the analysis period. This will delay the start of current buildup in TH_1. In addition, EC_2 is triggered to cut off current in TH_2. During delay period TF the equations are, simply,

$$D(1) = 0$$
$$D(2) = -RL\ X(2)/LL$$
$$D(3) = -RL\ X(3)/LL$$

At the conclusion of delay T_F both TH_1 and TH_2 are turned on, and system operation returns to the initial set of three equations.

What major changes should we make in the program of Appendix 18E to modify it for UPS analyses? First, remove all sine-wave-related equations. Introduce instead V_1, V_2, and V_3 equations based on the voltage drops across the R_L elements. Second, introduce a set of statements, based on Table 8-3, to direct the program to successive analysis time segments of the UPS output waveform. Third, develop equation sets similar to those developed above for each of the remaining operating segments. Introduce them into the program. Fourth, simplify the tabular output format of the program. Following a printout of an initial cycle of tabular information, automatically shift to a plotting function, and display the line-to-line ac voltage for one phase.

Analysis Results

Appendix 18U has been modified in the manner above. The parameters of the system are fundamentally the same as in Appendix 18E. New initial conditions have been introduced to establish the UPS as a system that has approximately 75 kVA at 208-V delta/120-V wye capabilities. TF is initially set at zero in the program.

Now load Appendix 18U and request a tabular output display. A printout similar to that of Figure 8-24 should result. This should be followed by a plot similar to Figure 8-25. Note that for the initial conditions selected:

1. Figure 8-24 shows that the average rms value of the three line-to-neutral voltages is somewhat in excess of 120 V and that the average rms value of the three line-to-line voltages is somewhat in excess of 208 V. The initial conditions were selected to ensure that the regulation function possessed a reasonable operating zone. For example, the excess voltage capability accommodates situations where reduced values of V_{DC} are available as the battery bank discharges.

2. Figure 8-25 illustrates that the ac waveform is symmetrical. In addition, it can be seen that the leakage inductances of the power transformer introduce noticeable exponential rise and fall times to the waveform.

3. In this analysis, L has been set to zero, but if a finite value were introduced, the exponential times of Fig. 8-25 would increase. Similarly, if an inductive component were added to each R_L (i.e., to represent a more typical system load), an increase in the rise and fall times would likewise result.

4. In addition to the plot, a tabular value of line-to-line voltage is printed out along the edge of Fig. 8-25. If a harmonic analysis is required, these values could be used as input data for Appendix 7.

Output Voltage Control Analysis

To illustrate how voltage control can be obtained with the UPS delay function, we can modify the value of T_F in program line 13 to 0.0005. Another run shows the following comparison between the two cases:

Case 1: Line to neutral = 130.71 V; line to line = 226.40 V

Case 2: Line to neutral = 122.25 V: line to line = 211.90 V

Tabular Response For An Uninterruptable 6 Pulse System

Sec.	TS	V1	V2	V3	I(1)	I(2)	I(3)	I(4)	I(5)	I(6)	VDC	X(1)	x(2)	X(3)
.00000	1	0	162	-162	0	281	0	0	0	281	176	0	281	-281
.00020	1	92	74	-167	160	289	0	0	0	450	176	160	129	-289
.00040	1	132	34	-166	229	288	0	0	0	518	176	229	59	-288
.00060	1	149	16	-165	259	286	0	0	0	545	176	259	27	-286
.00080	1	156	7	-164	272	284	0	0	0	556	176	272	12	-284
.00100	1	160	3	-163	277	283	0	0	0	560	176	277	6	-283
.00120	1	161	2	-162	279	282	0	0	0	561	176	279	3	-282
.00140	1	162	1	-162	280	282	0	0	0	562	176	280	1	-282
.00160	1	162	0	-162	281	281	0	0	0	562	176	281	1	-281
.00180	1	162	0	-162	281	281	0	0	0	562	176	281	0	-281
.00200	1	162	0	-162	281	281	0	0	0	562	176	281	0	-281
.00220	1	162	0	-162	281	281	0	0	0	562	176	281	0	-281
.00240	1	162	0	-162	281	281	0	0	0	562	176	281	0	-281
.00260	1	162	0	-162	281	281	0	0	0	562	176	281	0	-281
.00280	3	166	-56	-110	384	0	0	0	97	287	176	287	-97	-190
.00300	3	167	-116	-50	492	0	0	0	202	289	176	289	-202	-87
.00320	3	165	-142	-23	534	0	0	0	247	287	176	287	-247	-40
.00340	3	164	-154	-11	552	0	0	0	267	285	176	285	-267	-18
.00360	3	163	-158	-5	558	0	0	0	275	283	176	283	-275	-8
.00380	3	163	-160	-2	561	0	0	0	278	282	176	282	-278	-4
.00400	3	162	-161	-1	562	0	0	0	280	282	176	282	-280	-2
.00420	3	162	-162	-0	562	0	0	0	281	281	176	281	-281	-1
.00440	3	162	-162	-0	562	0	0	0	281	281	176	281	-281	-0
.00460	3	162	-162	-0	562	0	0	0	281	281	176	281	-281	-0
.00480	3	162	-162	-0	562	0	0	0	281	281	176	281	-281	-0
.00500	3	162	-162	-0	562	0	0	0	281	281	176	281	-281	-0
.00520	3	162	-162	-0	562	0	0	0	281	281	176	281	-281	-0
.00540	5	133	-164	31	285	0	54	0	339	0	176	231	-285	54
.00560	5	61	-167	106	290	0	183	0	473	0	176	106	-290	183
.00580	5	28	-166	138	288	0	239	0	527	0	176	49	-288	239
.00600	5	13	-164	152	285	0	263	0	549	0	176	22	-285	263
.00620	5	6	-163	157	284	0	273	0	557	0	176	10	-284	273
.00640	5	3	-163	160	283	0	278	0	560	0	176	5	-283	278
.00660	5	1	-162	161	282	0	280	0	562	0	176	2	-282	280
.00680	5	1	-162	162	282	0	281	0	562	0	176	1	-282	281
.00700	5	0	-162	162	281	0	281	0	562	0	176	0	-281	281
.00720	5	0	-162	162	281	0	281	0	562	0	176	0	-281	281
.00740	5	0	-162	162	281	0	281	0	562	0	176	0	-281	281
.00760	5	0	-162	162	281	0	281	0	562	0	176	0	-281	281
.00780	5	0	-162	162	281	0	281	0	562	0	176	0	-281	281
.00800	5	0	-162	162	281	0	281	0	562	0	176	0	-281	281

Figure 8-24 Appendix 18U analysis of a nonsynchronous three-phase bridge UPS.

Figure 8-24 Appendix 18U analysis of a nonsynchronous three-phase bridge UPS (*cont.*).

```
.00820  7   -76   -90  166   0    0  421  132  289   0  176  -132  -157   289
.00840  7  -125   -41  166   0    0  506  217  289   0  176  -217   -72   289
.00860  7  -146   -19  165   0    0  540  254  287   0  176  -254   -33   287
.00880  7  -155    -9  164   0    0  554  269  284   0  176  -269   -15   284
.00900  7  -159    -4  163   0    0  559  276  283   0  176  -276    -7   283
.00920  7  -161    -2  163   0    0  561  279  282   0  176  -279    -3   282
.00940  7  -161    -1  162   0    0  562  280  282   0  176  -280    -1   282
.00960  7  -162    -0  162   0    0  562  281  281   0  176  -281    -1   281
.00980  7  -162    -0  162   0    0  562  281  281   0  176  -281    -0   281
.01000  7  -162    -0  162   0    0  562  281  281   0  176  -281    -0   281
.01020  7  -162    -0  162   0    0  562  281  281   0  176  -281    -0   281
.01040  7  -162    -0  162   0    0  562  281  281   0  176  -281    -0   281
.01060  7  -162    -0  162   0    0  562  281  281   0  176  -281    -0   281
.01080  9  -166    56  110   0   97  287  384    0   0  176  -287    97   190
.01100  9  -167   116   50   0  202  289  492    0   0  176  -289   202    87
.01120  9  -165   142   23   0  247  287  534    0   0  176  -287   247    40
.01140  9  -164   154   11   0  267  285  552    0   0  176  -285   267    18
.01160  9  -163   158    5   0  275  283  558    0   0  176  -283   275     8
.01180  9  -163   160    2   0  278  282  561    0   0  176  -282   278     4
.01200  9  -162   161    1   0  280  282  562    0   0  176  -282   280     2
.01220  9  -162   162    0   0  281  281  562    0   0  176  -281   281     1
.01240  9  -162   162    0   0  281  281  562    0   0  176  -281   281     0
.01260  9  -162   162    0   0  281  281  562    0   0  176  -281   281     0
.01280  9  -162   162    0   0  281  281  562    0   0  176  -281   281     0
.01300  9  -162   162    0   0  281  281  562    0   0  176  -281   281     0
.01320  9  -162   162    0   0  281  281  562    0   0  176  -281   281     0
.01340 11  -133   164  -31   0  339    0  285    0  54  176  -231   285   -54
.01360 11   -61   167 -106   0  473    0  290    0 183  176  -106   290  -183
.01380 11   -28   166 -138   0  527    0  288    0 239  176   -49   288  -239
.01400 11   -13   164 -152   0  549    0  285    0 263  176   -22   285  -263
.01420 11    -6   163 -157   0  557    0  284    0 273  176   -10   284  -273
.01440 11    -3   163 -160   0  560    0  283    0 278  176    -5   283  -278
.01460 11    -1   162 -161   0  562    0  282    0 280  176    -2   282  -280
.01480 11    -1   162 -162   0  562    0  282    0 281  176    -1   282  -281
.01500 11    -0   162 -162   0  562    0  281    0 281  176    -0   281  -281
.01520 11    -0   162 -162   0  562    0  281    0 281  176    -0   281  -281
.01540 11    -0   162 -162   0  562    0  281    0 281  176    -0   281  -281
.01560 11    -0   162 -162   0  562    0  281    0 281  176    -0   281  -281
.01580 11    -0   162 -162   0  562    0  281    0 281  176    -0   281  -281
.01600 11    -0   162 -162   0  562    0  281    0 281  176    -0   281  -281
```

Line To Neut. RMS: V1 = 129.733   V2 = 131.2293   V3 = 131.1759   VLNavg = 130.7127
Line To Line RMS: V13= 225.5091   V21= 225.6026   V32= 228.0917   VLLavg = 226.4011

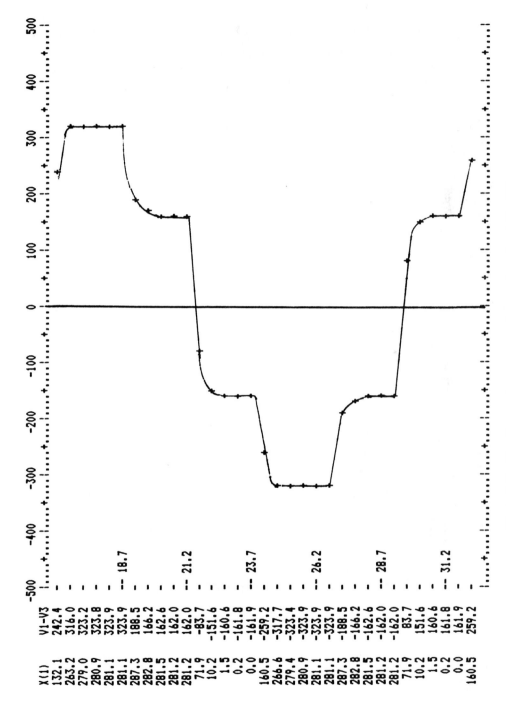

Figure 8-25 Line–line ac voltage on nonsynchronous three-phase bridge UPS.

Chapter 9

Low-Level Pulsing Techniques

Two fundamental approaches are possible: low-level pulsing and high-level pulsing. Selection of a pulsing approach requires investigation and definition of a number of system-related considerations. These include:

1. The type of load that is to be pulsed. For example, is it a solid state, a linear beam, a crossed field, a magnetron, and so on, approach?

2. The nature of the pulses to be generated. For example, is a train of pulses of constant time duration repeated at a constant pulse repetition rate required, or is a specialized train of pulses having variable durations and/or variable interpulse intervals required?

3. RF generator stability as a function of pulse generator stability. For example, in the case of a pulsed amplifier, what number of degrees of phase deviation will be introduced in the RF output wave relative to RF exciting wave when the dc applied to the RF generator's electrodes vary as a function of time during each pulse and/or from pulse to pulse?

4. The availability of inherent pulsing electrodes within the RF generator. For example, if a linear beam tube is to be pulsed, does it have a pulsing electrode or must it be beam pulsed?

In some cases, consideration of the above and other RF system-related factors will lead to the selection or a low-level pulser design approach. In Chapter 10 we review applications where consideration of the factors led to the selection of high-level pulsing configurations. For either the low- or high-level pulsing approach, *transient analyses* are essential.

Analysis of Low-Level Pulsing Systems

In this chapter low-level pulsing is intended to indicate the power level at which the pulse generating elements function. The RF generators themselves will usually be operating at high RF power output levels. Two specific arrangements will be used to illustrate low-level pulsing:

1. A solid-state phased array system wherein a large number of solid-state RF amplifiers have their output pulses controlled by externally generated low-level RF drive pulses supplied to each amplifier. However, all amplifiers will have their electrode voltages supplied from a single relatively large dc power source.

2. A high-powered traveling-wave-tube amplifier system that has its power amplification capabilities activated by low-level gate pulses applied to its beam control grid. Its RF generating capabilities are activated by low-level externally generated RF pulses applied to its RF input port (during the beam-activated period of the grid pulse). The beam accelerating voltage for the TWT is typically supplied from a high-power-level dc voltage source.

The common element in each of these systems is the *high-level dc source*, which for analysis purposes can be viewed as a dc energy source connected to a transient load. Variations in the amplitude of the voltage supplied by the source's storage elements are largely responsible for the phase pushing produced in either RF generation system.

9-1 PHASED ARRAY PULSING SYSTEM

Figure 9-1a shows the block diagram for the small phased array transmitter and radiating system we discussed in Chapter 6. In this example, let us assume that the eight solid-state RF amplifier assemblies have (1) their dc voltages supplied from a 50-kHz dc–dc power source similar to the one we analyzed in Chapter 8; (2) an electrode operating voltage level of 50 V dc; (3) their interpulse voltage held within 0.5 V during each pulse; (4) a peak pulse current requirement of 125 A (for each RF amplifier); (5) system operation at 10-μs pulse duration and 10-kHz pulse repetition rate; and (6) an average 1.52-m (5-ft) physical distance between the 50-V electrode source and each amplifier.

System Element Selection

For our initial analysis, the circuit configuration shown in Fig. 9-1b will be used. The dc–dc power supply of Appendix 19 will require the following modifications. Transformer

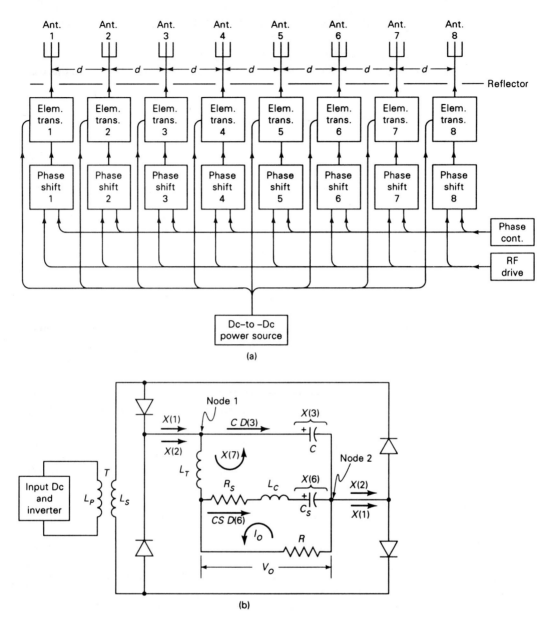

Figure 9-1 Phased array dc power/pulsed energy source: (a) phase array block diagram; (b) network diagram.

T will require a step-down ratio of approximately 8:1. This modification is made by dividing the present secondary inductance L_S by the turns ratio squared (i.e., $L_S = 0.005/64 = 7.8125 \times 10^{-5}$ H). Since a turn ratio of somewhat less than 8:1 is required, the 500-V source is controlled by adjustment of the input voltage to the system. In this case reducing the previous 500-V input level to 460 V provides the approximate output voltage level needed.

Capacitor C_S is typically determined by using the differential equation

$$C_S = I_P \frac{dt}{dv} \qquad (9\text{-}1)$$

where I_P is the peak pulse current (i.e., 1000 A for the eight amplifier assemblies) dt is the pulse duration of 10×10^{-6} s, and dv is the desired pulse regulation of 0.5 V. When these values are substituted in Eq. (9-1), the approximate value of C_S is 0.02 F.

Resistor R_S in Fig. 9-1b represents the equivalent series resistance (ESR) of the capacitor. High-capacity storage capacitors are typically of the aluminum electrolytic type. Capacitor technical literature defines representative values of ESR. The ESR voltage drop results in reduced phased array amplifier voltage, and ESR power dissipation heating limits the amount of pulse energy that can be obtained from a capacitor of given size. The value of R_S for our representative capacitor is 0.006 Ω.

Supply output capacitor C is assigned a relatively low value of 100 μF. Its primary function is to maintain a low ripple level from the rectified output of the 50-kHz inverter. A low ripple level assists in obtaining reasonably smooth power supply switching and rectifying element current waveforms.

Inductance L_T defines the inductance of the physical connections between the common power supply and the RF amplifiers installed on the array antenna system. For our average connection lengths, a value of 1 μH has been assigned.

Resistor R represents the dc load presented by the amplifiers (dc supply voltage/peak pulse current drawn by eight amplifier assemblies = 0.05 Ω). When the amplifiers are in their nonpulsed mode, R is assumed to have a bleeder resistance value of 100 Ω.

Appendix 21 is used for this analysis. It lists the system element revisions above between lines 60 and 66.

Initial Analysis Program

We first have to review the equations originally developed for the 50-kHZ inverter in Appendix 19. The inverter, transformer, and rectifier elements of the power supply will operate as before; hence the equations listed on program lines 1101–1104, 1111–1114, 1121–1124, and 1131–1134 can be used without change.

The load pulsing portion of the circuit requires reference to the two current summation nodes and the two meshes shown on Fig. 9-1b. If the currents at node 1 are summed,

$$C\,D(3) + X(7) = X(1) + X(2)$$

In program format this becomes

$$D\,(3) = (X(1) + X(2) - X(7))/C \qquad (9\text{-}2)$$

The mesh voltage equation around the I_O loop can be written as

$$R\ IO - X(6) - RS\ CS\ D(6) = 0$$

In program format this becomes

$$IO = (X(6) + RS\ CS\ D(6))/R \qquad (9\text{-}3)$$

If a summation is made at node 2 and all terms are multiplied by R, the result is

$$R\ IO + R\ C\ D(3) + R\ CS\ D(6) = R\ (X(1) + X(2))$$

When the equivalent value of $X(1) + X(2)$ is substituted, this becomes

$$R\ IO - R\ X(7) - R\ CS\ D(6) = 0$$

If this equation is subtracted from Eq. (9-3) and the result is put in program format, we have

$$D(6) = (R\ X(7) - X(6))/((R + RS)\ CS) \qquad (9\text{-}4)$$

The voltages around mesh 7 are

$$LT\ D(7) + RS\ CS\ D(6) + X(6) - X(3)$$

In program format this becomes

$$D(7) = (X(3) - X(6) + RS\ CS\ D(6))/LT \qquad (9\text{-}5)$$

Equations (9-2), (9-4), and (9-5) have been entered as program lines 1106, 1116, 1126, and 1136 of Appendix 21. Since we are now dealing with seven differential equations, N in line 30 becomes 7. A revised version of Eq. (9-3) has been used in lines 2010 and 2012 of the print subroutine.

Initial Analysis Results

The objective of the initial analysis is to determine the approximate steady-state operating conditions for the system and to establish the value of rms current associated with the energy storage elements. A set of approximate initial conditions have therefore been entered in lines 50–59. Line 36 has been set to allow six pulse cycles to elapse before these steady-state values are assessed.

If Appendix 21 is loaded and executed for a series of six pulses the analysis will show the following. *First*: V_O during the sixth pulse period has a droop value of somewhat over 0.6 V, and its average voltage is 49.42 V. *Second*: During the sixth pulse period, the current discharge level from CS has a value of approximately 910 A. If we assume that this is a rectangular current, its rms value would be $I_{pk}/DU = 910/\sqrt{0.1} = 288$ A, a power loss of around 498 W. This is in excess of representative 0.02-F, 50-V storage capacitor ratings. A requirement for a multiple-capacitor arrangement is therefore indicated. *Third*: If the 100 printout values for sixth-pulse-period voltage $X(3)$ are averaged, a figure of 54.946 V results—somewhat in excess of the design value of 50 V. *Fourth*: If the 100 printout values for sixth-pulse-period current $X(7)$ are averaged, a figure of 98.683 A results—somewhat below the design value of 100 A. *Fifth*: If the power product of the average values of the last two steps is calculated, a figure of 5422 W results—

about 10% in excess of the design value of 5000 W. *Sixth*: If the average power delivered to the RF amplifiers is calculated (i.e., using the average load voltage during the pulse squared divided by the pulsed load resistance), its value will be about 4885 W. By subtracting step 6 from step 5, we obtain 537 W, a more inclusive capacitor heating loss than the step 2 figure.

Revised System Configuration

The preceding analysis is for an idealized system. To demonstrate this, a plot of pulse voltage can be run if the value of T_2 on program line 34 is changed to 2×10^{-8}, T_3 on line 34 is changed to 12×10^{-6}, a <RUN> command is issued, and a P is entered to answer the plot question, and Fig. 9-2 will result. It shows that the voltage instantaneously drops from its initial dc value to its pulse voltage value at $T = 0$s. Similarly, the voltage instantaneously returns to its dc value at $T = 10$ μs. This is because the model treated the RF amplifiers as though they were ideal resistance loads. In fact, they have both inductive and capacitive components associated with their resistance values. Furthermore, any actual storage capacitor also contains internal inductance as well as its ESR. The next analysis will incorporate these inductive and capacitive elements into the pulsed load circuit.

We noted that the rms current associated with the storage capacitor was too great for a single 0.02-F unit. Assume that individual capacitors, located physically as close as practical to each of the eight RF amplifiers, are used. Even then an internal inductance on the order of 1 nH will be introduced into the composite pulse circuit. This value will be used in our model.

Now consider the RF amplifiers. Typically, a group of bypass capacitors (i.e., comprised of units encompassing a significant range of capacitance values) are used with the stages of each amplifier assembly to ensure the overall amplifier chain's stability. Assume that for the composite pulse circuit this capacitance has a value of 1 μF in our model.

The inverter and transformer circuit used for our revised circuit configuration will remain the same as in the previous analysis. Figure 9-3 illustrates how the new pulse circuit components have been added to previous model's diagram as elements L_C and C_B. Modified program equations will now have to be established for $D(6)$, $D(7)$, $D(8)$, and $D(9)$.

Revised System Equations

If a summation is performed at node B of the revised system of Fig. 9-3,

$$CS\ D(6) + X(8) = X(7) \tag{9-6}$$

In program format this is

$$D(6) = (X(7) - X(8))/CS \tag{9-7}$$

If the voltage equation is written around mesh 7, we have

$$(LT + LC)\ D(7) + X(6) + RS\ X(7) - LC\ D(8) - RS\ X(8) - X(3) = 0$$

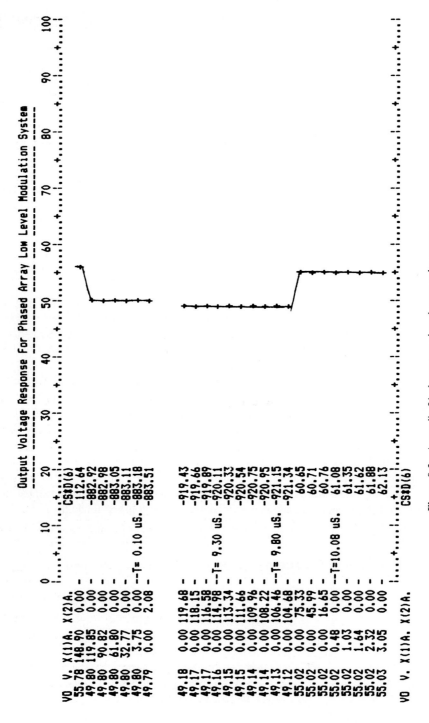

Figure 9-2 Appendix 21 elementary phased array voltage response.

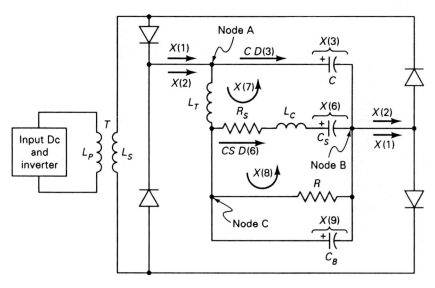

Figure 9-3 Phased array dc power/pulsed source: revised.

Around mesh 8 we can write

$$-LC\, D(7) - X(6) - RS\, X(7) + LC\, D(8) + RS\, X(8) + X(9) = 0$$

If these two equations are added and put in program format,

$$D(7) = (X(3) - X(9))/LT \tag{9-8}$$

By substitution we can obtain

$$D(8) = (X(6) + LC\, D(7) + RS\, X(7) - RS\, X(8) - X(9))/LC \tag{9-9}$$

A summation at node C gives

$$D(9) = (X(8) - X(9)/R)/CB \tag{9-10}$$

Appendix 22 lists the program for the modified system. Lines 1106, 1116, 1126, and 1136 show the altered equations for $D(6)$–$D(9)$. Because of the revised circuit configuration, numerical analysis stability requires that smaller integration steps be taken in lines 34, 1111, and 1131. Otherwise, the program is almost identical to that of Appendix 21 except that line 30 has been changed to N = 9 to accommodate the revised equations, and the new load circuit elements that have been added to the listing.

Revised Analysis Results

If you load Appendix 22 into your computer, issue a <RUN> command, and then enter P for a plot, a display similar to the abbreviated version shown in Fig. 9-4 will develop on your printer. Its significant features are the following. *First*: A clearly defined oscillatory response is present when the pulse starts at $T = 0$ s. Somewhat over 0.3 μs is required before its effects are damped out. This oscillation degrades each RF amplifier's performance capabilities during this period. After the decay is complete, the rest of the pulse is

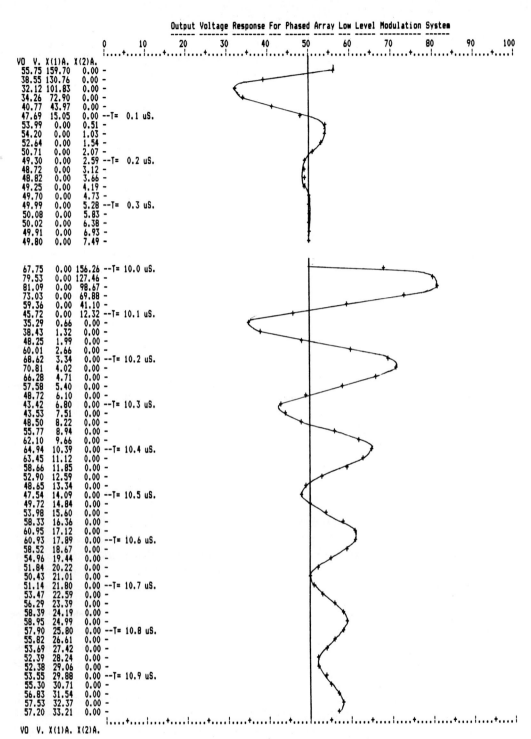

Figure 9-4 Appendix 22 modified phased array voltage response.

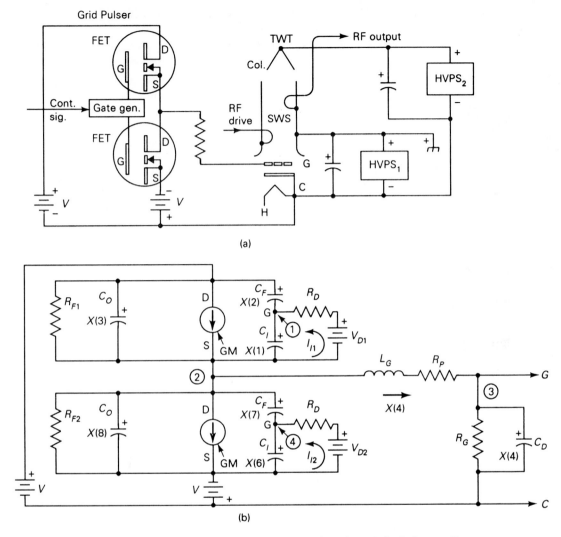

Figure 9-5 Traveling-wave-tube grid pulser configuration: (a) circuit diagram; (b) network diagram.

substantially identical to that of Appendix 21. *Second*: Based on the turn-on characteristics observed, similar oscillations are inevitable when the RF drive to the amplifiers is removed at $T = 10$ μs. Their amplitudes are significantly higher because the circuit now has less damping present. Voltage peaks of greatly increased amplitude will occur when critical combinations of L_C and C_B are present. Excessive oscillation amplitudes should definitely be reduced, perhaps by the addition of some nonlinear resistance across the loads, to minimize any potential damage to the RF amplifier transistors. The turn-off oscillation clearly does not affect performance since the system is no longer functioning as an RF amplifier when its RF drive is removed.

9-2 GRID PULSED TRAVELING-WAVE AMPLIFIER TUBE SYSTEM

Figure 9-5a shows a simplified circuit diagram for a widely used linear beam tube RF power amplifier arrangement. In this example a depressed collector, grid-controlled traveling-wave tube is used. $HVPS_1$ provides a high voltage between cathode C and the slow-wave structure (SWS) to accelerate the beam through the structure (as described in Chapter 6). This supply operates at a relatively low current level.

$HVPS_2$ delivers a lower level of high voltage between the cathode and the collector. This supply operates at a relatively high current level. Because of the depressed level of $HVPS_2$, the input power required to operate the tube is reduced, and the tube's efficiency is significantly improved compared to that obtained with nondepressed collector operation (where the collector operates at ground potential).

Typical TWTs of recent design use *shadow grids* to control the linear beam current. Representative grid voltage characteristics, for amplifiers rated in the 50- to 100-kW peak RF power output level range, show that the grid should be pulsed to a value of around 500 V negative with respect to the cathode in order to cut off beam current, and that it should be pulsed to a value around 500 V positive with respect to the cathode to turn on the beam current fully.

An outstanding advantage of shadow grids is that they have extremely low values of interception current (in contrast with earlier grid pulsed tubes, which often intercepted as much as 10% of the beam current). As a result, a relatively simple single-pole single-throw switch type of grid pulser is often used.

Assume that we have a specification requirement that defines a shadow grid pulsed TWT RF generating system. The system must provide 0.25-μs RF pulse durations. Our objective is to investigate the performance capabilities of one of the grid pulser design candidates. This approach is shown in Fig. 9-5a. Two FETs are configured to provide single-pole single-throw grid switching action. The pulser functions in response to commands from the control signal line.

The pulser and its associated circuits are physically located on a floating platform attached to the cathode of the TWT. Since the slow-wave structure of the tube is designed for operation at ground potential, the cathode typically operates in the region of 25 to 40 kV below ground reference potential.

The low-level control signals for the pulser frequently take the form of on and off triggers that are generated at ground potential. The triggers must therefore be delivered to the pulser through isolation elements (e.g., pulse transformers, laser beams, etc.) that are capable of providing the high-voltage insulation required between the grounded signal source and the floating platform.

If we assume that the TWT is initially turned off, the lower FET will have its gate turned on. The TWT's grid will then be effectively connected to the negative battery between the FET's source and the TWT's cathode.

When an on control trigger actuates the gate generator, it pulses on the gate of the upper FET and pulses off the gate of the lower FET. This causes the upper FET to turn on, and the TWT's grid will swing from its initial negative level to the effective level of the positive battery connected between the TWT's cathode and the drain of the upper FET.

An off gate causes the reverse action to occur in the two FETs, and the TWT will be turned off again. For this analysis the voltage level of each battery will be considered to be 500 V.

System Element Definition

Now let us proceed to define the characteristics of the candidate design for the grid pulser. Figure 9-7b shows a network diagram that defines some of the major pulser elements. Each FET circuit includes the elements labeled C_I, C_F, C_O, R_{F1} or R_{F2}, a current generator, and a drive source V_{D1} or V_{D2} along with its impedance R_D. Some details are as follows.

FET interelectode capacitance C_I exists between the gate and source electrodes. A check of 1000-V FET literature (the FETs must be capable of withstanding 1000 V between their drain and source elements) indicated representative C_I values of 4000 pF for these devices. The value of C_F, the FET's gate-to-drain transfer capacitance, lies in a representative region around 70 pF. C_O, the drain-to-source capacitance, falls within a range of values in the 250-pF region. The representative figures noted above are the values that will be used in our analysis.

Resistors R_{F1} and R_{F2} represent the reverse diode connected between the source and drain in power FET devices. Assume that when the drain is positive with respect to the source the resistance value is 1 MΩ, and when the polarity reverses the value is 0.25 Ω.

The current generators characterize the transconductance capabilities of the FET devices. The voltage across C_I controls the current delivered by the generator. Transconductance, G_M, has been assigned a representative value of 3.5 S.

Voltages V_{D1} and V_{D2} represent the source voltage levels of the gate generator. When an ON control signal trigger is applied, V_{D1} is programmed to rise in a linear manner from 0 to 5 V over a period of 40 n, and then hold that value until an OFF command is applied. Similarly, V_{D2} is programed to fall from 5 to 0 V during the same time period and hold that value. These two voltage levels reverse in identical fashion when an OFF trigger is applied. Because of the high value of the C_I elements, quite low driver source impedances, R_D, of 2.5 Ω are used to achieve acceptable short durations of gate pulse rise and fall times.

Capacitor C_D represents the grid-to-cathode capacitance associated with the TWT and its connecting leads. A value of 100 pF has been assigned to this capacitance.

Resistor R_G represents the interception current element of the TWT grid. A representative value of 100,000 Ω will be used.

Inductance L_G defines the connecting lead inductance between the pulser and the TWT grid. A value of 0.5 μH has been assigned to this lead inductance.

Resistor R_P has been added in series with the grid circuit. Even though the rise and fall times of the pulser are not instantaneous, the grid circuit elements can introduce superimposed oscillations of an objectionable magnitude on the flat top of the turned-on pulse—unless damping is introduced. A damping value of 150 Ω has been used in this application.

System Equations

With this background established, our next step concerns development of the analysis equations. The mesh associated with grid circuit current $X(4)$ can be defined as follows:

$$X(3) + LG\ D(4) + RP\ X(4) + X(5) - V = 0$$

In program format this becomes

$$D(4) = (V - X(3) - X(5) - RP\ X(4))/LG \tag{9-11}$$

If a current summation is made at node 3,

$$CG\ D(5) + X(5)/RG = X(4)$$

In program format this is

$$D(5) = (X(4) - X(5)/RG)/CG \tag{9-12}$$

For the driver circuit equation around mesh I_{11},

$$RD\ I1 + X(1) - VD1 = 0$$

In program format,

$$I1 = (VD1 - X(1))/RD \tag{9-13}$$

And similarly,

$$I2 = (VD2 - X(6))/RD \tag{9-14}$$

Now consider the mesh comprised of C_I, C_D, and C_O in the upper FET. We can write the derivatives around this loop as

$$D(1) + D(2) = D(3) \tag{9-15}$$

Next note that the sum of $D(3) + D(8)$ must always equal zero—because this voltage sum is connected across two series-connected 500-V batteries. Hence we can write

$$D(3) = -D(8) \tag{9-16}$$

And for the lower FET C_I, C_D, C_O mesh,

$$D(6) + D(7) = -D(3) \tag{9-17}$$

For the current summation at node 1 we have

$$CI\ D(1) - CF\ D(2) = I1 \tag{9-18}$$

And for a similar summation at node 4,

$$CI\ D(6) - CF\ D(7) = I2 \tag{9-19}$$

Finally, consider the current summation at node 2,

$$X(3)/RF1 + CO\ D(3) + GM1\ X(1) + CF\ D(2)$$
$$= X(4) + X(8)/RF2 + CO\ D(8) + GM2\ X(6) + CFF\ D(7) \tag{9-20}$$

This equation can be solved for $D(3)$ if the following preliminary solutions are made: Solve Eqs. (9-18) and (9-15) in terms of $D(2)$ and $D(3)$ to obtain

$$D(2) = (CI\ D(3) - II1)/(CI + CF) \tag{9-21}$$

Solve Eqs. (9-19) and (9-17) in terms of $D(7)$ and $D(3)$ to obtain

$$D(7) = -(CI\ D(3) + II2)/(CI + CF) \tag{9-22}$$

Now substitute Eqs. (9-16), (9-21), and (9-22) into (9-20) to obtain, in program format,

$$D(3) = ((CF/(CI + CF))\ (II1 - II2) - X(3)/RF1 + X(8)/RF2 \\ - GM1\ X(1) + GM2\ X(6) + X(4))/(2CO + 2CI\ CF/(CI + CF)) \tag{9-23}$$

The remaining equations can then be readily expressed as follows:

$$D(2) = (CI\ D(3) - II2)/(CI + CF) \tag{9-24}$$

$$D(1) = D(3) - D(2) \tag{9-25}$$

$$D(7) = -(CI\ D(3) + II2)/(CI + CF) \tag{9-26}$$

$$D(6) = -(D(3) + D(7)) \tag{9-27}$$

$$D(8) = -D(3) \tag{9-28}$$

Analysis Program

Equations (9-11), (9-12), and (9-23)–(9-28) are the complete set of differential equations needed for the analysis. They are listed as lines 1015–1090 of our new analysis program, Appendix 24. Supporting algebraic Eqs. (9-13) and (9-14) are listed as lines 1005 and 1010. Some further statements required for this analysis, via Appendix 24, follow.

Line 20 revises N for this system's eight-differential-equation format. Lines 32–36 are assigned analysis time increments consistent with the program execution requirements for the circuit. Lines 51–62 are the element values developed from the network diagram discussion. Lines 70–77 list the off-mode initial conditions for the system at $T = 0$. Lines 110–116 define the ramped values of V_{D1} and V_{D2} during the FET switching periods.

Lines 152–158 and 192–198 define the transconductance of the FETs as they traverse their drain voltage/drain current/gate voltage performance curves. Idealized characteristics are assumed wherein the FETs will cut off drain current when their gate voltages fall below zero. They will conduct drain currents of $G_{M1}\ X(3)$ or $G_{M2}\ X(6)$ when their values are positive. However, when the drain voltage falls below 10 V, the values of G_{M1} and G_{M2} fall linearly toward zero as the drain voltage approaches zero. If reverse voltage is applied from drain to source, lines 157 and 197 change R_{F1} or R_{F2} from 1 MΩ to 0.25 Ω.

Lines 2012 and 2015 print out each of the X analysis values and the drain current values for each FET when a detailed tabular analysis is requested.

TWT Grid Pulser Integration Time = .00025 Microseconds --

Microsec.	X(1)	X(2)	X(3)	X(4)	X(5)	X(6)	X(7)	X(8)	FET1 Cur.	FET2 Cur.
0.000	0.000	1000.000	1000.000	-0.005	-499.250	5.000	-5.000	0.000	0.000	0.000
0.010	0.884	997.168	998.052	0.001	-499.071	4.116	-2.169	1.948	3.092	2.806
0.020	2.699	986.793	989.492	0.033	-497.176	2.301	8.207	10.507	9.447	8.053
0.030	3.337	917.440	920.777	0.291	-483.551	1.663	77.560	79.222	11.681	5.819
0.040	3.404	828.583	831.986	0.616	-436.453	1.596	166.416	168.012	11.914	5.586
0.050	3.426	739.383	742.809	0.781	-365.052	1.574	255.616	257.190	11.992	5.508
0.060	3.435	650.281	653.717	0.847	-282.807	1.565	344.717	346.282	12.024	5.476
0.070	3.439	561.203	564.642	0.873	-196.352	1.561	433.795	435.356	12.037	5.463
0.080	3.440	472.133	475.574	0.883	-108.328	1.560	522.865	524.424	12.042	5.458
0.090	3.441	383.069	386.510	0.887	-19.729	1.559	611.929	613.488	12.044	5.456
0.100	3.441	294.010	297.451	0.889	69.080	1.559	700.987	702.546	12.045	5.455
0.110	3.441	204.957	208.398	0.891	157.962	1.559	790.040	791.599	12.045	5.455
0.120	3.442	115.909	119.350	0.892	246.868	1.558	879.088	880.647	12.046	5.454
0.130	3.442	26.867	30.308	0.892	335.779	1.558	968.131	969.689	12.046	5.454
0.140	4.197	-1.703	2.494	0.670	418.289	0.803	996.700	997.503	3.663	2.812
0.150	4.683	-3.779	0.904	0.312	465.473	0.317	998.776	999.093	1.482	1.108
0.160	4.876	-4.531	0.345	0.130	485.875	0.124	999.528	999.652	0.589	0.435
0.170	4.951	-4.816	0.135	0.054	494.046	0.049	999.813	999.862	0.234	0.170
0.180	4.981	-4.926	0.054	0.024	497.237	0.019	999.923	999.942	0.095	0.067
0.190	4.993	-4.969	0.023	0.012	498.472	0.007	999.966	999.973	0.041	0.026
0.200	4.997	-4.986	0.011	0.008	498.948	0.003	999.983	999.986	0.020	0.010
0.210	4.999	-4.992	0.006	0.006	499.132	0.001	999.989	999.990	0.011	0.004
0.220	5.000	-4.995	0.005	0.005	499.203	0.000	999.992	999.992	0.008	0.002
0.230	5.000	-4.996	0.004	0.005	499.231	0.000	999.993	999.992	0.007	0.001
0.240	5.000	-4.996	0.004	0.005	499.241	0.000	999.993	999.992	0.006	0.000
0.250	5.000	-4.996	0.003	0.005	499.245	0.000	999.993	999.992	0.006	0.000

0.260	5.000	0.003	0.005	499.247	0.000	999.993	999.992	0.006	0.000
0.270	5.000	0.003	0.005	499.247	0.000	999.993	999.992	0.006	0.000
0.280	5.000	0.003	0.005	499.247	0.000	999.993	999.992	0.006	0.000
0.290	5.000	0.003	0.005	499.247	0.000	999.993	999.992	0.006	0.000
0.300	5.000	0.003	0.005	499.247	0.000	999.993	999.992	0.006	0.000
0.310	5.000	0.003	0.005	499.247	0.000	999.993	999.992	0.006	0.000
0.320	5.000	0.003	0.005	499.247	0.000	999.993	999.992	0.006	0.000
0.330	5.000	0.003	0.005	499.247	0.000	999.993	999.992	0.006	0.000
0.340	5.000	0.003	0.005	499.247	0.000	999.993	999.992	0.006	0.000
0.350	5.000	0.003	0.005	499.247	0.000	999.993	999.992	0.006	0.000
0.360	5.000	0.003	0.005	499.247	0.000	999.993	999.992	0.006	0.000
0.370	5.000	0.003	0.005	499.247	0.000	999.993	999.992	0.006	0.000
0.380	5.000	0.003	0.005	499.247	0.000	999.993	999.992	0.006	0.000
0.390	5.000	0.003	0.005	499.247	0.000	999.993	999.992	0.006	0.000
0.400	5.000	0.003	0.005	499.247	0.000	999.993	999.992	0.006	0.000
0.410	4.154	1.846	-0.001	499.085	0.846	997.304	998.150	2.684	2.961
0.420	2.349	10.050	-0.031	497.267	2.651	987.294	989.945	8.222	9.278
0.430	1.666	77.050	-0.282	484.279	3.334	919.612	922.945	5.831	11.669
0.430	1.666	77.050	-0.282	484.279	3.334	919.612	922.945	5.831	11.669
0.440	1.597	165.782	-0.610	437.997	3.403	830.810	834.213	5.589	11.911
0.450	1.574	254.962	-0.778	367.010	3.426	741.607	745.033	5.509	11.991
0.460	1.565	344.055	-0.846	284.932	3.435	652.504	655.939	5.476	12.024
0.470	1.561	433.129	-0.873	198.540	3.439	563.426	566.865	5.463	12.037
0.480	1.560	522.198	-0.883	110.540	3.440	474.355	477.796	5.458	12.042
0.490	1.559	611.262	-0.887	21.949	3.441	385.291	388.732	5.456	12.044
0.500	1.559	700.320	-0.889	-66.857	3.441	296.232	299.673	5.455	12.045
0.510	1.559	789.373	-0.891	-155.737	3.441	207.179	210.620	5.455	12.045
0.520	1.558	878.421	-0.892	-244.643	3.442	118.131	121.572	5.454	12.046
0.530	1.558	967.463	-0.892	-333.554	3.442	29.088	32.530	5.454	12.046
0.540	0.822	996.610	-0.681	-416.608	4.178	-1.617	2.561	2.877	3.745
0.550	0.324	999.067	-0.319	-464.691	4.676	-3.749	0.927	1.134	1.517

Figure 9-6 Tabular printout for traveling-wave-tube grid pulser.

221

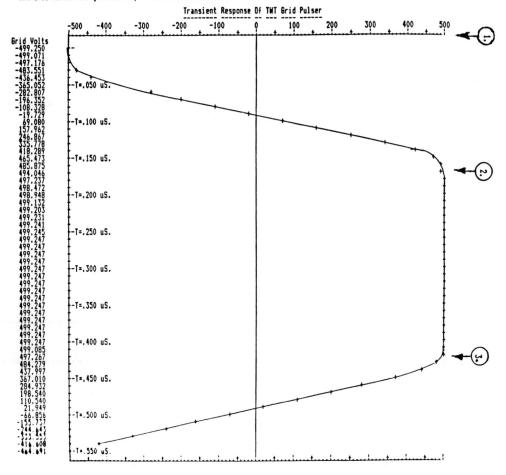

Figure 9-7 Appendix 24 TWT grid pulser voltage response.

Analysis Results

Now load Appendix 24 into your computer and enter a <RUN> command. A tabular listing similar to Fig. 9-6 should develop on your printer. Several points of interest are:

    1. The value of $X(5)$, the grid-to-cathode voltage of the TWT, can be seen to change rather slowly as the ramped gates are applied to the FETs during the first 40 ns of the run. Following this, the voltage changes in 80- to 90-V increments as the grid voltage rises. At about 150 ns into the run as the voltage approaches its flat-top value, the increments become smaller. The grid substantially reaches its final voltage value at 170 ns.

    2. The top remains completely flat until 400 ns into the run. At that time program lines 113 and 116 cause the turn off ramp to execute. The absolute value of the rate of

222   Low–Level Pulsing Techniques   Chap. 9

fall in grid voltage can be seen to be substantially the same as that noted for the rate of rise above.

3. Another point of interest concerns the drain currents. Their maximum values might be expected to be the result of the highest product of the V_D and G_M values (e.g., $5 \times 3.5 = 17.5$ A). However, it can be seen that during either the rise or fall time of the pulse, this current does not exceed about 12 A. On the other hand, the sum of the two FET currents during the major portion of the rise and fall times has a value of 17.5 A. The two drain current values are consequences of complex interactions that occur between of the six interelectrode capacitances (and other circuit operating factors) during the turn-on and turn-off periods.

4. The pulse characteristics of the system can be assessed more easily if a response plot is requested. A <RUN> command followed by a P response is entered into the program, a display similar to Figure 9-7 will develop on your printer. To clarify a point noted previously, three arrows are shown along the time display. Arrow 1 denotes where the command signal requests the pulser system to turn on. Arrow 2 denotes a reasonable point in time at which injection of RF drive into the TWT's input port should occur—to initiate RF power generation by the system. Arrow 3 denotes where two system actions should occur: (1) where the RF drive signal to the TWT should cease; and (2) where the system control signal line should issue a pulser OFF command. Hence the required RF pulse duration of approximately 0.25 µs is generated. For longer or shorter pulse durations, arrow 3 would be moved to an appropriate time relative to arrow 1.

A number of the values selected for this analysis were the result of several trial runs through the program. You are encouraged to try alternate values for some components (e.g., R_D, L_G, C_D, and R_P) so that you can determine the effect of their variations on system performance. If you try to speed up the rise and fall times significantly, some modifications in T_1 and the ramping on V_{D1} and V_{D2} will probably be needed.

Chapter 10

High-Level Pulsing Techniques

High-level pulsers can be considered to fall into either of two fundamental categories: those that use *hard switches* and those that use *soft switches*. In either case the switch(s) directly control pulse modulation within a power electronics system. This is typically accomplished by pulsing the major electrode (e.g., the anode of a magnetron or the collector of a bipolar transistor) of one or more of the RF oscillator and/or RF amplifier stages of the system.

Traditional hard-tube modulators represent the most elementary form of high-level pulsing configurations. The term *hard tube* designates a vacuum tube, used as a series switch, whose open or closed characteristics are controlled by means of a low-level voltage gate applied between its grid and cathode.

The same rationale can be applied to solid-state hard switches. In solid-state hard modulators, appropriate gates are applied to the base/gate of the bipolar/FET transistors used as the pulser's series power switches.

Hard-Tube Modulators

In their traditional form, hard-tube modulators can be described in terms of five functional areas: an energy storage capacitor, a vacuum-tube switching circuit, an RF power generator circuit, a dc power supply (HVPS), and a charging circuit, as shown in the simplified diagram of Fig. 10-1a. During the system's interpulse period the hard tube acts as an open switch. It is cut off by a negative bias applied to its grid.

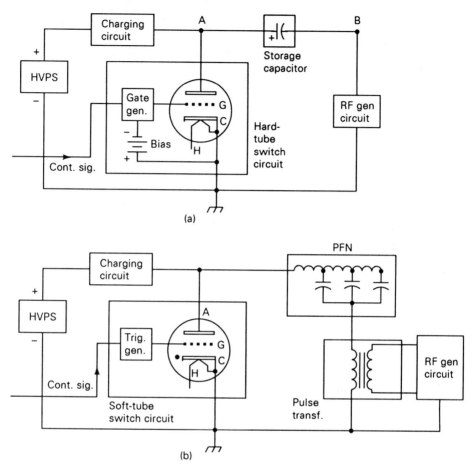

Figure 10-1 Two traditional high-level modulator approaches: (a) simplified hard-tube modulator circuit; (b) simplified soft-tube modulator circuit.

When a pulse is required from the RF generator, a control signal commands the gate generator to inject a positive gate pulse in series with the cutoff bias. This initiates the closing action of the hard-tube switch. In the configuration shown in Fig. 10-1a, when the closing action is complete, a negative high-voltage pulse is developed across the RF generator.

When the specified pulse duration has occurred, the control signal commands the gate generator to initiate the opening action of the hard-tube switch. This opening action is compete when full cutoff bias is applied to the hard tube and when the pulse is terminated by the overall pulser circuit. The system then returns to its interpulse status.

Pulse voltage droop occurs because the energy that is delivered to the RF oscillator/amplifier is removed from the storage capacitor during each pulse. Hence it is necessary to restore the capacitor's lost energy, by means of the power supply recharging path, during the ensuing interpulse period.

If a positive load polarity hard-tube modulator is required for a system, the diagram can readily be modified as follows. The vacuum-tube switching arrangement should have the lead presently attached to ground moved to connect to node B, and the storage capacitor should have the lead presently attached to node B connected to ground instead. The hard-tube modulator is a simple and highly effective arrangement that has been extensively used in radar pulsing systems for many decades.

Soft-Tube Modulators

Traditional soft-tube modulators typically use gas-filled elements, such as *thyratrons* or *ignitrons*, as their switches. *Silicon-controlled rectifiers* are representative of the solid-state approach to circuits that use soft switches. For either the tube of the solid-state case, in its open mode, the switch effectively assumes a high-impedance status. When the switch is triggered, it assumes a very low impedance status.

Soft switches normally possess no direct control means by which they can be returned to their open status. For this reason, they typically operate in conjunction with an arrangement of inductive and capacitive elements known as a pulse-forming network (PFN). The PFN stores only enough energy for a single pulse.

A traditional soft-tube modulator, shown in Fig. 10-1b, can be described in terms of six functional areas: a PFN, a soft-tube switching circuit, a pulse transformer, an RF generator circuit, a dc power supply, and a charging circuit. During the system's interpulse period, the soft switch assumes its high-impedance status. The dc power supply stores energy in the PFN, via the charging circuit, during this period.

When a high-level output pulse is required, a control signal commands the trigger generator to turn on the soft tube. The soft tube then rapidly (typically within a fraction of a microsecond) assumes its very low impedance status. It discharges the energy from the PFN (via the pulse transformer) into the RF generator in the form of a pulse of the desired duration and amplitude.

At the conclusion of the pulse, nearly all of the energy is dissipated, and the soft tube can therefore assume its open mode again. The HVPS and charging circuit then proceed to restore the PFN's energy. Either a positive or a negative load pulse polarity system can be provided by reversing the secondary connections of the pulse transformer.

This circuit is normally referred to as a *line-type* pulser or line-type modulator because the artificial line principles associated with PFNs relate closely to those of transmission lines. Just as with hard-tube modulators, line-type pulsers have been extensively used in radar pulsing applications for many decades.

Both hard-tube and line-type pulsers require transient analyses to evaluate their load pulse response stability capabilities accurately. A representative example of each type is presented in the following sections.

10-1 HARD-SWITCH PULSER FOR A MAGNETRON

In this section we consider the design of an updated version of a hard-tube modulator that will be used to pulse a magnetron oscillator. Historically, hard-switch pulsers for magnetrons were among the first high-level pulsing techniques utilized for radar transmitters during the infancy of radar. Despite all the years of technical progress in the field of electronic RF generators, magnetron oscillators are still widely used today.

Consider a current RF power generator requirement for a 50- to 100-kW superhigh-frequency short pulse radar system that specified only a moderately high level of RF stability. A coaxial magnetron oscillator would certainly be one representative candidate for the application. Furthermore, if the transmitter were required to provide frequent changes in pulse duration, and if in addition, frequent variations in the interpulse periods were required, a hard-switched modulator would clearly represent a logical candidate for the pulser. This analysis is based on a system scenario in which these two candidates have been selected for a new transmitter design.

System Design Rationale

Figure 10-2a shows a circuit diagram for a hard-switch pulser possessing only one tube—the magnetron itself. A representative magnetron for the superhigh-frequency region could have the following major characteristics:

Peak power output	75 kW
Pulse duration	0.08–1.5 µs
Duty	0.0012 maximum
Frequency (band center)	16.25 GHz
Tunability (via a mechanical drive)	250 MHz
Peak anode voltage (typical)	14 kV
Peak anode current (typical)	15 A
Rate of rise of anode voltage	50–160 kV/µs
Frequency pushing factor	150 kHz/A
Frequency pulling factor at 1.5 VSWR	8 MHz
Heater supply (typical)	7.8 V

The hard switch for the candidate design consists of 20 series-connected 1000-V FETs (similar to the type used in the grid pulser analysis in Chapter 9). These elements are configured along the left side of Fig. 10-2a.

Even when a well-matched set of FETs is selected for an application such as this, there are certain to be performance variations between individual units. For this reason, a safety factor of about 33% has been introduced in the drain-to-source voltage ratings (i.e., 20 rather than 15 FETs have been used in the string). In practical circuit applications, voltage divider elements would be connected across each of the FETs to ensure that nearly equal voltages exist across them.

One method of gating 20 switches, each operating with their elements at different voltage levels with respect to ground, can be provided by a toroidal pulse transformer arrangement similar to that shown in the figure. If the FETs are physically configured so that a common single-turn primary winding, L_P, is passed through a small toroidal transformer on each FET assembly (i.e., with an appropriate level of insulation between L_P and LS_1 through LS_{20}), a relatively simple gating system can be implemented. The secondary of each transformer drives its FET via resistors R_1 through R_{20}, in a manner

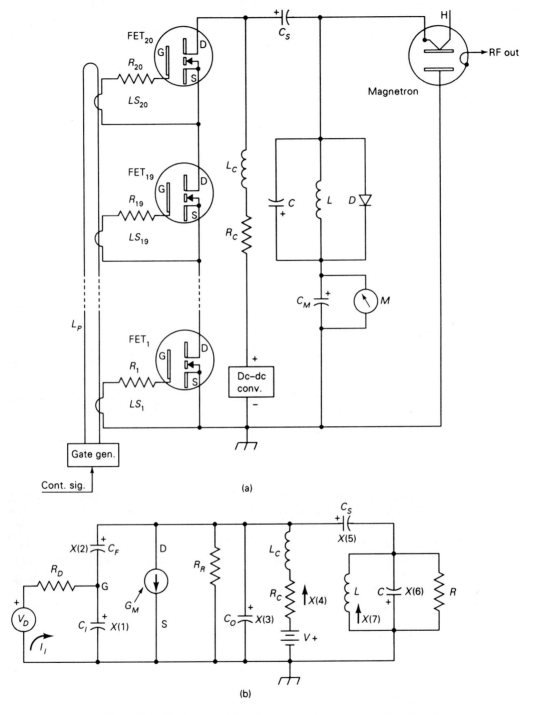

Figure 10-2 Hard-tube modulator for magnetron transmitter: (a) solid-state pulsar circuit for the magnetron oscillator; (b) network diagram for magnetron pulsing system.

similar to the grid pulser arrangement described in the preceding section. A control signal initiates the start and stop action of the gate generator that drives L_P.

A storage capacitor for the system, C_S, should have a capacitance value sufficient to limit the pulse droop to the level set by system requirements. It will be recalled that a magnetron operates in a manner similar to a biased diode; that is, it can be defined as a high-impedance until its operating region is approached—in this case around 14 kV. Above this level it can be considered to be a relatively constant resistance whose value decreases as the voltage across the tube is raised.

A first approximation of the value for C_S can be obtained if the voltage droop is taken as 1 or 2% of the magnetron's operating voltage, the current is taken as the rated operating value for the tube, and the droop time is taken as the maximum pulse duration. These values can then be substituted into Eq. (9-1). A more refined value can be established later after reviewing the analysis results.

The charging circuit for C_S shown in Fig. 10-2a consists essentially of a dc–dc converter similar to the one described in earlier examples (but implemented with a higher-ratio step-up transformer), R_C, L_C, and charging diode D. Other types of high-voltage power supplies could, of course, be used in place of the dc–dc unit.

In any event, the supply must have a dc voltage rating somewhat greater than the operating voltage required by the magnetron. It must have an average dc current rating consistent with system losses (i.e., the losses due to the magnetron's efficiency and the pulser circuit losses).

R_C and L_C serve the primary function of preventing the dc source from being short circuited during the pulse period. Traditionally, R_C and L_C have often been provided by the inherent characteristics of a conventional inductively wound power resistor.

Charging diode D must be rated to withstand the full magnetron voltage; hence for the solid-state rectifier shown, multiple series elements are required, and considerations similar to those noted for the FET string's voltage division properties should be exercised.

Next let us consider the C, L, and D elements. C represents an inescapable amount of stray capacitance associated with the elements that comprise the circuit (e.g., internal anode-to-cathode capacitance in the magnetron, stray capacitance from C_S and its connections to ground, capacitance of the magnetron's filament heating transformer to ground, etc.). Typically, C has a value in the region of 100 pF.

The pulse fall time is a function of the product of the high resistance of the magnetron's biased diode characteristic and C. This can be a significant number of microseconds in duration.

One method of greatly reducing the fall time is to add *tail clipping inductance L* to the circuit. It in conjunction with C forms an oscillatory circuit that causes the waveform to follow a portion of a damped sine-wave path during the period when C is transferring its stored energy into L. Were it not for the presence of the charging diode, this oscillation would continue to result in a high positive voltage swing at the magnetron's cathode after the voltage across C passed through zero. The diode prevents this; hence it is more correctly referred to as a backswing clipper diode rather than a charging diode.

Finally, consider M and C_M. M is a standard milliammeter that provides a qualitative gage of magnetron operation in terms of its average current. C_M serves the dual function of providing a low-impedance return to ground for the tail clipper circuit, and of providing a charge/discharge current averaging device for the milliammeter.

Network Diagram and Analysis Program

Figure 10-2b shows the network diagram to be used for our system analysis. This pulser analysis is developed in Appendix 25, a revised version of the grid pulser program we used in our last analysis. The rationale for the elements used in this program follows.

C_I, C_F, and C_O are interelectrode capacitances of a single equivalent FET that will be used to model the 20-FET series chain. These values were established by dividing each of the single unit values by N_S, the number of switches in series. They are listed as program lines 50–52.

Resistance R_D is the gate circuit resistance, R_R is the leakage/reverse diode resistance of the equivalent FET, and R is the backswing clipper diode forward resistance of the network. The single-unit R_D and R_R values are each multiplied by N_S. These are listed on lines 55 and 56.

Resistance R is the clipper diode resistance and is assumed to have the same value as diode action produced in R_R (when positive magnetron cathode voltage backswing occurs across R). This value is defined initially in program line 58 and redefined in lines 114, 116, and 118 during magnetron conduction.

Transconductance G is the transconductance of the equivalent FET and is defined as the single unit G divided by N_S on line 63.

Voltage V_D is the drive voltage for the equivalent FET and is defined an N_S times the single-FET drive voltage. Its initial value is set at zero in line 70.

The operating conditions assumed for this analysis are that storage capacitor charging has just been completed at $T = 0$, and at that instant the equivalent FET turn-on begins (the initial conditions for the X analysis values in lines 70–76 are also established based on this rationale, and the presence of a HVPS voltage of 15,000 V, listed on line 62). This action takes place in the form of ramped gate pulses in lines 112, 114, 116, and 118. Line 116 defines magnetron resistance during pulse rise time, and 118 defines its resistance during the pulse's RF generation period.

Based on previously described considerations C_S was established as 0.02 µF in line 53, and C as 100 pF in line 54. Inductive charging resistor R_C is set for 10,000 Ω in line 57, and its inductance L_C as 60 mH in line 59. Finally, a trial value of clipping inductor L is set as 1.5 mH in line 60.

System Equations

Now consider the analysis equation development. An algebraic equation for control mesh II can be written as

$$\text{II RD} + X(1) - \text{VD} = 0$$

which becomes, in program format,

$$\text{II} = (\text{VD} - X(1))/\text{RD} \tag{10-1}$$

Next, the differential voltages around the mesh composed of C_O, C, and C_S can be written as

$$D(3) + D(6) = D(5) \tag{10-2}$$

A summation of currents at node A gives

$$X(7) + C\,D(6) + X(6)/R - CS\,D(5) = 0$$

If these two equations are solved for $D(5)$ in terms of $D(3)$, we obtain

$$D(5) = (C\,D(3) - X(7) - X(6)/R)/(CS + C) \qquad (10\text{-}3)$$

If the differential voltages around the mesh composed of C_I, C_F, and C_O is written, we obtain

$$D(2) + D(1) = D(3) \qquad (10\text{-}4)$$

If a current summation is made at node G, we have

$$CF\,D(2) + II = CI\,D(1)$$

If these two equations are solved for $D(2)$ in terms of $D(3)$, the result is

$$D(2) = (CI\,D(3) - II)/(CF + CI) \qquad (10\text{-}5)$$

Now we can sum the currents at node D to obtain

$$CF\,D(2) + GM\,X(1) + X(3)/RR + CO\,D(3) + CS\,D(5) = 0$$

If the values of $D(5)$ from Eq. (10-3) and $D(2)$ from Eq. (10-5) are substituted into this equation and it is solved for $D(3)$, we obtain our key analysis equation,

$$D(3) = (X(4) - GM\,X(1) - X(3)/RR + CF\,II/(CF + CI) + CS\,(X(7)$$
$$+ X(6)/R)/(CS + C))/(CO + CF\,CI/(CF + CI) + C\,CS/(CS + C)) \qquad (10\text{-}6)$$

Then, if Eq. (10-5) follows this equation in the listings, $D(2)$ is defined. If Eq. (10-4) is rewritten and follows $D(2)$ in the listings, we can define $D(1)$ as

$$D(1) = D(3) - D(2) \qquad (10\text{-}7)$$

If we write the equation around the $X(4)$ mesh, it can be solved directly as

$$D(4) = (V - X(3) - RC\,X(4))/LC \qquad (10\text{-}8)$$

If Eq. (10-3) is placed next in the listing, $D(5)$ is defined. If Eq. (10-2) is rearranged and listed next, we can define $D(6)$ as

$$D(6) = D(5) - D(3) \qquad (10\text{-}9)$$

Finally, $D(6)$ can be written directly as

$$D(7) = X(6)/L \qquad (10\text{-}10)$$

The seven differential equations above, together with algebraic Eq. (10-1), constitute the complete analysis set listed as equation subroutine program lines 1005–1040 of Appendix 25. Line 20 of the program was changed to handle seven equations and to define NS, the number of series elements in the hard-switch string. Except for some bookkeeping modification of the printout subroutines, the new program structure is now complete.

Analysis Results

If you now load Appendix 25 into your computer and follow this with a <RUN> command, a tabular printout similar to Fig. 10-3 will develop on your printer. Some key results follow.

Column X(6) shows us that the magnetron reaches its biased on value of 14,000 V in about 0.14 μs and that the rate of rise of this voltage is about 93 kV/μs. It is important to make sure that the rate of rise of voltage across the magnetron falls within the range of values given on its data sheet. This in turn makes certain that the generated oscillations start in the tube's proper internal RF operating mode on each pulse.

The rise time of the pulse reaches its maximum value at about 0.23 μs into the run—at which time the 20 switches have been almost completely turned on and C has been almost completely charged up.

Pulse droop then begins to occur and continues until the drive signal reaches its turn-off point at $T = 0.4$ μs. The droop is caused by current delivered to two loads: the intended current, which is delivered to the magnetron (Mag. Cur. column), and the current, which builds up in tail-clipping inductance L [X(7) column].

The fall time, first, can be seen to start [X(6) column] at about $T = 0.4$ μs and to reach zero at about 0.895 μs. This results in a fall time of about two to three times the rise time. Second, if you look at $X(7)$ during this period of time it can be seen that it builds up as expected during the transfer of energy from C during the fall time of the pulse. Third, a system trade-off factor becomes evident now. If we want to obtain a faster fall time, we will have to use a smaller value of L. However, as noted previously, this will cause a greater droop to occur, and a larger value of storage capacitor C_S will have to be used. An optimum set of values could be established through the use of several computer runs.

At time values greater than 0.85 μs, the last column, which is now diode current rather than magnetron current, shows that the backswing clipper is performing its role as anticipated.

The tail clipping inductance in conjunction with the backswing diode are performing their intended function—to minimize noise generation from the magnetron during the receiving time of the radar system. One of the most effective ways of minimizing magnetron noise generation is to hold the tube's cathode at a small positive bias relative to ground during the interpulse period of the system.

A plot of magnetron voltage often makes it somewhat easier to visualize system operating conditions. If you issue a RUN command and follow this with a P response to the ensuing question, a printout similar to Fig. 10-4 will appear on your printer. It clearly shows the pulse rise and fall time responses.

From a system operating point of view: At $T = 0$, a control signal would initiate pulsing action and rise time starts; at $T = 0.4$ μs a pulse termination command is issued by the control signal and the pulse fall time starts. In this example an RF pulse of about 0.25 μs is produced by the magnetron. Simply repositioning the location of the second control point will permit adjustment of the pulse duration. It is important to avoid exceeding the pulse duration ratings of the magnetron when setting the turn-off control signal timing positions.

Another area that requires careful attention when dealing with pulse trains involving

```
Integration Step = .00025 Microseconds --

         Response Of A Hard Switched Type Modulator For A Magnetron
         ---------------------------------------------------------------

Microsec.  X(1)      X(2)       X(3)       X(4)     X(5)       X(6)     X(7)     FET Cur.  Mag.Cur.
0.000      0.000     15000.0    15000.0    0.0008   15000.00   0.0      0.0000   0.0000    0.0000
0.020      49.919    14362.3    14412.2    0.0008   14997.07   584.9    0.0021   8.7359    0.0058
0.040      76.658    12270.3    12347.0    0.0013   14986.76   2639.8   0.0228   13.4151   0.0264
0.060      78.953    9912.5     9991.4     0.0026   14974.95   4983.5   0.0736   13.8167   0.0498
0.080      79.250    7544.7     7624.0     0.0046   14963.00   7339.0   0.1558   13.8688   0.0734
0.100      79.417    5192.4     5271.8     0.0075   14950.97   9679.2   0.2692   13.8980   0.0968
0.120      79.608    2861.5     2941.1     0.0111   14938.93   11997.9  0.4138   13.9313   0.1200
0.140      84.073    828.7      912.8      0.0154   14926.56   14013.8  0.5888   14.7128   14.0331
0.160      95.927    600.1      696.0      0.0201   14910.63   14214.6  0.7768   16.7872   14.5217
0.180      97.763    364.9      462.6      0.0248   14893.86   14431.3  0.9679   17.1085   15.0782
0.200      98.600    208.4      307.1      0.0296   14876.81   14569.8  1.1613   17.2549   15.4480
0.220      99.165    114.9      214.1      0.0344   14859.60   14645.6  1.3561   17.3539   15.6552
0.240      99.827    96.6       196.4      0.0392   14842.47   14646.1  1.5515   17.1579   15.6581
0.260      99.987    97.8       197.8      0.0440   14825.27   14627.6  1.7466   17.3026   15.6074
0.280      100.010   99.4       199.4      0.0488   14807.93   14608.6  1.9415   17.4493   15.5557
0.300      100.065   105.4      205.4      0.0536   14790.46   14585.1  2.1362   17.5113   15.4921
0.320      100.183   126.4      226.6      0.0583   14772.96   14546.4  2.3304   17.5320   15.3881
0.340      100.275   158.1      258.4      0.0630   14755.46   14497.1  2.5240   17.5480   15.2566
0.360      100.340   197.2      297.6      0.0677   14737.96   14440.4  2.7170   17.5596   15.1069
0.380      100.388   241.7      342.1      0.0724   14720.46   14378.4  2.9091   17.5679   14.9450
0.400      100.423   290.1      390.6      0.0770   14702.96   14312.3  3.1004   17.5740   14.7750
0.420      46.955    636.7      683.7      0.0816   14688.61   14004.9  3.2892   8.2171    14.0182
0.440      8.213     823.3      831.5      0.0860   14684.88   13853.3  3.4754   1.4372    0.1385
```

Figure 10-3   Appendix 25 tabular analysis of a hard-switch pulser for a magnetron.

0.460	4.863	1302.1	1307.0	0.0904	14683.55	13376.5	3.6571	0.8510	0.1338
0.480	4.776	1841.7	1846.5	0.0946	14682.38	12835.9	3.8319	0.8359	0.1284
0.500	4.953	2407.3	2412.2	0.0985	14681.14	12268.9	3.9993	0.8668	0.1227
0.520	5.143	2995.0	3000.2	0.1023	14679.89	11679.7	4.1589	0.8999	0.1168
0.540	5.324	3603.6	3608.9	0.1058	14678.59	11069.6	4.3106	0.9317	0.1107
0.560	5.496	4231.8	4237.3	0.1092	14677.23	10439.9	4.4540	0.9617	0.1044
0.580	5.657	4878.3	4884.0	0.1123	14675.82	9791.8	4.5889	0.9899	0.0979
0.600	5.807	5542.1	5547.9	0.1152	14674.42	9126.5	4.7151	1.0162	0.0913
0.620	5.946	6221.7	6227.7	0.1178	14672.95	8445.2	4.8322	1.0405	0.0845
0.640	6.074	6916.0	6922.1	0.1202	14671.46	7749.4	4.9402	1.0629	0.0775
0.660	6.190	7623.5	7629.7	0.1224	14669.90	7040.2	5.0388	1.0833	0.0704
0.680	6.295	8343.0	8349.3	0.1243	14668.34	6319.0	5.1279	1.1016	0.0632
0.700	6.387	9073.1	9079.5	0.1260	14666.78	5587.3	5.2073	1.1178	0.0559
0.720	6.468	9812.4	9818.9	0.1274	14665.21	4846.3	5.2768	1.1319	0.0485
0.740	6.536	10559.5	10566.0	0.1286	14663.58	4097.5	5.3365	1.1439	0.0410
0.760	6.592	11313.1	11319.6	0.1295	14661.94	3342.3	5.3861	1.1537	0.0334
0.780	6.636	12071.6	12078.2	0.1302	14660.30	2582.1	5.4256	1.1613	0.0258
0.800	6.667	12833.7	12840.4	0.1306	14658.66	1818.3	5.4549	1.1668	0.0182
0.820	6.686	13598.0	13604.7	0.1308	14657.02	1052.3	5.4741	1.1701	0.0105
0.840	6.693	14363.0	14369.7	0.1307	14655.38	285.7	5.4830	1.1712	0.0029
0.860	2.102	14678.4	14680.5	0.1304	14654.37	-26.1	5.4835	0.3679	-5.2228
0.880	0.304	14681.8	14682.1	0.1300	14654.33	-27.8	5.4831	0.0532	-5.5559
0.900	0.045	14682.4	14682.4	0.1297	14654.43	-28.0	5.4827	0.0078	-5.6033
0.920	0.007	14682.6	14682.6	0.1294	14654.58	-28.0	5.4824	0.0013	-5.6096
0.940	0.002	14682.7	14682.8	0.1291	14654.74	-28.0	5.4820	0.0003	-5.6099
0.960	0.001	14682.9	14682.9	0.1287	14654.90	-28.0	5.4816	0.0002	-5.6093
0.980	0.001	14683.0	14683.1	0.1284	14655.05	-28.0	5.4812	0.0002	-5.6086
1.000	0.001	14683.2	14683.2	0.1281	14655.21	-28.0	5.4808	0.0002	-5.6079

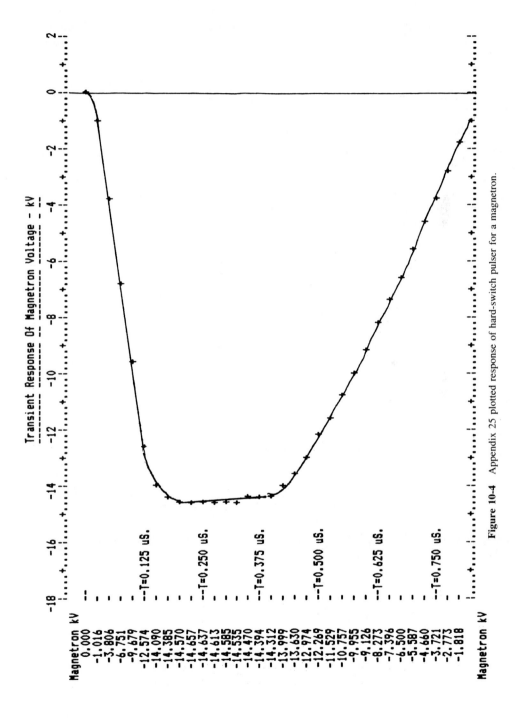

Figure 10-4 Appendix 25 plotted response of hard-switch pulser for a magnetron.

variable pulse durations and nonconstant interpulse spacings concerns the voltage regulation capabilities of the charging circuit. When operating conditions of this type are encountered, a simplified system model involving the charging circuit, bulk discharge periods, and the pulse train spacing intervals is recommended. For some pulser designs it will be found that the required degree of regulation can be obtained by means of dc–dc converter gating control alone. If such is not the case, a more comprehensive regulation system will have to be modeled and analyzed.

10-2 LINE PULSER SYSTEM ELEMENTS

The key element of any line pulser circuit is the pulse-forming network. Other elements that require careful consideration are (1) the pulse load impedance presented by the RF generator circuit (including the characteristics of associated pulse transformer assemblies), (2) the interrelated characteristics of the charging circuit and the inverse pulse energy circuit of the system, and (3) the characteristics of the high-voltage power supply circuit. In the next several sections we review these areas.

Pulse-Forming Network Properties

The general properties of PFNs can be deduced rather easily from the properties of conventional transmission lines. A wave traveling on a lossless transmission line has a velocity of

$$v = \frac{1}{(LC)^{1/2}} \tag{10-11}$$

where velocity is measured in units of line length per second, L is the number of henries of series inductance per unit length of line, and C is the number of farads of shunt capacitance per unit length of line.

For an isolated wave traveling on the line,

$$\frac{e}{i} = \left(\frac{L}{C}\right)^{1/2} = Z_0 \tag{10-12}$$

where e is the instantaneous voltage between the two line conductors, i is the instantaneous current at the point of observation, and Z_0 is the characteristic impedance of the line. When the isolated wave reaches the end of the line, a portion of it will be reflected back down the line in accordance with

$$\Gamma = \frac{Z_L - Z_0}{Z_L + Z_0} \tag{10-13}$$

where Γ is the *reflection coefficient* and Z_L is the load impedance.

Three special load impedance cases of an isolated pulse wave traveling on a line are shown in Fig. 10-5. In each case three points in time are shown: first, noted as 1, the pulse prior to reaching the load; second, noted as 2, the pulse halfway through its reflection; and third, noted as 3, the pulse after reflection.

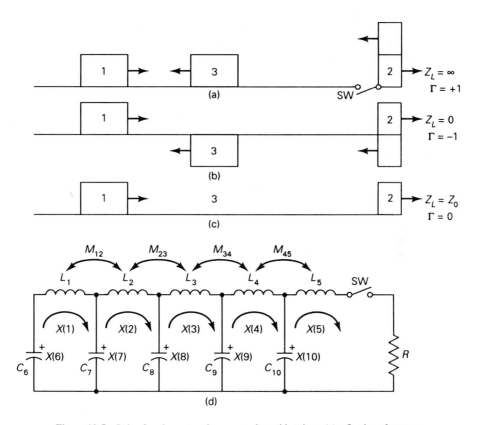

Figure 10-5 Pulse-forming networks—general considerations: (a) reflection of rectangular pulse, $Z_L = \infty$; (b) reflection of rectangular pulse, $Z_L = 0$; (c) reflection of rectangular pulse, $Z_L = Z_0$; (d) network diagram for type E pulse-forming network.

In the open-circuit load case of Fig. 10-5a, Eq. (10-13) indicates that Γ has a value of $+1$. During reflection the load voltage will double. In the short-circuit load case of Fig. 10-5b, the reflection coefficient is -1. The load voltage is zero during reflection. With the matched load case of Fig. 10-5c, Γ is zero. There is no reflection because all of the pulse's energy is absorbed in Z_L.

To illustrate how a pulse can be generated, consider a case where a conventional transmission line is used as a PFN. Refer to the open-circuit load case of Fig. 10-5a, time period 2. Now assume that at this instant in time we are able to isolate this "charged" length from the remainder of the line (e.g., by opening switch SW at this instant). This isolated line segment behaves in a fashion similar to that of a line that is charged to a voltage V_C by a charging source. If the source is disconnected when the line is charged (thus leaving both ends of the line open circuited), the line can be considered to have waves traveling in both directions on it, just as in the development of Fig. 10-5a. When a wave reaches either open end of the line, it experiences a positive reflection, it returns down the line, and as a result, the dual wave motion on the line continues.

Because the line has been assumed lossless, it will retain the same charged potential of V_C between the two conductors. However, because the waves are traveling in opposite directions, the amplitude of these traveling waves must be equal to $V_C/2$, as indicated in Fig. 10-5a.

Say that the length of the charged line segment is L_L. Now assume that we change the value of the load to Z_0. The traveling wave will then start to discharge into the load. Because of the two waves traveling in opposite directions, the line acts as though its length is $2L_L$. The product of the pulse duration and the velocity of the wave on the line, v, then must equal $2L_L$. This can be written as follows:

$$T_P = 2L_L(LC)^{1/2} \tag{10-14}$$

where T_P is the pulse duration in seconds. The velocity factor portion of this equation depends only on the dielectric used in the line. If it is air, $v = 3 \times 10^8$ m/s, and if we wanted to generate a 1-μs pulse,

$$L_L = \frac{T_P v}{2} = \frac{10^{-6}(3 \times 10^8)}{2} = 150 \text{ m}$$

The ungainly line length calculated in this representative example demonstrates why conventional transmission lines are rarely used for PFNs in practical pulsers.

Figure 10-5d illustrates the configuration of elements used for a five-section type E PFN. It is one of a number of artificial transmission line arrangements that were developed during World War II when a vital need first existed for compact forms of pulse energy storage in line-type pulsers.

The type E network probably achieved its popularity because all of its capacitors have the same value, a decided asset from a manufacturing point of view. It is, however, necessary to use unequal values of self-inductance for some of the network inductors (L_1–L_5), and these inductors must be arranged in such a manner that proper values of mutual inductance (M_{12}–M_{45}) exist between them. This requirement tends to complicate the network design process.

A cookbook approach for type E PFN design was documented during its early development stages. In general, it states:

1. Wind a continuous air-core solenoid in such a way that its total inductance is

$$L_T = \frac{T_P Z_0}{2} \tag{10-15}$$

2. The total network capacitance required is

$$C_T = \frac{T_P}{2Z_0} \tag{10-16}$$

Divide this value equally between the number of sections in the network.

3. Position the solenoid taps to obtain equal inductance for all except the end sections. The end sections should have 20 to 30% more self-inductance.
4. Select a coil section length/diameter ratio that will result in mutual inductance values that are 15% of the self-inductance of the center sections.

Five-Section PFN Equation and Program Development

E. A. Guillemin[9] pioneered most of the theoretical and mathematical background for lumped-parameter networks. His work showed that the quality of a lumped-parameter pulse network was a function of the number of sections used. As the number of sections increased, typically, the mathematical analysis complexity increased. Hence, until the wide use of digital computer analyses came about in recent years, this complexity tended to discourage the use of mathematical studies. Instead, the general PFN design approach consisted of (1) using the cookbook rules, (2) developing a hardware model of the network in the laboratory, and (3) finalizing the design through experimental efforts.

As our first step toward developing a fairly comprehensive five-section line-type pulser computer program, refer to the network diagram of Fig. 10-5d. We can write the network's five mesh equations as follows:

$$L5\ D(5) + M45\ D(4) - X(10) + R\ X(5) = 0 \quad (10\text{-}17)$$

$$M45\ D(5) + L4\ D(4) + M34\ D(3) - X(9) + X(10) = 0 \quad (10\text{-}18)$$

$$M34\ D(4) + L3\ D(3) + M23\ D(2) - X(8) + X(9) = 0 \quad (10\text{-}19)$$

$$M23\ D(3) + L2\ D(2) + M12\ D(1) - X(7) + X(8) = 0 \quad (10\text{-}20)$$

$$M12\ D(2) + L1\ D(1) - X(6) + X(7) = 0 \quad (10\text{-}21)$$

This set of equations can be readily solved by using the Gauss elimination method. We can first eliminate $D(5)$ by solving Eqs. (10-17) and (10-18) to obtain

$$K1\ D(4) + K2\ D(3) = K3$$

where

$$K1 = M45^2 - L5\ L4$$

$$K2 = -L5\ M34$$

$$K3 = M45\ (X(10) - R\ X(5)) - L5\ (X(9) - X(10))$$

Then this equation is solved with Eq. (10-19) to eliminate $D(4)$ to obtain

$$K4\ D(3) + K5\ D(2) = K6$$

where

$$K4 = M34\ K2 - K1\ L3$$

$$K5 = -K1\ M23$$

$$K6 = M34\ K3 - K1\ (X(8) - X(9))$$

Next this equation is solved with Eq. (10-20) to eliminate $D(3)$ and obtain

$$K7\ D(2) + K8\ D(1) = K9$$

where

$$K7 = M23\ K5 - K4\ L2$$

$$K8 = -K4\ M12$$

$$K9 = M23\ K6 - K4\ (X(7) - X(8))$$

Finally, this equation is solved with Eq. (10-21) to obtain

$$D(1) = (M12\ K9 - K7\ (X(6) - X(7))/(M12 - K7\ L1) \qquad (10\text{-}22)$$

By direct substitution into the original equations, we obtain

$$D(2) = (X(6) - X(7) - L1\ D(1))/M12 \qquad (10\text{-}23)$$
$$D(3) = (X(7) - X(8) - M12\ D(1) - L2\ D(2))/M23 \qquad (10\text{-}24)$$
$$D(4) = (X(8) - X(9) - M23\ D(2) - L3\ D(3))/M34 \qquad (10\text{-}25)$$
$$D(5) = (X(9) - X(10) - M34\ D(3) - L4\ D(4))/M45 \qquad (10\text{-}26)$$

The remaining equations can be written and solved directly from the current summations at the network's *LC* nodes as follows:

$$D(6) = -X(1)/C6 \qquad (10\text{-}27)$$
$$D(7) = (X(1) - X(2))/C7 \qquad (10\text{-}28)$$
$$D(8) = (X(2) - X(3))/C8 \qquad (10\text{-}29)$$
$$D(9) = (X(3) - X(4))/C9 \qquad (10\text{-}30)$$
$$D(10) = (X(4) - (X(3))/C10 \qquad (10\text{-}31)$$

Appendix 26 contains the program we will use to analyze the five-section type E network. Equations (10-22)–(10-31) are listed as equation subroutine program lines 1020–1065. The equations determined for K_1, K_2, K_4, K_5, K_7, and K_8 are listed on lines 64–69, and the equations for K_3, K_6, and K_9 are listed on lines 1005–1015.

Analysis Results

It is convenient to set up the program for the five-section type E network analysis through the use of a set of standard reference values (1-s pulse duration T_P, a 1-Ω characteristic impedance Z_0, and a 1-Ω resistive load value Z_L). These circuit element values have been used in the program. They are listed on lines 50–60 and on line 63. The network is assumed to be initially charged to 2 V, and the initial conditions for all capacitors are listed on lines 60 and 61. Line 20 lists 10 as the number of differential equations required for analysis solutions.

Now if you load Appendix 26 into your computer and issue a <RUN> command, a tabular response similar to Fig. 9-6 will develop on your printer. Several significant response features are as follows.

Pulse rise and fall times. Rise time, typically expressed between the 10 and 90% levels of the wave, requires approximately 5% of the 1-s value used for T_P. Similarly, the pulse fall time between the 90 and 10% levels requires approximately 12% of T_P. If shorter rise and fall times are required, more sections will be required for the network.

Integration Step = .005 Seconds

Response Of Type E Five Section PFN

Sec.	x(5)	x(4)	x(3)	x(2)	x(1)	x(10)	x(9)	x(8)	x(7)	x(6)	Vout
0.00	0.000	0.000	0.000	0.000	0.000	2.00	2.00	2.00	2.00	2.00	0.000
0.05	0.871	-0.050	-0.005	0.003	-0.000	1.69	2.04	2.00	2.00	2.00	0.871
0.10	1.075	0.361	-0.087	0.010	0.000	1.20	1.94	2.03	1.99	2.00	1.075
0.15	1.008	0.880	0.034	-0.047	0.008	0.97	1.56	2.05	2.00	2.00	1.008
0.20	0.973	1.077	0.491	-0.094	-0.006	0.99	1.14	1.87	2.05	2.00	0.973
0.25	1.000	1.003	0.953	0.105	-0.062	1.02	0.97	1.45	2.05	2.01	1.000
0.30	1.011	0.956	1.068	0.590	-0.061	1.00	1.01	1.06	1.82	2.05	1.011
0.35	0.997	1.008	0.949	1.025	0.173	0.99	1.03	0.96	1.38	2.04	0.997
0.40	0.996	1.018	0.935	1.094	0.648	1.00	0.98	1.06	1.00	1.82	0.996
0.45	1.005	0.970	1.052	0.932	1.073	1.00	0.98	1.07	0.93	1.33	1.005
0.50	0.995	0.982	1.066	0.904	1.097	0.98	1.04	0.97	1.06	0.71	0.995
0.55	0.984	1.047	0.956	1.050	0.695	1.00	1.03	0.95	1.03	0.21	0.984
0.60	1.000	1.037	0.950	1.028	0.193	1.03	0.97	1.02	0.68	-0.03	1.000
0.65	1.022	0.963	1.062	0.644	-0.079	1.02	0.98	0.94	0.22	-0.05	1.022
0.70	1.011	0.980	0.978	0.155	-0.086	0.99	1.02	0.58	-0.05	0.01	1.011
0.75	0.995	1.056	0.538	-0.086	-0.002	1.00	0.89	0.15	-0.07	0.03	0.995
0.80	1.020	0.912	0.053	-0.061	0.039	1.01	0.49	-0.05	-0.01	0.02	1.020
0.85	1.028	0.440	-0.107	0.005	0.032	0.82	0.07	-0.03	0.02	-0.00	1.028
0.90	0.877	-0.031	-0.006	0.009	0.009	0.39	-0.06	0.01	0.03	-0.02	0.877
0.95	0.535	-0.117	0.039	0.015	-0.016	-0.07	0.01	0.00	0.02	-0.01	0.535
1.00	0.167	0.125	-0.031	0.030	-0.027	-0.26	0.02	0.01	-0.01	-0.00	0.167
1.05	-0.029	0.289	0.000	-0.011	-0.008	-0.16	-0.12	0.04	-0.03	0.01	-0.029
1.10	-0.033	0.179	0.176	-0.065	0.014	0.00	-0.21	-0.03	0.00	0.01	-0.033
1.15	0.019	-0.007	0.254	0.012	-0.002	0.05	-0.13	-0.18	0.03	0.00	0.019
1.20	0.027	-0.048	0.105	0.199	-0.020	0.01	0.00	-0.23	-0.04	0.01	0.027
1.25	0.006	0.004	-0.064	0.269	0.047	-0.01	0.02	-0.10	-0.17	0.01	0.006
1.30	-0.000	0.005	-0.042	0.109	0.184	-0.01	-0.03	0.05	-0.22	-0.06	-0.000
1.35	-0.001	-0.027	0.065	-0.083	0.240	-0.01	-0.01	0.05	-0.10	-0.18	-0.001
1.40	-0.013	0.004	0.046	-0.076	0.102	-0.02	0.04	-0.05	0.06	-0.28	-0.013
1.45	-0.015	0.054	-0.055	0.057	-0.134	0.01	0.02	-0.05	0.05	-0.27	-0.015
1.50	0.008	0.018	-0.042	0.058	-0.257	0.03	-0.04	0.03	-0.11	-0.16	0.008

Figure 10-6 Appendix 26 tabular analysis of a five-section type E network.

Pulse duration. It is evident that the useful pulse duration will be somewhat shorter than T_P. If additional flat top is required, the value used for T_P in Eqs. (10-15) and (10-16) should be set for a higher value in the initial calculations.

Overshoot/undershoot. The pulse exhibits an overshoot of approximately 7.5% on its leading edge and approximately 3.5% just prior to its trailing edge. For the remainder of the body of the pulse, it generally stays within ± 1.5% of its normal flat-top value. A number of factors influence these values—such as modified section L values, modified section-to-section M values, and/or an increased number of network sections will improve the response—if system requirements dictate a greater degree of pulse top flatness.

Energy dissipation. Substantially all of the stored energy is dissipated in the load resistor. This is indicated by the pulse's very small back swing residue after the trailing edge reaches zero. Pulse-forming networks are seldom operated with Z_L exactly equal to Z_0. Since plotted computer outputs can be used very effectively to evaluate mismatched load operating conditions, issue another RUN command, and then follow this with a P. A matched load output similar to that shown in Fig. 10.7a should develop on your printer—a plotted duplicate of the tabular response shown in the last column of Fig. 10-6. The V_{out} values are printed adjacent to the time scale.

Figure 10.7b illustrates the response of the network if the value of R (i.e., in program line 63) is changed to 2 Ω, thereby producing a positive match. If values of Z_0 = 1 and Z_L = 2 are substituted into Eq. (10-13), Γ = 0.333.

This checks out quite well with the plot. It shows the flat-top response to average in the 1.333-V region (as we would predict from the simple positive reflection coefficient wave rationale shown in Fig. 10.5a). Similarly, the second step of the pulse, with values of $T > 1$, would be expected to have an average flat-top value in the 1.333 × 0.333 = 0.444 V region. It can be seen that the second step pulse does indeed fall within this region.

Figure 10.7c shows the response you can obtain if the network is connected to a negative match (if you change program line 63 to a value of 0.5 Ω) and another plot is requested. The value of Γ = -0.333 for this case, hence we would expect an average flat-top value of 1 -0.333 or 0.667 V. The plot confirms this value. The second pulse value would be expected to have a value of 0.667 × -0.333 = -0.222 V. Again the plot's average value confirms this calculation.

In typical applications, a line-type pulser is operated with some degree of negative matching (and usually with Z_L much closer to Z_0 than was the case in the demonstration case of Fig. 10.7c). This mode of operation ensures that an inverse voltage will be applied to the modulator's soft switch immediately following the main pulse. This causes the switch to return rapidly to its open-circuit mode, and in turn, this permits PFN charging for the subsequent pulse to begin as soon as possible.

10-3 LINE PULSER WITH PULSE TRANSFORMER AND NONLINEAR RF AMPLIFIER LOAD

Pulse transformers provide the medium through which the modulator's PFN and soft-switch transfer stored pulse energy into the load impedance presented by the RF generator. Typically, the pulse transformer's windings provide a significant step-up voltage ratio. The impedance presented to the line-type pulser at the primary winding (due to steady-state transformation of the RF load impedance) is in accordance with

$$Z_P = \frac{Z_L}{TR^2} \qquad (10\text{-}32)$$

where Z_L is the effective load impedance of the RF load, and TR is the step-up voltage ratio of the transformer.

This expression provides steady-state rationale, but from a transient point of view, Z_L is variable. Also, the transformer introduces inherent pulse-related quantities that significantly affect the transformation waveforms. A quantitative assessment of such a system requires a network diagram that more completely defines the pulse transformer and RF load characteristics.

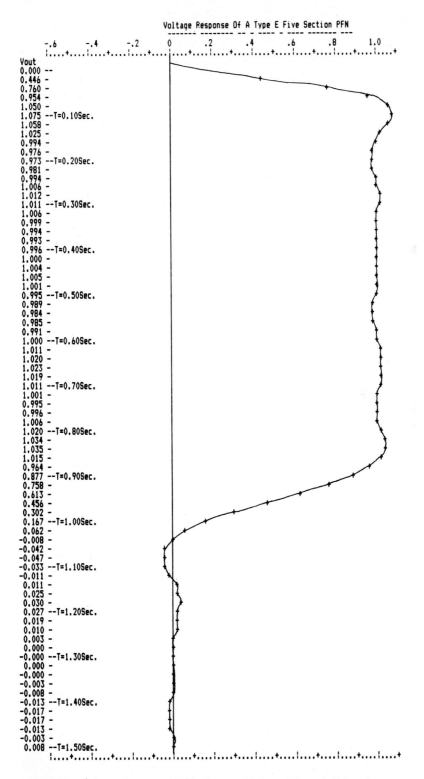

Figure 10-7 Plots of five-section PFN response with three different load impedances: (a) $Z_L = Z_0$; (b) $Z_L = 2Z_0$; (c) $Z_L = 0.5Z_0$.

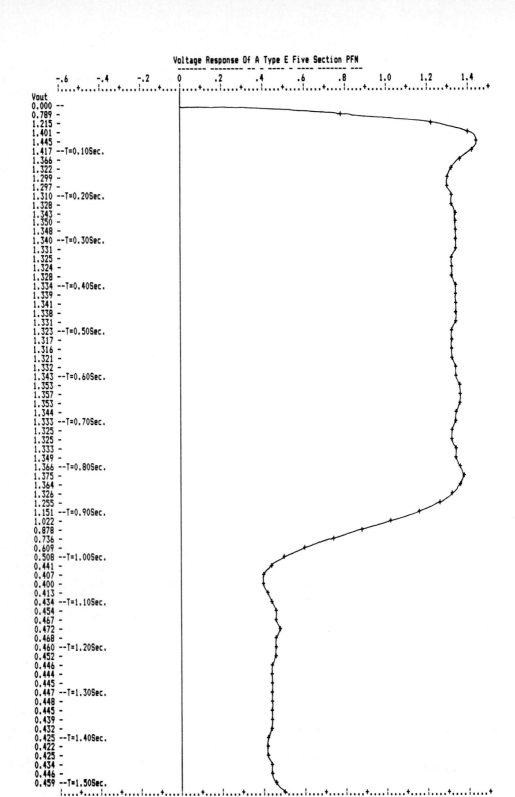

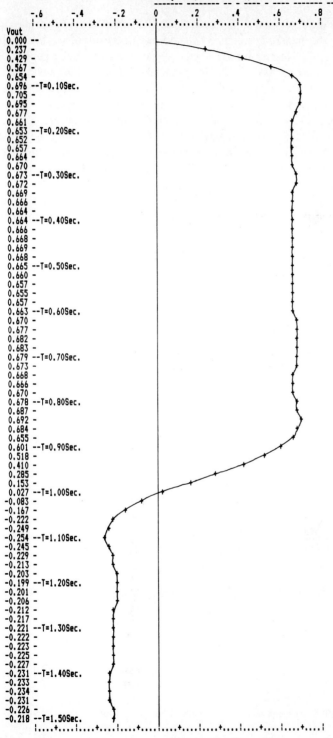

Sec. 10-3 Line Pulser with Pulser Transformer

Network Diagram

Figure 10-8 shows the network diagram we will use to demonstrate the principles involved. For this example, the standard reference pulse-forming network of the last analysis will be used so that the relative response characteristics of the two cases can easily be compared. Also, we will use a ratio of 1:1 in the pulse transformer to further aid the comparison. The primary and secondary windings are designated L_P and L_S and the mutual coupling as M—in the same manner as for our power supply analyses.

A new and significant addition to the model is C. This element represents a summation of the turn-to-turn and layer-to-layer capacitances associated with the transformer's windings (plus all associated load circuit capacitances connected across the winding). It, in conjunction with the leakage inductance of the windings, forms an LC circuit which is an inherent, although invisible part of every transformer.

The relative values of these two elements are of particular importance in a pulse transformer because they affect the voltage rise time and overshoot characteristics obtainable. The steady-state RF load impedance is ideally made equal to

$$Z_L = \left(\frac{L}{C}\right)^{1/2} \tag{10-33}$$

where L is the equivalent transformer leakage inductance and C is the equivalent transformer shunt capacitance.

If an ideal PFN (whose characteristic impedance matches the steady-state Z_P value of the pulse transformer) is used to drive the system, the rise time and overshoot characteristics of the transformer will be found to be nearly optimum. Hence the leakage inductance in conjunction with the transformer's equivalent capacitance are figures of major significance to both the system equipment designer and the pulse transformer designer.

From a practical point of view, actual systems do not have the aforementioned ideal matched load and/or ideal PFN characteristics. Nevertheless, the transformer's values of C and L are fundamental elements that require definition on the network diagram.

Z_L typically possesses a nonlinear value. For example,

$$I_L = KV_L^{1.5} \tag{10-34}$$

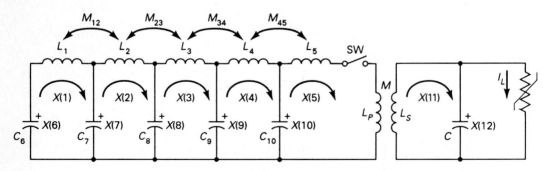

Figure 10-8 Network diagram for type E PFN/pulse transformer/nonlinear load.

defines a typical function of load current versus load voltage, where K is the perveance of the RF amplifier. Most relatively high-powered linear beam tube applications are associated with perveance values in the region 10^{-6} to 3×10^{-6}. Other RF amplifiers have different values that are functions of both their electronic amplification principles and their structural materials. In our network diagram the load is shown to have a saturating load impedance. It is defined in terms of load current I_L.

System Equations

With the diagram rationale defined, next consider the system equations. The $D(1)$–$D(4)$ mesh relationships and the $D(6)$–$D(10)$ node relationships remain the same as in our previous analysis. The revised mesh 5 equation is now

$$-M\ D(11) + (LP + L5)\ D(5) + M45\ D(4) - X(10) = 0 \qquad (10\text{-}35)$$

and the mesh 11 expression is

$$LS\ D(11) - M\ D(5) + X(12) = 0 \qquad (10\text{-}36)$$

If these two equations are solved to eliminate $D(11)$, we have

$$S1\ D(5) + S2\ D(4) = S3$$

where

$$S1 = LS\ (L5 + LP) - M^2$$
$$S2 = LS\ M45$$
$$S3 = LS\ X(10) - M\ X(12)$$

Now if this equation is solved with Eq. (10-18) to eliminate $D(5)$, revised values of K_1, K_2, and K_3 can be established as follows:

$$K1 = M45\ S2 - S1\ L4$$
$$K2 = -S1\ M34$$
$$K3 = M45\ S3 - S1\ (X(9) - X(10))$$

If we assume that the load resistor is defined by the relation shown in Eq. (10-34), current I_L can be written as

$$IL = PV\ X(12)^{1.5} \qquad (10\text{-}37)$$

where PV is the perveance, and the $D(12)$ can be written directly as

$$D(12) = (X(11) - IL)/C \qquad (10\text{-}38)$$

Program Modifications

Appendix 27, a revised version of Appendix 26, will be used for this analysis. The modified new values of S_1 and S_2 have been written on lines 71 and 72, and the revised values of K_1 and K_2 are on lines 73 and 74. The revised value of K_3 is on line 1005, and the new value of S_3 is on line 1017. The pulse value of I_L from Eq. (10-37) is listed on line 1080. $D(11)$, from Eq. (10-36), is listed on line 1070, and $D(12)$, from Eq. (10-38), is listed on line 1085.

Line 1075 has also been added. It causes I_L to assume a value of 0 when the voltage across the load assumes a backswing status. For some types of RF amplifiers this would not be a valid procedure (e.g., solid-state devices are often shunted with reverse diodes), but for a tube-type amplifier the approach shown is reasonable.

The pulse transformer and load elements are listed between lines 64 and 70. In pulse transformers, the primary inductance, L_P, should be selected to have the lowest practical value. This in turn causes the leakage inductance to achieve its lowest value.

We now encounter a trade-off area. Too low a value of primary inductance (i.e., across the load) can introduce excessive droop on the pulse. For this example a value of 10 H was selected. This value holds the droop to less than 10%. For our 1:1 ratio transformer, L_S will have the same value. Both are listed on line 64.

A representative value of coefficient of coupling in a pulse transformer of 0.999 is listed on line 65, and its associated value of mutual inductance is calculated on line 66. The PFN's value of Z_0 is listed on line 67, and the transformer's leakage inductance is calculated on line 68.

With this background established, the desired value of C for the transformer is calculated on line 69. Finally, the amplifier's perveance is listed on line 70. Its value was selected so that the load on the transformer would be equal to the value of R (i.e., used in the previous example) when the output voltage of the transformer crosses the 1-V level.

Analysis Results

Load your disk, issue a <RUN> command, and request a tabular printout. A display similar to Fig. 10-9 should develop on your printer. If this printout is compared to Fig. 10-6, the relative effects of introducing a pulse transformer and nonlinear load can be assessed. The following are several items of interest.

Rise and fall times. If V_{out} of Fig. 10-6 is compared to $X(12)$ of Fig. 10-9, it can be seen that the rise time is slightly greater with the pulse transformer. Similarly, the fall time is slightly greater.

Overshoot/undershoot. The initial overshoot is somewhat lower with the pulse transformer, and the main body of the pulse has somewhat lower peak-to-peak ripple variations.

Flat top. The flat top of the pulse for the transformer analysis case has a definite droop. This can be assessed if $X(5)$, the primary current, is compared with $X(11)$, the secondary current. The rise in $X(5)$ is countered with a droop in $X(11)$.

If you issue another <RUN> command and follow this with a P command, a plot of the output voltage similar to Fig. 10-10 should develop on your printer. If this plot is compared with Fig. 10-7a, the general characteristics above can be observed.

Integration Step = .005 Seconds

Output Response Of 5 Section PFN, Pulse Transformer, And Non-linear Load

Sec.	x(5)	x(4)	x(3)	x(2)	x(1)	x(10)	x(9)	x(8)	x(7)	x(6)	x(11)	x(12)
0.00	0.000	0.000	0.000	0.000	0.000	2.00	2.00	2.00	2.00	2.00	0.00	0.000
0.05	0.836	-0.054	-0.003	0.002	-0.000	1.71	2.04	2.00	2.00	2.00	0.83	0.679
0.10	1.093	0.329	-0.083	0.010	0.000	1.22	1.95	2.02	2.00	2.00	1.09	1.027
0.15	1.049	0.860	0.020	-0.042	0.007	0.96	1.59	2.05	2.00	2.00	1.04	1.037
0.20	0.987	1.099	0.460	-0.095	-0.004	0.96	1.15	1.89	2.05	2.00	0.97	0.989
0.25	0.997	1.038	0.944	0.085	-0.058	1.01	0.95	1.48	2.05	2.01	0.97	0.980
0.30	1.018	0.967	1.097	0.562	-0.065	1.00	0.98	1.07	1.84	2.05	0.99	0.992
0.35	1.013	1.005	0.982	1.022	0.152	0.98	1.02	0.94	1.41	2.04	0.98	0.990
0.40	1.010	1.027	0.939	1.126	0.623	0.99	0.98	1.03	1.00	1.84	0.97	0.982
0.45	1.018	0.988	1.045	0.964	1.072	0.99	0.97	1.06	0.90	1.36	0.98	0.984
0.50	1.015	0.993	1.079	0.905	1.134	0.97	1.02	0.97	1.03	0.73	0.97	0.981
0.55	1.004	1.058	0.981	1.040	0.744	0.98	1.02	0.94	1.03	0.20	0.95	0.970
0.60	1.015	1.059	0.960	1.047	0.218	1.01	0.96	1.00	0.70	-0.06	0.96	0.970
0.65	1.042	0.988	1.067	0.690	-0.089	1.01	0.95	0.94	0.23	-0.08	0.98	0.984
0.70	1.044	0.993	1.008	0.188	-0.106	0.97	1.00	0.59	-0.07	-0.02	0.98	0.987
0.75	1.027	1.074	0.589	-0.087	-0.008	0.97	0.89	0.15	-0.12	0.01	0.96	0.974
0.80	1.045	0.958	0.085	-0.076	0.050	0.98	0.49	-0.08	-0.05	-0.00	0.97	0.975
0.85	1.068	0.501	-0.108	0.006	0.046	0.81	0.05	-0.08	-0.00	-0.03	0.99	0.991
0.90	0.951	0.005	-0.010	0.027	0.018	0.37	-0.12	-0.03	0.00	-0.05	0.86	0.940
0.95	0.604	-0.104	0.058	0.035	-0.012	-0.11	-0.05	-0.03	-0.01	-0.05	0.51	0.735
1.00	0.165	0.145	0.005	0.046	-0.023	-0.32	-0.03	-0.03	-0.05	-0.04	0.07	0.379
1.05	-0.090	0.315	0.040	0.005	-0.001	-0.19	-0.17	-0.01	-0.07	-0.03	-0.18	-0.038
1.10	0.003	0.165	0.210	-0.042	0.024	-0.00	-0.24	-0.09	-0.05	-0.04	-0.09	-0.450
1.15	0.253	-0.032	0.255	0.044	0.011	-0.05	-0.13	-0.24	-0.03	-0.05	0.16	-0.340
1.20	0.244	0.052	0.059	0.224	0.000	-0.21	-0.03	-0.26	-0.11	-0.05	0.16	0.118
1.25	0.037	0.227	-0.068	0.249	0.071	-0.20	-0.13	-0.10	-0.23	-0.06	-0.05	0.097
1.30	-0.021	0.160	0.100	0.055	0.189	-0.08	-0.25	-0.01	-0.25	-0.14	-0.11	-0.166
1.35	0.100	-0.022	0.280	-0.067	0.199	-0.06	-0.17	-0.14	-0.12	-0.25	0.01	-0.306
1.40	0.171	0.005	0.148	0.082	0.042	-0.16	-0.02	-0.28	-0.04	-0.32	0.08	-0.152
1.45	0.097	0.181	-0.087	0.253	-0.117	-0.19	-0.06	-0.19	-0.17	-0.30	0.01	-0.015
1.50	0.038	0.182	-0.025	0.119	-0.100	-0.11	-0.22	-0.02	-0.36	-0.23	-0.05	-0.085

Figure 10-9 Appendix 27 tabular analysis of five-section PFN/pulse transformer/nonlinear load.

The magnified nature of the backswing is a new feature that stands out. This is due to the fact that in the transformer case there is no resistive loading during backswing, while in the case of the PFN only the load was present at all times. For our present analysis case, the backswing could easily be corrected by means or a reverse diode and an oscillation damping resistor.

These program elements have been selected to produce a demonstration pulse waveform of generally acceptable quality. Nevertheless, there is room for improved waveform characteristics. The analysis equations have been set up in such a way that they can be readily revised to accommodate a modified number of PFN sections, the addition of circuit loss elements, modification to desired pulse lengths, modified transformer characteristics, and so on. You are encouraged to modify the program in any way that assists in determining the specific capabilities of the pulser you have to design or investigate. That was the main reason for developing this program.

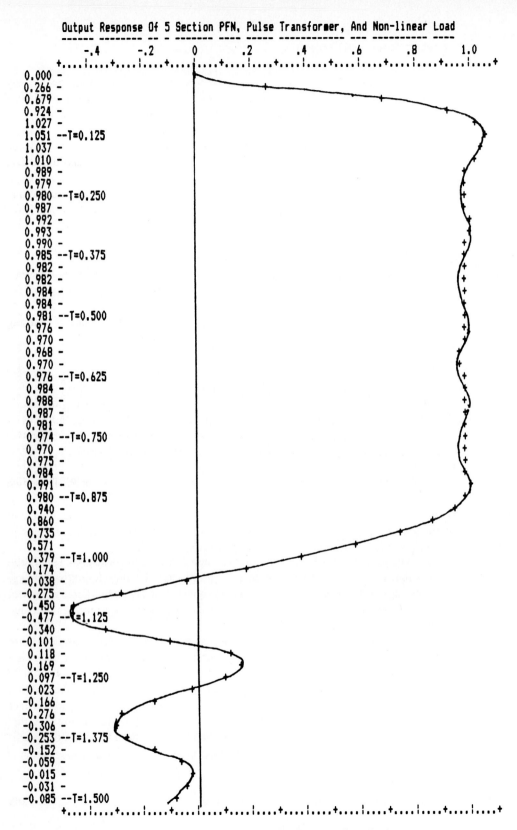

Figure 10-10 Appendix 27 V_{out} of five-section PFN/pulse transformer/nonlinear load.

Line-Type Pulser Charging and Inverse Principles

In this section we analyze one of the most frequently used techniques for repetitively recharging a PFN. In addition, a situation closely associated with PFN charging—the effect produced by the inverse voltage that typically remains on the PFN after each pulse—is reviewed, and a typical method of controlling this voltage is analyzed.

Figure 10-11a encompasses these two areas by means of an extension to our previous network diagram. The upper group of elements, consisting of V, L, R, and diode D_C, describe a PFN charging technique known as dc resonant charging. For this initial assessment, assume that there is no inverse voltage across the PFN at $T = 0$. Also assume that the pulse has just concluded and that the soft switch has returned to its high-impedance status. Voltage V will then charge capacitors C_6–C_{10} of the PFN via a path that includes L, R, D_C, L_1–L_5, and L_P.

Typical interpulse periods vastly exceed the pulse duration of a radar system; hence L will have a value of inductance an order of magnitude greater than the other inductors noted. As a result, the other inductors are frequently omitted to simplify the analysis. We will consider that only L is present in our analysis.

If the complete interpulse period is used to recharge the PFN, the value of charging inductor L, is

$$L = \frac{1}{(4\pi \cdot \text{PRR})^2 \cdot C} \qquad (10\text{-}39)$$

If the value of series resistance R is zero, the PFN will be charged to a voltage of exactly 2 times V during the charging period. Practical inductors are usually constructed using iron-core coils. Therefore, they have combined resistance/core losses, and R always has a finite value. As a result, the PFN voltage will always be somewhat less than 2 V when the initial conditions noted are present.

The inverse voltage discharge circuit consists of diode D_I and resistor R_I. These two elements are connected as a front of the line clipper in this case. Sometimes they are connected across C_6, and the circuit is referred to as an end-of-the-line clipper. In either location they serve the important function of discharging the network's inverse energy before a significant portion of the charging period has elapsed.

The simplified circuit shown in Fig. 10-11b will allow us to assess the capabilities of a dc resonant charging/inverse voltage circuit by means of Appendix 28. The mesh equation for the $X(1)$ current path is

$$L\,D(1) + R\,X(1) + X(2) - V = 0$$

which becomes, in program format,

$$D(1) = (V - R\,X(1) - X(2))/L \qquad (10\text{-}40)$$

It is listed as line 1010 of Appendix 28.

The node equation can be written directly as

$$D(2) = (X(1) - \text{II})/C \qquad (10\text{-}41)$$

It is listed as line 1020 of the appendix. An intermediate line, 1015, simulates the action of inverse diode D_I. Similarly, lines 154 and 194 simulate the action of charging diode D_C.

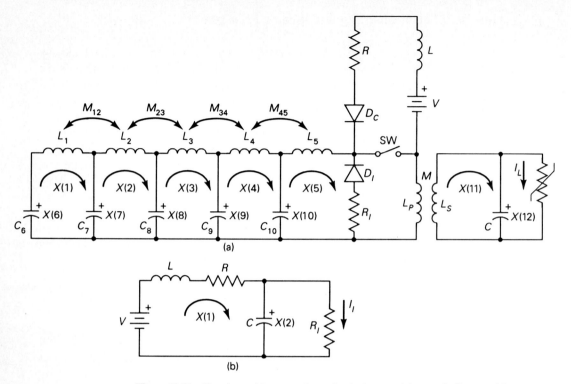

Figure 10-11 Charging and inverse voltage circuit elements: (a) network diagram with charging and inverse circuits added; (b) revised network diagram for charging and inverse circuits.

For this example we will charge the standard network used in the previous two examples. Assume that a 1000-s charging time is involved. The value of L will then be 222,684 H, and it is listed on line 50. If a typical inductor Q of 20 is used, the value of R would be about 35 Ω, and this is listed on line 54. The total network capacitance C is listed on line 52 as 0.455 F, and it has been assigned a 20% inverse voltage level of 0.2 V on line 60. Inverse resistor R_I has been chosen arbitrarily to produce a relatively short RC time constant value by using the 44-Ω figure on line 56. Finally, the power supply voltage V has been assigned a value of 1 V in line 58.

Now if you load Appendix 28 and issue a <RUN> command, a tabular response similar to the one shown in Fig. 10-12 will develop on your printer. A <RUN> followed by a P command will give the plot shown in Fig. 10-12b. These displays show:

1. Charging current X(1) is essentially a half sinusoidal wave that starts at 0, peaks at about halfway through the charge, and falls to zero at full charging voltage.
2. The charging voltage wave starts at the initial inverse voltage level of -0.2 V, returns rapidly on an RC time constant curve to zero volts, and then charges up to almost double voltage on a $[1 - \cos(\omega t)]$ type of waveform path.

D-C Resonant Charging And Inverse Diode Circuit

Sec.	x(1)	x(2)
0	.00000	-0.200
20	.00010	-0.076
40	.00019	-0.026
60	.00028	-0.003
80	.00037	0.012
100	.00046	0.030
120	.00054	0.052
140	.00063	0.078
160	.00071	0.107
180	.00078	0.140
200	.00085	0.176
220	.00092	0.215
240	.00099	0.257
260	.00105	0.302
280	.00111	0.350
300	.00116	0.399
320	.00121	0.452
340	.00125	0.506
360	.00129	0.562
380	.00132	0.619
400	.00135	0.678
420	.00137	0.738
440	.00139	0.798
460	.00140	0.860
480	.00140	0.921
500	.00140	0.983
520	.00140	1.045
540	.00139	1.106
560	.00137	1.167
580	.00135	1.227
600	.00132	1.285
620	.00129	1.343
640	.00125	1.399
660	.00121	1.453
680	.00116	1.505
700	.00111	1.555
720	.00106	1.603
740	.00100	1.648
760	.00093	1.690
780	.00087	1.730
800	.00080	1.767
820	.00073	1.800
840	.00065	1.830
860	.00057	1.857
880	.00049	1.881
900	.00041	1.900
920	.00033	1.917
940	.00024	1.929
960	.00016	1.938
980	.00007	1.943
1000	.00000	1.945
1020	.00000	1.945
1040	.00000	1.945
1060	.00000	1.945
1080	.00000	1.945
1100	.00000	1.945

Figure 10-12 Plot and listing for revised charging and inverse network.

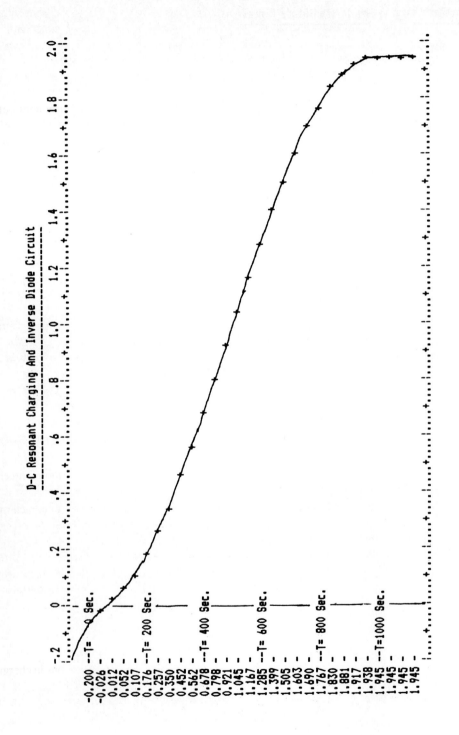

You are encouraged to introduce a number of variable operating conditions into the program to demonstrate the wide range of charging characteristics possible with this system. For example, one of the original reasons for using an inverse circuit was to stabilize line-type pulser performance under arcing load operating conditions. An inverse voltage almost equal to the full charging voltage appears on the PFN when this occurs, and if the inverse circuit does not respond fairly rapidly, a charging voltage significantly in excess of twice the power supply voltage will occur on the subsequent pulse. This might well cause an unstable situation to develop wherein additional arcing results.

10-4 MTI PERFORMANCE VERSUS DC CHARGING STABILITY

In most systems the pulse repetition frequency is not related harmonically to the ac power source frequency. As a result, dc resonant charging introduces nonuniform loading on the dc power supply (e.g., loading may occur either during the commutation period or during the normal conduction period). This will cause the power supply voltage to vary from pulse to pulse, and this causes the resonant charging voltage to change from pulse to pulse as well.

The following will illustrate a technique for determining the power supply voltage versus RF phase stability characteristics of a beam pulsed MTI transmitter (i.e., one that uses a line-type pulser with dc resonant charging). The representative MTI system in this example consists of the following:

1. A 5-MW klystron that has a beam interactive structure 25 rad long.
2. A three-phase full-wave dc power source (rated at 10 kV/20 A) that delivers energy to the system's dc resonant charging elements (we will use the power supply program of Appendix 18C).
3. A single delay MTI processor that operates with a constant pulse repetition frequency of 1 kHz.
4. A modulator PFN that has network capacitance value of 1 μF and a characteristic impedance of 5 Ω. When the network is switched it produces high-power-level pulses that have peak amplitudes of about 20 MW and time durations of about 10 μs. These pulses then pass through a pulse transformer that has a step-up ratio of about 1:15. The transformer's secondary voltage provides beam voltage pulses to the klystron. (Note: The value of PFN capacitance is the only essential item needed for our analysis; the other details describe more fully the representative MTI system.)

The quantitative value of the PFN's fully charged voltage is directly related to the klystron's pulse voltage—since we are dealing with a linear system. Variations in charging voltage from pulse to pulse are therefore the first-order quantities of interest. In this analysis we assume that there will be no pulse-to-pulse variations in the remaining elements of the MTI system.

Network Diagram, Equations, and Program

Figure 10-13 shows the network diagram used to define the computer program. The analysis will simply combine the features of two programs, Appendix 18C and Appendix 28. The composite, Appendix 29, will be structured as follows.

1. The program will begin ($T = 0$) at a point in time where an assumed rectangular beam pulse has just been completed. For this initial condition, the inverse voltage of the PFN is assumed to be -925 V at this instant. This is shown on line 52 of Appendix 29.

2. To simplify the program, all subsequent pulses are assumed to be have a duration of zero (since the high-speed pulse duration is such a small percentage of the relatively slow-moving inverse and network charging voltage functions). Also, the program will assign a representative inverse voltage value of 5% of the peak charging voltage of the network to each succeeding pulse. Line 112 defines the rationale used.

3. Capacitor CN (listed on line 72) will be used alone to represent the PFN. This greatly increases the speed of program execution without significantly affecting the accuracy of the analysis (by eliminating a large number of damped secondary network reflections of small amplitude and time duration that would otherwise be superimposed on the relatively long-duration network charging wave).

4. In accordance with Fig. 10-13, the charging inductor, L_{CH}, is assigned a value of 0.10132 H on line 70 [using Eq. (10-39)]. Its series resistance, R_{CH}, is assigned a typical value of 16 Ω on line 74. On line 76 the inverse discharge resistor R_{IN} has a representative value assigned as 20 Ω.

5. Three new equations are required for the analysis. They can be derived directly from Appendix 28 and the network diagram of Fig. 10-13 as

$$D(7) = (X(6) - RCH\ X(7) - X(8))/LCH \quad (10\text{-}42)$$

$$II = X(8)/RIN \quad (10\text{-}43)$$

$$D(8) = (X(7) - II)/CN \quad (10\text{-}44)$$

These equations, along with the inverse diode operation rationale, are listed as lines 401 and 402.

6. With regard to the three-phase power supply, the elements shown on lines 58–64, and the peak supply voltage of line 66, are the same as those used in Appendix 18C. The three power supply phase voltages are defined on line 165, and the initial conditions for the analysis are given on lines 50–55.

7. The power supply equation subroutine is listed on lines 1000–1124. It has a few minor changes that make it compatible with the currents associated with the node formed by C and L_C in Fig. 10-13. The current summation at this node can be written directly to provide revised equation $D(6)$. It is

$$D(6) = (X(5) - X(7))/C \quad (10\text{-}45)$$

This revision is shown in lines 1017, 1037, 1057, 1077, 1097, and 1117.

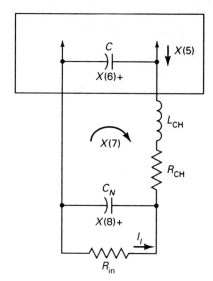

Figure 10-13 Three-phase FW power supply modification for dc resonant charging/inverse clipper load.

8. Other modifications to Appendix 29 include a change in line 30 to accommodate an eight-equation analysis, a change in line 52 to define the value of charging voltage $X(8)$ at the start of the analysis, and several miscellaneous analysis print/plot statement changes.

Analysis Results

If you load Appendix 29 and issue a <RUN> command, the resulting tabular response will list successive resonant charging voltages the VCH column. Variations in peak charging voltages exist during early charging periods due to inexact initial conditions selected (i.e., the long-time-constant currents in the inductive and capacitive elements of the three-phase power supply stabilize much less rapidly than do those of the short-time-constant charging circuit). If the last 10 charging periods are averaged, the pulse-to-pulse variation is approximately 2.8 V and the peak charging level is 18,861 V.

Now let us examine pulse-to-pulse phase variations in the klystron. The following linear beam tube phase pushing relationship is often used in analyses:

$$d\Phi = \frac{dV\Phi}{2V}$$

where Φ is the phase length of the tube (assume that $\Phi = 25$ rad), V is the beam voltage, and dV is the voltage variation.

To a close approximation, the beam voltage is proportional to the charging voltage of the PFN. Hence for the average steady-state charging voltage variation noted above, and for a 25-rad-long klystron we can write

$$d\Phi = \frac{2.8 \times 25}{2 \times 18{,}861} = 0.00186 \text{ rad} = 0.106 \text{ deg}$$

In Chapter 11 we examine RF generator systems in terms of improvement factors. RF phase pushing values (e.g., such as $d\Phi = 0.106$ deg calculated above) represent essential data used in the determination of MTI system performance capabilities.

Example 10-1

The last analysis was idealized. For example, three-phase prime power lines are seldom exactly balanced as in Appendix 29. We can modify the program to demonstrate the effects of a 5% source voltage unbalance simply by introducing the following two changes in line 165: $V_1 = 0.95\, V_P \sin(A_1 + K)$: $V_3 = 1.05\, V_P \sin(A_3 + K)$. Then one line is 5% high, one line is 5% low, and one line is at the original value. If we make a computer run using these changes and use the last 10 charging voltage cycles of the run, we obtain

average $dV = 6.1$ V   average $V_{CH} = 18{,}859$ V   average $d\Phi = 0.232$ deg

Example 10-2

A system operating problem occurs in single-cancellation MTI systems when a moving target has a Doppler frequency equal to (or at a multiple of) the radar system's PRF. A blind speed occurs, and the target vanishes from the radar's indicator. For example, if the MTI system had an operating frequency of 1000 MHz and a PRF of 1000 Hz, the first blind speed would occur if the target had a radial velocity of 538 km/h (334 mi/h) based on Eq. (6-15).

A technique that has been found effective in minimizing the blind speed problem requires the use of staggered PRF. In contrast to our preceding constant PRF analyses, staggered PRFs involve two, three, four, or more pulses. Let us use Appendix 29 to analyze a triple stagger with 800-, 1000-, and 1200-μs interpulse periods.

One simple and reasonably accurate program modification is to add two more statements similar to line 112, and arrange to have the program address the statements in sequence to set up the three interpulse spacings. L_{CH} will be modified to resonantly charge at the highest PRF, and T_2 will be modified to ensure that the peak charging wave values are printed out. To summarize the changes:

```
34 T2=.0002
36 T3=.016
70 LCH=.064845
110 IF PRF=0 AND TR>.0008-T1/2 THEN VCH=X(8): TR=0
   : X(8)=-.05*VCH: PRF=1
112 IF PRF=1 AND TR>.001-T1/2 THEN VCH=X(8): TR=0
   : X(8)=-.05*VCH: PRF=2
114 IF PRF=2 AND TR>.0012-T1/2 THEN VCH=X(8): TR=0
   : X(8)=-.05*VCH: PRF=0
```

Now if we run the program and average the last nine peak charging voltages, we have

average $dV = 102.3$ V   average $V = 18{,}803$ V   , $d\Phi = 3.90$ deg

Staggered PRF operation often degrades a conventional system's MTI beyond specified performance capability limits. If this is the case, alternate courses of action include: (1) modification of the system design to include a high-voltage power supply voltage regulator, or (2) if such a regulator is not feasible/practical, investigate the design for an entirely different power supply/pulser approach.

Chapter 11

Analysis of an MTI System's Performance Capabilities

In Chapter 10 we investigated the pulse-to-pulse phase stability that could be obtained from a beam pulsed klystron power amplifier. We found that the relative pulse-to-pulse phase stability was a function of both the source interface characteristics and the form of loading impressed on the high-voltage dc power supply (e.g., constant PRF loading, constant PRF loading but with an unbalanced prime power source, and staggered PRF loading). However, an analysis tool was missing in these investigations—a quantitative yardstick that would allow us to define performance capabilities in terms of how well or how poorly the equipment was doing its job. In the next section we examine a technique that is frequently used for this purpose. With this background we then analyze/determine the performance capabilities of more versatile forms of RF power generators.

11-1 RADAR CLUTTER AND SYSTEM IMPROVEMENT FACTORS

In Chapter 6 it was noted that fixed objects on the ground—buildings, towers, mountains, etc.—interfere with the detection of moving aircraft targets when using a conventional radar system. These objects produce *clutter*. MTI radar systems will, under ideal circumstances, remove these clutter echos from the systems indicator display and allow it to present only the moving targets that exist in the surveyed volume of space.

More demanding sources of motion-related radar clutter exist as well: for example,

TABLE 11-1 RF POWER GENERATOR INSTABILITY LIMITATIONS

Pulse-to-pulse instability	Limit on improvement factor
1. Transmitter frequency	$I = 20 \log[(1/\pi \, \Delta f \, t)]$
2. Stalo or Coho frequency	$I = 20 \log[(1/2\pi \, \Delta f \, T)]$
3. Power amplifier phase shift	$I = 20 \log(1/\Delta\Phi)$
4. Coho locking	$I = 20 \log(1/\Delta\Phi)$
5. Pulse timing	$I = 20 \log[t/(\sqrt{2}\Delta t \, \sqrt{B \cdot t})]$
6. Pulse width	$I = 20 \log[t/(\Delta PW \, \sqrt{B \cdot t})]$
7. Pulse amplitude	$I = 20 \log(A/\Delta A)$

where
Δf = interpulse frequency change
t = transmitted pulse length
T = transmission time to and from target
$\Delta\Phi$ = interpulse phase change
Δt = time jitter
$B \cdot t$ = time–bandwidth product of pulse compression system ($B \cdot t$ = unity for uncoded pulses)
ΔPW = pulse width jitter
A = pulse amplitude, volts
ΔA = interpulse amplitude change

the surface of surrounding land (e.g., with tree, vehicle, etc., motion present); the surface of nearby bodies of water (e.g., with lake, sea, etc., motion/waves present); natural atmospheric reflections from rain and cloud motion; man-made atmospheric reflections (e.g., from chaff, consisting of drifting metallic elements distributed in space, usually introduced for military target masking), and so on. Based on consideration of these and other factors, a system's baseline clutter reference level can be determined. This is the reference level for a set of MTI quantitative capability yardsticks—yardsticks called *improvement factors*.

An improvement factor, I, is a power ratio. It is defined as

$$I = \frac{R_o}{R_i} \qquad (11\text{-}1)$$

where R_o is the power ratio of the target to the clutter at the output interface of the system, and R_i is the power ratio of the target to the clutter at the input interface of the receiver averaged over all target speeds. How does this relate to RF power generators? Table 11-1 lists the typical forms of RF power generator instability and the limit that each places on the system improvement factor [10]. The following example examines these instabilities.

Example 11-1

A 3000-MHz RF generator, transmitting an uncoded 2-μs pulse, has a requirement that no single system instability will limit the MTI improvement factor attainable at a range of 100 nautical mi (160.93 km) to less than 50 dB (a voltage ratio of 316:1).

In the following we consider this example in terms of the instabilities defined in Table 11.1 (i.e., addressing only RF power amplifier systems—items 1 and 4 in the table are for pulsed oscillator systems).

Stalo or Coho Frequency

In the second expression note that the interpulse frequency change period is determined by the elapsed time required for 100 mi (160.93 km) round-trip detection by the radar. RF wave travel time is 12.36 μs for each round-trip nautical mile. Then, in accordance with this expression, the MTI system's stalo and coho (see Fig. 6-23) must have pulse-to-pulse frequency changes of less than

$$\Delta f = \frac{1}{2\pi \times 316 \times 100 \times 12.36 \times 10^{-6}} = 0.4 \text{ Hz}$$

Transmitter Phase Shift

In accordance with the third expression, the pulse-to-pulse phase shift in the RF amplifier system must be less than

$$\Delta \phi = \frac{1}{316} = 0.00316 \text{ rad} = 0.018 \text{ deg}$$

Pulse Timing Jitter

In accordance with the fifth expression, the pulse-to-pulse timing jitter must be less than

$$\Delta t = \frac{2 \times 10^{-6}}{316 \cdot 2^{1/2} \cdot 1^{1/2}} = 4.5 \times 10^{-9} \text{ s}$$

Pulse-Width Jitter

In accordance with the sixth expression, the pulse-to-pulse width jitter must be less than

$$\Delta PW = \frac{2 \times 10^{-6}}{316 \cdot 1^{1/2}} = 6 \times 10^{-9} \text{ s}$$

Pulse Amplitude Change

In accordance with the seventh expression, the pulse-to-pulse amplitude change must be less than

$$\frac{\Delta A}{A} = \frac{1}{316} = 0.00316 = 0.3\%$$

Some Additional Phase Improvement Factor Examples

Let us now determine the pulse-to-pulse phase improvement factors of the klystron systems that we analyzed in Examples 10-1 and 10-2:

1. For the constant PRF/balanced line case we determined that the average pulse-to-pulse phase variation was 0.00186 rad. Then from the third expression in Table 11-1, we would have an improvement factor of 54.61 dB.
2. For the constant PRF/unbalanced line case the average pulse-to-pulse phase variation was 0.00404 rad. The improvement factor has now fallen to 47.87 dB.
3. For the staggered PRF case the average pulse-to-pulse phase variation was 0.0680 rad. Our improvement factor has now plunged to 23.35 dB.

Most current MTI radar systems require operating modes similar to or more demanding than the staggered PRF case described above. In Section 10-4 we noted two courses of action to satisfy these demanding cases: (1) either use of a well-regulated high power supply for the beam pulser, or (2) use of an alternate technical approach. In the following sections of this chapter we investigate several alternate technical approaches capable of providing improved system performance capabilities.

11-2 DESIGN APPROACH FOR A HIGH-STABILITY RF AMPLIFIER

Consider a burst pulse radar RF power amplifier with the following preliminary technical specification:

1. Frequency of operation   3.1–3.5 GHz
2. Peak power output   100 kW
3. Pulse burst   Four 1-μs pulses
4. Burst pulse separation   5–10 μs
5. Burst PRF   1000 pulses/s
6. Pulse droop (maximum)   10 V
7. Droop recovery time   Before next pulse in burst
8. MTI improvement factor   50 dB

Proposed System Configuration

A technical approach compatible with the burst pulse system requirements above is shown in Fig. 11-1. It is a variation of the grid-pulsed, depressed collector traveling-wave tube circuit analyzed in Chapter 9. Our analyses in Chapter 10 illustrated that a dc power supply operating directly from a public utility has an insufficient high recharging rate for high-PRF applications. This problem has been overcome in the critical TWT beam voltage area of Fig. 11-1 through the use of a dc–dc converter. Its switching frequency can be made compatible with the system's PRF recharging requirements.

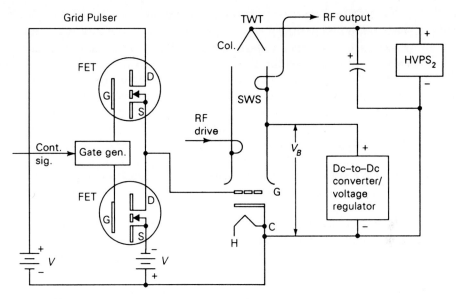

Figure 11-1 TWT power amplifier with depressed collector and beam regulation.

Assume that the system will use a representative 3-GHz 100-kW peak power level TWT: beam voltage 25 kV, collector voltage 15 kV, and -500 to $+500$ V grid swing. The beam current and the body current during a pulse have values of 15 and 0.75 A, respectively. The tube's RF gain is 50 dB.

The following are representative phase-and amplitude-sensitivity figures for a tube of this type:

PHASE SENSITIVITY

Beam voltage	40° for 1% voltage change
Grid voltage	7.5° for 1% voltage change
Collector voltage	0.5° for 1% voltage change
RF drive power	6° for 1 dB change

AMPLITUDE SENSITIVITY

Beam voltage	0.4 dB for 1% voltage change
Grid voltage	0.3 dB for 1% voltage change

Phase and Amplitude Budget

Based on our review of Table 11-1, we know that to achieve a 50-dB improvement factor, the value of $\Delta\Phi$ must be less than 0.00316 rad or 0.18 deg. A typical stability analysis approach starts with documentation of a phase budget. We will allocate appropriate portions of the 0.18-deg overall phase pushing figure to each of the phase areas indicated above. These individual $\Delta\Phi$ values will then be summed up to provide a worst-case analysis. In some cases the values are summed in rms fashion.

Sec. 11-2   Design Approach for a High–Stability RF Amplifier

TWT collector. Most of the dc energy supplied to the TWT amplifier's electrodes is consumed at the collector (i.e., 15,000 V × 14.25 A or 213.75 kW of peak power per pulse). Since this area's voltage pushing value is small, an unregulated 60-Hz power supply is typically used for the dc source. To support this point of view, note that the burst PRF is 1000 pulses/s. Therefore, the klystron analysis in Section 10-4 provides relevant information on the stability we could expect in this application. Based on that analysis it appears reasonable to predict better than 15-V burst-to-burst stability. Then (15 V/150 V) × 0.5 deg = 0.05 deg is a reasonable goal for a similar supply in this phase pushing area.

TWT grid. Based on our analysis of the grid pulser system in Chapter 9, and based on the advanced state of the art for regulated 500-V dc power supplies, it is reasonable to predict that a very high degree of stability can be achieved on a pulse-to-pulse basis in this area. Therefore, it seems reasonable to allocate 0.01 deg, equivalent to about 0.09 V pulse to pulse, to this area of the budget. This value should provide sufficient overall margin to allow for long-term characteristic variations in both the power supply and the switching circuit elements.

TWT RF drive. Assume that as a result of extensive system background and hardware design expertise available in this equipment area—which includes a low-power RF driver amplifier (several watts of peak power output), associated electrode power sources, and so on—the pulse-to-pulse phase pushing can be held below 0.02 deg.

TWT beam voltage. The remainder of the overall phase budget—0.08 deg (0.5 V) pulse to pulse—will be allocated to this critical area.

TWT amplitude. The amplitude improvement factor is below the previously determined 0.3% level (it is 0.166% based on the grid and beam voltage pulse-to-pulse departures noted above). Therefore, it makes no contribution to the phase and amplitude budget.

TWT RF signal and pulse gating. Based on the advanced state of the art in low-level generator technology, we will assume that the waveform operating package can readily accommodate all of the 50-dB improvement factor requirements associated with pulse-to-pulse RF frequency and pulse timing/control areas.

The budget clearly demonstrates that the beam voltage source is the area of prime concern because (1) it operates at a high voltage level, and (2) it has a high phase pushing figure. Therefore, in the analyses that follow we will be concerned fundamentally with the capabilities of several 25-kV dc regulated power source candidates.

11-3 DC-TO-DC POWER SUPPLY WITH A DC VOLTAGE SOURCE

Several previous analyses of dc–dc power supply systems were performed for systems designed to provide relatively low levels of dc output voltage, typically for the electrodes of solid-state power amplifier systems. In the analyses that follow, similar techniques will be investigated to establish these capabilities for high-voltage power supply applications. However, the models to be used will be more exact than those used in the low-voltage analyses. These improved analysis techniques are readily adaptable to any other type of system—low voltage or high voltage—that may be encountered.

System Diagram and Element Selection

   1. Figure 11-2a shows a circuit diagram of the power supply we consider as our first candidate for the TWT beam voltage source. The circuit defines a full-wave bridge dc–dc configuration. Battery V is the low-voltage source, and in these analyses we will assign it a value of 1000 V.

   2. Four 1000-V FETs—the same type as we used in the TWT pulser in Chapter 9—will be used as the inverter switches. For this analysis the inverter operating frequency will be 50 kHz.

   3. Transformer T is in effect a 10-µs pulse transformer. It has a step-up turns ratio of 1:30. The transformer's secondary winding connects to the input terminals of a bridge rectifier. Energy storage capacitor C_S connects across the dc output terminals of the bridge. R_B represents the load presented by the tube's body. Its value is pulsed on and off to simulate the pulse template.

   4. Inverter operation is controlled by the gate generators. The triggers from the control signal line initiate gate generator output in accordance with the following sequence. FET_1 and FET_2 are gated on in unison to provide one half-cycle of the ac wave applied to the primary of T. When the gates applied to this set of FETs terminate after 10 µs, FET_3 and FET_4 are gated on in unison to provide the alternate polarity half-cycle during the next 10-µs period—and so on—in normal operation.

   5. A quite simple form of voltage regulation will be utilized for this approach. When a system voltage sensor detects the fact that the power supply has charged up energy storage capacitor C_S to a value equal to or greater than 25,000 V, it will instantaneously send a command to the inverter circuit to cease charging. The rationale for accomplishing the regulation command action is to generate an overriding control signal that will terminate the gates supplied to the inverter FETs. We are in effect introducing a form of pulse width modulation when we introduce this approach.

   To summarize, this power supply concept is based on (1) using a high-frequency inverter to obtain a high storage capacitor recharging rate, and (2) using pulse width modulation to further enhance the regulation capabilities of the supply. A determination of the extent to which this approach is effective will be established by our first computer analysis.

Analysis Program Structure

Figure 11-2b details the network diagram used for the circuit analysis. The program of Appendix 31 documents the diagram's circuit element values as follows. Lines 51–53 list the input capacitance C_I, the feedback capacitance C_F, and the output capacitance C_O,

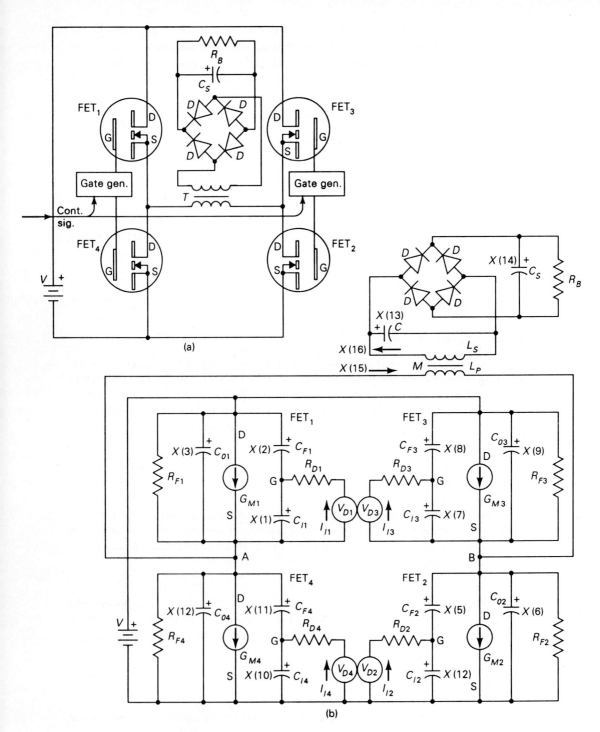

Figure 11-2 TWT beam voltage dc–dc regulator with a voltage source: (a) circuit diagram; (b) network diagram.

respectively, for each of the inverter FETs. Lines 56 and 57 define the primary and secondary inductance of the inverter transformer. Line 58 defines the mutual inductance of the two windings. Line 54 defines the transformer's shunt capacitance. Line 55 lists C_S, the pulse energy storage capacitor. This value is sufficient to limit the droop of each of the 1-µs pulses to 10 V.

Line 59 lists the resistance of R_B—the TWT's body loading value. When a 1-µs pulse is in progress R_B has a value of 33,333 Ω. This absorbs 0.75 A of TWT body interception current. During the pulsed-off periods, a value of 50 MΩ is assigned. The transitions are controlled by lines 1102–1110.

The first entry on line 60 lists the value of R_D, the series resistance associated with each gate drive circuit. The second set of entries list the values of R_{F1}, R_{F2}, R_{F3}, and R_{F4}—the blocking resistances of the FET reverse diodes. Program lines 1180 and 1220 modify these resistance values to 1 Ω when a reverse diode conduction is required.

Line 61 defines the input dc voltage, V, as 1000 V. Line 62 defines the maximum FET transconductance as G. The specific value of G_M for each set of FETs depends on the inverter operating conditions that exist at any point in time. Lines 1170, 1200, 1210, 1240, and 1250 determine these specific values. Line 20 denotes 17 differential equations are involved. Lines 70–77 list the initial conditions for the voltages and currents involved.

The dc–dc power supply status represented by the initial conditions is as follows: first, the system is in operation at $T = 0$; second, the 1-µs value (pulsed on value of 33,333 Ω) of load resistor R_B has just been established at $T = 0$, and this value will remain in effect until $T = 10^{-6}$; third, the inverter has just concluded the negative high-voltage output half-cycle and has received the command to start switching to the positive half-cycle; and fourth, the inverter is not delivering energy to storage capacitor C_S at $T = 0$.

System Equations

The systems equations can be developed as follows. First, the FET gate drive algebraic equations are as follows:

$$II1 = (VD1 - X(1))/RD \qquad (11\text{-}2)$$

$$II2 = (VD2 - X(6))/RD \qquad (11\text{-}3)$$

$$II3 = (VD3 - X(9))/RD \qquad (11\text{-}4)$$

$$II4 = (VD4 - X(12))/RD \qquad (11\text{-}5)$$

These four equations are listed in the equations subroutine of Appendix 30 on lines 1006–1012.

For each of the FETs, the sum of the derivatives of capacitor voltage around each of their respective loops is zero; hence we can write

$$D(3) = D(1) + D(2) \qquad (11\text{-}6)$$

$$D(6) = D(4) + D(5) \qquad (11\text{-}7)$$

$$D(9) = D(7) + D(8) \qquad (11\text{-}8)$$

$$D(12) = D(10) + D(11) \qquad (11\text{-}9)$$

Also, the following equations are based on the fact that indicated derivative sums are connected across a constant, V:

$$D(3) + D(12) = 0 \tag{11-10}$$

$$D(6) + D(9) = 0 \tag{11-11}$$

Current summations at the four FET G nodes give

$$II1 = CI\ D(1) - CF\ D(2) \tag{11-12}$$

$$II2 = CI\ D(4) - CF\ D(5) \tag{11-13}$$

$$II3 = CI\ D(7) - CF\ D(8) \tag{11-14}$$

$$II4 = CI\ D(10) - CF\ D(11) \tag{11-15}$$

Next, if a current summation is made at node A, the following can be written:

$$X(3)/RF1 + CO\ D(3) + GM1\ X(1) + CF\ D(2) = $$
$$X(13) + X(12)/RF4 + CO\ D(12) + GM4\ X(10) + CF\ D(11) \tag{11-16}$$

Also, Eqs. (11-6) and (11-12) can be combined to give

$$D(2) = (CI\ D(3) - II1)/(CI + CF) \tag{11-17}$$

and Eqs. (11-9) and (11-15) combined give

$$D(11) = -(CI\ D(3) + II4)/(CI + CF) \tag{11-18}$$

When Eqs. (11-10), (11-17), and (11-18) are substituted into Eq. (11-16), the solution for $D(3)$ is

$$D(3) = (X(15) + X(12)/RF4 - X(3)/RF1 + GM4\ X(10)$$
$$- GM1\ X(1) + K1(II1 - II4))/K2 \tag{11-19}$$

The constants introduced for equation simplification are

$$K1 = CF/(CI + CF) \tag{11-20}$$

$$K2 = (2\ CO + 2\ CI\ K1) \tag{11-21}$$

$$K3 = CI + CF \tag{11-22}$$

Equation (11-19) is listed in the equations subroutine of Appendix 31 on line 1014. The remaining equations associated with the current summation at node A follow directly from the above as

$$D(2) = (CI\ D(3) - II1)/K3 \tag{11-23}$$

$$D(1) = D(3) - D(2) \tag{11-24}$$

$$D(12) = -D(3) \tag{11-25}$$

$$D(11) = -(CI\ D(3) + II4)/K3 \tag{11-26}$$

$$D(10) = -(D(3) + D(11)) \tag{11-27}$$

These equations are in the form listed on program lines 1016–1024.

Now if a similar current summation and solution approach is performed at node B, the following equations will result:

$$D(9) = (-X(15) + X(6)/RF2 - X(9)/RF3 + GM2\ X(4) \\ - GM3\ X(7) + K1(I I3 - I I2))/K2 \quad (11\text{-}28)$$

$$D(8) = (CI\ D(9) - I I3)/K3 \quad (11\text{-}29)$$

$$D(7) = D(9) - D(8) \quad (11\text{-}30)$$

$$D(6) = -D(9) \quad (11\text{-}31)$$

$$D(5) = -(CI\ D(9) + I I2)/K3 \quad (11\text{-}32)$$

$$D(4) = -(D(9) + D(5)) \quad (11\text{-}33)$$

These six equations are listed in the program's equation subroutine as lines 1026–1036.

If mesh equations are written around the loops associated with X(15) and X(16), we have

$$LP\ D(15) - M\ D(16) + X(3) + X(6) - V = 0 \\ -M\ D(15) + LS\ D(16) + X(13) = 0$$

When these two equations are solved, we obtain

$$D(15) = (LS\ (V - X(6) - X(3)) - M\ X(13)/(LP\ LS - M^2) \quad (11\text{-}34)$$

$$D(16) = (M\ D(15) - X(13))/LS \quad (11\text{-}35)$$

These two equations are listed as program lines 1002 and 1004.

If the absolute value of X(13) is less than X(14), the following two high-voltage node equations can be written:

$$D(13) = X(16)/C \quad (11\text{-}36)$$

$$D(14) = -X(14)/(RB\ CS) \quad (11\text{-}37)$$

The program will then proceed. Program line 1038 lists the two equations above and a conditional statement. Hence, if the BASIC conditional statement,

$$\text{IF } (X(13)) = > X(14)\ \text{AND SGN } (X(13)) = \text{SGN } (X(16))$$

prevails, an alternate approach is required.

The alternate approach is based on the fact that C and C_S will be charged in parallel (i.e., energy will be stored in C_S). In this case we need only calculate one of the derivatives, D(14), and using BASIC terminology we can write

$$X(13) = \text{SGN } (X(13))\ X(14) \quad (11\text{-}38)$$

$$D(14) = (\text{ABS } (X(16) - X(14)/RB)/(C + CS) \quad (11\text{-}39)$$

$$D(13) = \text{SGN } (D(13))\ D(14) \quad (11\text{-}40)$$

These three equations are listed on line 1046.

An initial analysis was conducted with the program set up in this manner. The analysis showed that the system operated outside its voltage variation budget figure of

0.5 V. It was rationalized that this problem could be corrected if the drive to all four FETs was switched off during any pulse generation period.

To accomplish this, note line 1116. A new variable, D_R, denotes the status of the storage capacitor's voltage. If $D_R = 0$, the voltage across C_S is equal to or greater than 25,000 V or a pulse is in progress. If $D_R = 1$, the voltage across C_S is less than 25,000 V.

In addition, an integration period change subroutine has also been added starting at line 1300. It functions to speed up the execution of the program by increasing T_1 to five times the initially programed value when $D_R = 0$.

The reasoning behind this approach is that when $X(14)$ is greater than 25,000 V, the rate of change in the system variables occurs at a relatively slow rate, hence no significant degree of system computational accuracy is lost. For the program to function properly with the speed-up subroutine, program lines 32, 1038, 1040, and 1042 have been modified to accommodate the alternate slow/fast system operating modes.

Analysis Results

If you load Appendix 31 into your computer, execute a <RUN> command, and follow this with a P command, a secondary winding voltage response plot [i.e., $X(13)$] similar to Fig. 11-3 will develop on your printer. In addition, a tabulation of the power supply output voltage [i.e., $X(14)$] is shown along the edge of the plot.

These two displays allow us to assess (1) if the power supply is performing within its budgeted pulse-to-pulse voltage variation value of 0.5 V, and (2) it allows us to assess how well the regulation control modification is doing its job. The following is a summary of these two areas.

1. The beam voltage values from Fig. 11-3, shown below, indicate that any initial pulse voltage, any final pulse voltage, or any pulse droops check out within 0.5 V of one another for the three cases. Therefore, the alternate regulation approach appears to be effective.
2. Termination of inverter action is evident at each of the voltage departures from the ±25-kV lines in the plotted display of inverter switching voltage. In effect, these departures are formed by a stored energy interchange between, on the one hand, the transformer and capacitor C, and on the other hand, the FET capacitances (i.e., after the FET gates are cut off). In some cases capacity coupling, via C_F, causes the FETs to turn on slightly during the energy transfer.

Pulse	Intial voltage	Final voltage	Droop
1	25,000.01	24,992.86	7.15
2	25,000.34	24,992.93	7.41
3	25,000.30	24,992.88	7.42
4	25,000.34		

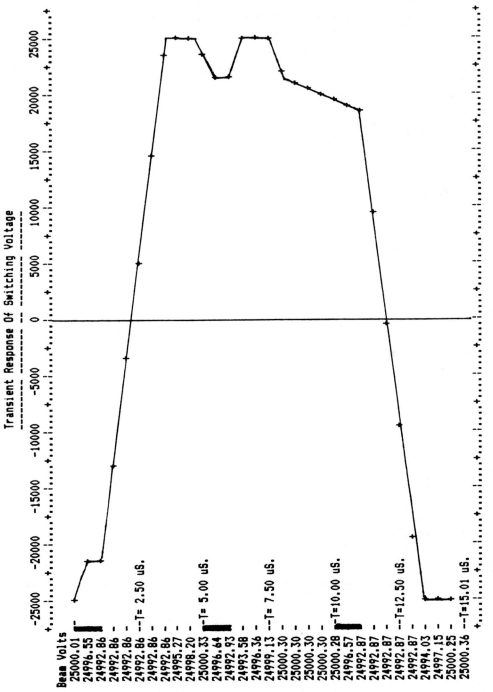

Figure 11-3 Appendix 31 plot of dc–dc regulator/voltage source/inverter clanking.

3. A superimposed high-frequency oscillation will also be detected if a fine-grain observation is performed. This oscillation occurs due to a resonant circuit involving the leakage inductance and the previously noted capacitances.

11-4 DC-TO-DC POWER SUPPLY WITH A CURRENT SOURCE

Another fundamental technique found in dc–dc power supply circuit repertoires employs a current source in place of the voltage source we have used in our previous analyses. Figure 11-4a shows the circuit diagram of this arrangement. The diagram is identical to Fig. 11-2a except for the source symbol.

Modifications and Equation Development

Typically, the constant-current source used in these power supplies is a fairly large inductor, as indicated in the network diagram of Fig. 11-4b. Why would anyone become involved with the additional complexity associated with a current source? One reason is that the optimum source voltage required to operate a dc–dc supply varies as a function of the instantaneous point in time observed during the ac cycle of the inverter. For example, consider the period of time where the inverter's output voltage wave is reversing polarity. From a qualitative point of view we can see that the dc voltage required to operate the system need only be sufficient to supply the required primary voltage plus the active switch voltage drops of the system. With a voltage source system the full level of voltage is present at all times. Therefore, the switching elements must operate at quite high dissipation levels while passing through this region.

On the other hand, with a current-fed system the source voltage appears across the energy storage inductor. It acts to adjust itself to a voltage level just adequate to support the inverter's primary side voltage drops. As a result, significantly reduced switch heating commonly results. Perhaps the easiest way to establish some quantitative background on a current-fed system is to perform a simplified computer analysis. Our initial analysis is based on the diagram of Fig. 11-7b. The same general approach that we used in the voltage-fed systems will be used to set up the current-fed analysis equations. For example, program lines 1000–1010 present the same FET gate drive equations used for the voltage source system.

For this analysis, in place of energy storage inductor L_G, a constant-current source has been set up by definition, $I_S = 16.5$ A, in program line 78. If a summation is made at the junction encompassing current I_S, the currents at the D node of FET_1, and the currents at the D node of FET_3, we can write

$$IS = X(3)/RF1 + CO\ D(3) + GM1\ X(1) + CF\ D(2) \\ + X(9)/RF3 + CO\ D(9) + GM3\ X(7) + CF\ D(8) \quad (11\text{-}41)$$

In similar fashion we can sum up the currents at node A,

$$X(15) = X(3)/RF1 + CO\ D(3) + GM\ X(1) + CF\ D(2) \\ - X(12)/RF4 - CO\ D(12) - GM\ X(10) - CF\ D(11) \quad (11\text{-}42)$$

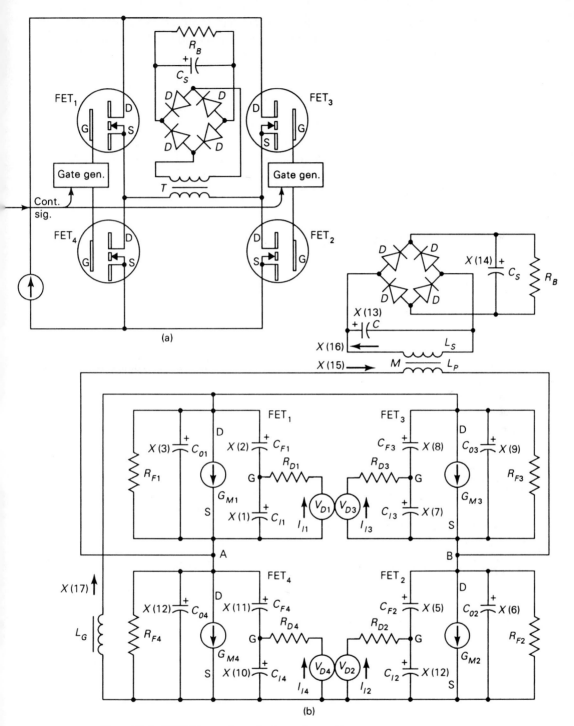

Figure 11-4 TWT beam voltage dc–dc regulator with a current source: (a) circuit diagram; (b) network diagram.

Sec. 11-4   DC–to–DC Power Supply with a Current Source   273

It will be recalled from the voltage source analysis that $X(7) = X(10)$, $D(9) = D(12)$, and $D(11) = D(8)$. By making the substitutions and adding Eq. (11-41) and (11-42), we obtain

$$X(3)/RF1 + CO\ D(3) + GM\ X(1) + CF\ D(2) = (IS + X(15))/2$$

Also from the voltage source analysis we have the equation $D(2) = (CI\ D(3) - I11)/K3$. If this substitution is made in the above, the resulting equation simplifies to

$$D(3) = ((IS + X(15))/2 - X(3)/RF1 - GM1\ X(1) \\ + K1\ I11)/(CO + CI\ K1) \tag{11-43}$$

This equation, along with several associated equations derived previously in the voltage source analysis, are listed as program line 1052.

If Eq. (11-42) is subtracted from Eq. (11-41) we have

$$X(9)/RF3 + CO\ D(9) + GM3\ X(7) + CF\ D(8) \\ = (IS - X(15))/2 \tag{11-44}$$

By following the same procedure as used in solving for $D(3)$ we will obtain

$$D(9) = ((IS - X(15)) - X(3)/RF1 - GM3\ X(7) \\ + K1\ I11)/(CO + CI\ K1) \tag{11-45}$$

This equation, along with several associated equations derived previously in the voltage source analysis, are listed as program line 1054. Lines 1052 and 1054 therefore define the current-fed system's derivatives $D(1)$–$D(12)$.

By writing a mesh equation including FET_1, FET_3, and the transformer, we have

$$LP\ D(15) - M\ D(16) + X(3) - X(9) = 0$$

The secondary equation is the same as for the voltage source case,

$$-M\ D(15) + LS\ D(16) + X(13) = 0$$

If the two equations above are solved, we obtain

$$D(15) = (LS\ (X(9) - X(3)) - M\ X(13))/(LP\ LS - M^2) \tag{11-46}$$

$$D(16) = (M\ D(15) - X(13))/LS \tag{11-47}$$

These two equations are listed on program lines 1062 and 1064.

Finally, the equations for $D(13)$ and $D(14)$ will be the same as for the voltage source analysis, since that portion of the circuit was not changed. They are listed on program lines 1068 and 1072. We now have a complete set of equations for our first constant-current-source analysis.

The program was initially evaluated in this form. A major flaw was uncovered— when the program removes the gate drive from all FETs their drain-to-source voltage rises above the 1000-V maximum rating for the devices. To correct this situation, the design approach was changed to the voltage-limited current source shown in the diagram of Fig. 11-5b. It functions as follows. First, the voltage-limited current source uses four elements. L_G is an energy storage inductor, V is a voltage-limiting power supply (the same supply that we used in the voltage source case), D_L is a voltage limiting diode, and D_F is a *free wheeling* diode. Second, the current path via freewheeling diode D_F and storage inductor L_G provides the identical current source route that we used in Fig. 11.4. This path will be utilized when two simultaneous system conditions prevail—when the voltage across L_G is less than V, and when the current required by the inverter (the current supplied by L_G) is equal to or greater than I_S. Third, when the inverter load current falls below current I_S, the voltage across L_G will start to rise. When this voltage's level exceeds the voltage of V, the excess current from L_G will be diverted into V. If we were unaware of the circuit being used, an external observation at the circuit's terminals would indicate that it was a voltage source. Hence the four circuit elements will function as a simple voltage-limited current source.

We will use the program of Appendix 33 to analyze this configuration. The two modes of operation will be defined by string variable M\$ = "I" for current source operation and M\$ = "V" for voltage source operation. The modes will be handled as two subroutine paths. The first subroutine (where M\$ = "I") is almost complete. We need only introduce a conditional voltage sensing statement that will transfer to the program to the M\$ = "V" mode when the inverter voltage is greater than V. Referring to Fig. 11-5b, the voltage-sensing statement will use the sum of the voltages across $X(3)$ and $X(12)$ as the conditional value that is compared to V. Line 1060 defines this conditional statement. The current subroutine of Appendix 33 is then complete.

The voltage subroutine, used when M\$ = "V", requires us to place four currents under observation. From their status we can establish when to transfer back to the program's current subroutine. The currents are storage current I_S, transformer primary current $X(15)$, current passing through FET_1—I_1, and the current passing through FET_3—I_2. The equations for I_1 and I_2 can be written as

$$I1 = CO\ D(3) + CF\ D(2) + GM1\ X(1) + X(3)/RF1 \qquad (11\text{-}48)$$

$$I2 = CO\ D(9) + CF\ D(8) + GM3\ X(4) + X(9)/RF3 \qquad (11\text{-}49)$$

These two equations are listed as program lines 1020 and 1022 of the voltage subroutine.

The analysis equations when using the voltage path are identical to those we developed for previous voltage source systems. The equations for derivatives $D(1)$–$D(16)$ are listed in compressed form on program lines 1016, 1018, and 1028–1036. Either of two conditional statements will transfer the subroutine's mode from the voltage path to the current path. The rationale for these statements is as follows: FIRST: A stable transfer to the current mode can be made if transformer primary current $X(15)$ is following a decreasing amplitude path as it passes through a value of I_S, and at the same point in time, the value of $I_1 + I_2$ is greater than that of I_S. SECOND: A stable transfer to the current mode can also be made if transformer primary current $X(15)$ is following an increasing amplitude path as it passes through a value of I_S, and at this same point in time, the value of $I_1 - I_2$ is less than I_S. These statements are written in conditional form in program lines 1024 and 1026.

This completes the voltage subroutine entries for the program. Two additional programming details are: *line 1113* has been added to change the drive voltage on all FETs to zero volts during a pulse—if the voltage across the secondary of the transformer is greater than 20,000 V. *A printout subroutine column* has been added to indicate if the voltage or current mode is in use.

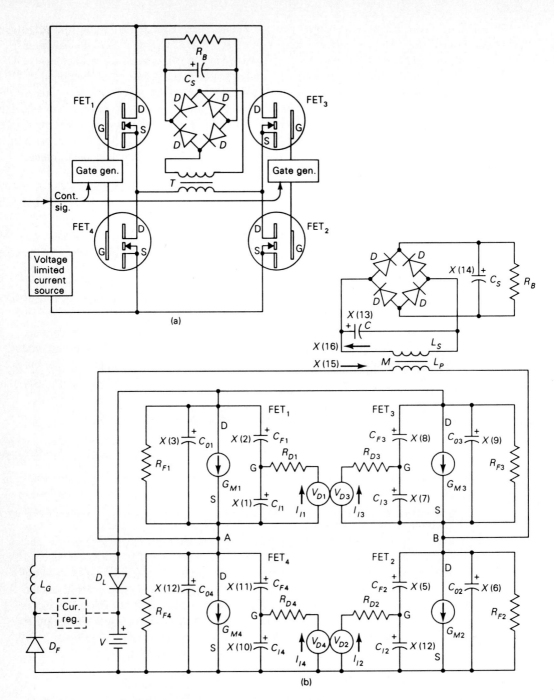

Figure 11-5 TWT beam voltage regulator with voltage-limited current source: (a) circuit diagram; (b) network diagram.

The rationale for the estimated initial conditions listed between lines 70 and 77 is as follows: (1) the transformer is in the final stages of completing a negative voltage half-cycle; (2) the storage capacitor is fully charged and the inverter is in its cutoff mode; and (3) the first body current-loading pulse will occur between $T = 1$ and $T = 2$ μs.

Analysis Results

Now if you load Appendix 33 and then enter a <RUN> command, a display similar to Fig. 11-6 will begin to develop on your printer. The objective of this run is to assess how the system is functioning.

1. For the period $T = 0$ through 1.0 μs we can see that the system has nearly steady-state status other than for a small damped oscillation that can be detected across the FET capacitances. This oscillation occurs when stored energy is interchanged between the transformer winding inductances and the FET capacitances (i.e., the FETs have been driven into their cutoff mode of operation during this period).

2. For the period $T = 1.0$ through 2.0 μs pulse loading occurs. TWT body voltage droops from 25,000.30 V to 24,992.49 V during the pulse. Also, as in the voltage source system case, the inverter remains in its cutoff mode during the pulse.

3. For the period $T = 2.0$–4.6 μs voltage $X(13)$ polarity reversal is taking place across the transformer secondary. Note that the active mode of the inverter is now in effect, and operation alternates between the voltage and current modes. When the current mode is active, the circuit works to decrease FET voltage drops, and hence the heat dissipation in the FETs is reduced. Although the current mode is not as predominant as with constant-current operation, it exercises a moderating effect on system operation.

4. For the period $T = 4.7$–6.0 μs the inverter is recharging the pulse storage capacitor of the power supply. Output voltage $X(14)$ settles in at essentially the same initial voltage level as for the previous pulse.

5. For the period $T = 6.0$–7.0 μs the second TWT body voltage pulse occurs. It droops from 25,000.33 V to 24,992.52 V. Based on this short run the system's degree of pulse-to-pulse stability significantly exceeds our requirements.

11-5 STEADY-STATE DC-TO-DC POWER SUPPLY ANALYSIS

We have no assurance that the dc–dc power supply system described in Section 11-4 will continue to operate in the manner indicated by the short computer run—other than by conducting a steady-state run over a longer period of time. Let us conduct such a run over a time period encompassing the next six TWT pulses. If we select our initial conditions at the conclusion of the second pulse of the previous run (i.e., at 7.0 μs), it seems likely that minimum startup perturbations will occur since the inverter is then in its cutoff mode.

Based on the rationale above, the following program line modifications need to be inserted in Appendix 33:

```
34 T2=.0000005
34 T3=.000037
38 T=.000007:Q=T
70X(1)=.2:X(2)=171.4:X(3)=171.6
71X(4)=.2:X(5)=171.4:X(6)=171.6
72X(7)=-.2:X(8)=854.3:X(9)=854.1
73X(10)=-2:X(1)=854.3:X(12)=854.1
74X(13)=22320.9
75X(14)=24992.52
76X(15)=1.2
77X(16)=.035
```

Now if we load the program, issue a <RUN> command, and follow this with a P, the plot shown in Fig. 11-7 will develop. In a very brief summary, it shows that:

1. A composite of the first eight 25,000-V TWT pulses (i.e., from both runs) shows that they have initial values that are within 0.05 V of one another. The final values of the eight pulses are also within 0.04 V of one another.
2. The pre-pulse and post-pulse inverter cutoff periods are functioning properly.
3. Beam voltage stability exceeds the 0.5-V level of dV budgeted for 50-dB MTI performance. Based on this analysis, the beam voltage improvement factor is in excess of 70 dB.
4. The system performance capability present during this analysis appears to be marginal (e.g., there has been barely enough time available to recharge the storage capacitor C_S following each inverter voltage reversal). An improved safety margin could probably be achieved by using a transformer with a lower internal value of shunt capacitance C. This would increase the transformer's rate of change of voltage during each reversal and thereby provide a longer time period for storage capacitor charging.

Tabulated Characteristics Of A D-C To D-C Regulator With A Voltage Limited Current Source

uSec.	Mode	X(1)	X(2)	X(3)	X(4)	X(5)	X(6)	X(7)	X(8)	X(9)	X(10)	X(11)	X(12)	X(13)	X(14)	X(15)	X(16)
0.000	V	0.0	900.0	900.0	0.0	900.0	900.0	0.0	100.0	100.0	0.0	100.0	100.0	-21300.00	25000.30	0.00	0.000
0.100	V	-0.1	836.2	836.1	-0.1	836.2	836.1	0.1	163.8	163.9	0.0	163.8	163.9	-21379.12	25000.30	-0.91	-0.030
0.200	V	0.1	834.9	835.0	0.1	834.9	835.1	-0.1	165.1	165.0	-0.1	165.1	165.0	-21386.32	25000.30	0.73	0.026
0.300	V	-0.1	888.4	888.5	0.0	888.4	888.5	-0.0	111.6	111.5	-0.0	111.6	111.5	-21307.92	25000.30	0.16	0.007
0.400	V	-0.1	846.4	846.3	-0.1	846.4	846.3	-0.1	153.6	153.7	-0.0	153.6	153.7	-21352.09	25000.30	-0.73	-0.022
0.500	V	0.0	835.9	835.9	0.0	835.9	835.9	-0.1	164.1	164.1	-0.0	164.1	164.1	-21363.40	25000.30	0.45	0.018
0.600	V	-0.1	878.4	878.4	0.0	878.4	878.4	-0.1	121.6	121.6	-0.0	121.6	121.6	-21295.77	25000.30	0.24	0.011
0.700	V	-0.1	852.5	852.5	-0.1	852.5	852.5	-0.0	147.5	147.5	-0.0	147.5	147.5	-21313.39	25000.30	-0.56	-0.015
0.800	V	0.0	837.9	837.9	0.0	837.9	837.9	-0.0	162.1	162.1	-0.0	162.1	162.1	-21322.56	25000.30	-0.25	-0.013
0.900	V	-0.1	870.1	870.2	-0.1	870.1	870.2	-0.1	129.9	129.8	-0.0	129.9	129.8	-21263.84	25000.30	0.25	0.014
1.000	V	-0.1	855.6	855.6	-0.1	855.6	855.6	0.1	144.4	144.4	-0.1	144.4	144.4	-21261.54	25000.30	-0.41	-0.008
1.100	V	0.0	840.1	840.1	-0.1	840.1	840.1	-0.1	159.9	159.9	-0.0	159.9	159.9	-21212.60	24998.75	0.12	-0.010
1.200	V	0.0	863.6	863.6	-0.0	863.6	863.6	-0.0	136.4	136.4	-0.0	136.4	136.4	-21212.60	24998.75	0.24	-0.015
1.300	V	-0.0	856.6	856.5	-0.0	856.6	856.5	0.0	143.4	143.5	-0.0	143.4	143.5	-21195.40	24997.97	-0.30	-0.002
1.400	V	-0.0	841.9	841.9	-0.0	841.9	841.9	0.1	158.1	158.1	-0.0	158.1	158.1	-21190.82	24997.18	-0.03	0.009
1.500	V	-0.0	858.3	858.3	-0.0	858.3	858.3	-0.0	141.7	141.7	-0.0	141.7	141.7	-21142.60	24996.40	-0.20	0.016
1.600	V	-0.0	856.0	856.0	-0.0	856.0	856.0	-0.0	144.0	144.0	-0.0	144.0	144.0	-21114.18	24995.62	-0.20	0.003
1.700	V	-0.0	843.1	843.1	-0.0	843.1	843.1	0.0	156.9	156.9	-0.0	156.9	156.9	-21100.84	24994.84	-0.02	-0.009
1.800	V	-0.0	854.0	854.0	-0.0	854.0	854.0	-0.0	146.0	146.0	-0.0	146.0	146.0	-21054.41	24994.06	-0.17	-0.016
1.900	V	-0.0	854.4	854.4	-0.0	854.4	854.4	-0.0	145.6	145.6	-0.0	145.6	145.6	-21017.32	24993.28	-0.13	-0.007
2.000	V	-0.0	843.7	843.7	-0.0	843.7	843.7	0.0	156.3	156.3	-0.0	156.3	156.3	-20995.06	24992.50	-0.05	0.010
2.100	V	0.9	329.3	333.2	3.9	329.3	333.2	0.9	666.8	666.8	0.4	666.4	666.8	-20681.33	24992.49	9.93	0.343
2.200	V	8.2	770.7	779.0	8.2	770.7	779.0	0.0	3.7	3.7	-0.0	-4.4	3.7	-18347.26	24992.49	27.94	0.944
2.300	I	4.9	1025.1	1029.9	4.9	1025.1	1029.9	0.1	-4.4	-4.2	0.1	-4.4	-4.2	-15885.53	24992.49	16.23	0.554
2.400	I	4.3	681.8	686.1	4.3	681.8	686.1	0.7	338.9	339.6	-0.7	338.9	339.6	-14608.19	24992.49	9.77	0.339
2.500	I	5.1	505.1	510.2	5.1	505.1	510.2	-0.1	515.6	515.5	-0.1	515.6	515.5	-13114.93	24992.49	18.77	0.639
2.600	I	5.5	820.8	826.3	5.5	820.8	826.3	-0.5	199.9	199.4	-0.5	199.9	199.4	-10799.59	24992.49	20.69	0.704
2.700	I	4.5	739.9	744.5	4.5	739.9	744.5	0.5	280.8	281.2	0.3	280.8	281.2	-9019.60	24992.49	12.41	0.428
2.800	I	4.7	472.5	477.2	4.7	472.5	477.2	0.3	548.2	548.5	-0.3	548.2	548.5	-7651.74	24992.49	14.53	0.499
2.900	I	5.4	566.5	571.9	4.9	566.5	571.9	-0.4	454.2	453.8	-0.4	454.2	453.8	-5665.92	24992.49	20.77	0.707
3.000	I	4.9	688.0	692.9	4.9	688.0	692.9	0.1	332.7	332.8	-0.3	332.7	332.8	-3575.03	24992.49	16.24	0.556
3.100	V	4.6	490.0	494.6	4.6	490.0	494.6	0.4	530.7	531.1	-0.4	530.7	531.1	-2033.71	24992.49	13.32	0.459
3.200	V	5.1	397.0	402.1	5.1	397.0	402.1	-0.1	623.6	623.6	-0.2	623.6	623.6	-340.25	24992.49	18.14	0.620
3.300	V	5.2	544.9	550.1	5.2	544.9	550.1	-0.3	475.8	475.6	-0.3	475.8	475.6	1763.30	24992.49	18.81	0.642
3.400	I	4.7	482.1	486.8	4.7	482.1	486.8	0.3	538.6	538.9	0.3	538.6	538.9	3571.98	24992.49	14.57	0.501
3.500	V	4.8	330.9	335.7	4.8	330.9	335.7	0.3	689.8	690.0	0.3	689.8	690.0	5176.75	24992.49	15.75	0.540
3.600	V	5.2	364.0	369.2	5.2	364.0	369.2	-0.2	656.5	656.5	-0.2	656.7	656.5	7103.21	24992.49	18.97	0.647

Figure 11-6   Appendix 33 analysis of dc–dc regulator/voltage-limited current source.

3.700 V	4.9	412.9	417.8	0.1	607.8	607.9	0.1	9086.25	24992.49	16.68	0.571
3.800 I	4.8	298.7	303.5	0.2	722.0	722.2	0.2	10781.76	24992.49	15.05	0.516
3.900 V	5.0	231.3	236.3	0.0	789.4	789.4	0.0	12540.55	24992.49	17.46	0.596
4.000 V	5.1	288.8	293.9	-0.1	731.9	731.8	-0.1	14520.77	24992.49	18.09	0.616
4.100 V	4.9	252.7	257.6	0.1	768.0	768.1	0.1	16367.59	24992.49	15.86	0.542
4.200 I	4.9	157.2	162.1	-0.1	863.5	863.6	-0.1	18081.71	24992.49	16.14	0.551
4.300 V	5.1	146.4	151.4	-0.1	874.7	874.6	-0.1	19945.68	24992.49	17.95	0.610
4.400 V	5.0	166.4	171.4	0.0	854.3	854.3	0.0	21876.59	24992.49	17.14	0.583
4.500 I	4.9	102.6	107.4	-0.1	918.1	918.3	-0.1	23658.19	24992.49	15.94	0.542
4.600 V	5.0	66.1	71.1	-0.0	954.6	954.6	-0.0	24992.62	24992.62	16.60	0.563
4.700 V	5.0	68.5	73.5	0.0	952.2	952.2	0.0	24993.21	24993.21	17.16	0.582
4.800 V	5.0	85.4	90.4	-0.0	949.0	949.0	-0.0	24993.80	24993.80	17.61	0.596
4.900 V	5.0	81.0	85.9	-0.0	935.3	935.3	-0.0	24994.38	24994.38	17.30	0.595
5.000 V	5.0	72.4	77.4	-0.0	939.7	939.7	-0.0	24994.97	24994.97	17.44	0.584
5.100 V	5.0	79.1	84.1	0.0	948.3	948.3	0.0	24995.55	24995.55	17.44	0.588
5.200 V	5.0	83.4	88.4	0.0	941.6	941.5	0.0	24996.14	24996.14	17.44	0.594
5.300 V	5.0	76.2	81.2	-0.0	937.3	937.3	-0.0	24996.73	24996.73	17.37	0.587
5.400 V	5.0	75.2	80.2	-0.0	944.5	944.5	-0.0	24997.31	24997.31	17.57	0.584
5.500 V	5.0	82.2	87.2	-0.0	945.5	945.5	-0.0	24997.90	24997.90	17.55	0.590
5.600 V	5.0	79.5	84.5	-0.0	938.5	938.5	-0.0	24998.48	24998.48	17.39	0.588
5.700 V	5.0	75.3	80.3	-0.0	941.2	941.1	-0.0	24999.07	24999.07	17.47	0.582
5.800 V	5.0	526.0	528.0	-0.0	945.4	945.4	-0.0	24999.66	24999.66	13.28	0.584
5.900 V	2.0	526.0	528.0	0.9	499.6	497.7	0.9	25000.23	25000.23	-13.56	0.444
6.000 V	-1.8	336.5	334.7	-1.9	689.2	691.0	-1.9	24550.61	25000.33	-13.56	-0.451
6.100 V	0.4	15.9	16.3	-0.4	1009.8	1009.8	-0.4	23148.84	24998.56	-10.65	-0.355
6.200 V	0.2	6.8	7.0	-0.2	1018.9	1018.7	-0.2	22338.27	24998.77	-4.46	-0.149
6.300 V	0.2	30.8	31.0	-0.2	994.9	994.6	-0.2	22205.19	24997.99	1.87	0.061
6.400 V	-0.4	230.9	231.1	-0.4	794.8	794.6	-0.4	22467.55	24997.21	1.31	0.042
6.500 V	-0.1	125.0	124.6	0.4	900.7	901.1	0.4	22341.05	24996.43	-2.58	-0.088
6.600 V	-0.1	46.2	46.3	-0.1	979.5	979.4	-0.1	22210.42	24995.65	1.00	0.030
6.700 V	-0.2	196.6	196.6	-0.2	829.1	829.1	-0.2	22396.51	24994.87	1.32	0.040
6.800 V	-0.3	141.1	140.9	-0.3	884.6	884.8	-0.3	22327.36	24994.09	-1.87	-0.066
6.900 V	-0.0	62.9	62.9	-0.3	962.8	962.8	-0.3	22198.07	24993.31	0.42	0.009
7.000 V	0.2	171.4	171.6	-0.2	854.3	854.1	-0.2	22320.90	24992.52	-1.20	-0.035
7.100 V	4.9	11.2	16.1	-0.7	1010.3	1009.6	-0.7	22477.94	24992.52	3.91	0.124
7.200 V	5.0	31.4	36.4	-0.0	989.3	989.3	-0.0	23137.68	24992.52	8.61	0.280
7.300 V	5.0	44.9	50.0	-0.0	975.7	975.7	-0.0	24211.29	24992.52	11.72	0.383
7.400 V	5.0	52.3	57.3	-0.0	968.4	968.3	-0.0	24992.67	24992.67	13.41	0.439
7.500 V	5.0	57.4	62.4	-0.0	963.3	963.3	-0.0	24993.15	24993.15	14.59	0.477

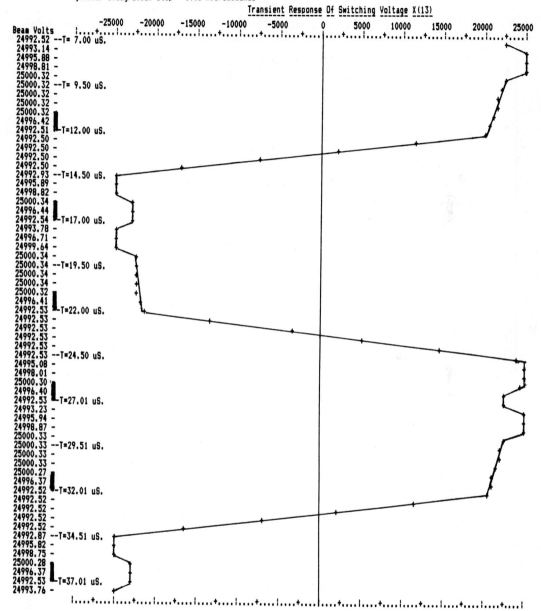

Figure 11-7 Appendix 33 steady-state analysis of dc–dc regulator/voltage-limited current source.

Sec. 11-5 Steady–State DC-to-DC Power Supply Analysis

Chapter 12

BASIC-to-C Conversion

Two objectives of this book are to provide a source of power electronics background and to demonstrate some power electronics/RF power generator analysis techniques. Stated another way, the book is intended to provide you with a means of obtaining technical insight. Your personal problem—in conjunction with a modified or unmodified version of one of the programs in conjunction with your personal computer—will provide the capability of establishing enormous amounts of insight into your specific and unique technical problem.

12.1 BASIC PROGRAMMING

Up to this point in the book the BASIC language has been used for all programs. Why? Because it is in common use worldwide, it is easy to comprehend and is available on nearly all computers. As a result, using BASIC typically requires no additional study of programming and no additional financial expenditure on your part.

It may have been evident to you that BASIC provides rather slow solution printouts on some of the analyses we have performed. That is because BASIC is an interpreter language. It reads a line of your program, it converts it into machine language commands, and then the machine commands are executed by the computer. Following that, the interpreter proceeds line by line through the program in the same fashion. If a loop is present in the program—as is typically the case—this tedious procedure is repeated for each line and on each pass through the loop. From a positive point of view, when a BASIC program modification is necessary—which is also typically the case when performing analyses—programs can be edited very easily/rapidly. You then simply issue a RUN command, and your revised analysis is in progress.

Each BASIC program in the appendices is intended to represent a point of departure for a modified program that will solve your specific technical problems. These modifications require only a modest amount of skill in BASIC programming on your part.

12.2 C LANGUAGE PROGRAMMING

If you possess (or develop) some additional computer expertise, you may find it beneficial to convert some of the BASIC programs used in this book into another language, such as C. Why C? It is a powerful and flexible compiler language that is widely used by professional programmers. Because C programs are compiled before they are executed, your computer functions continuously in machine language. Hence for a given analysis, a C program will execute considerably more rapidly than a BASIC program.

Let us assume that you are familiar with BASIC programming but want to learn what C can do for you. A typical question might be: How can a BASIC analyses be performed in C? To perform a BASIC-to-C conversion, first, you would develop and enter a set of statements into a C editor program (i.e., in much the same manner as you develop statements in a BASIC editor program) to generate an initial C program called the source code. Second, you would enter this program into a C compiler program that functions to generate a C object file. (*Note*: If programming errors are encountered during object file generation, diagnostics are displayed, and the source code must be reedited before proceeding.) When an acceptable object file is generated, the system automatically proceeds to a C linker—functioning in conjunction with a stored C library—to generate an executable C file.

From this brief description it is apparent that you will require several new computer-related additions if you want to perform analyses in C—a set of C programming instructions plus C editor, C compiler, C linker, and C library software. How can these things be acquired? One effective approach is to purchase a C instruction manual and the necessary C software disks. [*Note*: The American National Standards Institute (ANSI) has established "ANSI Standard C" to ensure uniformity in C compilers.] Peter Aitken's recent book [11] provides within its covers very fundamental and useful C programming instructions, examples, and floppy disk software. If you have followed through with the BASIC programming background developed in this book thus far, a C instruction manual/disk approach should supply all of the additional background you need to get started in C programming.

12.3 BRIEF REVIEWS OF BASIC-TO-C CONVERSIONS

Several converted C source code programs—similar to our most frequently used BASIC programs—will be reviewed briefly. While these programs utilize only the most fundamental features of the C language, they will clearly illustrate a number of similarities and differences that exist in the two languages.

FOURIER.C

This is one of the C source code programs on your disk. The contents of this steady-state program are also listed in the appendices under FOURIER.C. It performs a graphical Fourier analysis in the same manner as the BASIC program listed in Appendix 7. For the sake of simplicity, FOURIER.C does not have all the bells and whistles present in Appendix 7, but it accomplishes the same end result. We use it to illustrate some of the major features of C programming. If you look at the FOURIER.C listing you will note that:

1. C does not use line numbers for its statements.

2. The first line is a remark statement. In BASIC a remark line starts with the letters REM. In C a remark starts with /* and terminates with */ symbols.

3. The second and third lines state that the analysis program must include the standard input and output <stdio.h> and mathematics <math.h> functions from the C library.

4. All variables and constants must be documented before they are executed in a C program (i.e., in our six C program conversions all variables and constants will be declared before any analysis statements occur). For example, the fourth line declares Y as a symbolic constant with a value of 500, the fifth line declares integer variables n, n7, n8 etc., and floating-point variables t, a, t3, etc). Note that all C statements are terminated with a semicolon.

5. The sixth line lists seven floating-point array statements Each array has a storage value of 500 (e.g., float f[Y] is an array that can store up to Y floating point values in its 500 elements).

6. Line 7 defines functions—not present in BASIC. A major objective in C is to break the programming code into simplified independent sections known as functions. Different types of functions are available in C, but all consist of three parts. In this program the first part is the word void and a space, the second part is a description (i.e., wording selected by the programmer), and the third part is the word void in parentheses. This particular function arrangement very closely resembles a BASIC subroutine (e.g., void load-calculate(void) contains the statements needed (a) to *load* waveform ordinates, and (b) to *calculate* the values of the harmonic values of the wave's Fourier series. Words in a function description must be separated by the dash format shown.

7. The word/symbols main() followed by { defines the point at which program statement execution starts. Just as in Appendix 7, the FOURIER.C program user must define the number of graphical ordinates, the lowest and highest number of the Fourier harmonics required in the analysis, and the period of the wave in seconds. Note that printf is similar to PRINT in BASIC (i.e., in C a −n symbol denotes a line feed). A scanf statement in C is similar to an INPUT statement in BASIC [e.g., the program's first scanf denotes that a decimal integer, %d, should be entered to define the number of ordinates n, &n, etc.]. Other definitive inputs are entered in similar fashion using printf and scanf format. Variables n, n7, n8, and t are the same variables used in Appendix 7.

8. The C symbols used during analysis are either identical or similar to those of BASIC. Symbols +, −, *, and / are the same. The BASIC FOR-NEXT has a more complex *for* format and no *next* listing in C (i.e., all statements to be processed are placed between { } symbols). BASIC's IF-THEN-ELSE is similar but somewhat more complex in C (i.e., several of C's relational operators are equal, = =, not equal, ! =, and, &&, or, ::, and not, !). BASIC's power symbol,ˆ, is replaced by pow( ) in C. PI is stored in the <math.h> function and does not require initial definition. BASIC's trigonometrical functions are similar to C's, but C has a superior form of ATAN function called atan2 (e.g., it is correct over the full 2π radians scale). By using this limited primer of symbols, most of the FOURIER.C programming should be understandable in the description that follows, but reference to a manual of C programming instructions is advised before you initially try your own programming.

9. The main() program's next statement, load-calculate(), transfers us to the void load-calculate(void) function. In response to lines 2–6 of the function, the user enters values for the waveform's ordinate amplitudes. Except for the differences noted in item 8, the statements should be self explanatory since they closely follow APPENDIX 7. After entering the ordinates, the statements that follow execute harmonic calculations. In-progress calculations are then stored in arrays. At the conclusion of the function we are returned to the main() program.

10. The main program's printout() statement next transfers us to the void print-out(void) function. The statements used for displaying analysis results parallel those established in Appendix 7. At the conclusion of this function we return to the main() program where the } symbol terminates execution of the program.

The comparison above, despite its abbreviated nature, illustrates the close relationship that exists between C and BASIC programming. Now let us explore how C functions after a debugged program much as FOURIER.C has been prepared.

Assume that you have an IBM (or IBM-compatible) PC with two disk drives and that you have loaded Peter Aitken's C software [11]. The disk in drive 1 will contain the active programs (e.g., editor, compiler, etc.). The disk in drive 2 will contain source programs—and sufficient available space for storing new programs. Assume that your C system is operational and that the source code of drive 2 is active. Then simply remove the C disk from drive 1 and insert the disk supplied with this book. Type < copy a:fourier.c b:fourier.c > and press the Enter key. Within a few seconds the source code will be transferred to drive 2. Request a drive 2 directory. It should now list FOURIER C 2348, indicating that your program has been stored using 2348 bytes.

The next step is to compile your source code. Simply enter < ZTC FOURIER.C > and press the Enter key. A number of subsequent in-process printouts will be displayed on your monitor, but after a short time the display will return to B>. This indicates a successful compile cycle has been completed. A new drive 2 directory check will display FOURIER OBJ 7561 and FOURIER EXE 24596, showing that the object and execute files have been entered on the source disk.

To execute the program simply type <FOURIER> and press the Enter key. After a delay of a few seconds a program screen printout will request waveform ordinate input information from your keyboard. For purposes of comparison, enter the same pulse data

```
Enter the following information from the keyboard:

1. Number ordinates (number must be odd);<Enter>
   Number = 65
2. Lowest harmonic;space;highest harmonic,<Enter>.
   Lowest harmonic = 1, highest harmonic = 17
3. Period of waveform in seconds.
   Period = 1.000000

Type waveform ordinate;<Enter>
        Ordinate 0 = 100.000000
        Ordinate 1 = 100.000000
        Ordinate 2 = 100.000000
        Ordinate 3 = 100.000000
        Ordinate 4 = 100.000000
        Ordinate 5 = 100.000000
        Ordinate 6 = 100.000000
        Ordinate 7 = 100.000000
        Ordinate 8 = 100.000000
        Ordinate 9 = 100.000000
        Ordinate 10 = 100.000000
        Ordinate 11 = 100.000000
        Ordinate 12 = 100.000000
        Ordinate 13 = 100.000000
        Ordinate 14 = 100.000000
        Ordinate 15 = 100.000000
        Ordinate 16 = 50.000000
        Ordinate 17 = 0.000000
        Ordinate 18 = 0.000000
        Ordinate 19 = 0.000000
        Ordinate 20 = 0.000000
        Ordinate 21 = 0.000000
        Ordinate 22 = 0.000000
        Ordinate 23 = 0.000000
        Ordinate 24 = 0.000000
        Ordinate 25 = 0.000000
        Ordinate 26 = 0.000000
        Ordinate 27 = 0.000000
        Ordinate 28 = 0.000000
        Ordinate 29 = 0.000000
        Ordinate 30 = 0.000000
        Ordinate 31 = 0.000000
        Ordinate 32 = 0.000000
        Ordinate 33 = 0.000000
        Ordinate 34 = 0.000000
        Ordinate 35 = 0.000000
        Ordinate 36 = 0.000000
        Ordinate 37 = 0.000000
        Ordinate 38 = 0.000000
        Ordinate 39 = 0.000000
        Ordinate 40 = 0.000000
        Ordinate 41 = 0.000000
        Ordinate 42 = 0.000000
        Ordinate 43 = 0.000000
        Ordinate 44 = 0.000000
        Ordinate 45 = 0.000000
        Ordinate 46 = 0.000000
        Ordinate 47 = 0.000000
        Ordinate 48 = 50.000000
        Ordinate 49 = 100.000000
        Ordinate 50 = 100.000000
        Ordinate 51 = 100.000000
        Ordinate 52 = 100.000000
        Ordinate 53 = 100.000000
        Ordinate 54 = 100.000000
        Ordinate 55 = 100.000000
        Ordinate 56 = 100.000000
        Ordinate 57 = 100.000000
        Ordinate 58 = 100.000000
        Ordinate 59 = 100.000000
        Ordinate 60 = 100.000000
        Ordinate 61 = 100.000000
        Ordinate 62 = 100.000000
        Ordinate 63 = 100.000000
        Ordinate 64 = 100.000000
```

Figure 12-1 C analysis using compiled FOURIER.C.

```
--- Calculations In Progress ---
Harmonic 1 = 63.662006 @ 0.000000 deg.
Harmonic 2 = 0.000000 @ 0.000000 deg.
Harmonic 3 = 21.221546 @ 180.000000 deg.
Harmonic 4 = 0.000000 @ 0.000000 deg.
Harmonic 5 = 12.736629 @ 0.000000 deg.
Harmonic 6 = 0.000000 @ 0.000000 deg.
Harmonic 7 = 9.106499 @ 180.000000 deg.
Harmonic 8 = 0.000000 @ 0.000000 deg.
Harmonic 9 = 7.099934 @ 0.000000 deg.
Harmonic 10 = 0.000000 @ 0.000000 deg.
Harmonic 11 = 5.838091 @ 180.000000 deg.
Harmonic 12 = 0.000000 @ 0.000000 deg.
Harmonic 13 = 4.986133 @ 0.000000 deg.
Harmonic 14 = 0.000000 @ 0.000000 deg.
Harmonic 15 = 4.392012 @ 180.000000 deg.
Harmonic 16 = 0.000000 @ 0.000000 deg.
Harmonic 17 = 3.981643 @ 0.000000 deg.
```

Harmonic	Voltage	V.Angle	Amperes	A.Angle	Power
1	63.662006	0.000000	63.662006	0.000000	2026.425537
2	0.000000	0.000000	0.000000	0.000000	0.000000
3	21.221546	180.000000	21.221546	180.000000	225.177017
4	0.000000	0.000000	0.000000	0.000000	0.000000
5	12.736629	0.000000	12.736629	0.000000	81.110855
6	0.000000	0.000000	0.000000	0.000000	0.000000
7	9.106499	180.000000	9.106499	180.000000	41.464161
8	0.000000	0.000000	0.000000	0.000000	0.000000
9	7.099934	0.000000	7.099934	0.000000	25.204529
10	0.000000	0.000000	0.000000	0.000000	0.000000
11	5.838091	180.000000	5.838091	180.000000	17.041655
12	0.000000	0.000000	0.000000	0.000000	0.000000
13	4.986133	0.000000	4.986133	0.000000	12.430761
14	0.000000	0.000000	0.000000	0.000000	0.000000
15	4.392012	180.000000	4.392012	180.000000	9.644885
16	0.000000	0.000000	0.000000	0.000000	0.000000
17	3.981643	0.000000	3.981643	0.000000	7.926742

```
      Rms voltage = 70.330833
      Rms current = 70.330833
     Volt-amperes = 4946.426270
            Power = 4946.426270
     Power factor = 1.000000
```

as was used in Fig. 5-3. Figure 12-1 shows your end result, a C program printout that concurs with the BASIC printout of Fig. 5.4.

TRANS LI.C

This steady-state C program, present on your disk and in the appendix, is a conversion encompassing the essential features of the BASIC program listed as Appendix 5. The fundamental equations required for transmission line analysis, Eqs. (4-16) and (4-17), were developed in Chapter 4. These equations, defining voltage and current at any point along a line, can also be written in their equivalent hyperbolic function form as

$$v_{si} = v_l[\cosh(Pd) + \frac{z_0}{z_l}\sinh(Pd)] \qquad (12\text{-}1)$$

$$i_{si} = i_l[\cosh(Pd) + \frac{z_l}{z_0}\sinh(Pd)] \qquad (12\text{-}2)$$

Since C's <math.h> function also contains hyperbolic functions, these equations are used in TRANS-LI.C. This allows us to perform a cross-check on Appendix 5, where exponential functions were used in the BASIC approach.

For a summary of the main features of TRANS-LI.C, refer to its listing in the appendix. We note that:

1. The definition portion of the program utilizes the same procedures as those outlined in FOURIER.C.
2. In the main() portion of the program, printf and scanf statements are used to read/enter data on transmission line characteristic and operating conditions, in substantially the same way as outlined in FOURIER.C.
3. Following data entry, main() routes the program to the void hyperbolic(void) function. It solves for the line's input voltage and current and returns to main(). The program is then routed to the void printout(void) function to display the input/output parameters of the line and is then returned to main(). In main() a separation line and an incremental analysis heading are printed out, a for-next loop is established, and for each loop step the program is routed to the void hyperbolic(void) and void print_tab(void) functions. The loop then calculates and prints out a table of voltages, currents, and impedances along the line at increments defined by the for-next loop. At the conclusion of this operation, a separation line is printed, the symbol } of main() is encountered, and the program terminates.

If a compiled program is executed using the same data as in the analysis of Fig. 4-3, a printout similar to Fig. 12-2 will result. C's hyperbolic and BASIC's exponential analyses concur.

SSANAL.C

This C program, present on your disk and in the appendix, is a conversion of the steady-state circuit analysis program listed in BASIC as Appendix 2. All significant features of the BASIC program have been retained. For reference, follow the SSANAL.C program in the appendix. It shows that:

1. The program's introductory portion is similar but more extensive than in the two previous C programs.
2. The first part of main() uses printf and scanf statements to load the source voltage level and printf statements to provide instructions for loading system elements and nodes. Next, it loads system elements into arrays via an infinite for-next loop, in conjunction with void element(void), void load_r(void), void load_c(void),

```
            Enter the following values from the keyboard:

1. Freq.in MHz:space:load voltage:space:load resis.:space:load react.;<Enter>.
   f = 9.348000, vl = 450.000000, rl = 300.000000, xl = 0.000000.

2. Line Z0:space,dB loss/100 ft:space:prop.const.:space:length in ft:<Enter>.
   z0 = 600.000000, db = 0.250000, pc = 0.950000, l = 50.000000.

   Load impedance = 300.000000 @ 0.000000 deg.
  Load refl. coef. = 0.333333 @ 180.000007 deg.
        Load vswr = 2.000000
     Load voltage = 450.000000 @ 0 deg.
     Load current = 1.500000 @ 0.000000 deg.
                  = 1.500000 - j 0.000000
    Source voltage = 462.995758 @ 179.999980 deg.
                  = -462.995758 + j 0.000180
    Source current = 1.510946 @ 179.999993 deg.
                  = -1.510946 + j 0.000000
```

Ft To Load	Voltage	Current	Impedance	Angle deg.
0.000000	450.000000	1.500000	300.000000	0.000000
5.000000	511.545258	1.446410	353.665503	23.713827
10.000000	643.991638	1.293631	497.817166	35.305595
15.000000	776.935608	1.074840	722.838417	35.208765
20.000000	869.875916	0.858303	1013.483402	23.487737
25.000000	903.260498	0.760810	1187.235099	0.000007
30.000000	871.238770	0.862134	1010.560410	-23.338388
35.000000	779.983948	1.080955	721.569206	-34.823836
40.000000	649.495361	1.301251	499.131386	-34.728207
45.000000	520.740356	1.455496	357.775288	-23.116461
50.000000	462.995758	1.510946	306.427702	-0.000014

Figure 12-2 C analysis using compiled TRANS LI.C.

void__l(void), void__f(void), and/or load__n(void) functions. The printf and scanf statements in void element(void) prints out an element list.

3. After the last element is loaded, the for-next loop in main() is terminated and the p, q, and or q1 arrays are loaded by accessing void load__p__q__q1(void). Program execution starts as main() next accesses the void i__o__sweep(void) function. This function initially uses printf and scanf statements to define/print out analysis nodes, frequency analysis range, frequency increments, and column titles for the analysis solutions. Following this, under control of a for-next loop, the function accesses

```
            Enter system elements from keyboard as follows.
1. Type:<source voltage value><Enter>
   Source voltage = 480.000000.
2. Type element symbol:<R L C or F><Enter>
3. For R L or C type:<first node#><space><second node#><space><value><Enter>
   Value is in ohms, hy., or uF. After final element entry type:<E><Enter>
4. Type F input similarly except nodes are for G S D and value is GM.
L4 1 2 0.000489 hy.
R4 2 3 0.011520 ohms.
R1 3 0 9.216000 ohms.
L2 3 4 0.007334 hy.
R2 4 0 3.686400 ohms.
C3 3 5 479.705994 uF.
R3 5 0 7.372800 ohms.
End of list

Type analysis nodes:<input node><space><output node><Enter>
Input node = 1 and output node = 3.
Type sweep values in hz:<low><space><high><space><delta><Enter>
Lowest = 60.000000 Hz, highest = 60.000000 Hz, increment = 1.000000 Hz.


   Freq=Mhz    V=volts     Phase=deg.     Real V       Imag. V     Gain=dB
   0.000060    471.293936  -3.785995      470.265402   -31.119544  -0.158988

For an analysis at another circuit node enter Y.
```

Figure 12-3 C analysis using compiled SSANAL.C.

the void solutions(void) function at each frequency, then accesses the void determinant(void) function, and so on.

4. When the solution is completed at the current analysis frequency, the void i_o_sweep(void) function accesses void print(void) and displays the values. The for-next loop continues until analyses at all sweep frequencies are complete. The program then returns to main(), where } terminates execution.

If a compiled program is executed and loaded with the same information as used in 60-Hz analysis circuit of Fig. 2-6, a printout similar to Fig. 12.3 will result. A cross-check of the node 3 voltage and phase angle in C and BASIC shows that the programs concur.

SSANAL2.C

You will recall that a variation of Appendix 2 was derived as Appendix 4. The objective of Appendix 4 was to minimize potential errors often associated with entering a large number of elements and nodes from the keyboard. This problem was overcome by storing fixed circuit elements and nodes in the program and using a READ statement to access the stored DATA during execution.

SSANAL2.C has the same objective as Appendix 4, but in addition, it permits entering variable elements from the keyboard. This permits you to run a series of analyses in relatively rapid fashion, without a need for extensive keyboard manipulation.

Reference to the appendix listing for SSANAL2.C shows that:

1. The first portion of the program is substantially the same as SSANAL.C except that the function, void fixed_el(void), was added to store fixed element and node information.
2. The fixed element function is accessed in the first step of the main() section (i.e., C does not have READ–DATA procedures). It was also necessary to make some revisions to the void elements(void) function to accommodate loading the fixed elements.

SSANAL2.C has fixed values for an OTH-B transmitter bandpass filter circuit stored. If a compiled program is executed, and if the load variable—a resistance of 50 Ω connected between node 7 and 0 is inserted from the keyboard—a printout similar to Fig. 12.4 will result. A comparison of this printout with the first analysis printout of Fig. 6-18 shows that the C and BASIC analyses concur.

OPAMP.C

Now let us examine several BASIC transient analysis program conversions to C. OPAMP.C is stored on your disk and listed in the appendix. It closely parallels the operational amplifier model described in BASIC Appendix 14.

If you refer to OPAMP.C in your appendix, you will note that:

1. The initial portion of the program is similar to those reviewed previously but of reduced scope. However, note that initial conditions have been introduced.
2. The main() portion of the program first accesses function void solutions(void), solves the four differential equations, stores the results, and returns to main(). Next, column headings are printed, void solutions(void) is accessed, initial conditions x[] values are printed out, and the program returns to main().
3. At this point main() enters an infinite for-next loop that contains the Euler–Cauchy algorithm that we used for all of our BASIC differential equation solutions. Perhaps it would have been more in the character of C to list this routine as a function; nevertheless, in either location it proceeds in normal fashion to solve for all x[] values at each t1 increment of time.
4. When sets of successive t1 increments (i.e., po) total to t2, the routine accesses void solutions(void), prints out the current x[] values, and returns to main(). When total elapsed time t reaches t3, a break statement occurs, the infinite loop is terminated, } is encountered, and program execution is terminated.

```
The fixed circuit element are:              Enter variables from the keyboard as follows.
                                            1. Type:<source voltage value><Enter>
F g=1 s=0 d=2 gm=1.000000 mhos.                Source voltage = 1.000000.
R 2 0 1600.000000 ohms.                     2. Type element symbol:<C L R or X>;<space>;<Enter>.  Then type:<first node#>;
C 2 0 0.000430 uF.                             <space><second node#><space><value><Enter>. Note: value = uf.,hy. or ohms.
L 2 3 0.000001 hy.                          3. After final element entry type:<E><Enter>
L 3 0 0.000001 hy.                             r 7 0 50.000000 ohms.
C 3 0 0.000006 uF.                             End of list.
L 3 8 0.000003 hy.
C 8 0 0.000010 uF.                          Type analysis nodes:<input node><space><output node><Enter>
L 8 4 0.000003 hy.                          Input node = 1 and output node = 7.
C 4 0 0.000006 uF.                          Type sweep values in hz:<low><space><high><space><delta><Enter>
L 5 0 0.000000 hy.                          Lowest = 4500000.000000 Hz, highest = 7000000.000000 Hz, increment = 250000.000000 Hz.
C 4 5 0.000114 uF.
C 5 0 0.002390 uF.                          Freq=Mhz    V=volts     Phase=deg.     Real V       Imag. V      Gain=dB
C 5 6 0.000130 uF.                          4.500000    14.147531   -168.009796    -13.838876   -2.939071    23.013613
C 6 0 0.000006 uF.                          4.750000    93.099983    94.684631     -7.603585   92.788967    39.378992
L 6 9 0.000003 hy.                          5.000000    95.980934    -9.100090     94.772876  -15.180308    39.643699
C 9 0 0.000010 uF.                          5.250000    97.970978   -80.230614     16.624006  -96.550271    39.821949
L 9 7 0.000003 hy.                          5.500000    98.447617  -137.689514    -72.802791  -66.269803    39.864104
L 7 0 0.000001 hy.                          5.750000    99.122551   166.009842    -96.182305   23.963394    39.923449
C 7 0 0.001317 uF.                          6.000000    94.261826   115.423096    -40.466529   85.133729    39.486717
                                            6.250000    99.144562    64.845840     42.141915   89.742427    39.925378
                                            6.500000    99.146286     4.996666     98.769507    8.635421    39.925529
                                            6.750000    99.440361   -64.561790     42.713360  -89.799523    39.951254
                                            7.000000    42.080654  -160.105209    -39.569242  -14.319797    32.481650

                                            For an analysis at another circuit node enter Y.

                                            Figure 12-4   C analysis using compiled SSANAL2.C.
```

If a compiled program is executed, the tabular response shown in Fig. 12-5a will result. A comparison of the x[4] output voltage column with the Appendix 14 upper scale display in Fig. 7-10 shows the BASIC and C concur.

OPAMPP.C

Up to this point in our BASIC-to-C conversions, all solution printouts have been presented in tabular form. It appears that C does not include the simplistic TAB() function technique used for our line printer plotting of BASIC programs. There are undoubtedly many approaches possible to display response plots in C, but the one used to convert OPAMP.C to a plotted response is stored on your disk as OPAMPP.C and listed in the appendix. If you refer to the appendix, you will note that:

1. The initial part of the program is similar to OPAMP.C except that <stdlib>, three floating-point variables, three integer variables, and two functions have been added.
2. The main() portion of the program is similar to OPAMP.C except that voltage scale numerics are initially printed out, void vscale(void) is accessed, an incremental scale is printed, we return to main(), and then void plot(void) is accessed.
3. Function void plot(void) plots x[4] through the use of another function called malloc() (i.e., abbreviation for memory allocation function). A C manual is suggested for malloc() details, but for this brief discussion it serves to allocate 80 tabular spaces of memory for each line feed. Next, a set of if-else statements are encountered that print out time-scale increments, and at 0, 5, 10, and so on, increments, print out time-scale values. Next, we encounter a loop in which the variable *count* cycles through tabular positions. When the value of x[4] corresponds to the value of *count*, a + symbol is plotted. Another C technique, pointers, is used to implement this plotting function. Again a C manual is suggested for background in this area. Next we return to main().
4. The initial pass through void plot(void) registers t = 0 information. Following this, void plot(void) is only accessed in increments of t2, and when t3 is reached the program plots the value, returns to main, accesses plot vscale(void), prints a incremental scale, and returns to main() where the program terminates.

If this program is compiled and executed, a plotted response similar to Fig. 12-5b will result. It concurs with the BASIC plot of Fig. 7-10.

3PHFWPS.C

This program is for transient analysis of a three-phase full-wave dc power supply. It is converted from the BASIC program of Appendix 18C. The fundamental calculation routine used is similar to OPAMP.C, but 3PHFWPS.C is significantly more involved because of rectifier commutation considerations and a higher number of analysis equations. A snap-on analysis is programmed—that is, the dc output filter capacitor has no initial

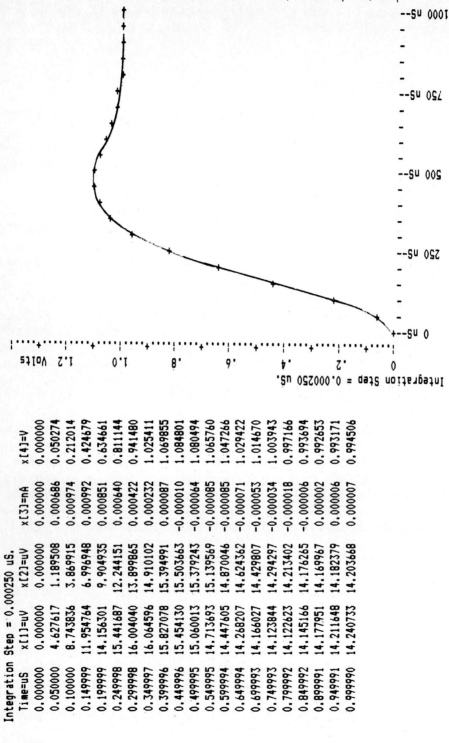

Integration Step = 0.000250 uS.				
Time=uS	x[1]=uV	x[2]=uV	x[3]=nA	x[4]=V
0.000000	0.000000	0.000000	0.000000	0.000000
0.050000	4.627617	1.189508	0.000686	0.050274
0.100000	8.743836	3.869915	0.000974	0.212014
0.149999	11.954764	6.996948	0.000992	0.424679
0.199999	14.156301	9.904935	0.000851	0.634661
0.249998	15.441687	12.244151	0.000640	0.811144
0.299998	16.004040	13.899865	0.000422	0.941480
0.349997	16.064596	14.910102	0.000232	1.025411
0.399996	15.827078	15.394991	0.000087	1.069855
0.449996	15.454130	15.503663	-0.000010	1.084801
0.499995	15.060013	15.379243	-0.000064	1.080494
0.549995	14.713693	15.139569	-0.000085	1.065760
0.599994	14.447605	14.870046	-0.000085	1.047266
0.649994	14.268207	14.624362	-0.000071	1.029422
0.699993	14.166027	14.429807	-0.000053	1.014670
0.749993	14.123844	14.294297	-0.000034	1.003943
0.799992	14.122623	14.213402	-0.000018	0.997166
0.849992	14.145166	14.176265	-0.000006	0.993694
0.899991	14.177951	14.169967	0.000002	0.992653
0.949991	14.211648	14.182379	0.000006	0.993171
0.999990	14.240733	14.203668	0.000007	0.994506

Figure 12-5 C analysis using compiled OPAMP.C and OPAMPP.C.

charge; hence worst-case operating conditions occur throughout the power supply system during the early portion of the analysis.

If you refer to the appendix listing of 3PHFWPS.C, you will note that:

1. The variables, arrays, and functions have been revised to coincide with the system elements present in the APPENDIX 18C circuit configuration.
2. The statements in main() have been minimized. Functions void heading(void), void timing(void), void equations(void), and void printout(void) are accessed sequentially to do the following, respectively: print out analysis column headings, determine in which of the six rectification sectors the power supply is operating (and assign a ts value), calculate the values of the six differential equations for the system and store the values, and print out the initial condition x[] values.
3. Function main() then enters an infinite for-next loop and accesses void integration(void). It follows the Euler–Cauchy routine's second-order analysis procedure, and after determining the status of rectifier commutation in final five steps, it returns to main().
4. A t2 and t3 routine noted in conjunction with OPAMP.C is again followed for printouts, and when t3 is reached the program terminates execution.

If the program is compiled and executed, a tabular display similar to Fig. 12.6 will result. An identical display does not appear in the text, but a separate run of Appendix 18C indicates that BASIC AND C concur.

12PULSE.C

This transient analysis program is for a 12-pulse bridge dc power supply. It has been converted from BASIC APPENDIX 18G—perhaps the most complex transient analysis program presented in the text. Although rather long, the program follows substantially the same procedure as outlined in 3PHFWPS.C, except that there are now two dc supplies in series, one using a delta rectification circuit and one using a wye rectification circuit. If you refer to the discussion of the circuit in Section 8-5 and Appendix 18G, you should be readily able to follow through the program listed in the appendix as 12PULSE.C.

Instead of the statement "If this program is compiled and executed . . . ," in this case *you* will have the opportunity to run the program. How can this be done if you do not have C software? The answer to that question points out one of the other advantages noted for C—an executable C program is "portable." The program 12PULSE EXE 25562 on your disk can be run on your computer if it is an IBM PC or compatible type. Simply turn your computer on, and when DOS has been fully activated, insert your disk into the active drive, type < 12PULSE >, and press the Enter key. Within a few seconds a printout similar to Fig. 12-7 will appear on your screen. If you want a hard copy and your printer is active, type < 12PULSE>PRN > and Enter. For a cross-check, refer to the tabular portion of Fig. 8-12 and you will note that BASIC and C concur.

It is important to recognize that this or any other C EXECUTE program is relatively inflexible. If you want to change any program statement, it will be necessary to revise all

Time	ts	i[1]	i[2]	i[3]	i[4]	i[5]	i[6]	x[5]	vin	vout	v[1]	v[2]	v[3]
0.0000	1	0.0	0.1	0.0	0.0	0.0	0.1	0.1	0.1	0.0	3023	3023	-6046
0.0010	1	14.3	0.0	0.0	0.0	0.0	14.3	4.1	10562.6	34.3	4797	789	-5586
0.0020	1	12.4	0.0	0.0	0.0	0.0	12.4	9.3	10217.3	165.2	5840	-1565	-4275
0.0030	2	13.1	0.0	0.0	0.0	2.1	11.0	13.9	8627.4	388.3	5994	-3681	-2314
0.0040	2	19.4	0.0	0.0	0.0	19.4	0.0	17.9	9611.0	684.7	5236	-5236	0
0.0050	2	22.5	0.0	0.0	0.0	22.5	0.0	22.4	9490.0	1054.0	3681	-5994	2314
0.0060	3	17.3	0.0	7.8	0.0	25.1	0.0	26.2	8438.3	1490.4	1565	-5840	4275
0.0070	3	0.0	0.0	31.2	0.0	31.2	0.0	29.5	9613.6	1977.7	-789	-4797	5586
0.0080	3	0.0	0.0	32.9	0.0	32.9	0.0	33.2	9090.0	2516.2	-3023	-3023	6046
0.0090	4	0.0	0.0	34.9	16.1	18.8	0.0	36.1	8229.4	3098.6	-4797	-789	5586
0.0100	4	0.0	0.0	40.7	40.7	0.0	0.0	38.6	9721.0	3708.7	-5840	1565	4275
0.0110	5	0.0	1.8	39.1	40.9	0.0	0.0	41.3	8851.5	4348.1	-5994	3681	2314
0.0120	5	0.0	27.2	14.4	41.7	0.0	0.0	43.2	7875.1	5007.3	-5236	5236	0
0.0130	5	0.0	46.7	0.0	46.7	0.0	0.0	44.8	9737.9	5671.7	-3681	5994	-2314
0.0140	6	0.0	45.9	0.0	39.3	0.0	6.7	46.4	8752.0	6344.9	-1565	5840	-4275
0.0150	6	0.0	45.1	0.0	4.3	0.0	40.8	47.1	7371.9	7014.7	789	4797	-5586
0.0160	6	0.0	49.1	0.0	0.0	0.0	49.1	48.0	9487.8	7670.7	3023	3023	-6046
0.0170	1	14.5	33.2	0.0	0.0	0.0	47.7	48.5	8528.8	8316.7	4797	789	-5586
0.0180	1	49.1	0.0	0.0	0.0	0.0	49.1	48.1	8959.4	8938.7	5840	-1565	-4275
0.0190	2	48.6	0.0	0.0	0.0	1.2	47.4	48.3	9170.6	9533.9	5994	-3681	-2314
0.0200	2	46.4	0.0	0.0	0.0	25.2	21.2	47.7	8142.6	10102.4	5236	-5236	-0
0.0210	2	49.0	0.0	0.0	0.0	49.0	0.0	46.8	9891.4	10631.0	3681	-5994	2314
0.0220	3	40.1	0.0	5.7	0.0	45.8	0.0	46.1	9004.1	11125.2	1565	-5840	4275
0.0230	3	4.0	0.0	38.6	0.0	42.6	0.0	44.6	7596.7	11578.9	-789	-4797	5586
0.0240	3	0.0	0.0	44.5	0.0	44.5	0.0	43.3	9772.3	11985.2	-3023	-3023	6046
0.0250	4	0.0	0.0	41.0	13.6	27.4	0.0	41.9	8715.5	12351.5	-4797	-789	5586
0.0260	4	0.0	0.0	41.8	41.8	0.0	0.0	39.8	9824.8	12667.7	-5840	1565	4275
0.0270	5	0.0	1.0	37.2	38.2	0.0	0.0	38.4	9319.8	12938.1	-5994	3681	2314
0.0280	5	0.0	24.4	10.4	34.8	0.0	0.0	36.3	8270.7	13163.3	-5236	5236	0
0.0290	5	0.0	36.1	0.0	36.1	0.0	0.0	34.3	10299.0	13336.6	-3681	5994	-2314
0.0300	6	0.0	31.6	0.0	26.1	0.0	5.5	32.4	9066.5	13468.8	-1565	5840	-4275
0.0310	6	0.0	31.1	0.0	0.0	0.0	31.1	30.0	9673.2	13552.9	789	4797	-5586
0.0320	6	0.0	28.5	0.0	0.0	0.0	28.5	28.3	9844.5	13593.0	3023	3023	-6046
0.0330	1	13.5	11.3	0.0	0.0	0.0	24.9	26.1	8731.2	13594.4	4797	789	-5586
0.0340	1	26.0	0.0	0.0	0.0	0.0	26.0	24.1	10586.1	13551.1	5840	-1565	-4275
0.0350	2	21.3	0.0	0.0	0.0	1.1	20.3	22.3	9264.7	13475.7	5994	-3681	-2314
0.0360	2	20.7	0.0	0.0	0.0	20.7	0.0	20.1	9649.0	13363.3	5236	-5236	0
0.0370	2	19.3	0.0	0.0	0.0	19.3	0.0	18.6	10307.6	13218.1	3680	-5994	2314
0.0380	3	9.8	0.0	5.8	0.0	15.5	0.0	16.8	8989.4	13048.4	1565	-5840	4275
0.0390	3	0.0	0.0	16.9	0.0	16.9	0.0	15.1	10627.0	12848.0	-789	-4796	5586
0.0400	3	0.0	0.0	13.2	0.0	13.2	0.0	13.9	9684.5	12629.6	-3023	-3023	6046
0.0410	4	0.0	0.0	12.0	12.0	0.0	0.0	12.2	9415.0	12391.3	-4797	-789	5586
0.0420	4	0.0	0.0	12.2	12.2	0.0	0.0	11.3	10543.9	12134.9	-5840	1565	4275
0.0430	5	0.0	1.2	7.5	8.8	0.0	0.0	10.2	9122.8	11870.6	-5994	3681	2313
0.0440	5	0.0	10.6	0.0	10.6	0.0	0.0	9.1	10469.8	11592.9	-5236	5236	-0
0.0450	5	0.0	8.5	0.0	8.5	0.0	0.0	8.6	10111.1	11311.7	-3680	5994	-2314
0.0460	6	0.0	6.2	0.0	0.0	6.2	7.7	8798.2	11028.7	-1564	5840	-4276
0.0470	6	0.0	8.5	0.0	0.0	8.5	7.3	10576.7	10741.6	790	4796	-5586
0.0480	6	0.0	6.0	0.0	0.0	6.0	7.1	9493.9	10461.9	3023	3023	-6046
0.0490	1	7.4	0.0	0.0	0.0	7.4	6.6	10061.0	10184.7	4797	789	-5586
0.0500	1	7.2	0.0	0.0	0.0	7.2	6.7	10366.3	9915.1	5840	-1565	-4275

Figure 12-6 C analysis using compiled 3PHFWPS.C.

Time	y ts	d ts	i [1]	i [2]	i [3]	i [4]	i [5]	i [6]	i [7]	i [8]	i [9]	i [10]	i [11]	i [12]	x [5]	dc vout
0.0000	6	8	0	19	0	0	0	19	0	22	0	0	0	7	20	19243.89
0.0002	1	8	1	18	0	0	0	19	0	22	0	0	0	22	20	19243.81
0.0004	1	8	3	16	0	0	0	19	0	22	0	0	0	22	20	19243.83
0.0006	1	8	7	12	0	0	0	19	0	22	0	0	0	22	20	19243.92
0.0008	1	8	12	7	0	0	0	19	0	21	0	0	0	21	20	19244.01
0.0010	1	8	19	0	0	0	0	19	0	20	0	0	0	20	20	19243.97
0.0012	1	8	21	0	0	0	0	21	0	20	0	0	0	20	20	19243.78
0.0014	1	8	22	0	0	0	0	22	0	19	0	0	0	19	20	19243.68
0.0016	1	9	22	0	0	0	0	22	1	18	0	0	0	19	20	19243.72
0.0018	1	9	22	0	0	0	0	22	4	15	0	0	0	19	20	19243.86
0.0020	1	9	21	0	0	0	0	21	9	10	0	0	0	19	20	19244.06
0.0022	1	9	21	0	0	0	0	21	14	4	0	0	0	19	20	19244.21
0.0024	1	10	20	0	0	0	0	20	20	0	0	0	0	20	20	19244.17
0.0026	1	10	19	0	0	0	0	19	22	0	0	0	0	22	20	19244.00
0.0028	2	10	19	0	0	0	0	19	22	0	0	0	0	22	20	19243.90
0.0030	2	10	19	0	0	0	2	17	22	0	0	0	0	22	20	19243.90
0.0032	2	10	19	0	0	0	5	14	22	0	0	0	0	22	20	19244.00
0.0034	2	10	19	0	0	0	10	9	21	0	0	0	0	21	20	19244.12
0.0036	2	10	19	0	0	0	17	2	21	0	0	0	0	21	20	19244.15
0.0038	2	10	21	0	0	0	21	0	20	0	0	0	0	20	20	19243.97
0.0040	2	10	22	0	0	0	22	0	19	0	0	0	0	19	20	19243.82
0.0042	2	11	22	0	0	0	22	0	19	0	0	0	1	18	20	19243.81
0.0044	2	11	22	0	0	0	22	0	19	0	0	0	3	16	20	19243.91
0.0046	2	11	22	0	0	0	22	0	19	0	0	0	7	12	20	19244.09
0.0048	2	11	21	0	0	0	21	0	19	0	0	0	12	7	20	19244.27
0.0050	2	12	20	0	0	0	20	0	19	0	0	0	19	7	20	19244.30
0.0052	2	12	20	0	0	0	20	0	21	0	0	0	21	0	20	19244.14
0.0054	2	12	19	0	0	0	19	0	22	0	0	0	22	0	20	19244.00
0.0056	3	12	18	0	1	0	19	0	22	0	0	0	22	0	20	19243.97
0.0058	3	12	15	0	4	0	19	0	22	0	0	0	22	0	20	19244.03
0.0060	3	12	10	0	9	0	19	0	21	0	0	0	21	0	20	19244.15
0.0062	3	12	4	0	14	0	19	0	21	0	0	0	21	0	20	19244.22
0.0064	3	12	0	0	20	0	20	0	20	0	0	0	20	0	20	19244.11
0.0066	3	12	0	0	22	0	22	0	19	0	0	0	19	0	20	19243.94
0.0068	3	13	0	0	22	0	22	0	19	0	0	0	19	0	20	19243.89
0.0070	3	13	0	0	22	0	22	0	17	0	2	0	19	0	20	19243.96
0.0072	3	13	0	0	22	0	22	0	14	0	5	0	19	0	20	19244.13
0.0074	3	13	0	0	21	0	21	0	9	0	10	0	19	0	20	19244.32
0.0076	3	13	0	0	21	0	21	0	2	0	17	0	19	0	20	19244.42
0.0078	3	14	0	0	20	0	20	0	0	0	21	0	21	0	20	19244.31
0.0080	3	14	0	0	19	0	19	0	0	0	22	0	22	0	20	19244.14
0.0082	4	14	0	0	19	1	18	0	0	0	22	0	22	0	20	19244.07
0.0084	4	14	0	0	19	3	16	0	0	0	22	0	22	0	20	19244.11
0.0086	4	14	0	0	19	7	12	0	0	0	22	0	22	0	20	19244.21
0.0088	4	14	0	0	19	12	7	0	0	0	21	0	21	0	20	19244.31
0.0090	4	14	0	0	19	19	0	0	0	0	20	0	20	0	20	19244.28
0.0092	4	14	0	0	21	21	0	0	0	0	20	0	20	0	20	19244.10
0.0094	4	14	0	0	22	22	0	0	0	0	19	0	19	0	20	19244.01
0.0096	4	15	0	0	22	22	0	0	0	0	19	1	18	0	20	19244.06
0.0098	4	15	0	0	22	22	0	0	0	0	19	4	15	0	20	19244.21
0.0100	4	15	0	0	21	21	0	0	0	0	19	9	10	0	20	19244.41

Figure 12-7 C analysis using compiled 12PULSE.C.

three C programs (i.e., modify the source code and then proceed through all steps of the compile process). To accomplish this you will, of course, need C software.

12.4 CONCLUDING THOUGHT

Regardless of the programming technique you select for use in your future analyses—BASIC, C, or another—your new programs will increase the rate at which your technical insight grows. Perhaps Shakespeare had technical insight in mind when he wrote: "See first that the design is wise and just: that ascertained, pursue it resolutely; do not for one repulse forego the purpose that you resolved to effect."

Bibliography

1. Richard C. Dorf, *Introduction to Electric Circuits*. New York: Wiley, 1989, pp. 12–14.
2. Donald G. Fink, *Standard Handbook for Electrical Engineers*, 11th ed. New York: McGraw-Hill, 1978, pp. 13-3 and 13-4.
3. Bert K. Erickson, "Network Analysis on the Personal Computer," *RF Design*, Vol. 9. No. 12 (December 1986), pp. 29–34.
4. *OTH-B WCRS Transmit Site Prime Power System*. General Electric Co., Syracuse, N.Y., internal reference document, June 1989, pp. 11–24.
5. Michael Karchmar, "A Simple Guide to Specifying Ferrite Circulators," *Microwaves & RF*, Vol. 26, Nos. 12 and 13 (November and December 1987).
6. Jack Browne, "Tubes Continue the Chase for Power, Gain and Bandwidth," *Microwaves & RF*, Vol. 29, No. 3 (March 1990).
7. Carl Eichenauer, *Transient System Analysis on a Personal Computer*. New York: Wiley, 1988, pp. 166–169.
8. W. G. Wright *Numerical Integration*, General Electric Co. Technical Information Series No. DF-71-DS-1, July 1971.
9. G. N. Glasoe and J. V. Lebacaz, *Pulse Generators*. New York: McGraw-Hill, 1948, pp. 189–207.
10. Merrill I. Skolnik (editor in chief), *Radar Handbook*. New York: McGraw-Hill, 1970, pp. 17-46 to 17-49.
11. Peter Aitken, *Learning C*. Carmel, Ind.: Howard W. Sams/Macmillan Computer Publishing, 1991.

Appendix A

BASIC and C Software Operation

The disk provided with this book includes thirty-two programs written in Microsoft™ BASIC, thirty-four programs written in ANSI Standard C source code and 1 program written in Zortech™ C execute code. These programs are for use with IBM™ XT, AT, or compatible computers.

All of the BASIC programs are briefly reviewed in the text. To execute them you will require Microsoft™ BASIC software (version 2.0 or higher)—not supplied with this disk. The following briefly describes how to load the programs into BASIC and how to execute them. Before running the programs you should refer to the list at the end of the BASIC description. It describes the contents of each program and references the chapter in which the program is described.

The programs listed in C source code cannot be executed until they are compiled. To compile them, you will require ANSI Standard C compiler software—not supplied with this disk. The C program operation section describes how to compile the programs and how to execute them.

Some background associated with C programming and some elementary BASIC to C conversions are briefly discussed in Chapter 12. One compiled C program is also included to illustrate C program transportability.

BASIC PROGRAM OPERATION

1. Load your BASIC or BASICA system (Microsoft version 2.0 or higher) into drive A of the computer. When loading is complete the screen should indicate <Ok>.
2. Insert this disk into drive A.
3. To load a program from the disk press F3. The screen should indicate <LOAD">. Then key in the number of the APPENDIX and press Enter key.
4. When loading is complete the screen should indicate <Ok>. Then press F2 to start program execution.
5. If questions are asked, respond from your keyboard. For some questions execution continues as soon as you enter a response. For others you must press the Enter key following the response.
6. To stop the program during execution simultaneously press the Ctrl and Break keys. The screen should indicate <Break in xxx> (where xxx is the line number at which interruption took place). On the next line it will indicate <Ok>.
7. If you want to continue the same program press F5. <CONT> will appear on the next line and execution will continue from where it was interrupted.
8. Whenever the screen indicates <Ok>, the same program can be started over again by pressing F2.
9. Whenever the screen indicates <Ok> and you want a different program, proceed as in step 3 except key in the new APPENDIX number and Enter.
10. Whenever the screen indicates <Ok> and you want to return to DOS, type SYSTEM and then press Enter. If you want to run additional BASIC programs, it is now necessary to start from step 1.
11. If you want a hard copy of a program that has been loaded, check that your printer is turned on. Type LLIST and then press Enter key.

BASIC PRINTER OPERATION

There is, unfortunately, no common system of functional codes used by printer manufacturers. Therefore you may find that the positioning of your printed displays are not consistent with those shown in the text. To correct this difficulty, check the codes given in your printer's manual against the following list (that was used to document the programs), and revise the programs for proper positioning.

1. CHR$(27);"@";—Resets printer to "power on" status.
2. CHR$(27);"A";CHR$(n);—Sets line feed to n/72 inches.
3. CHR(27);"G";—Sets double strike print mode.
4. CHR(15);—Condensed print (17 characters per inch).
5. CHR(27);"B";CHR$(2);—Elite print (12 characters per inch).
6. CHR(27);"M";CHR(n)—Indents left margin n characters.

7. CHR(27);"Q";CHR(n)—Sets right margin at n characters.
8. WIDTH "LPT1:",n—(BASIC) Sets line length to n characters.

Most of the programs are structured to provide printed output only. If you want monitor output instead, modify the program by changing <LPRINT> to <PRINT> wherever it appears.

BASIC Program Cross References

Appendix		Chapter
1	Parallel to Series Conversion of Circuit Elements	2
2	Steady State Analysis from Keyboard	2
3	Delta to Y and Y to Delta Circuit Conversions	3
4	Three Phase System Analysis from READ Statements	3
5	Transmission Line Analysis	4
6	Rectangular Pulse Spectrum Analysis	5
7	Graphical Fourier Analysis (Manual or DATA List Input)	5
8	Not Programmed	
9	Bessell Functions of the First Kind	6
10	Not Programmed	
11	DC Motor Acceleration Analysis Program	7
12	Transient Numerical Integration Program	7
13	Simplified Power Supply Transient Analysis	7
14	Operational Amplifier Transient Analysis	7
15	Band Pass Amplifier Transient Analysis	8
16	DC Power Supply Transient Analysis—1-Phase Center Tap	8
17	DC Power Supply Transient Analysis—1-Phase Bridge	8
18	DC Power Supply Transient Analysis—3-Phase Bridge	8
18C	3 Phase Full Wave Power Supply, RC Snubber, Revised Commutation, And Expanded Plot	8
18D	Controlled 3-Phase Bridge Converter for Shunt Motor	8
18E	Controlled 3-Phase Bridge Inverter for Shunt Generator	8
18G	12 Pulse Full Wave Power Supply	8
18U	Non-synchronized 3-Phase Y-Delta UPS Driving Resistance Load	8
19	DC-DC Converter Power Supply Transient Analysis	8
20	Not Programmed	
21	Phased Array Pulsing System Analysis	9
22	Phased Array Pulsing System Analysis (modified)	9
23	Not Programmed	
24	Grid Pulser Analysis Program	9
25	Hard Switched Analysis Program	10
26	Pulse Forming Network Analysis Program	10
27	Line Type Pulser Analysis Program	10
28	DC Resonant Charge/Inverse Diode Analysis	10
29	3-Phase Full Wave Power Supply, DC Resonant Charging, and Inverse Diode	10
30	Not Programmed	
31	MTI Stability of TWT RF Power Generator—Voltage Source	11
32	Not Programmed	
33	MTI Stability of TWT RF Power Generator—Volt/Current Source	11

C PROGRAM OPERATION

The fundamental principles of C program compiling and execution are outlined in Chap. 12. Since there are a number of different C compilers available on the market, only a general description will be attempted at this point. Your ANSI Standard C Compiler disk instruction manual should be consulted for full information. However, the essential process is as follows:

1. Load your ANSI Standard C Compiler program into your computer.
2. To run one of the source code programs supplied with this book copy the program from the book's disk onto your ANSI Standard C Source disk.
3. Request your ANSI Standard C system to compile. It will then typically proceed though a temporary storage file phase, an object file development phase, a linking phase, and finally to an execute file development phase. These phases are normally listed on your computer screen as they occur. When completed, a DOS symbol (e.g. >B) will be displayed on the screen.
4. To execute the program type its source code and press Enter.
5. To originate a new program, request the editor function from your ANSI Standard C software. Then proceed to type in the new program. When it is completed, save it. Then proceed with compiling as described in step 3.
6. Frequently there are errors in a new program's structure. If they are present, during the object file development phase a number of error messages will be printed out on your screen. You must then proceed back to the editor phase and make corrections. When an acceptable program structure is obtained the complete set of phases will be completed.
7. After a compiled program is available execute it as described in step 4. However, as with all new computer programs, be sure to check that it is executing as you intended (i.e., that it is calculating the correct answers). At this point in time it is often useful to have a BASIC counterpart solution with which to compare the new program's solutions.
8. If you want a hard copy of a program's source code enter the DOS mode and check that your printer is turned on. The following keyboard input is required <type xxxxxx.c>prn> where xxxxxx.c is the program's source code. Then press the Enter key.

C Program Listings

Coding	Description	BASIC version	Chapter
FOURIER.C	Graphical Fourier Analysis	Appendix 7	12
TRANS_LI.C	Transmission Line Analysis	Appendix 5	12
SSANAL.C	Steady State Circuit Analysis	Appendix 2	12
SSANAL2.C	Bandpass Amplifier Analysis	Appendix 15	12
OPAMP.C	Transient Analysis of Opamp.	Appendix 14	12
OPAMPP.C	Plotted Trans. Response of Opamp.	Appendix 14	12
3PHFWPS.C	Analysis of 3-Phase Bridge Power Supply	Appendix 18C	12

12PULSE.C	Analysis of 12-Pulse Bridge Power Supply	Appendix 18G	12
12PULSE.EXE	Compiled Version of 12PULSE.C		12

Additional C. Source Code on Disk

Coding	Description	BASIC version
SER_PAR.C	Series to Parallel Conversions	Appendix 1
DEL_Y.C	Delta to Y Circuit Conversions	Appendix 3
3PHANAL.C	Three-Phase Steady State Analysis	Appendix 4
REC_PUL.C	Rectangular Pulse Fourier Analysis	Appendix 6
BESSEL.C	Bessel Function Numerical Analysis	Appendix 9
DCMOT.C	DC Motor Acceleration Analysis	Appendix 11
TRANS.C	Fundamental Transient Analysis	Appendix 12
PSTRANS.C	Elementary DC Power Supply Transients	Appendix 13
BPTRANS.C	Bandpass Amplifier Transient Response	Appendix 15
1PHCTPS.C	1 Phase Center Tap Power Supply Transients	Appendix 16
1PHBRPS.C	1 Phase Bridge Power Supply Transients	Appendix 17
P_3PHFW.C	Preliminary 3 Phase Full Wave Transients	Appendix 18
DCMOT.C	Controlled 3-Phase Motor Power Supply	Appendix 18D
C3PHGEN.C	Controlled 3-Phase Generator Power Supply	Appendix 18E
C3PHUPS.C	3-Phrase Uninterruptible Power Supply	Appendix 18U
DC_DCPS.C	50 kHz DC—DC Converter Analysis	Appendix 19
PH_ARP1.C	Prelim. Phased Array Power Analysis	Appendix 21
ARRAY2.C	Revised Phased Array Power Analysis	Appendix 22
GR_PULS.C	A TWT Grid Pulser Transient Analysis	Appendix 24
HARD_SW.C	A Hard Switch Pulser Analysis	Appendix 25
5SECPFN.C	5 Section Pulse Forming Network	Appendix 26
LINETYP.C	A Line Type Pulser System Analysis	Appendix 27
DCR_ID.C	Dc Resonant Charging with Inverse Diode	Appendix 28
3PDCRID.C	3 Phase Power Supply with Dc Resonant Charging and Inverse Diode	Appendix 29
V_TWTPS.C	TWT Regulated Power—Voltage Source	Appendix 31
L_TWTPS.C	TWT Regulated Power—Current Source	Appendix 33

Appendix B

BASIC and C Listings

```
10 REM APPENDIX 1 -- Parallel And Series Conversions Of Circuit Elements --
20 CLS:PRINT"                Parallel And Series Element Conversions
                 ----------------------------------------"
30 INPUT"1. Are two resistors or two reactors to be combined(Enter Y or N)";A$:I
F A$="Y" OR A$="y" THEN INPUT"2. Are the elements in series or parallel(Enter S
 or P)";A$:GOTO 80
40 INPUT"2. Is conversion series to parallel or parallel to series(Enter SP or P
S)";A$
50 IF A$="SP" OR A$="sp" THEN INPUT"3. Enter value of series resistance in ohms"
;R:INPUT"4. Enter value of series reactance in ohms";X:N=R^2+X^2:PRINT"5. Equiva
lent parallel resistance =";N/R;"ohms":PRINT"6. Equivalent parallel reactance ="
;N/X;"ohms"
60 IF A$="PS" OR A$="ps" THEN INPUT"3. Enter value of parallel resistance in ohm
s";R:INPUT"4. Enter value of parallel reactance in ohms";X:D=(X/R)+(R/X):PRINT"5
. Equivalent series resistance =";X/D;"ohms":PRINT"6. Equivalent series reactanc
e =";R/D;"ohms"
70 GOTO 90
80 INPUT"3. Enter the value of the first element in ohms";A:INPUT"4. Enter the v
alue of the second element in ohms.";B:IF A$="P" OR A$="p" THEN PRINT"5. The res
ultant value =";A*B/(A+B);"ohms" ELSE PRINT"5. The resultant value =";A+B;"ohms"
90 INPUT"   Do you want to perform another analysis(Y or N)";A$:IF A$="Y" OR A$=
"y" THEN PRINT"----------------------------------------------------------------
--------":GOTO 30
100 END



10 REM APPENDIX 2 -- This Program Performs Steady State Network Analyses --
20 REM ***** THE FOLLOWING STATEMENTS SELECT THE TYPE OF PROGRAM OUTPUT DISPLAY
*****
30 CLS:PRINT:PRINT:INPUT"Do you want to review program functions and types of an
alysis (enter Y or N)";A$:GOSUB 3080:IF A$="Y" OR A$="y" THEN GOSUB 2240
40 CLEAR ,49152!,3000:CLS:PRINT:INPUT"Enter P if you want a printer display or M
 if you want a monitor display";A$:GOSUB 3060:IF A$="P" OR A$="p" THEN P=1 ELSE
GOTO 90
50 INPUT"Your printer must be on to run this program.  Is it on (enter Y or N)";
A$:GOSUB 3080
60 IF P=1 THEN GOTO 70 ELSE END
70 LPRINT CHR$(27);"@";CHR$(27);"A";CHR$(8);CHR$(27);"B";CHR$(2);CHR$(27);"6";CH
R$(27);"M";CHR$(10);CHR$(27);"N";CHR$(9);
80 REM ***** THE FOLLOWING STATEMENTS DEFINE THE SYSTEM MODEL USED FOR THE ANALY
SIS *****
90 CLS:PRINT"        The Input Information For This Analysis Is As Follows:
         --- ----- ----------- --- ---- -------- -- -- --------"
100 PRINT"     Note: Separate all multiple entry information with commas."
110 IF P=1 THEN LPRINT"      The Input Information For This Analysis Is As Follow
s:"
115 IF P=1 THEN LPRINT"      --- ----- ----------- --- ---- -------- -- -- ------
--"
120 PRINT:INPUT "              1. Number of network nodes =";Y
130 IF P=1 THEN LPRINT"                 1. Number of network nodes =";Y
140 DIM A(Y,Y),B(Y,Y),P(Y,Y),Q(Y,Y),B1(Y,Y),Q1(Y,Y)
150 PI=3.14159654#:LGTEN=20/LOG(10):CT=1:VS=1
160 CT=CT+1:PRINT USING"              ##.  ";CT;
170 IF P=1 THEN LPRINT USING"              ##.  ";CT;
180 INPUT"Enter element symbol ";Z$
190 E$=Z$:Z$=LEFT$(Z$,1)
200 IF Z$="r" OR Z$="R" THEN GOSUB 450 ELSE IF Z$="c" OR Z$="C" THEN GOSUB 480 E
LSE IF Z$="l" OR Z$="L" THEN GOSUB 510 ELSE IF Z$="a" OR Z$="A" THEN GOSUB 610 E
```

Appendix B

```
LSE IF Z$="f" OR Z$="F" THEN GOSUB 540 ELSE IF Z$="n" OR Z$="N" THEN GOSUB 570
210 IF Z$="x" OR Z$="X" THEN GOSUB 400 ELSE IF Z$="e" OR Z$="E" THEN GOTO 220 EL
SE PRINT"    Your symbol is in error.  Please enter it again.";:INPUT " ";Z$:GOTO
 190
220 IF Z$="e" OR Z$="E" THEN PRINT"                             End of list.":PRINT"-----
--------------------------------------------------------------------------"
230 IF P=1 THEN LPRINT"End of list":LPRINT"---------------------------------------
----------------------------------------"
240 REM ***** THE FOLLOWING STATEMENTS SELECT THE TYPE OF ANALYSIS *****
250 INPUT" o Do you want the time delay list (enter Y of N)";A$:GOSUB 3080:IF A$
="Y" OR A$="y" THEN ND=0 ELSE ND=1
260 IF P=0 AND ND=0 THEN GOTO 660
270 IF P=1 AND ND=0 THEN LPRINT"        --- This Is A Voltage Analysis With  Time D
elay ---":GOTO 660
280 INPUT" o Do you want an impedance analysis (enter Y of N)";A$:GOSUB 3080:IF
A$="Y" OR A$="y" THEN QR=1 ELSE QR=0
290 IF P=0 AND QR=1 THEN GOTO 660
300 IF P=1 AND QR=1 THEN LPRINT"              --- This Is An Impedance Analysis ---"
:GOTO 660
310 INPUT" o Do you want a branch power analysis (enter Y or N)";A$:GOSUB 3080:I
F A$="Y" OR A$="y" THEN BP=1 ELSE BP=0
320 IF P=1 AND BP=1 THEN LPRINT"This Is A Branch Power Analysis Where:"
330 IF BP=1 THEN INPUT"   (1) Enter the source voltage";VS:IF P=1 THEN LPRINT"
 (1) Source voltage =";VS
340 IF BP=1 THEN INPUT"   (2) Enter the frequency";G:H=G:D=1:IF P=1 THEN LPRINT"
   (2) Frequency =";G;"Hz"
350 IF BP=1 THEN INPUT"   (3) Enter the first node number of branch";F:E=1:IF P=
1 THEN LPRINT"   (3) The first branch node is";F:GOTO 680
360 IF BP=1 THEN GOTO 680
370 IF P=1 THEN LPRINT"              --- This Is A Voltage Analysis ---"
380 GOTO 660
390 REM ***** THE FOLLOWING STATEMENTS DEFINE ELEMENT NODES AND LOAD ANALYSIS RO
UTINE *****
400 INPUT"                    Nodes and ohms are";I,J,V
410 IF P=1 THEN GOSUB 2490
420 INPUT"                    At what frequency was reactance measured";FX
430 IF V=>0 THEN V=V/(2*PI*FX):GOTO 530
440 V=ABS(1000000!/(2*PI*FX*V)):GOTO 500
450 INPUT"                    Nodes and ohms are";I,J,V
460 IF P=1 THEN GOSUB 2490
470 V=1/V:GOSUB 1240:GOTO 160
480 INPUT"                    Nodes and microfarads are";I,J,V
490 IF P=1 THEN GOSUB 2490
500 V=V*.000001:GOSUB 1320:GOTO 160
510 INPUT"                    Nodes and henries are";I,J,V
520 IF P=1 THEN GOSUB 2490
530 V=-1/V:GOSUB 1380:GOTO 160
540 INPUT"                    Nodes G,S,D and A/V are";K,J,I,V
550 IF P=1 THEN GOSUB 2500
560 L=J:GOTO 650
570 INPUT"                    Nodes B,E,C, beta and B-E Ohms are";K,J,I,B1,V
580 IF P=1 THEN GOSUB 2510
590 L=I:I=K:V=1/V:GOSUB 1240
600 I=L:L=J:GOTO 640
610 PRINT"                    Nodes IN+,IN-,OUT+,OUT-, Gain and Ohms are";
620 INPUT K,L,J,I,B1,V:IF P=1 THEN GOSUB 2520
630 V=1/V:GOSUB 1240
640 V=B1*V
```

```
650 GOSUB 1440:GOTO 160
660 REM ***** THE FOLLOWING STATEMENTS COMPUTE SYSTEM INPUT/OUTPUT RATIOS *****
670 INPUT" o Enter the input and output nodes for this analysis";E,F
680 FOR I=0 TO N:FOR J=0 TO N
690 P(I,J)=A(I,J):Q1(I,J)=B1(I,J)
700 Q(I,J)=B(I,J):NEXT J:NEXT I
710 IF BP=1 THEN GOTO 840 ELSE GOTO 810
720 REM ***** THE FOLLOWING STATEMENTS DETERMINE REVISED ANALYSIS DATA *****
730 PRINT"For this revised run do you want:"
740 INPUT" o Time delay list Y/N";A$:GOSUB 3080
750 IF A$="Y" OR A$="y" THEN ND=0 ELSE ND=1
755 IF A$="Y" OR A$="y" THEN GOTO 780
760 INPUT" o Impedance format Y/N";A$:GOSUB 3080
770 IF A$="Y" OR A$="y" THEN QR=1 ELSE QR=0
780 INPUT" o New input/output nodes Y/N";A$:GOSUB 3080
790 IF A$="Y" OR A$="y" THEN 800 ELSE 820
800 INPUT" o Input & output nodes are";E,F
810 IF P=1 THEN LPRINT"              Input and output nodes are";E;",";F
820 PRINT" o Enter frequency values in hz: ";:INPUT"lowest, highest, increment";
G,H,D
830 REM ***** THE FOLLOWING STATEMENTS CONTROL ANALYSIS SOLUTION SEQUENCE *****
840 IF QR=0 AND BP=0 THEN GOSUB 2540
850 IF QR=0 AND P=1 AND BP=0 THEN GOSUB 2550
860 IF QR THEN GOSUB 2740
870 IF QR AND P=1 THEN GOSUB 2780
880 IF D<0 THEN F2=-D ELSE F2=1+(H-G)/D
890 IF D<0 THEN D=-((H/G)^(1/(-D-1)))
900 F1=G:FOR I1=1 TO F2
910 I2=0:IF ND OR QR THEN 930
920 FOR I2=0 TO 1
930 W=2*PI*F1:D1=E:D2=F:GOSUB 2040
940 V=B1:U=D2
950 IF (-1)^(E+F)>0 THEN 970
960 U=U-180
970 D1=E:D2=E:GOSUB 2040
980 IF V=0 OR B1=0 THEN 1000
990 V=V/B1:DB=LGTEN*LOG(V)
1000 U=U-D2:IF U>180 THEN U=U-360
1010 IF U <-180 THEN U=U+360
1020 IF QR THEN 1050
1030 IF P=0 AND I2=0 AND BP=0 THEN GOSUB 2580
1040 IF P=1 AND I2=0 AND BP=0 THEN GOSUB 2650
1050 IF QR AND P=0 THEN 2840
1060 IF QR AND P=1 THEN 2940
1070 DU=U1-U:U1=U:IF ND THEN 1110
1080 F1=F1*(1+.0001):NEXT I2
1090 U2=DU/U3/360*10000!
1100 IF P=1 AND BP=0 THEN 2720 ELSE 2640
1110 IF P=1 AND BP=0 THEN  LPRINT
1120 IF P=0 AND BP=0 THEN PRINT
1130 IF D<0 THEN F1=-U3*D ELSE F1=U3+D
1140 NEXT I1:IF P=1 AND BP=0 THEN GOSUB 3040
1150 IF BP=1 THEN GOSUB 3100:IF CTR=1 THEN GOTO 840
1160 CTR=0:VP=0:AP=0:INPUT"Do you want to continue (enter Y or N)";A$:PRINT:GOSU
B 3080:IF A$="N" OR A$="n" THEN END
1170 IF BP=0 THEN GOTO 1200 ELSE IF A$="Y" OR A$="y" AND BP=1 THEN INPUT"    (3)
Enter the first node number of branch";F:E=1
1180 IF P=1 THEN LPRINT"This Is A Revised Branch Power Analysis Where:":LPRINT"
```

```
          (3) The first branch node is";F
1190 GOTO 840
1200 IF A$="Y" OR A$="y" THEN GOTO 730
1210 END
1220 REM ***** THE FOLLOWING SET OF SUBROUTINES OBTAIN SOLUTIONS FOR EACH ANALYS
IS FREQUENCY STEP *****
1230 REM --- Form element matrices ---
1240 IF I=0 THEN 1280
1250 A(I,I)=A(I,I)+V
1260 IF J=0 THEN 1290
1270 A(I,J)=A(I,J)-V:A(J,I)=A(J,I)-V
1280 A(J,J)=A(J,J)+V
1290 IF I>N THEN N=I
1300 IF J>N THEN N=J
1310 RETURN
1320 IF I=0 THEN 1360
1330 B(I,I)=B(I,I)+V
1340 IF J=0 THEN 1290
1350 B(I,J)=B(I,J)-V:B(J,I)=B(J,I)-V
1360 B(J,J)=B(J,J)+V
1370 GOTO 1290
1380 IF I=0 THEN 1420
1390 B1(I,I)=B1(I,I)+V
1400 IF J=0 THEN 1290
1410 B1(I,J)=B1(I,J)-V:B1(J,I)=B1(J,I)-V
1420 B1(J,J)=B1(J,J)+V
1430 GOTO 1290
1440 IF I<>0 AND K<>0 THEN A(I,K)=A(I,K)+V
1450 IF J<>0 AND K<>0 THEN A(J,K)=A(J,K)-V
1460 IF I<>0 AND L<>0 THEN A(I,L)=A(I,L)-V
1470 IF K>N THEN N=K
1480 IF L>N THEN N=L
1490 GOTO 1290
1500 REM --- Compute Determinant ---
1510 IF N>1 THEN 1530
1520 D1=A(N,N):D2=B(N,N):RETURN
1530 D1=1:D2=0:K=1
1540 L=K
1550 S=ABS(A(K,K))+ABS(B(K,K))
1560 FOR I=K TO N
1570 T=ABS(A(I,K))+ABS(B(I,K))
1580 IF S>=T THEN 1600
1590 L=I:S=T
1600 NEXT I
1610 IF L=K THEN 1690
1620 FOR J=1 TO N
1630 S=-A(K,J)
1640 A(K,J)=A(L,J)
1650 A(L,J)=S
1660 S1=-B(K,J)
1670 B(K,J)=B(L,J):B(L,J)=S1
1680 NEXT J
1690 L=K+1
1700 FOR I=L TO N
1710 S1=A(K,K)*A(K,K)+B(K,K)*B(K,K)
1720 S=(A(I,K)*A(K,K)+B(I,K)*B(K,K))/S1
1730 B(I,K)=(A(K,K)*B(I,K)-A(I,K)*B(K,K))/S1
1740 A(I,K)=S:NEXT I
```

```
1750 J2=K-1
1760 IF J2=0 THEN 1820
1770 FOR J=L TO N
1780 FOR I=1 TO J2
1790 A(K,J)=A(K,J)-A(K,I)*A(I,J)+B(K,I)*B(I,J)
1800 B(K,J)=B(K,J)-B(K,I)*A(I,J)-A(K,I)*B(I,J)
1810 NEXT I:NEXT J
1820 J2=K:K=K+1
1830 FOR I=K TO N
1840 FOR J=1 TO J2
1850 A(I,K)=A(I,K)-A(I,J)*A(J,K)+B(I,J)*B(J,K)
1860 B(I,K)=B(I,K)-B(I,J)*A(J,K)-A(I,J)*B(J,K)
1870 NEXT J:NEXT I
1880 IF K<> N THEN 1540
1890 L=1
1900 J2=INT(N/2)
1910 IF N=2*J2 THEN 1950
1920 L=0
1930 D1=A(N,N)
1940 D2=B(N,N)
1950 FOR I=1 TO J2
1960 J=N-I+L
1970 S=A(I,I)*A(J,J)-B(I,I)*B(J,J)
1980 S1=A(I,I)*B(J,J)+A(J,J)*B(I,I)
1990 T=D1*S-D2*S1
2000 D2=D2*S+D1*S1
2010 D1=T
2020 NEXT I
2030 RETURN
2040 N1=N:N=N-1:I=0
2050 FOR K=1 TO N
2060 IF K<>D1 THEN 2080
2070 I=1
2080 J=0
2090 FOR L=1 TO N
2100 IF L<>D2 THEN 2120
2110 J=1
2120 A(K,L)=P(K+I,L+J)
2130 B(K,L)=W*Q(K+I,L+J)+Q1(K+I,L+J)/W
2140 NEXT L:NEXT K
2150 GOSUB 1510
2160 N=N1
2170 B1=SQR(D1*D1+D2*D2)
2180 IF D1=0 AND D2=0 THEN 2220
2190 IF D1<0 AND D2=0 THEN 2210
2200 D2=360/PI*ATN(D2/(B1+D1)):RETURN
2210 D2=180
2220 RETURN
2230 REM ***** THE FOLLOWING SUBROUTINES DOCUMENT GENERAL PROGRAM INFORMATION **
***
2240 REM Program Information Subroutine
2250 CLS:PRINT"    Keyboard Data Input Is Used To Perform Steady State Analyses
As Follows:       -------- ---- ----- -- ---- -- ------- ------ ----- ---------
-- --------"
2260 PRINT"1. Program input is a 1 volt ac source connected between node 1 and n
ode 0.":PRINT
2270 PRINT"2. First draw your circuit diagram.  Label all circuit elements using
 only the     prefix letters shown on the following list.  The letters can be fo
```

Appendix B   313

llowed   with additional numbers or letters to further define elements. Th
en proceed"
2280 PRINT"   to label all circuit nodes on the diagram. The number of nodes on
 the   diagram is your first entry item for the program."
2290 PRINT:PRINT"3. Next, enter all elements into the program using the node seq
uences and the   value definitions shown below:
2300 PRINT"   o R (Resistor) from node #, to node #, ohms value"
2310 PRINT"   o C (Capacitor) from node #, to node #, microfarads value"
2320 PRINT"   o L (Inductor) from node #, to node #, henries value"
2330 PRINT"   o X (Reactor) from node #, to node #, ohms value"
2340 PRINT"   o F (Field Effect Transistor) with nodes Gate,Source,Drain,and Gai
n value"
2350 PRINT"   o N (NPN Transistor) with nodes Base,Emit,Coll,and Beta,B-E ohms v
alues"
2360 PRINT"   o A (Operational Amp.) with nodes +IN,-IN,+OUT,-OUT,and Gain,Ohms
out values"
2370 PRINT"   o E (Exit), used to indicate all elements have been entered."
2380 PRINT"---
---------- For a copy turn on printer. Press Shift and PrtSc keys. To continue
 press F5.":END
2390 CLS:PRINT"   Types Of Program Analyses And Displays
                              ----- -- ------- -------- --- --------"
2400 PRINT"1. VOLTAGE VS. FREQUENCY: In this type of analysis the frequency of
 the 1 volt   source is stepped between the lower and upper frequency limits usi
ng the   frequency limits and the increments you specify. Voltage and dB r
esponse "
2410 PRINT"   between the input and output nodes is calculated/listed for each f
requency.   Either uniform or logarithmic frequency increments can be used. T
ime delay   between the two nodes (at each step) will be calculated/listed if
 requested."
2420 PRINT:PRINT"2. IMPEDANCE VS. FREQUENCY: In this type of analysis the progr
am proceeds in   the same frequency stepping manner described above, but the
impedance at a   specified node with respect to the reference node is now cal
culated/listed.
2430 PRINT:PRINT"3. BRANCH POWER: This type of analysis uses 1 above to determi
ne the node vol- tages at each end of a branch. Dividing this voltage differ
ence by branch   impedance gives current. Power, VA, and power factor are ca
lculated";
2440 PRINT"/listed."
2450 PRINT:PRINT"4. A monitor display of the desired output versus frequency can
 be listed.":PRINT
2460 PRINT"5. A hard copy of the desired output versus frequency can be printed.

---------- For a copy turn on printer. Press Shift and PrtSc keys. To continue
 press F5.":END
2470 CLS:RETURN
2480 REM ***** THE FOLLOWING SUBROUTINES PROVIDE ELEMENT DEFINITION PRINTOUTS **
***
2490 LPRINT   E$;" =";I;",";J;",";V:RETURN
2500 LPRINT   E$;" =";K;",";J;",";I;",";V:RETURN
2510 LPRINT   E$;" =";K;",";J;",";I;",";B1;",";V:RETURN
2520 LPRINT   E$;" =";K;",";L;",";J;",";I;",";B1;",";V:RETURN
2530 REM ***** THE FOLLOWING SUBROUTINES DEVELOP FORMAT FOR THE MONITOR AND / OR
 THE PRINTER *****
2540 PRINT   "   FREQ.   VOLTS   DB   PHASE   DELAY":RETUR
N
2550 PRINT   "   --- Note: Results are being printed ---"
2560 LPRINT   "   FREQ.   VOLTS   DB   PHASE   DELAY"

```
2570 LPRINT            "  ------------------------------------------------
":RETURN
2580 U3=F1:PRINT USING"          ##.##^^^^";F1;
2590 IF V=0 OR B1=0 THEN 2610
2600 PRINT USING" #####.##";V*VS;:GOTO 2620
2610 PRINT"    inf";
2620 PRINT" ";:PRINT USING"###.##";DB;
2630 PRINT" ";:PRINT USING"####.#";U;:RETURN
2640 PRINT" ";:PRINT USING"##.#^^^^";U2:GOTO 1130
2650 U3=F1:LPRINT     USING "      ##.##^^^^";F1;
2660 LPRINT      "   ";:IF V=0 OR B1=0 THEN 2690
2670 LPRINT USING"######.##";V*VS;
2680 LPRINT USING"####.##";DB;:GOTO 2700
2690 LPRINT      "   inf";
2700 LPRINT      "  ";
2710 LPRINT      USING" ####.#";U;:RETURN
2720 LPRINT      " ";:LPRINT    USING "##.##^^^^";U2
2730 GOTO 1130
2740 PRINT     "          The input or transfer impedance is:"
2750 PRINT     "      FREQ.        REAL";
2760 PRINT     "        IMAG.     AMPL.";
2770 PRINT     "    ANGLE":RETURN
2780 PRINT          "      --- Note: Results are being printed ---"
2790 LPRINT    "          The input or transfer impedance is:"
2800 LPRINT    "      FREQ.        REAL";
2810 LPRINT    "        IMAG.     AMPL.";
2820 LPRINT    "    ANGLE"
2830 LPRINT         "  ------------------------------------------------
":RETURN
2840 U3=F1:U=(180+U)*PI/180
2850 PRINT     USING"   ##.##^^^^";F1;
2860 PRINT     " ";
2870 PRINT     USING"#######.###";V*COS(U);
2880 PRINT     " ";
2890 PRINT     USING"#######.###";V*SIN(U);
2900 PRINT     " ";
2910 PRINT     USING"#######.###";V;
2920 PRINT     " ";
2930 PRINT     USING"####.##";U*180/PI:GOTO 1130
2940 U3=F1:U=(180+U)*PI/180
2950 LPRINT    USING" ##.##^^^^";F1;
2960 LPRINT    " ";
2970 LPRINT    USING"#######.###";V*COS(U);
2980 LPRINT    " ";
2990 LPRINT    USING"#######.###";V*SIN(U);
3000 LPRINT    " ";
3010 LPRINT    USING"#######.###";V;
3020 LPRINT    " ";
3030 LPRINT    USING"####.##";U*180/PI:GOTO 1130
3040 LPRINT         "  ------------------------------------------------
":RETURN
3050 REM ***** THE FOLLOWING SUBROUTINE CHECKS LETTER INPUTS FROM THE KEYBOARD *
****
3060 IF A$="P" OR A$="p" OR A$="M" OR A$="m" THEN RETURN ELSE PRINT"Your symbol
is in error.  Please enter it again.";
3070 INPUT" ";A$:GOTO 3060
3080 IF A$="Y" OR A$="y" OR A$="N" OR A$="n" THEN RETURN ELSE PRINT"Your symbol
is in error.  Please enter it again.";
```

```
3090 INPUT" ";A$:GOTO 3080
3100 REM ***** THE FOLLOWING SUBROUTINE CALCULATES BRANCH POWER INFORMATION ****
*
3110 CTR=CTR+1:V1=V*VS:AV1=U*PI/180
3120 IF CTR=2 THEN GOTO 3160
3130 INPUT"   (4) Enter the second node number of the branch";F:E=1:IF P=1 THEN
 LPRINT"   (4) The second branch node is";F
3140 IF F>0 THEN VP=V1:AP=AV1:RETURN
3150 R=V1*COS(AV1):X=V1*SIN(AV1):VN=SQR(R^2+X^2):GOSUB 3290:AVN=A:PRINT"   (5) B
ranch voltage =";VN;"@";AVN*180/PI;"Degrees":CTR=2:GOTO 3170
3160 R=-V1*COS(AV1)+VP*COS(AP):X=-V1*SIN(AV1)+VP*SIN(AP):VN=SQR(R^2+X^2):GOSUB 3
290:AVN=A:PRINT"   (5) Branch voltage =";VN;"@";AVN*180/PI;"Degrees"
3170 IF P=1 THEN LPRINT"   (5) Branch voltage =";VN;"@";AVN*180/PI;"degrees"
3180 INPUT"   (6) Enter branch resistance value in ohms =";R
3190 IF P=1 THEN LPRINT"   (6) Branch resistance =";R;"Ohms"
3200 INPUT"   (7) Enter branch reactance symbol (C,L,X), comma, value =";E$,X
3210 IF P=1 THEN LPRINT"   (7) Branch ";:IF E$="C" OR E$="c" THEN LPRINT"capacit
ance = "; ELSE IF E$="L" OR E$="l" THEN LPRINT"inductance ="; ELSE LPRINT"reacta
nce =";
3220 IF P=1 THEN LPRINT X;:IF E$="C" OR E$="c" THEN LPRINT"uF" ELSE IF E$="L" OR
 E$="l" THEN LPRINT"H" ELSE LPRINT"Ohms"
3230 IF E$="C" OR E$="c" THEN X=-1000000!/(2*PI*6*X):GOTO 3250
3240 IF E$="L" OR E$="l" THEN X=2*PI*6*X
3250 ZD=SQR(R^2+X^2):GOSUB 3290:AZD=A:A=VN/ZD:AA=AVN-AZD:PRINT"   (8) Branch cur
rent =";A;"@";AA*180/PI;"Degrees"
3260 IF P=1 THEN LPRINT"   (8) Branch current =";A;"@";AA*180/PI;"degrees"
3270 PRINT"   (9) Branch volt-amperes =";VN*A:PRINT"   (10) Branch power =";VN*A*
COS(AVN-AA);"W":PRINT"   (11) Branch power factor =";COS(AVN-AA)
3280 IF P=1 THEN LPRINT"   (9) Branch volt-amperes =";VN*A:LPRINT"   (10) Branch
 power =";VN*A*COS(AVN-AA);"W":LPRINT"   (11) Branch power factor =";COS(AVN-AA):L
PRINT"----------------------------------------":RETURN
3290 REM ***** THE FOLLOWING SUBROUTINE DEFINES TANGENTS IN ALL QUADRANTS *****
3300 IF R>0 THEN A=ATN(X/R):RETURN
3310 IF R<0 THEN A=ATN(X/R)+PI:RETURN
3320 IF X>0 THEN A=PI/2:RETURN
3330 IF X<0 THEN A=-PI/2:RETURN
3340 A=0:RETURN




10 REM APPENDIX 3  -- Delta To Y Or Y To Delta Circuit Conversion --
20 CLS:PRINT:PRINT:INPUT"    Enter P if you want a printer display or M if you w
ant a monitor display";A$:GOSUB 610:IF A$="P" OR A$="p" THEN P=1 ELSE GOTO 60
30 PRINT:INPUT"   Your printer must be on to run this program.  Is it on (enter
 Y or N)";A$:PRINT:GOSUB 630
40 IF A$="Y" OR A$="y" THEN GOTO 50 ELSE END
50 LPRINT CHR$(27);"@";CHR$(27);"G";CHR$(27);"B";CHR$(3);CHR$(27);"M";CHR$(10);C
HR$(27);"N";CHR$(10);
60 CLS:PI=3.141592654#
70 PRINT"              Three Phase Circuit Conversion
              ----- ----- ------- ----------":PRINT
80 IF P=1 THEN LPRINT"              Three Phase Circuit Conversion
              ----- ----- ------- ----------":LPRINT
90 INPUT"1. Do you want a delta to Y or a Y to delta conversion (enter DY or YD)
";A$:GOSUB 560:PRINT:IF P=1 THEN LPRINT"1. This is a ";:IF A$="dy" OR A$="DY" TH
EN LPRINT"delta to Y circuit conversion." ELSE LPRINT"Y to delta circuit convers
ion."
100 INPUT"2. What is the power source frequency in hz";F:PRINT:IF P=1 THEN LPRIN
```

```
T:LPRINT"2. The power source frequency is";F;"Hz."
110 PRINT"3. Enter each circuit branch as a resistive element and a reactive ele
ment. The      reactive element can be a capacitor of C farads, an inductor of L h
enries or     a known reactance of + or - X ohms (if branch reactance is 0 enter
 X,0);"
120 IF P=1 THEN LPRINT:LPRINT"3. The ";:IF A$="DY" OR A$="dy" THEN LPRINT"delta
circuit elements are as follows:" ELSE LPRINT"Y circuit elements are as follows:
"
130 REM Definition Of Original Circuit Elements
140 FOR A=1 TO 3
150 PRINT"     Branch";A;:INPUT"resistance value in ohms =";V:R(A)=V:IF P=1 THEN L
PRINT"    Branch";A;"resistance =";V;"Ohms"
160 PRINT"     Branch";A;:INPUT"reactance (C, L, X), a comma, and its value =";E$,
V
170 IF E$="C" OR E$="c" THEN C(A)=V:X(A)=-1/(2*PI*F*C(A)):IF P=1 THEN LPRINT"
Branch";A;"capacitance =";V;"F (";X(A);"Ohms)" ELSE 200
180 IF E$="L" OR E$="l" THEN L(A)=V:X(A)=2*PI*F*L(A):IF P=1 THEN LPRINT"     Branc
h";A;"inductance =";V;"H (";X(A);"Ohms)" ELSE 200
190 IF E$="X" OR E$="x" THEN X(A)=V:IF P=1 THEN LPRINT"     Branch";A;"reactance =
 ";V;"Ohms ";:X=V:GOSUB 480
200 GOSUB 590:NEXT A:IF A$="YD" OR A$="yd" THEN GOTO 270
210 REM Delta To Y Calculations
220 A1=ATN(X(1)/R(1)):A2=ATN(X(2)/R(2)):A3=ATN(X(3)/R(3)):Z1=SQR(R(1)^2+X(1)^2):
Z2=SQR(R(2)^2+X(2)^2):Z3=SQR(R(3)^2+X(3)^2)
230 ZD=SQR((R(1)+R(2)+R(3))^2+(X(1)+X(2)+X(3))^2):AD=ATN((X(1)+X(2)+X(3))/(R(1)+
R(2)+R(3)))
240 ZY1=Z1*Z3/ZD:AY1=A1+A3-AD:R1=ZY1*COS(AY1):X1=ZY1*SIN(AY1)
250 ZY2=Z1*Z2/ZD:AY2=A1+A2-AD:R2=ZY2*COS(AY2):X2=ZY2*SIN(AY2)
260 ZY3=Z2*Z3/ZD:AY3=A2+A3-AD:R3=ZY3*COS(AY3):X3=ZY3*SIN(AY3):GOTO 330
270 REM Y To Delta Calculations
280 Z1=SQR(R(1)^2+X(1)^2):A1=ATN(X(1)/R(1)):Z2=SQR(R(2)^2+X(2)^2):A2=ATN(X(2)/R(
2)):Z3=SQR(R(3)^2+X(3)^2):A3=ATN(X(3)/R(3))
290 ZN1=Z1*Z2:AN1=A1+A2:ZN2=Z2*Z3:AN2=A2+A3:ZN3=Z3*Z1:AN3=A3+A1:R=ZN1*COS(AN1)+Z
N2*COS(AN2)+ZN3*COS(AN3):X=ZN1*SIN(AN1)+ZN2*SIN(AN2)+ZN3*SIN(AN3):ZN=SQR(R^2+X^2
):GOSUB 500:AN=A
300 ZD1=ZN/Z3:AD1=(AN-A3):R1=ZD1*COS(AD1):X1=ZD1*SIN(AD1)
310 ZD2=ZN/Z1:AD2=(AN-A1):R2=ZD2*COS(AD2):X2=ZD2*SIN(AD2)
320 ZD3=ZN/Z2:AD3=(AN-A2):R3=ZD3*COS(AD3):X3=ZD3*SIN(AD3)
330 REM Solution Printout
340 PRINT:PRINT"4. The equivalent ";:IF A$="YD" OR A$="yd" THEN PRINT"delta circ
uit elements are as follows:" ELSE PRINT"Y circuit elements are as follows:"
350 IF P=1 THEN LPRINT:LPRINT"4. The equivalent ";:IF A$="YD" OR A$="yd" THEN LP
RINT"delta circuit elements are as follows:" ELSE LPRINT"Y circuit elements are
as follows:"
360 PRINT"     R1 =";R1:IF P=1 THEN LPRINT"     R1 = ";R1
370 PRINT"     X1 =";X1;:X=X1:GOSUB 450:IF P=1 THEN LPRINT"     X1 =";X1;:GOSUB 480
380 PRINT"     R2 =";R2:IF P=1 THEN LPRINT"     R2 = ";R2
390 PRINT"     X2 =";X2;:X=X2:GOSUB 450:IF P=1 THEN LPRINT"     X2 =";X2;:GOSUB 480
400 PRINT"     R3 =";R3:IF P=1 THEN LPRINT"     R3 = ";R3
410 PRINT"     X3 =";X3;:X=X3:GOSUB 450:IF P=1 THEN LPRINT"     X3 =";X3;:GOSUB 480
420 PRINT"     ------------------------------------------------------------------
------"
430 IF P=1 THEN LPRINT"    -------------------------------------------------------
-----------":LPRINT
440 INPUT"    Do you want to perform another conversion (enter Y or N)";A$:GOSUB
 630:IF A$="Y" OR A$="y" THEN GOTO 60 ELSE END
450 REM Element Value Subroutine
460 IF X<0 THEN PRINT"(";ABS(1/(2*PI*F*X));"F)":RETURN
```

Appendix B

```
470 PRINT"(";X/(2*PI*F);"H)":RETURN
480 IF X<0 THEN LPRINT"(";ABS(1/(2*PI*F*X));"F)":RETURN
490 LPRINT"(";X/(2*PI*F);"H)":RETURN
500 REM Tangent Subroutine
510 IF R>0 THEN A=ATN(X/R):RETURN
520 IF R<0 THEN A=ATN(X/R)+PI:RETURN
530 IF X>0 THEN A=PI/2:RETURN
540 IF X<0 THEN A=-PI/2:RETURN
550 A=0:RETURN
560 REM Symbol Check Subroutine
570 IF A$="DY" OR A$="dy" OR A$="YD" OR A$="yd" THEN RETURN ELSE PRINT:PRINT"   Your symbol is in error.  Please check and run the program again."
580 PRINT:END
590 IF E$="L" OR E$="l" OR E$="C" OR E$="c" OR E$="X" OR E$="x" THEN RETURN ELSE PRINT:PRINT"   Your symbol is in error.  Please check and run the program again."
600 PRINT:END
610 IF A$="P" OR A$="p" OR A$="M" OR A$="m" THEN RETURN ELSE PRINT"   Your symbol is in error.  Please enter it again.";
620 INPUT"    ";A$:GOTO 610
630 IF A$="Y" OR A$="y" OR A$="N" OR A$="n" THEN RETURN ELSE PRINT"   Your symbol is in error.  Please enter it again.";
640 INPUT"    ";A$:GOTO 630


10 REM APPENDIX.4  -- Three Phase System Analysis From READ Statements --
20 CLEAR ,49152!,3000
30 CLS:GOSUB 2190:LPRINT CHR$(27);"@";CHR$(27);"B";CHR$(3);CHR$(27);"A";CHR$(9);CHR$(27);"Q";CHR$(92);CHR$(27);"G";CHR$(27);"M";CHR$(10);CHR$(27);"N";CHR$(8);
40 CLEAR:PRINT:INPUT"1. This program determines the values of three phase circuit voltages and         currents.  What is the power source frequency in hz";F1: IF F1=0 THEN PRINT "   Please enter a finite number.":GOTO 40
50 PRINT:INPUT"2. What is the generator's line to line voltage";VLL:PRINT"   The generator's line to neutral voltage =";:VPH=VLL/SQR(3):PRINT VPH:PRINT
60 INPUT"3. Is the load circuit delta, Y, or combined delta-Y (D,Y,C)";Q$:PRINT
70 IF Q$="d" OR Q$="D" THEN PRINT"4. Enter your value for each circuit element listed in lines 1990 through 2070."
80 IF Q$="c" OR Q$="C" THEN PRINT"4. Enter your value for each circuit element listed in lines 1990 through 2150."
90 IF Q$="y" OR Q$="Y" THEN PRINT"4. Enter your value for each circuit element listed in lines 1990 through 2090."
100 PRINT:PRINT"5. If the required circuit values have been entered press F5 key.":PRINT:STOP
110 NODE=7:E=1:F=7
120 IF Q$="d" OR Q$="D" THEN LPRINT "Three phase system analysis with a delta connected load.":Y=12:UL=13
130 IF Q$="y" OR Q$="Y" THEN LPRINT "Three phase system analysis with a Y connected load.":Y=14:UL=15
140 IF Q$="c" OR Q$="C" THEN LPRINT "Three phase system analysis with a delta and Y connected load.":Y=17:UL=21
150 LPRINT:LPRINT"Line-Line V =";VLL;": Line-Neut. V =";VPH;": F =";F1;"Hz"
160 LPRINT:LPRINT"-- System Elements Are As Follows: --":LPRINT
170 DIM A(Y,Y),B(Y,Y),P(Y,Y),Q(Y,Y),B1(Y,Y),Q1(Y,Y),R(20),X(20)
180 PI=3.141592654#:LGTEN=20/LOG(10)
190 REM--THE FOLLOWING STATEMENTS READ THE SYSTEM ELEMENTS INTO THE PROGRAM--
200 FOR A=1 TO 3
210 READ Z$,K,L,J,I,B1,V:GOSUB 290:Z$=LEFT$(Z$,1):IF Z$="A" THEN S$="A"
220 NEXT A
```

```
230 FOR EL = 1 TO UL
240 READ Z$,I,J,V
250 GOSUB 290
260 NEXT EL
270 LPRINT:GOTO 390
280 REM -- THE FOLLOWING SUBROUTINE LOADS THE ELEMENT MATRICES --
290 E$=Z$:Z$=LEFT$ (Z$,1):IF EL=>5 THEN CT=CT+1
300 IF Z$="R" THEN LPRINT E$;", ";I;", ";J;", ";V:R(CT)=V:V=1/V:GOSUB 810:RETURN
310 IF Z$="C" THEN LPRINT E$;", ";I;", ";J;", ";V:X(CT)=-1000000!/(2#PI#F1#V):V=
V#.000001:GOSUB 890:RETURN
320 IF Z$="L" THEN LPRINT E$;", ";I;", ";J;", ";V:X(CT)=2#PI#F1#V:V= -1/V:GOSUB
950:RETURN
330 IF Z$="X" THEN LPRINT E$;", ";I;", ";J;", ";V:IF V=>0 THEN X(CT)=V: V=V/(2#P
I#F1):V=-1/V:GOSUB 950:RETURN
340 IF Z$="X" AND V<0 THEN X(CT)=V:V=ABS(1/(2#PI#F1#V)):GOSUB 890:RETURN
350 IF Z$="A" THEN LPRINT E$;", ";K;", ";L;", ";J;", ";I;", ";B1;", ";V:GOSUB 25
10:V= 1/V:GOSUB 810:V=B1#V:GOSUB 1010:RETURN
360 IF Z$="F" THEN LPRINT E$;", ";K;", ";J;", ";I;", ";V:L=J:GOSUB 1010:RETURN
370 LPRINT"--- An improper element symbol was entered. Please correct data list
. ---":END
380 IF Z$="N" THEN LPRINT E$;", ";K;", ";J;", ";I;", ";B1;", ";V:L=I:I=K::V=1/V:
GOSUB 810:I=L:L=J:V=B1#V:GOSUB 1010:RETURN
390 PRINT:REM --- Compute I/O Ratios ---
400 FOR I=0 TO N:FOR J=0 TO N
410 P(I,J)=A(I,J):Q1(I,J)=B1(I,J)
420 Q(I,J)=B(I,J):NEXT J:NEXT I
430 LPRINT"Input and output nodes are ";E;", ";F:GOTO 450
440 LPRINT"New input and output nodes are ";E;", ";F
450 GOSUB 1810
460 W=2#PI#F1:D1=E:D2=F:GOSUB 1620
470 V=B1:U=D2
480 IF (-1)^(E+F)>0 THEN 500
490 U=U-180
500 D1=E:D2=E:GOSUB 1620
510 IF V=0 OR B1=0 THEN 530
520 V=V/B1
530 U=U-D2:IF U>180 THEN U=U-360
540 IF U <-180 THEN U=U+360
550 GOSUB 1840
560 GOSUB 2250:IF NODE=10 AND Y=14 THEN NODE=13:GOTO 600
570 IF NODE=10 AND Y=17 THEN NODE=13:GOTO 600
580 IF NODE=10 AND Y=12 THEN GOTO 610
590 IF NODE>10 THEN GOTO 610
600 T=1:F=NODE:GOTO 440
610 LPRINT"   Line To Line Voltage 1-2 =";V12;"@";AV1;"Degrees"
620 LPRINT"   Line To Line Voltage 2-3 =";V23;"@";AV2;"Degrees"
630 LPRINT"   Line To Line Voltage 3-1 =";V31;"@";AV3;"Degrees"
640 LPRINT"          Line Current 1 =";IL1;"@";AI1;"Degrees"
650 LPRINT"          Line Current 2 =";IL2;"@";AI2;"Degrees"
660 LPRINT"          Line Current 3 =";IL3;"@";AI3;"Degrees"
670 IF Y=14 THEN GOTO 710
680 LPRINT"         Delta Current 1-2 =";I12;"@";AI12;"Degrees"
690 LPRINT"         Delta Current 2-3 =";I23;"@";AI23;"Degrees"
700 LPRINT"         Delta Current 3-1 =";I31;"@";AI31;"Degrees":IF Y=12 THEN GOT
O 790
710 LPRINT" Line 1 To Neutral Voltage =";V1N;"@";AV1N;"Degrees"
720 LPRINT" Line 2 To Neutral Voltage =";V2N;"@";AV2N;"Degrees"
730 LPRINT" Line 3 To Neutral Voltage =";V3N;"@";AV3N;"Degrees"
```

```
740 LPRINT" Neutral To Ground Voltage =";VNY;"@";AVNY;"Degrees":IF Y=14 THEN GOT
O 780
750 LPRINT"         Y Current In Leg 1 =";I1N;"@";AI1N;"Degrees"
760 LPRINT"         Y Current In Leg 2 =";I2N;"@";AI2N;"Degrees"
770 LPRINT"         Y Current In Leg 3 =";I3N;"@";AI3N;"Degrees"
780 LPRINT" Neutral To Ground Current =";ILN;"@";AILN;"Degrees"
790 LPRINT"-------------------------------------------------------------
":LPRINT:INPUT"Do you want to run another analysis (Y/N)";Q$:IF Q$="Y" OR Q$="y"
 THEN GOTO 40 ELSE END
800 REM --- Form element matrices ---
810 IF I=0 THEN 850
820 A(I,I)=A(I,I)+V
830 IF J=0 THEN 860
840 A(I,J)=A(I,J)-V:A(J,I)=A(J,I)-V
850 A(J,J)=A(J,J)+V
860 IF I>N THEN N=I
870 IF J>N THEN N=J
880 RETURN
890 IF I=0 THEN 930
900 B(I,I)=B(I,I)+V
910 IF J=0 THEN 860
920 B(I,J)=B(I,J)-V:B(J,I)=B(J,I)-V
930 B(J,J)=B(J,J)+V
940 GOTO 860
950 IF I=0 THEN 990
960 B1(I,I)=B1(I,I)+V
970 IF J=0 THEN 860
980 B1(I,J)=B1(I,J)-V:B1(J,I)=B1(J,I)-V
990 B1(J,J)=B1(J,J)+V
1000 GOTO 860
1010 IF I<>0 AND K<>0 THEN A(I,K)=A(I,K)+V
1020 IF J<>0 AND L<>0 THEN A(J,L)=A(J,L)+V
1030 IF J<>0 AND K<>0 THEN A(J,K)=A(J,K)-V
1040 IF I<>0 AND L<>0 THEN A(I,L)=A(I,L)-V
1050 IF K>N THEN N=K
1060 IF L>N THEN N=L
1070 GOTO 860
1080 REM --- Compute Determinant ---
1090 IF N>1 THEN 1110
1100 D1=A(N,N):D2=B(N,N):RETURN
1110 D1=1:D2=0:K=1
1120 L=K
1130 S=ABS(A(K,K))+ABS(B(K,K))
1140 FOR I=K TO N
1150 T=ABS(A(I,K))+ABS(B(I,K))
1160 IF S>=T THEN 1180
1170 L=I:S=T
1180 NEXT I
1190 IF L=K THEN 1270
1200 FOR J=1 TO N
1210 S=-A(K,J)
1220 A(K,J)=A(L,J)
1230 A(L,J)=S
1240 S1=-B(K,J)
1250 B(K,J)=B(L,J):B(L,J)=S1
1260 NEXT J
1270 L=K+1
1280 FOR I=L TO N
```

```
1290 S1=A(K,K)*A(K,K)+B(K,K)*B(K,K)
1300 S=(A(I,K)*A(K,K)+B(I,K)*B(K,K))/S1
1310 B(I,K)=(A(K,K)*B(I,K)-A(I,K)*B(K,K))/S1
1320 A(I,K)=S:NEXT I
1330 J2=K-1
1340 IF J2=0 THEN 1400
1350 FOR J=L TO N
1360 FOR I=1 TO J2
1370 A(K,J)=A(K,J)-A(K,I)*A(I,J)+B(K,I)*B(I,J)
1380 B(K,J)=B(K,J)-B(K,I)*A(I,J)-A(K,I)*B(I,J)
1390 NEXT I:NEXT J
1400 J2=K:K=K+1
1410 FOR I=K TO N
1420 FOR J=1 TO J2
1430 A(I,K)=A(I,K)-A(I,J)*A(J,K)+B(I,J)*B(J,K)
1440 B(I,K)=B(I,K)-B(I,J)*A(J,K)-A(I,J)*B(J,K)
1450 NEXT J:NEXT I
1460 IF K<> N THEN 1120
1470 L=1
1480 J2=INT(N/2)
1490 IF N=2*J2 THEN 1530
1500 L=0
1510 D1=A(N,N)
1520 D2=B(N,N)
1530 FOR I=1 TO J2
1540 J=N-I+L
1550 S=A(I,I)*A(J,J)-B(I,I)*B(J,J)
1560 S1=A(I,I)*B(J,J)+A(J,J)*B(I,I)
1570 T=D1*S-D2*S1
1580 D2=D2*S+D1*S1
1590 D1=T
1600 NEXT I
1610 RETURN
1620 N1=N:N=N-1:I=0
1630 FOR K=1 TO N
1640 IF K<>D1 THEN 1660
1650 I=1
1660 J=0
1670 FOR L=1 TO N
1680 IF L<>D2 THEN 1700
1690 J=1
1700 A(K,L)=P(K+I,L+J)
1710 B(K,L)=W*Q(K+I,L+J)+Q1(K+I,L+J)/W
1720 NEXT L:NEXT K
1730 GOSUB 1090
1740 N=N1
1741 Q1=ABS(D1):Q2=ABS(D2):IF Q1>1E+18 OR Q2>1E+18 THEN PRINT"D1 =";D1;": D2 =";
D2:Q1=Q1/1E+12:Q2=Q2/1E+12:B1=1E+12*SQR(Q1^2+Q2^2):GOTO 1760
1750 B1=SQR(D1*D1+D2*D2)
1760 IF D1=0 AND D2=0 THEN 1800
1770 IF D1<0 AND D2=0 THEN 1790
1780 D2=360/PI*ATN(D2/(B1+D1)):RETURN
1790 D2=180
1800 RETURN
1810 LPRINT     :LPRINT "      FREQ           REAL";
1820 LPRINT     "         IMAG        AMPL";
1830 LPRINT     "      ANGLE":LPRINT"---------------------------------------------
------------------------":RETURN
```

```
1840 U3=F1:U=(180+U)*PI/180
1850 LPRINT     USING"##.##^^^^";F1;
1860 LPRINT         "      ";
1870 LPRINT     USING"#######.###";V*COS(U);
1880 LPRINT         "      ";
1890 LPRINT     USING"#######.###";V*SIN(U);
1900 LPRINT         "      ";
1910 LPRINT     USING"#######.###";V;
1920 LPRINT         "      ";
1930 LPRINT     USING"####.###";U*180/PI:LPRINT"---------------------------------
---------------------------------":LPRINT:RETURN
1940 REM -- DATA INPUT FOR THE SYSTEM --
1950 DATA"A1",1,0,4,0,-1,.002304
1960 DATA"A2",2,0,5,0,2,.002304
1970 DATA"A3",3,0,6,0,2,.002304
1980 DATA"XC",1,2,-1.732050808
1990 DATA"RC",2,0,1
2000 DATA"XL",1,3,1.732050808
2010 DATA"RL",3,0,1
2020 DATA"X1",4,7,.027648
2030 DATA"X2",5,8,.027648
2040 DATA"X3",6,9,.027648
2050 DATA"R4",7,10,.4096
2060 DATA"X5",10,13,.3072
2070 DATA"R6",8,11,.4726
2080 DATA"X7",11,13,.3545
2090 DATA"R8",9,12,.5585
2100 DATA"X9",12,13,.4189
2110 DATA"R10",13,14,1E6
2120 DATA"X11",14,0,.027648
2130 DATA"R12",7,15,.4096
2140 DATA"X13",15,13,.3072
2150 DATA"R14",8,16,.4726
2160 DATA"X15",16,13,.3545
2170 DATA"R16",9,17,.5585
2180 DATA"X17",17,13,.4189
2190 REM Printer On/Tabular Output/Plotted Output Subroutine
2200 PRINT"Your printer must be on to run this program.  Is it on (Y or N)?"
2210 A$=INKEY$:IF A$="" THEN 2210
2220 IF A$="n" OR A$="N" THEN PRINT"For operation without a printer go to DOS.
Type INFORMATION.":END
2230 CLS:PRINT"                          PROGRAM IS IN PROGRESS
                            ------- -- -- --------"
2240 RETURN
2250 REM Line Voltage And Line Current Calculations
2260 IF NODE=7 THEN RV1=V*COS(U):XV1=V*SIN(U):RI1=VPH:XI1=0:R=RI1-RV1:X=XI1-XV1:
V1=SQR(R^2+X^2):GOSUB 2450:AV1=A
2270 IF NODE=7 THEN Z=SQR(R(1)^2+X(1)^2):R=R(1):X=X(1):GOSUB 2450:AZ=A:IL1=V1/Z:
AI1=(AV1-AZ)*180/PI
2280 IF NODE=8 THEN RV2=V*COS(U):XV2=V*SIN(U):RI2=-VPH/2:XI2=-VPH*SIN(PI/3):R=RI
2-RV2:X=XI2-XV2:V2=SQR(R^2+X^2):GOSUB 2450:AV2=A
2290 IF NODE=8 THEN Z=SQR(R(2)^2+X(2)^2):R=R(2):X=X(2):GOSUB 2450:AZ=A:IL2=V2/Z:
AI2=(AV2-AZ)*180/PI
2300 IF NODE=9 THEN RV3=V*COS(U):XV3=V*SIN(U):RI3=-VPH/2:XI3=VPH*SIN(PI/3):R=RI3
-RV3:X=XI3-XV3:V3=SQR(R^2+X^2):GOSUB 2450:AV3=A
2310 IF NODE=9 THEN Z=SQR(R(3)^2+X(3)^2):R=R(3):X=X(3):GOSUB 2450:AZ=A:IL3=V3/Z:
AI3=(AV3-AZ)*180/PI
2320 IF NODE=13 THEN RVNY=V*COS(U):XVNY=V*SIN(U):VNY=SQR(RVNY^2+XVNY^2):R=RVNY:X
```

```
=XVNY:GOSUB 2450:AVNY=A*180/PI
2330 IF NODE=13 THEN Z=SQR(R(10)^2+X(11)^2):R=R(10):X=X(11):GOSUB 2450:AZ=A*180/
PI:ILN=VNY/Z:AILN=AVNY-AZ
2340 IF NODE=8 THEN R=RV1-RV2:X=XV1-XV2:V12=SQR(R^2+X^2):GOSUB 2450:AV1=A*180/PI
:R=R(4):X=X(5):Z=SQR(R^2+X^2):GOSUB 2450:AZ=A*180/PI:I12=V12/Z:AI12=AV1-AZ
2350 IF NODE=9 THEN R=RV2-RV3:X=XV2-XV3:V23=SQR(R^2+X^2):GOSUB 2450:AV2=A*180/PI
:R=R(6):X=X(7):Z=SQR(R^2+X^2):GOSUB 2450:AZ=A*180/PI:I23=V23/Z:AI23=AV2-AZ
2360 IF NODE=9 THEN R=RV3-RV1:X=XV3-XV1:V31=SQR(R^2+X^2):GOSUB 2450:AV3=A*180/PI
:R=R(8):X=X(9):Z=SQR(R^2+X^2):GOSUB 2450:AZ=A*180/PI:I31=V31/Z:AI31=AV3-AZ
2370 IF NODE=13 THEN R=RV1-RVNY:X=XV1-XVNY:V1N=SQR(R^2+X^2):GOSUB 2450:AV1N=A*18
0/PI
2380 IF NODE=13 THEN R=RV2-RVNY:X=XV2-XVNY:V2N=SQR(R^2+X^2):GOSUB 2450:AV2N=A*18
0/PI
2390 IF NODE=13 THEN R=RV3-RVNY:X=XV3-XVNY:V3N=SQR(R^2+X^2):GOSUB 2450:AV3N=A*18
0/PI
2400 IF NODE=13 AND Y=17 THEN R=R(12):X=X(13):Z=SQR(R^2+X^2):GOSUB 2450:AZ=A*180
/PI:I1N=V1N/Z:AI1N=AV1N-AZ
2410 IF NODE=13 AND Y=17 THEN R=R(14):X=X(15):Z=SQR(R^2+X^2):GOSUB 2450:AZ=A*180
/PI:I2N=V2N/Z:AI2N=AV2N-AZ
2420 IF NODE=13 AND Y=17 THEN R=R(16):X=X(17):Z=SQR(R^2+X^2):GOSUB 2450:AZ=A*180
/PI:I3N=V3N/Z:AI3N=AV3N-AZ
2430 NODE=NODE+1
2440 RETURN
2450 REM Tangent Subroutine
2460 IF R>0 THEN A=ATN(X/R):RETURN
2470 IF R<0 THEN A=ATN(X/R)+PI:RETURN
2480 IF X>0 THEN A=PI/2:RETURN
2490 IF X<0 THEN A=-PI/2:RETURN
2500 A=0:RETURN
2510 REM Line Voltage And Source Resistor Subroutine
2520 B1=B1*VPH
2530 RNUM=RNUM+1
2540 R(RNUM)=V
2550 RETURN


10 REM APPENDIX 5   -- Transmission Line Analysis --
20 GOSUB 3110
30 LPRINT CHR$(27);"@";:LPRINT CHR$(27);"G";
40 LPRINT CHR$(27);"A";CHR$(9);CHR$(27);"B";CHR$(2);CHR$(27);"M";CHR$(12);
50 LPRINT CHR$(27);"Q";CHR$(92);CHR$(27);"N";CHR$(8);:WIDTH "LPT1:",80
60 PI=3.141592654#:KMF=2.54#12/100:KMK=KMF*5.28
70 LPRINT"     This Is A Steady State Analysis Of A Transmission Line With Losse
s"
80 LPRINT"    ---- -- - ------ ----- -------- -- - ------------ ---- ---- -----
-":LPRINT
90 INPUT"Is this a Utility line or an RF line application (U or R) ";L$:IF L$="U
" OR L$="u" OR L$="R" OR L$="r" THEN LPRINT"    Is this a Utility line or an RF
line application (U or R)? ";:LPRINT L$:IF L$="R" OR L$="r" THEN GOTO 280
100 IF L$="U" OR L$="u" THEN GOTO 110 ELSE PRINT"       ---Your symbol is in erro
r.  Please repeat.---":GOTO 90
110 INPUT"Is line information physical or electrical (P or E) ";I$:LPRINT"    Is
 line information physical or electrical (P or E)? ";:LPRINT I$:IF I$="E" OR I$=
"e" THEN GOSUB 3170: GOTO 430
120 IF I$="P" OR I$="p" THEN GOTO 130 ELSE PRINT"       ---Your symbol is in erro
r.  Please repeat.---":GOTO 110
130 INPUT"Are line dimensions in SI or NBS units (S or N) ";M$:LPRINT"    Are li
```

```
ne dimensions in SI or NBS units (S or N) ? ";:LPRINT M$:IF M$="s" OR M$="S" THE
N GOTO 150
140 INPUT"What is the diameter of the conductors in inches ";DI:LPRINT"    What
is the diameter of the conductors in inches? ";:LPRINT DI:GOTO 160
150 INPUT"What is the diameter of the conductors in cm ";DI:LPRINT"    What is t
he diameter of the conductors in cm? ";:LPRINT DI:DI=DI/2.54:IF M$="s" OR M$="S"
 THEN GOTO 170
160 INPUT"What is the center to center spacing of conductors in inches ";D:LPRIN
T"    What is the center to center spacing of conductors in inches? ";:LPRINT D:
GOTO 180
170 INPUT"What is the center to center spacing of conductors in cm ";D:LPRINT"
   What is the center to center spacing of conductors in cm ?";:LPRINT D:D=D/2.54
:IF M$="s" OR M$="S" THEN GOTO 190
180 INPUT"What is the series resistance of one conductor per mile ";RR:LPRINT"
   What is the series resistance of one conductor per mile? ";:LPRINT RR:GOTO 200
190 INPUT"What is the series resistance of one conductor per km ";RR:LPRINT"
What is the series resistance of one conductor per km? ";:LPRINT RR:RR=RR*KMK:IF
 M$="s" OR M$="S" THEN GOTO 210
200 INPUT"What is the shunt conductance of one conductor to neutral per mile ";G
:LPRINT"    What is the shunt conductance of one conductor to neutral per mile?
";:LPRINT G:GOTO 220
210 INPUT"What is the shunt conductance of one conductor to neutral per km ";G:L
PRINT"    What is the shunt conductance of one conductor to neutral per km? ";:L
PRINT G:G=G*KMK
220 INPUT" 1. The line's frequency of operation in Hz = ";FO:LPRINT" 1. The line
's frequency of operation in Hz = ";:LPRINT FO
230 C=3.883E-08/(.43429*LOG(2*D/DI)):L=(741.13*.43429*LOG(2*D/DI)+80.47)*.000001
:A=RR:B=2*PI*FO*L:Z=SQR(A^2+B^2):GOSUB 1240:AZ=AN:A=G:B=2*PI*FO*C:Y=SQR(A^2+B^2)
:GOSUB 1240:AY=AN:P=SQR(Z*Y):AP=(AZ+AY)/2:AA=P*COS(AP):BB=P*SIN(AP):ZO=SQR(Z/Y):
AZO=(AZ-AY)/2
240 RO=ZO*COS(AZO):XO=ZO*SIN(AZO):PRINT"       P =";P;"  AA =";AA;"  BB =";BB:PRIN
T"       ZO =";ZO;"  RO =";RO;"  XO =";XO:V=2*PI*FO/BB:KP=V/186280!:PRINT"       V =
";V;"  KP =";KP:LPRINT" 2. The line's wave velocity propagation factor =";KP
250 IF M$="s" OR M$="S" THEN GOTO 270
260 LPRINT" 3. The line's characteristic impedance = ";ZO:LPRINT" 4. The line's
 attenuation factor in nepers / mi =";AA:INPUT" 5. The line's length in mi = ";LM
:LPRINT" 5. The line's length in miles = ";:LPRINT LM:GOTO 430
270 LPRINT" 3. The line's characteristic impedance = ";ZO:LPRINT" 4. The line's
 attenuation factor in nepers / km =";AA*KMK:INPUT" 5. The line's length in km =
 ";LM:LPRINT" 5. The line's length in km = ";:LPRINT LM:LM=LM/KMK:GOTO 430
280 INPUT"Is this a parallel wire, coaxial, or waveguide line (P,C,or W) ";A$:IF
 A$="P" OR A$="p" OR A$="C" OR A$="c" OR A$="W" OR A$="w" THEN LPRINT"    Is thi
s a parallel wire, coaxial, or waveguide line? ";:LPRINT A$:GOTO 300
290 PRINT"       ---Your symbol is in error.  Please repeat.---":GOTO 280
300 INPUT"Are line dimensions in SI or NBS units (S or N) ";M$:LPRINT"    Are li
ne dimensions in SI or NBS units (S or N) ? ";:LPRINT M$
310 INPUT"Do you know the line's electrical characteristics (Y or N) ";B$:IF B$=
"Y" OR B$="y" OR B$="N" OR B$="n" THEN LPRINT"    Do you know the line's electri
cal characteristics? ";:LPRINT B$:IF B$="Y" OR B$="y" THEN GOTO 350 ELSE GOTO 33
0
320 PRINT"       ---Your symbol is in error.  Please repeat.---":GOTO 310
330 INPUT"Do you know the line's physical characteristics (Y or N) ";B1$:IF B1$=
"Y" OR B1$="y" OR B1$="N" OR B1$="n" THEN LPRINT"    Do you know the line's phys
ical characteristics? ";:LPRINT B1$
340 IF B1$="n" OR B1$="N" THEN PRINT"         --- Insufficient Information For Anal
ysis ---":END
350 INPUT" 1. The line's frequency of operation in MHz = ";FO:LPRINT" 1. The lin
e's frequency of operation in MHz = ";:LPRINT USING"#####.####";FO
```

```
360 IF B1$="y" OR B1$="Y" THEN GOSUB 1310:IF M$="s" OR M$="S" THEN GOTO 420 ELSE
 GOTO 410
370 INPUT" 2. The line's characteristic impedance value in ohms = ";Z0:LPRINT" 2
. The line's characteristic impedance value in ohms = ";:LPRINT Z0
380 INPUT" 3. The line's velocity propagation factor value = ";VP:LPRINT" 3. The
 line's velocity of propagation factor value = ";:LPRINT VP:IF M$="s" OR M$="S"
THEN GOTO 400
390 INPUT" 4. The line's attenuation in dB / 100 ft = ";AA:LPRINT" 4. The line's
 attenuation in dB / 100 ft = ";:LPRINT AA:GOTO 410
400 INPUT" 4. The line's attenuation in dB / 100 m = ";AA:LPRINT" 4. The line's
attenuation in dB / 100 m = ";:LPRINT AA:AA = AA*KMF:IF M$="s" OR M$="S" THEN GO
TO 420
410 INPUT" 5. The line's length value in ft = ";LF:LPRINT" 5. The line's length
value in ft = ";:LPRINT LF:GOTO 430
420 INPUT" 5. The line's length value in m = ";LF:LPRINT" 5. The line's length v
alue in m = ";:LPRINT LF:LF=LF/KMF
430 L=LF:INPUT"    Do you know the value of load impedance (Y,N)? ";Z$:IF Z$="Y"
 OR Z$="y" OR Z$="N" OR Z$="n" THEN LPRINT"    Do you know the value of load imp
edance? ";:LPRINT Z$:IF Z$="N" OR Z$="n" THEN GOSUB 1840:GOTO 490
440 IF Z$="Y" OR Z$="y" THEN GOTO 450 ELSE PRINT"      ---Your symbol is in erro
r.  Please repeat,---":GOTO 430
450 INPUT" 6. The load's series resistance value in ohms = ";RL:LPRINT" 6. The l
oad's series resistance value in ohms = ";:LPRINT RL
460 INPUT" 7. The load's series reactance value in ohms = ";XL:LPRINT " 7. The l
oads series reactance value in ohms =";:LPRINT XL
470 ZL=SQR(RL*RL+XL*XL):A=RL:B=XL:GOSUB 1240:ZA=AN
480 LPRINT"    The value of ZL = ";ZL;" @ ";ZA*180/PI;" Deg."
490 INPUT" 8. The source's series resistance value in ohms = ";RS:LPRINT" 8. The
 source's series resistance value in ohms = ";:LPRINT RS
500 INPUT" 9. The source's series reactance value in ohms = ";XS:LPRINT" 9. The
source's series reactance value in ohms = ";:LPRINT XS
510 INPUT"    Will analysis be referenced to load voltage or source voltage (S o
r L)";V$:IF V$="S" OR V$="s" OR V$="L" OR V$="l" THEN LPRINT"    Will analysis b
e referred to load voltage or source voltage (S or L)? ";:LPRINT V$: GOTO 530
520 PRINT"      ---Your symbol is in error.  Please repeat.---":GOTO 510
530 IF V$="S" OR V$="s" THEN INPUT"10. The source RMS voltage = ";VS:LPRINT"10.
The source rms voltage = ";:LPRINT VS:GOTO 550
540 IF V$="L" OR V$="l" THEN INPUT"10. The load RMS voltage = ";VL:LPRINT"10. Th
e load rms voltage = ";:LPRINT VL
550 LPRINT        "-----------------------------------------------------------
--------"
560 GOSUB 1090: LPRINT"   The Solutions At The Line's Sending And Receiving Ends
 Are:"
570            LPRINT"  --- --------- -- --- ------ ------- --- --------- ---- -
---"
580 LPRINT"  Input Current = ";:LPRINT USING "###.###";II;:LPRINT " @ ";:LPRINT
USING "####.###";AI*180/PI;:LPRINT " Deg.: Input Power = ";:LPRINT USING "######
#.###";IP;:LPRINT " Watts"
590 LPRINT"  Input Voltage = ";:LPRINT USING "#####.###";VI;:LPRINT " @ ";:LPRIN
T USING "####.###";(AV)*180/PI;:LPRINT " Deg."
600 LPRINT"  Input Impedance: RI = ";:LPRINT USING "####.##";RI;:LPRINT": XI = "
;:LPRINT USING "####.##";XI;:LPRINT": ZI = ";:LPRINT USING "####.##";ZI;:LPRINT"
 @";:LPRINT USING "####.###";AZ*180/PI;:LPRINT " Deg."
610 LPRINT" Output Current = ";:LPRINT USING "###.###";IL;:LPRINT " @ ";:LPRINT
USING "####.###";IA*180/PI;:LPRINT " Deg.: Output Power = ";:LPRINT USING "######
##.###";LP;:LPRINT " Watts"
620 LPRINT" Output Voltage = ";:LPRINT USING "######.###";VL;:LPRINT " @ ";:LPRI
NT USING "####.###";VA*180/PI;:LPRINT " Deg."
```

```
630 LPRINT" Power Attenuation = ";:LPRINT USING "###.###";4.343*LOG (LP/IP);:LPR
INT" dB";
640 LPRINT": Line Efficiency = ";:LPRINT USING "###.###";100*LP/IP;:LPRINT " %":
CT=1
650 LPRINT         "----------------------------------------------------------
--------":IF L$="U" OR L$="u" THEN END
660 INPUT"Is a table of solutions at intermediate line points required (Y,N)";A$
:IF A$="Y" OR A$="y" THEN LPRINT"Is a table of solutions at intermediate line po
ints required (Y,N)? ";:LPRINT A$:GOTO 720
670 IF A$="N" OR A$="n" THEN GOTO 680 ELSE PRINT"         ---Your symbol is in erro
r.  Please repeat.---":GOTO 660
680 INPUT"Is a plot of intermediate line voltage and current solutions required
(Y,N) ";C$:IF C$="Y" OR C$="y" THEN LPRINT"Is a plot of intermediate line voltag
e and current solutions required (Y,N)? ";:LPRINT C$:GOTO 720
690 IF C$="N" OR C$="n" THEN GOTO 700 ELSE PRINT"         ---Your symbol is in erro
r.  Please repeat.---":GOTO 680
700 INPUT"Is matching stub information required (Y,N) ";MS$:IF MS$="Y" OR MS$="y
" THEN LPRINT"Is matching stub information required (Y,N)? ";:LPRINT MS$:GOSUB 2
250:END
710 IF MS$="N" OR MS$="n" THEN END ELSE PRINT"         ---Your symbol is in error.
 Please repeat.---":GOTO 700
720 IF M$="s" OR M$="S" THEN GOTO 750
730 INPUT"How far from the load (in ft) is the initial solution required";L1:LPR
INT"How far from the load (in ft) is the initial solution required? ";:LPRINT L1
:INPUT"How far from the load (in ft) is the final solution required ";L2
740 LPRINT"How far from the load (in ft) is the final solution required? ";:LPRI
NT L2:GOTO 770
750 INPUT"How far from the load (in m) is the initial solution required";L1:LPRI
NT"How far from the load (in m) is the initial solution required? ";:LPRINT L1:L
1=L1/KMF:INPUT"How far from the load (in m) is the final solution required ";L2
760 LPRINT"How far from the load (in m) is the final solution required? ";:LPRIN
T L2:L2=L2*KMF:GOTO 780
770 INPUT"What point to point increments (in ft) are required ";ST:LPRINT"What p
oint to point  increments (in ft) are required? ";:LPRINT ST:IF C$="Y" OR C$="y"
  THEN GOSUB 1950: GOTO 870 ELSE GOTO 790
780 INPUT"What point to point increments (in m) are required ";ST:LPRINT"What po
int to point  increments (in m) are required? ";:LPRINT ST:ST=ST/KMF:IF C$="Y" O
R C$="y" THEN GOSUB 1950: GOTO 870 ELSE GOTO 830
790 LPRINT"-----------------------------------------------------------------
-"
800 LPRINT"Dist. To"
810 LPRINT TAB(1);"Load(ft.)";TAB(12);"Resistance";TAB(25);"Reactance";TAB(37);"
Current/Deg.";TAB(55);"Voltage/Deg."
820 LPRINT         "----------------------------------------------------------
--------":GOTO 870
830 LPRINT"-----------------------------------------------------------------
-"
840 LPRINT"Dist. To"
850 LPRINT TAB(1);"Load(m.)";TAB(12);"Resistance";TAB(25);"Reactance";TAB(37);"C
urrent/Deg.";TAB(55);"Voltage/Deg."
860 LPRINT         "----------------------------------------------------------
--------"
870 FOR LF = L1 TO L2 STEP ST
880 GOSUB 1090: II=IL*D:AI=(IA+DA)*180/PI:VI=II*ZI:AV=AI+(AZ*180/PI)
890 IF AI<0 THEN AI=360+AI
900 IF AV<0 THEN AV=360+AV
910 IF C$="Y" OR C$="y" THEN GOSUB 2060: GOTO 970
920 LPRINT TAB(1);:LPRINT USING"###.##";LF;
```

```
930 LPRINT TAB(13);:LPRINT USING"#####.###";RI;
940 LPRINT TAB(25);:LPRINT USING"#####.###";XI;
950 LPRINT TAB(37);:LPRINT USING"###.###";II;:LPRINT"@";:LPRINT USING"####.###";AI;
960 LPRINT TAB(55);:LPRINT USING"###.###";VI;:LPRINT"@";:LPRINT USING"####.###";AV
970 NEXT LF
980 IF C$="Y" OR C$="y" THEN GOTO 1000
985 LPRINT"Load vswr =";:RC=SQR((RL-Z0)*(RL-Z0)+XL*XL)/SQR((RL+Z0)*(RL+Z0)+XL*XL
):SW=(1+ABS(RC))/(1-ABS(RC)):LPRINT USING "###.###";SW;:LPRINT" : 1"
990 LPRINT        "----------------------------------------------------------
--------":LPRINT:END
1000 LPRINT "            |....+....|....+....|....+....|....+....|....+....|....
+....|....|....+....|....+....|....+....|....+....|....+"
1010 IP=VM/Z0
1020 FOR CO = 11 TO 111 STEP 10:LPRINT TAB(CO);:LPRINT USING"##.###";.01*(CO-11)
*VM/Z0;:NEXT CO:LPRINT ""
1030 LPRINT
1040 IF M$="s" OR M$="S" THEN GOTO 1060
1050 LPRINT"                     # = Intermediate Current On Line Vs. Distance In
 Feet From Load End":GOTO 1070
1060 LPRINT"                     # = Intermediate Current On Line Vs. Distance In Met
ers From Load End"
1070 LPRINT:LPRINT CHR$(27);"M";
1080 LPRINT:END
1090 REM       -- Calculation Subroutine --
1100 IF L$="U" OR L$="u" THEN A=AA*LM:B=BB*LM:AG=B:ZZ=A:GOTO 1120
1110 A=.1151*AA*LF/100:B=2*PI*LF*F0/(984*VP):AG=B:ZZ=A
1120 SR=(EXP(A)-EXP(-A))*COS(AG)/2:SI=(EXP(A)+EXP(-A))*SIN(AG)/2:SH=SQR(SR*SR+SI
*SI):A=SR:B=SI:GOSUB 1240:SA=AN:A=ZZ
1130 CR=(EXP(A)+EXP(-A))*COS(AG)/2:CI=(EXP(A)-EXP(-A))*SIN(AG)/2:CH=SQR(CR*CR+CI
*CI):A=CR:B=CI:GOSUB 1240:CA=AN
1140 NR=(ZL*CH/Z0)*COS(CA+ZA-AZ0)+SR:NI=(ZL*CH/Z0)*SIN(CA+ZA-AZ0)+SI:N=SQR(NR*NR
+NI*NI):A=NR:B=NI:GOSUB 1240:NA=AN
1150 DR=(ZL*SH/Z0)*COS(SA+ZA-AZ0)+CR:DI=(ZL*SH/Z0)*SIN(SA+ZA-AZ0)+CI:D=SQR(DR*DR
+DI*DI):A=DR:B=DI:GOSUB 1240:DA=AN
1160 ZI=Z0*N/D:AZ=AZ0+NA-DA:RI=ZI*COS(AZ):XI=ZI*SIN(AZ)
1170 IF CT=1 THEN RETURN
1180 Z=((RS+RI)*(RS+RI)+(XS+XI)*(XS+XI))^.5:A=RS+RI:B=XS+XI:GOSUB 1240:ANZ=AN
1190 IF V$="L" OR V$="l" THEN A=VL*CH*COS(CA)+(Z0*VL*SH/ZL)*COS(AZ0+SA-ZA):B=VL*
CH*SIN(CA)+(Z0*VL*SH/ZL)*SIN(AZ0+SA-ZA):VI=SQR(A^2+B^2):GOSUB 1240:AV=AN:II=VI/Z
I:AI=AV-AZ:VS=II*Z:AVS=AI+ANZ:IP=II*II*RI:VC=II*VI
1200 IF V$="L" OR V$="l" THEN IL=VL/ZL:IA=-ZA:LP=IL*IL*RL:CV=IL*VL:RETURN
1210 II=VS/Z:AI=AVS-ANZ:VI=II*ZI:AV=AI+AZ:IP=II*II*RI:VC=II*VI
1220 IL=II/D:IA=AI-DA:VL=IL*ZL:VA=IA+ZA:LP=IL*IL*RL:CV=IL*VL
1230 RETURN
1240 REM        -- Tangent Subroutine --
1250 IF A>0 THEN AN=ATN(B/A):RETURN
1260 IF A<0 THEN AN=ATN(B/A)+PI:RETURN
1270 IF B>0 THEN AN=PI/2:RETURN
1280 IF B<0 THEN AN=-PI/2:RETURN
1290 AN=0:RETURN
1300 END
1310 REM         -- Line Characteristic Calculation Subroutine --
1320 IF A$ = "W" OR A$ = "w" THEN GOTO 1680
1330 IF A$ = "C" OR A$ = "c" THEN GOTO 1510
1340 REM         -- Parallel Wire Line Calculation --
1350 IF M$="s" OR M$="S" THEN GOTO 1370
```

```
1360 INPUT "    What is the diameter of the conductors in inches ";DI:LPRINT"
    What is the diameter of the conductor in inches? ";:LPRINT DI:GOTO 1380
1370 INPUT "    What is the diameter of the conductors in cm ";DI:LPRINT"    Wha
t is the diameter of the conductor in cm? ";:LPRINT DI:DI=DI/2.54:IF M$="s" OR M
$="S" THEN GOTO 1390
1380 INPUT "    What is the center to center spacing of the conductors in inches
 ";D:LPRINT"    What is the center to center spacing of the conductor in inches?
 ";:LPRINT D:GOTO 1400
1390 INPUT "    What is the center to center spacing of the conductors in cm ";D
:LPRINT"    What is the center to center spacing of the conductor in cm? ";:LPRI
NT D:D=D/2.54
1400 INPUT "    What material, copper, aluminum, silver or brass is used (C,A,S,
B) ";ME$
1410 IF ME$="C" OR ME$="c" OR ME$="A" OR ME$="a" OR ME$= "S" OR ME$="s" OR ME$="
B" OR ME$="b" THEN LPRINT"    What material, copper, aluminum, silver, or brass
is used (C,A,S or B)? ";:LPRINT ME$:GOSUB 2190:GOTO 1430
1420 PRINT"      ---Your symbol is in error. Please repeat.---":GOTO 1400
1430 INPUT "    What is the power factor of the line's insulation ";PF:LPRINT"
    What is the power factor of the line's insulation? ";:LPRINT PF
1440 INPUT "    What is the dielectric constant of the line's insulation ";E:LPR
INT "    What is the dielectric constant of the line's insulation? ";:LPRINT E
1450 VP=1/SQR(E):LPRINT" 2. The line's wave velocity propagation factor = ";VP
1460 Z0=120*LOG(2*D/DI)/SQR(E)
1470 LPRINT " 3. The line's characteristic impedance in ohms = ";Z0
1480 AA=.00362*RR*SQR(F0)/(DI*.4343*LOG(2*D/DI))+100*PI*F0*PF/(984*VP*.1151):IF
M$="s" OR M$="S" THEN GOTO 1500
1490 LPRINT " 4. The line's attenuation in dB / 100 feet = ";AA:RETURN
1500 LPRINT " 4. The line's attenuation in dB / 100 m = ";AA/KMF:RETURN
1510 REM            -- Coaxial Line Calculation --
1520 IF M$="s" OR M$="S" THEN GOTO 1540
1530 INPUT "    What is the O.D. of the inner conductor in inches ";DI:LPRINT"
    What is the O.D. of the inner conductor in inches? ";:LPRINT DI:GOTO 1550
1540 INPUT "    What is the O.D. of the inner conductor in cm ";DI:LPRINT"    Wh
at is the O.D. of the inner conductor in cm? ";:LPRINT DI:DI=DI/2.54:GOTO 1560
1550 INPUT "    What is the I.D. of the outer conductor in inches ";D:LPRINT"
    What is the I.D. of the outer conductor in inches? ";:LPRINT D:GOTO 1570
1560 INPUT "    What is the I.D. of the outer conductor in cm ";D:LPRINT"    Wha
t is the I.D. of the outer conductor in cm? ";:LPRINT D:D=D/2.54
1570 INPUT "    What material, copper, aluminum, silver or brass is used (C,A,S,
B) ";ME$
1580 IF ME$="C" OR ME$="c" OR ME$="A" OR ME$="a" OR ME$= "S" OR ME$="s" OR ME$="
B" OR ME$="b" THEN LPRINT"    What material, copper, aluminum, silver, or brass
is used (C,A,S or B)? ";:LPRINT ME$:GOSUB 2190:GOTO 1600
1590 PRINT"      ---Your symbol is in error. Please repeat.---":GOTO 1570
1600 INPUT "    What is the power factor of the line's insulation";PF:LPRINT"
    What is the power factor of the line's insulation? ";:LPRINT PF
1610 INPUT "    What is the dielectric constant of the line's insulation ";E:LPR
INT "    What is the dielectric constant of the line's insulation? ";:LPRINT E
1620 VP=1/SQR(E):LPRINT" 2. The line's velocity propagation factor = ";VP
1630 Z0=60*LOG(D/DI)/SQR(E)
1640 LPRINT " 3. The line's characteristic impedance in ohms = ";Z0
1650 AA=.00362*RR*SQR(F0)*(1+D/DI)/(D*.4343*LOG(D/DI))+100*PI*F0*PF/(984*VP*.115
1):IF  M$="s" OR M$="S" THEN GOTO 1670
1660 LPRINT " 4. The line's attenuation in dB / 100 feet = ";AA:RETURN
1670 LPRINT " 4. The line's attenuation in dB / 100 m = ";AA/KMF:RETURN
1680 REM            -- Waveguide Line Calculation --
1690 IF M$="s" OR M$="S" THEN GOTO 1710
1700 INPUT"    What is the longer guide dimension, a, in inches ";A:LPRINT"    W
```

```
hat is the longer guide dimension, a, in inches? ";:LPRINT A:GOTO 1720
1710 INPUT"     What is the longer guide dimension, a, in cm ";A:LPRINT       What
 is the longer guide dimension, a, in cm? ";:LPRINT A:A=A/2.54:GOTO 1730
1720 INPUT"     What is the shorter guide dimension, b, in inches? ";B:LPRINT"
 What is the shorter guide dimension, b, in inches? ";:LPRINT B:GOTO 1740
1730 INPUT"     What is the shorter guide dimension, b, in cm? ";B:LPRINT"      Wha
t is the shorter guide dimension, b, in cm? ";:LPRINT B:B=B/2.54
1740 INPUT "     What material, copper, aluminum, silver or brass is used (C,A,S,
B) ";MS$
1750 IF MS$="C" OR MS$="c" OR MS$="A" OR MS$="a" OR MS$= "S" OR MS$="s" OR MS$="
B" OR MS$="b" THEN LPRINT"     What material, copper, aluminum, silver, or brass
 is used (C,A,S or B)? ";:LPRINT MS$:GOSUB 2190:GOTO 1770
1760 PRINT"        ---Your symbol is in error. Please repeat.---":GOTO 1740
1770 FC=11808/(2*A):LPRINT" 2. The line's cutoff frequency in MHz. = ";FC
1780 VP=SQR(1-(FC/F0)^2)
1790 Z0=120*PI*VP*B/A
1800 LPRINT " 3. The line's characteristic impedance in ohms = ";Z0
1810 A=(.129*RR/(A^1.5))*((A/(2*B))*(F0/FC)^1.5+(F0/FC)^-.5)/((F0/FC)^2-1)^.5:IF
 M$="s" OR M$="S" THEN GOTO 1830
1820 AA=A/.1151:LPRINT" 4. The line's attenuation in dB / 100 feet = ";AA:RETURN
1830 AA=A/.1151:LPRINT" 4. The line's attenuation in dB / 100 m = ";AA/KMF:RETUR
N
1840 REM              -- ZL Calculation Subroutine --
1850 INPUT"    What is the line's measured VSWR at the load end ";SW
1860 LPRINT"    What is the line's measured VSWR at the load end? ";:LPRINT SW
1870 INPUT "    What is the radian measurement from load to first Vmax toward so
urce ";RA:LPRINT       What is the radian measurement from load to first Vmax towa
rt source? ";:LPRINT RA
1880 RC=(SW-1)/(SW+1):X=RC*COS(2*RA):Y=RC*SIN(2*RA)
1890 A=1+X:B=Y:GOSUB 1240:NA=AN:N=Z0*SQR(A*A+B*B)
1900 A=1-X:B=-Y:GOSUB 1240:DA=AN:D=SQR(A*A+B*B)
1910 ZL=N/D:ZA=NA-DA:RL=ZL*COS(ZA):XL=ZL*SIN(ZA)
1920 LPRINT    " 6. The load's series resistance value in ohms = ";RL
1930 LPRINT    " 7. The load's series reactance value in ohms = ";XL
1940 RETURN
1950 REM -- Plotting Subroutine --
1960 INPUT"What scale multiplying factor is required ";MF:LPRINT"What scale mult
iplying factor is required? =";:LPRINT MF
1970 LPRINT CHR$(27);"@";CHR$ (15);CHR$(27);"A";CHR$(8);CHR$(27);"G":WIDTH"LPT1:
",135
1980 IF M$="s" OR M$="S" THEN GOTO 2000
1990 LPRINT"                     + = Intermediate  Voltages On  Line Vs. Distance
 In Feet From Load End: VSWR =";:GOTO 2010
2000 LPRINT"                     + = Intermediate Voltages On Line Vs. Distance In Me
ters From Load End: VSWR =";
2010 RC=SQR((RL-Z0)*(RL-Z0)+XL*XL)/SQR((RL+Z0)*(RL+Z0)+XL*XL):SW=(1+ABS(RC))/(1-
ABS(RC)):LPRINT USING "###.#";SW;:LPRINT : 1":LPRINT
2020 VM=VS*MF
2030 FOR CO=11 TO 111 STEP 10:LPRINT TAB(CO);:LPRINT USING"###.##";.01*(CO-11)*
VM;:NEXT CO:CO=0:LPRINT ""
2040 LPRINT "Amps @ Deg.  !....+....!....+....!....+....!....+....!....+....!..
..+....!....+....!....+....!....+....!....+....!....+ Volts @ Deg."
2050 RETURN
2060 CO=CO+1
2070 IF LF=L1 THEN GOTO 2100
2080 IF CO < 10 THEN LPRINT TAB(1);:LPRINT USING"#.###";II;:LPRINT"@";:LPRINT US
ING"####.#";AI;:LPRINT TAB(14);"-";:GOTO 2120
2090 IF M$="s" OR M$="S" THEN GOTO 2110
```

```
2100 LPRINT TAB(1);:LPRINT USING"#.###";II;:LPRINT"@";:LPRINT USING"####.#";AI;:
LPRINT TAB(14);"--";TAB(16);:LPRINT USING"###.##";LF;:CO=0:GOTO 2120
2110 LPRINT TAB(1);:LPRINT USING"#.###";II;:LPRINT"@";:LPRINT USING"####.#";AI;:
LPRINT TAB(14);"--";TAB(16);:LPRINT USING"###.##";LF/3.281;:CO=0
2120 PP=14+VI*100/VM
2130 PQ=14+II*ZO*100/VM
2140 IF INT(PQ+.5)=INT(PP+.5) THEN GOTO 2180
2150 IF PQ>PP THEN GOTO 2170
2160 LPRINT TAB(PQ);"*";TAB(PP);"+";TAB(121);:LPRINT USING"###.#";VI;:LPRINT"@";
:LPRINT USING"####.#";AV:RETURN
2170 LPRINT TAB(PP);"+";TAB(PQ);"*";TAB(121);:LPRINT USING"###.#";VI;:LPRINT"@";
:LPRINT USING"####.#";AV:RETURN
2180 LPRINT TAB(PP);"0";TAB(121);:LPRINT USING"###.#";VI;:LPRINT"@";:LPRINT USIN
G"####.#";AV:RETURN
2190 REM -- Relative Resistivity Subroutine --
2200 IF MS$ ="C" OR MS$="c" THEN RR=1:RETURN
2210 IF MS$ ="A" OR MS$="a" THEN RR=1.28:RETURN
2220 IF MS$ ="S" OR MS$="s" THEN RR=.97:RETURN
2230 IF MS$ ="B" OR MS$="b" THEN RR=1.51:RETURN
2240 RR=1:LPRINT"          Note: Copper has been used as a default material.":R
ETURN
2250 REM -- Matching Stub Subroutine --
2260 LPRINT"------------------------------------------------------------------
-"
2270 PRINT"          Matching Stub Calculations"
2300 PRINT"          -- --- ------ ---- ---"
2310 PRINT TAB(1);"% \/4";TAB(10);"Parallel Resistance";TAB(35);"Parallel Reacta
nce"
2320 PRINT TAB(1);"Length";TAB(10);"Main Line";TAB(35);"Main Line"
2330 PRINT"-----------------------------------------------------------"
2340 LQ=984*VP/(4*FO):PC=-2
2350 ST=.05*LQ:LF=-ST
2360 IF SS=1 THEN GOTO 2380
2370 P=RL:Q=XL:J=ZA:K=ZL
2380 LF=LF+ST
2390 PV=RO:PX=XO
2400 GOSUB 1090:R=RI:X=XI:GOSUB 3000:RO=RP:XO=XP
2410 IF LF>0 THEN GOTO 2440
2420 IF RO<ZO THEN S$="OC":GOTO 2440
2430 S$="SC"
2440 PRINT TAB(1);INT(100*LF/LQ+.5);TAB(10);INT(100*RO+.5)/100;TAB(35);INT(100*X
0+.5)/100
2450 IF S$="OC" THEN GOTO 2480
2460 IF RO<ZO THEN GOTO 2500
2470 GOTO 2490
2480 IF RO>ZO THEN GOTO 2500
2490 GOTO 2380
2500 PRINT"-----------------------------------------------------------"
2510 DE=ABS(ABS(PV)-ABS(RO)):IN=ABS(ABS(PV)-ABS(ZO)):SA=LF-(1-IN/DE)*ST:XX=PX+(X
0-PX)*IN/DE:IF SS=1 THEN GOSUB 3060:GOTO 2550
2520 IF M$="s" OR M$="S" THEN GOTO 2540
2530 LPRINT"-- The stub attachment point is ";INT(10000*SA+.5)/10000;" feet fro
m the load end. --":GOTO 2550
2540 LPRINT"-- The stub attachment point is ";INT(10000*SA*KMF+.5)/10000;" mete
rs from the load end. --"
2550 LPRINT"-- At this point the line's RP = ";INT(100*(PV+(RO-PV)*IN/DE)+.5)/10
0;", and its XP = ";INT(100*(PX+(XO-PX)*IN/DE)+.5)/100;" ohms. --"
2570 IF S$="OC" THEN GOTO 2600
```

```
2580 GOTO 2780
2590 REM -- Open Stub Subroutine --
2600 PRINT TAB(1);"% \/4";TAB(10);"Parallel Resistance";TAB(35);"Parallel Reacta
nce"
2610 PRINT TAB(1);"Length";TAB(10);"Open Stub";TAB(35);"Open Stub":LPRINT
              "--------------------------------------------------------":LF=-ST
2620 LF=LF+ST
2630 PV=R0:PX=X0:RL=1000000!:XL=-1000000!:ZL=1414000!:ZA=-PI/4
2640 GOSUB 1090:R=RI:X=XI:GOSUB 3000:R0=RP:X0=XP
2650 PRINT TAB(1);INT(100*LF/LQ+.5);TAB(10);INT(100*R0+.5)/100;TAB(35);INT(100*X
0+.5)/100
2660 IF (X0+XX)>0 THEN GOTO 2680
2670 GOTO 2620
2680 PRINT"--------------------------------------------------------"
2690 DE=ABS(ABS(PX)-ABS(X0)):IN=ABS(ABS(PX)-ABS(XX)):SA=LF-(1-IN/DE)*ST
2700 IF M$="s" OR M$="S" THEN GOTO 2720
2710 LPRINT"-- The length of the open stub is ";INT(SA*10000+.5)/10000;" feet. -
-":GOTO 2730
2720 LPRINT"-- The length of the open stub is ";INT(SA*KMF*10000+.5)/10000;" met
ers. --"
2730 LPRINT"-- The stub's input RP = ";INT(100*(PV+(R0-PV)*IN/DE)+.5)/100;", and
 its XP = ";INT(100*(PX+(X0-PX)*IN/DE)+.5)/100;" ohms. --"
2740 IF SS=1 THEN GOTO 2980
2750 LPRINT"--------------------------------------------------------"
2760 GOTO 2940
2770 REM -- Shorted Stub Subroutine --
2780 PRINT TAB(1);"% \/4";TAB(10);"Parallel Resistance";TAB(35);"Parallel Reacta
nce"
2790 PRINT TAB(1);"Length";TAB(10);"Shorted Stub";TAB(35);"Shorted Stub":LPRINT
              "--------------------------------------------------------":LF=
-ST
2800 LF=LF+ST
2810 PV=R0:PX=X0:RL=.000001:XL=.000001:ZL=1.414E-06:ZA=PI/4
2820 GOSUB 1090:R=RI:X=XI:GOSUB 3000:R0=RP:X0=XP
2830 PRINT TAB(1);INT(100*LF/LQ+.5);TAB(10);INT(100*R0+.5)/100;TAB(35);INT(100*X
0+.5)/100
2840 IF (X0+XX)>0 THEN GOTO 2860
2850 GOTO 2800
2860 PRINT"--------------------------------------------------------"
2870 DE=ABS(ABS(PX)-ABS(X0)):IN=ABS(ABS(PX)-ABS(XX)):SA=LF-(1-IN/DE)*ST
2880 IF M$="s" OR M$="S" THEN GOTO 2900
2890 LPRINT"-- The length of the shorted stub is ";INT(SA*10000+.5)/10000;" feet
. --":GOTO 2910
2900 LPRINT"-- The length of the shorted stub is ";INT(SA*KMF*10000+.5)/10000;"
meters. --
2910 LPRINT"-- The stub's input RP = ";INT(100*(PV+(R0-PV)*IN/DE)+.5)/100;", and
 its XP = ";INT(100*(PX+(X0-PX)*IN/DE)+.5)/100;" ohms. --"
2920 IF SS=1 THEN RETURN
2930 LPRINT"--------------------------------------------------------"
2940 INPUT"Is a source matching stub required (Y,N) ";Q$:IF Q$="Y" OR Q$="y" THE
N LPRINT"Is a source matching stub required (Y,N)? ";:LPRINT Q$:GOTO 2960
2950 IF Q$ = "N" OR Q$ = "n" THEN GOTO  2990 ELSE PRINT"      ---Your symbol is
in error.  Please repeat.---":GOTO 2940
2960 LPRINT"--------------------------------------------------------":PRINT:SS=
1
2970 RL=RS:XL=XS:ZL=SQR(RL*RL+XL*XL):A=RL:B=XL:GOSUB 1240:ZA=AN:CT=0:GOSUB 1090:
SS=1:CT=1:GOSUB 2310
2980 RL=P:XL=Q:ZA=J:ZL=K
```

Appendix B

```
2990 LPRINT"-----------------------------------------------------------":END
3000 REM -- Series To Parallel Conversion Subroutine --
3010 IF R=0 THEN RR=1000000!:GOTO 3030
3020 RP=(R*R+X*X)/R
3030 IF X=0 THEN XP=1000000!:GOTO 3050
3040 XP=(R*R+X*X)/X
3050 RETURN
3060 REM -- Source End Attachment Point Subroutine --
3070 IF M$="s" OR M$="S" THEN GOTO 3090
3080 LPRINT"-- The stub attachment point is ";INT(10000*SA+.5)/10000;" feet from
 source end. --":RETURN
3090 LPRINT"-- The stub attachment point is ";INT(10000*SA*KMF+.5)/10000;" meter
s from source end. --"
3100 RETURN
3110 REM Printer On/Tabular Output/Plotted Output Subroutine
3120 CLS
3130 INPUT"Your printer must be on to run this program.  Is it on (Y or N)";A$
3140 IF A$="n" OR A$="N" THEN PRINT"For operation without a printer go to DOS.
Type INFORMATION.":END
3150 CLS:PRINT"                       PROGRAM IS IN PROGRESS
                                      ------- -- -- --------"
3160 IF A$="Y" OR A$="y" THEN RETURN ELSE PRINT"        --- Your symbol is in erro
r.  Please repeat.---":GOTO 3130
3170 REM   -- Utility Line Electrical Data Subroutine --
3180 INPUT"Are measurements in km or mi (K or M) ";I$:LPRINT"    Are measurement
s in km or mi (K or M)? ";I$
3190 INPUT"What is the value of km or mi per unit length ";INC:LPRINT"    What i
s the number of km or mi per unit length? ";INC
3200 INPUT"What is the value of resistance per unit length ";RPI:LPRINT"    What
 is the value of resistance per unit length ";RPI
3210 INPUT"What is the value of reactance per unit length ";XPI:LPRINT"    What
is value of reactance per unit length? ";XPI
3220 INPUT"What is value of conductance per unit length ";GPI:LPRINT"    What is
 the value of conductance per unit length ";GPI
3230 INPUT"What is the value of susceptance per unit length ";BPI:LPRINT"    Wha
t is the value of susceptance per unit length? ";BPI
3240 INPUT"Are the above values expressed in terms of a base value (Y or N) ";BA
$:LPRINT"    Are the above values expressed in terms of a base value (Y or N)? "
;BA$:IF BA$="N" OR BA$="n" THEN KB=1:VA=1:VLL=1:GOTO 3260
3250 OV=VA:INPUT"Enter the line to line voltage, a comma, and the base VA ";VLL,
VA:LPRINT"    Enter the line to line voltage, a comma, and the base VA?";VLL;:LP
RINT",";VA:KB=VLL^2/VA:VA=OV
3260 INPUT" 1. The line's frequency of operation in Hz = ";F0:LPRINT" 1. The lin
e's frequency of operation in Hz = ";:LPRINT F0
3270 A=RPI*KB:B=XPI*KB:Z=SQR(A^2+B^2):GOSUB 1240:AZ=AN:A=GPI/KB:B=BPI/KB:Y=SQR(A
^2+B^2):GOSUB 1240:AY=AN:P=SQR(Z*Y):AP=(AZ+AY)/2:AA=P*COS(AP):BB=P*SIN(AP):Z0=SQ
R(Z/Y):AZ0=(AZ-AY)/2
3280 R0=Z0*COS(AZ0):X0=Z0*SIN(AZ0):PRINT"         P =";P;":  AA =";AA;":  BB =";BB:PRI
NT"     Z0 =";Z0;":  R0 =";R0;":  X0 =";X0:V=2*PI*F0*INC/BB:KP=V/300000!:PRINT"
  V =";V;":  KP =";KP:LPRINT" 2. The line's wave velocity propagation factor =";K
P
3290 LPRINT" 3. The line's characteristic impedance = ";Z0:LPRINT" 4. The line's
 attenuation factor in nepers / km. =";AA/INC:INPUT" 5. The line's length in inc
rements = ";LM:LPRINT" 5. The line's length in increments = ";:LPRINT LM:RETURN


10 REM APPENDIX.6 -- Rectangular Pulse Spectrum Analysis --
12 CLS:GOSUB 3500
```

```
15 GOSUB 3120:DU=TP#F:TR=TP/DU:IF CF=0 THEN GOSUB 3310:GOTO 25
20 LPRINT:LPRINT"Spectrum Of";HN; "Fourier Coefficients Of A";INT(1000#DU+.5)/10
00;"DU Rect. Pulse: TP =";TP;"Sec.":LPRINT
25 DIM C(500):PI=3.141592654#:TR=TP/DU:SC=-1:IF CF=1 THEN GOSUB 1000
30 FOR N = 1 TO HN
35 C(N)=2#DU#(SIN(N#PI#DU)/(N#PI#DU))
40 IF CF = 1 THEN GOSUB 1045:GOTO 50
45 GOSUB 3370
50 NEXT N:IF CF=1 THEN GOSUB 3100
55 IF CT=0 THEN GOSUB 3390:GOTO 65
60 LPRINT:LPRINT"-------------------------------------------------------------
----------------":LPRINT"f(t) Sum Of";HN;"Rect. Pulse Fourier Coef.: TP=";TP;":P
RR=";F;" DU=";INT(100000!#DU+.5)/100000!:LPRINT
65 SU=DU:W=2#PI#F
70 IF CT = 1 THEN GOSUB 2000
75 FOR T = (TR-TP) TO (TR+TP) STEP TP/20
80 FOR N = 1 TO HN STEP 1
85 K=C(N)#COS(N#W#T)
90 SU=SU+K
95 NEXT N:IF CT =1 THEN GOSUB 2015
100 IF CT=0 THEN GOSUB 3430
105 SU=DU
110 NEXT T:IF CT=1 THEN GOSUB 3100
120 END
1000 REM --- Spectrum Plot Subroutine ---
1005 LPRINT TAB(4);"-";INT(1000#.5#DU+.5)/1000;TAB(16);0;TAB(25);INT(1000#.5#DU+
.5)/1000;TAB(35);INT(1000#DU+.5)/1000;TAB(45);INT(1000#1.5#DU+.5)/1000;
1010 LPRINT TAB(55);INT(1000#2#DU+.5)/1000;TAB(65);INT(1000#2.5#DU+.5)/1000
1015 GOSUB 3100
1040 LPRINT TAB(1);"--F=0" ;:LPRINT TAB(16);"!";;PP=16.5+20:GOSUB 3040:LPRINT TA
B(71);TAB(73);INT(DU#1000+.5)/1000:RETURN
1045 SF=SF+1
1050 PP=(16  +20#C(N)/DU)
1055 LPRINT TAB(1);"-";:IF SF = 0 OR SF = 10 THEN LPRINT TAB(2);"-F=";F#N;:SF=0
1060 IF ABS(20#C(N)/DU) < .5 THEN LPRINT TAB(16);"+";TAB(71);"-";TAB(72);:LPRINT
 USING"#.####";C(N):RETURN
1065 IF PP <16 THEN GOSUB 3000: LPRINT TAB(16);"!";TAB(71);"-";TAB(72);:LPRINT U
SING"#.####";C(N):RETURN
1070 LPRINT TAB(16);"!";;:IF PP > 16 THEN GOSUB 3040:LPRINT TAB(71);"-";TAB(72);:
LPRINT USING"#.####";C(N):RETURN
1075 GOSUB 3000:LPRINT TAB(71);"-";TAB(72);:LPRINT USING"#.####";C(N):RETURN
2000 REM --- Coefficient Summation Subroutine ---
2005 LPRINT"     -.1  0   .1  .2  .3  .4  .5  .6  .7  .8  .9 1.0 1.1
 1.2"
2010 GOSUB 3100
2015 CO=CO+1
2020 IF CO=1 THEN RETURN
2030 SC=SC+1
2035 PP=(11  +50#SU)
2040 IF PP>17 THEN GOTO 2085
2045 IF PP<10.5 THEN GOTO 2070
2050 IF PP>10.5 AND PP<11.5 THEN LPRINT TAB(11);"+";:GOTO 2060
2055 LPRINT TAB(11);"-";TAB(PP);"+";
2060 IF SC=0 OR SC=10 THEN LPRINT TAB(PP+1);"--";TAB(PP+4);"T=";:LPRINT USING".#
#####";T;:SC=0:LPRINT TAB(61);"!";:TAB(71);"-";TAB(72);INT(10000#SU+.5)/10000:RET
URN
2065 LPRINT TAB(61);"!";TAB(71);"-";TAB(72)INT(10000#SU+.5)/10000:RETURN
2070 LPRINT TAB(PP);"+";
```

Appendix B

```
2075 IF SC=0 OR SC=10 THEN LPRINT TAB(11);"--";TAB(13);"T=";:LPRINT USING".#####
#";T;:SC=0:LPRINT TAB(61);"!";TAB(71);"-";TAB(72);INT(10000*SU+.5)/10000:RETURN
2080 LPRINT TAB(11);"-";TAB(61);"!";TAB(71);"-";TAB(72);INT(10000*SU+.5)/100000!
:RETURN
2085 IF SC=0 OR SC=10 THEN LPRINT TAB(11);"--";TAB(13);"T=";:LPRINT USING".#####
#";T;:SC=0:GOTO 2105
2095 LPRINT TAB(11);"-";
2100 IF PP>60.5 AND PP<61.5 THEN LPRINT TAB(61);"+";TAB(71);"-";:GOTO 2115
2105 IF PP<60.5 THEN LPRINT TAB(PP);"+";TAB(61);"!";TAB(71);"-";:GOTO 2115
2110 IF PP>61.5 THEN LPRINT TAB(61);"!";TAB(PP);"+";TAB(71);"-";
2115 IF SC=0 THEN LPRINT TAB(72);INT(10000*SU+.5)/10000:RETURN
2120 LPRINT TAB(72);INT(10000*SU+.5)/10000:RETURN
3000 REM --- Spectrum Negative Line Plot Subroutine ---
3010 LPRINT TAB(PP);"-";
3020 PP=PP+1:IF PP <15.5 THEN GOTO 3010
3030 RETURN
3040 REM --- Spectrum Positive Line Plot Subroutine ---
3050 QQ=17:LPRINT TAB(17);"-";
3060 QQ=QQ+1
3070 IF QQ = < PP THEN LPRINT TAB(QQ);"-";:GOTO 3060
3080 RETURN
3100 REM --- Trailing Edge Scale Subroutine ---
3110 LPRINT"!....!....!....!....!....!....!....!....!....!....!....!....!..
..!":RETURN
3120 REM --- Analysis Requirements Subroutine ---
3140 PRINT" --- This Is A Rectangular Pulse Analysis Program Which: ---":PRINT
3150 PRINT"      o Calculates, plots, and tabulates the Fourier coefficients."
3160 PRINT"      o Performs an inverse Fourier transform to reconstruct the pulse
."
3170 PRINT"      o Requires the following input information:":PRINT
3180 INPUT"      1. What is the pulse duration in seconds";TP
3190 INPUT"      2. What is the pulse repetition rate";F
3200 INPUT"      3. What is the number of harmonics desired";HN
3210 INPUT"      4. Is a coefficient vs. frequency display desired Y/N";A$
3220 IF A$="Y" OR A$="y" THEN CF=1
3230 IF CF=0 THEN PRINT"           --In lieu of displays, tabular listings will b
e given--"
3240 INPUT"      5. Is a coefficient vs. time display desired Y/N";A$
3250 IF A$="Y" OR A$="y" THEN CT=1
3260 IF CT=1 THEN GOTO 3280
3270 IF CT=0 THEN PRINT"           --In lieu of displays, tabular listings will b
e given--"
3280 INPUT"      6. Is your printer turned on Y/N";A$
3290 IF A$="Y" OR A$="y" THEN LPRINT CHR$(27);"@";CHR$(15);CHR$(27);"A";CHR$(8):
RETURN
3300 PRINT              "           --Printer must be turned on to continue--":GOT
O 3280
3310 REM --- Tabular Listing Subroutine ---
3320 LPRINT:LPRINT" ------------ List Of Fourier Coefficients And Their Frequenc
ies For: ---------":LPRINT
3330 LPRINT TAB(5);"TP=";TP;" Sec.";TAB(25);"TR=";INT(TR*100000!+.5)/100000!;" S
ec.";TAB(45);"PRR=";F;" HZ.";TAB(65);"DU=";INT(DU*100000!+.5)/100000!:LPRINT
3340 LPRINT TAB(30)"Frequency, Hz.";TAB(50)"Amplitude"
3350 LPRINT TAB(30)"---------    ---";TAB(50)"---------"
3360 IF N = 0 THEN LPRINT TAB(30);0;TAB(50);INT(1000000!*DU+.5)/1000000!:GOTO 33
80
3370 LPRINT TAB(30);INT (N*F+.5);TAB(50);INT(100000!*C(N)+.5)/100000!
3380 RETURN
```

```
3390 LPRINT:LPRINT"---------- Sum Of";HN;"Fourier Coefficients Versus Time For -
---------"
3400 LPRINT:LPRINT TAB(5);"TP=";TP;" Sec.";TAB(25);"TR=";INT(TR#100000!+.5)/1000
00!;" Sec.";TAB(45);"PRR=";F;" Hz.";TAB(65);"DU=";INT(DU#100000!+.5)/100000!:LPR
INT
3410 LPRINT TAB(30);"Time, Sec.";TAB(50);"Summation"
3420 LPRINT TAB(30);"----- ----";TAB(50);"---------"
3430 LPRINT"                                                  ";
3432 LPRINT USING".#######";T;
3434 LPRINT"              ";
3436 LPRINT USING"#.######";SU
3440 RETURN
3500 REM Printer On/Tabular Output/Plotted Output Subroutine
3510 CLS:PRINT"Your printer must be on to run this program.  Is it on (Y or N)?"
3520 A$=INKEY$:IF A$="" THEN 3520
3530 IF A$="n" OR A$="N" THEN PRINT"For operation without a printer go to DOS.
Type INFORMATION.":END
3560 CLS:PRINT"                        PROGRAM IS IN PROGRESS
                                        ------- -- -- --------"
3570 RETURN


10 REM APPENDIX.7 -- Graphical Fourier Analysis (Manual Or Data List Input) --
12 CLS
20 DIM F(500),C(500),D(500),E(500),A2(500),V(500),AV(500):PI=3.141592654#
30 REM -- Data Input --
40 GOSUB 100
50 REM -- Coefficient Evaluation --
60 GOSUB 1000
70 REM -- Printout Of Results --
80 GOSUB 300
90 END
100 REM -- This Is The Data Input Subroutine --
110 PRINT"---Fourier Series Coefficient Evaluation By Means Of Numerical Integra
tion---"
120 PRINT:PRINT"This program offers two approaches.  The choices and input data
required are:":PRINT
130 PRINT"  1. An approach that requires descrete pulse amplitude points as inpu
t data."
140 PRINT"  2. An approach using a mathematical description of the pulse as inpu
t data."
150 INPUT"     Which approach, 1 or 2, do you want to use";Z9
160 IF Z9=1 THEN GOTO 180
170 GOSUB 400:GOTO 210
180 INPUT"    Number of data points must be odd.  The number is";N
190 IF (N/2)<>INT((N/2)+.5)THEN GOTO 210
200 GOTO 180
210 INPUT"     The lowest harmonic desired is";N7
215 INPUT"     The highest harmonic desired is";N8
220 RETURN
300 REM -- This Is The Data Printout Subroutine --
310 INPUT"     Do you want a monitor or a printer copy M / P";A$
312 IF A$="P" OR A$="p" THEN GOSUB 3500:LPRINT CHR$(27);"@";CHR$(15);CHR$(27);"A
";CHR$(9);CHR$(27);"G";CHR$(27);"M";CHR$(10);CHR$(27);"N";CHR$(11);:WIDTH "LPT1:
",140
316 IF A$="P" OR A$="p" THEN GOSUB 1520 ELSE GOSUB 1920
322 INPUT"     Do you want to continue the analysis (Y/N)";B$
```

```
324 IF B$="y" OR B$="Y" THEN GOSUB 1000:GOSUB 300:GOTO 322
325 INPUT"     Do you want a volt-ampere / power analysis (Y/N)";F$
326 INPUT"     Do you want a monitor or a printer copy M / P";A$
327 IF F$="Y" OR F$="y" THEN GOSUB 4000
328 INPUT"     Do you want to continue the analysis (Y/N)";B$
329 IF B$="y" OR B$="Y" THEN GOSUB 1000:GOSUB 300:GOTO 325
330 END
400 REM --This Is The Mathematical Waveform Subrouine--
410 PRINT"    The waveforms as f(t) must be entered in lines 500-600."
500 Y=1
502 N=321
504 FOR TS=0 TO N-1
506 IF TS<10 THEN F(Y)=100:GOTO 560
508 IF TS>21 THEN GOTO 512
510 F(Y)=50*(1+COS(2*PI*(TS-10)/24)):GOTO 560
512 IF TS<298 THEN F(Y)=0:GOTO 560
514 IF TS>309 THEN F(Y)=100:GOTO 560
516 F(Y)=50*(1+COS((2*PI*(TS-298)/24)-PI))
560 PRINT"    Ordinate ";TS;" = ";F(Y)
570 Y=Y+1
580 NEXT TS
600 RETURN
1000 REM --Coefficient Analysis Subroutine--
1100 PASS=PASS+1
1101 IF Z9=1 THEN GOTO 1103
1102 GOTO 1150
1103 INPUT"     To insert data points manually, enter M.
                To read programmed data points, enter R.
                Which approach, M or R ";D$
1104 IF D$="R" OR D$="r" THEN GOTO 1110
1105 PRINT"    Data Points Are As Follows:"
1106 FOR I=1 TO N:PRINT"    Point "I-1;:INPUT" ---> ";F(I):NEXT I:GOTO 1150
1108 PRINT"    Data Points Are As Follows:"
1110 FOR I=1 TO N
1120 PRINT"    Point ";I-1;
1130 READ DP:F(I)=DP:PRINT F(I)
1140 NEXT I
1150 INPUT"    Signal Period ----->  ";T
1160 N3=N-1
1170 A=2*PI/T
1180 T3=T/N3
1190 A0=0
1200 FOR I=1 TO N3 STEP 2
1210 A0=A0+F(I)+4*F(I+1)+F(I+2)
1220 NEXT I
1230 A0=A0*T3/(3*T)
1240 IF ABS(A0)<.001 THEN A0=0
1245 PRINT"                    --- Calculations In Progress ---"
1250 FOR X=N7 TO N8
1280 C=0:D=0:E=0:A2=0
1290 T1=-T3
1300 FOR J=1 TO N3 STEP 2
1310 T1=T1+2*T3
1320 C=C+(F(J)*COS(X*A*(T1-T3)))+(4*F(J+1)*COS(X*A*T1))+(F(J+2)*COS(X*A*(T1+T3))
)
1330 D=D+(F(J)*SIN(X*A*(T1-T3)))+(4*F(J+1)*SIN(X*A*T1))+(F(J+2)*SIN(X*A*(T1+T3))
)
1340 NEXT J
```

```
1350 C=2*C*T3/(3*T):D=2*D*T3/(3*T)
1360 IF ABS(C)<.0001 THEN C=0
1370 IF ABS(D)<.0001 THEN D=0
1380 E=SQR(C*C+D*D)
1390 IF C>0 THEN GOTO 1470
1400 IF C<0 THEN GOTO 1402
1401 GOTO 1410
1402 IF D<0 THEN GOTO 1480
1410 IF C=0 THEN GOTO 1412
1411 GOTO 1420
1412 IF D>0 THEN GOTO 1460
1420 IF C=0 THEN GOTO 1422
1421 GOTO 1440
1422 IF D=0 THEN GOTO 1430
1423 GOTO 1440
1430 A2=0:GOTO 1500
1440 IF C=0 AND D<0 THEN GOTO 1450
1441 GOTO 1490
1450 A2=-90:GOTO 1500
1460 A2=90:GOTO 1500
1470 A2=(ATN(D/C))*180/PI:GOTO 1500
1480 A2=-180+(ATN(D/C))*180/PI:GOTO 1500
1490 A2=180+(ATN(D/C))*180/PI
1500 PRINT"                      Harmonic";X;" = ";E;" @ ";A2;" Deg."
1502 C(X)=C:D(X)=D:E(X)=E:A2(X)=A2
1505 IF PASS=1 THEN V(X)=E(X):AV(X)=A2(X):V0=A0
1510 NEXT X
1515 RETURN
1520 REM   --Numerical Analysis Result Printout--
1522 LPRINT"                         ---------------------------------------------
------------------------------"
1524 LPRINT"                      ---Numerical Integration Evaluation Of
Fourier Coefficients---"
1526 LPRINT"                           Number Of Ordinates = ";N
1528 LPRINT"                           D-C Term, A(0) = ";A0
1530 INPUT "     Is a data point / ordinate printout desired Y/ N =";A$
1532 IF A$="y" OR A$="Y" THEN GOSUB 1800
1534 LPRINT:LPRINT TAB(20);"HARMONIC";TAB(38)"COS TERM";TAB(55)"SIN TERM";TAB(73
);"MAGNITUDE";TAB(89)"PHASE";TAB(100)"PERCENT"
1536 LPRINT TAB(20);"--------";TAB(38)"--- ----";TAB(55)"--- ----";TAB(73);"----
-----";TAB(89);"-----";TAB(100);"--------"
1540 FOR I=N7 TO N8
1550 LPRINT TAB(22);:LPRINT USING"###";I;:LPRINT TAB(37);:LPRINT USING"######.##
";C(I);:LPRINT TAB(54);:LPRINT USING"######.##";D(I);:LPRINT TAB(73);:LPRINT USI
NG"######.##";E(I);:LPRINT TAB(87);:LPRINT USING"####.##";A2(I);
1551 LPRINT TAB(101);:LPRINT USING"###.##";E(I)*100/E(1):NEXT I
1560 F=1/T
1570 LPRINT:LPRINT TAB(20);"Fundamental Frequency = ";F;" Hertz":LPRINT
1580 RETURN
1800 REM --Data Point Printout Subroutine--
1820 LPRINT"                                              -- Data Ordinates -
-"
1822 LPRINT TAB(10);" Point #";TAB(20);" Ordinate";TAB(30);" Point #";TAB(40);
" Ordinate";TAB(50);" Point #";TAB(60);" Ordinate";TAB(70);" Point #";TAB(80);
" Ordinate";TAB(90);" Point #";TAB(100);" Ordinate"
1824 LPRINT TAB(10);" --------";TAB(20);" --------";TAB(30);" --------";TAB(40);
" --------";TAB(50);" --------";TAB(60);" --------";TAB(70);" --------";TAB(80);
```

Appendix B

```
" --------";TAB(90);"  -------";TAB(100);"  --------"
1830 FOR CO=1 TO N
1840 IF F(CO)=0 THEN GOTO 1860
1842 COUNT=COUNT+1
1850 IF COUNT=1 THEN LPRINT TAB(10);:LPRINT USING"     ###";CO-1;:LPRINT TAB(20);
:LPRINT USING"#####.###";F(CO);:GOTO 1860
1852 IF COUNT=2 THEN LPRINT TAB(30);:LPRINT USING"     ###";CO-1;:LPRINT TAB(40);
:LPRINT USING"#####.###";F(CO);:GOTO 1860
1853 IF COUNT=3 THEN LPRINT TAB(50);:LPRINT USING"     ###";CO-1;:LPRINT TAB(60);
:LPRINT USING"#####.###";F(CO);:GOTO 1860
1854 IF COUNT=4 THEN LPRINT TAB(70);:LPRINT USING"     ###";CO-1;:LPRINT TAB(80);
:LPRINT USING"#####.###";F(CO);:GOTO 1860
1855 LPRINT TAB(90);:LPRINT USING"     ###";CO-1;:LPRINT TAB(100);:LPRINT USING"#
####.###";F(CO)
1856 COUNT=0
1860 NEXT CO:COUNT=0
1870 LPRINT:RETURN
1920 REM  --Numerical Analysis Result Screen Display--
1922 PRINT"-----------------------------------------------------------------------
----------"
1924 PRINT"          ---Numerical Integration Evaluation Of Fourier Coefficients
---"
1926 PRINT"     Number Of Ordinates = ";N
1928 PRINT"          D-C Term, A(0) = ";A0
1930 INPUT "    Is a data point / ordinate printout desired Y/ N =";A$
1932 IF A$="y" OR A$="Y" THEN GOSUB 2000
1934 PRINT:PRINT TAB(10);"HARMONIC";TAB(20)"COS TERM";TAB(30)"SIN TERM";TAB(40);
"MAGNITUDE";TAB(50);"   PHASE";TAB(60)"PERCENT"
1936 PRINT TAB(10);"--------";TAB(20)"--- ----";TAB(30)"--- ----";TAB(40);"-----
----";TAB(50);"  -----";TAB(60);"-------"
1940 FOR I=N7 TO N8
1950 PRINT TAB(12);:PRINT USING"###";I;:PRINT TAB(20);:PRINT USING"#####.##";C(I
);:PRINT TAB(30);:PRINT USING"#####.##";D(I);:PRINT TAB(41);:PRINT USING"#####.#
#";E(I);:PRINT TAB(51);:PRINT USING"####.##";A2(I);
1951 PRINT TAB(61);:PRINT USING"###.##";E(I)*100/E(1):NEXT I
1960 F=1/T
1970 PRINT:PRINT TAB(20);"Fundamental Frequency = ";F;" Hertz":PRINT
1980 RETURN
2000 REM --Data Point Printout Subroutine--
2020 PRINT"              -- Data Ordinates --":PRINT
2022 PRINT TAB(2);"Pt.";TAB(8);"Ord.";TAB(15);"Pt.";TAB(21);"Ord.";TAB(28);"Pt."
;TAB(34);"Ord.";TAB(41);"Pt.";TAB(47);"Ord.";TAB(54);"Pt.";TAB(60);"Ord."
2024 PRINT TAB(2);"---";TAB(8);"----";TAB(15);"---";TAB(21);"----";TAB(28);"---"
;TAB(34);"----";TAB(41);"---";TAB(47);"----";TAB(54);"---";TAB(60);"----"
2030 FOR CO=1 TO N
2040 IF F(CO)=0 THEN GOTO 2060
2042 COUNT=COUNT+1
2050 IF COUNT=1 THEN PRINT TAB(2);:PRINT USING"###";CO-1;:PRINT TAB(6);:PRINT US
ING"#####.##";F(CO);:GOTO 2060
2052 IF COUNT=2 THEN PRINT TAB(15);:PRINT USING"###";CO-1;:PRINT TAB(19);:PRINT
USING"#####.##";F(CO);:GOTO 2060
2053 IF COUNT=3 THEN PRINT TAB(28);:PRINT USING"###";CO-1;:PRINT TAB(32);:PRINT
USING"#####.##";F(CO);:GOTO 2060
2054 IF COUNT=4 THEN PRINT TAB(41);:PRINT USING"###";CO-1;:PRINT TAB(45);:PRINT
USING"#####.##";F(CO);:GOTO 2060
2055 PRINT TAB(54);:PRINT USING"###";CO-1;:PRINT TAB(58);:PRINT USING"#####.##";
F(CO)
2056 COUNT=0
```

```
2060 NEXT CO:COUNT=0
2070 PRINT:RETURN
2100 DATA -7055,-7058,-7008,-6977,-6943,-6901,-6849,-6787,-6715,-6633,-6541
2110 DATA -6439,-6343,-5791,-5820,-5760,-5651,-5513,-5357,-5186,-5006,-4816
2120 DATA -4618,-4413,-4201,-3983,-3758,-3528,-3320,-3139,-2953,-2764,-2572
2130 DATA -2377,-2178,-1976,-1772,-1564,-1354,-1172,-279,95,423,727,1017,1297
2140 DATA 1571,1840,2106,2367,2625,2878,3127,3370,3609,3748,3910,4065,4216
2150 DATA 4360,4498,4629,4754,4871,4980,5082,5724,5955,6145,6301,6432,6546,6647
2160 DATA 6735,6813,6880,6936,6982,7017,7042,7055,7058,7007,6976,6942,6900,6848
2170 DATA 6786,6714,6632,6540,6438,6341,5793,5821,5759,5650,5511,5355,5184,5003
2180 DATA 4813,4615,4410,4198,3980,3755,3525,3319,3136,2950,2762,2570,2374,2176
2190 DATA 1974,1769,1561,1352,1169,282,-94,-425,-730,-1019,-1300,-1574,-1844
2200 DATA -2109,-2371,-2628,-2881,-3130,-3374,-3612,-3750,-3912,-4067,-4218
2210 DATA -4362,-4500,-4631,-4755,-4872,-4981,-5083,-5723,-5955,-6145,-6301
2220 DATA -6433,-6547,-6648,-6736,-6814,-6881,-6937,-6982,-7017,-7042,-7055
2300 DATA -36.1,-36.2,-36.2,-36.4,-36.6,-36.7,-36.8,-36.8,-36.9,-36.8,-36.8
2310 DATA -36.7,-36.5,-37.2,-38.0,-38.5,-38.7,-38.7,-38.7,-38.6,-38.5,-38.4
2320 DATA -38.2,-37.9,-37.6,-37.3,-36.9,-36.5,-36.0,-35.3,-34.1,-32.6,-30.7
2340 DATA -28.4,-25.8,-22.8,-19.4,-15.6,-11.5,-8.8,-7.4,-6.8,-6.2,-5.7,-5.2
2350 DATA -4.7,-4.2,-3.7,-3.2,-2.7,-2.1,-1.6,-1.1,-.6,0,.6,1.6,3.0,4.8,6.9,9.4
2360 DATA 12.2,15.4,18.8,22.6,26.7,30.1,31.3,32.2,32.9,33.4,33.9,34.3,34.7,35.1
2370 DATA 35.4,35.7,35.9,36.1,36.3,36.4,36.5,36.6,36.7,36.9,37.0,37.1,37.2,37.2
2380 DATA 37.2,37.1,37.0,36.8,37.6,38.4,38.8,39.0,39.0,39.0,39.0,38.8,38.7,38.5
2390 DATA 38.2,37.9,37.6,37.2,36.8,36.3,35.6,34.4,32.9,31.0,28.7,26.1,23.0,19.6
2400 DATA 15.9,11.7,9.1,7.7,7.1,6.5,6.0,5.5,5.0,4.5,4.0,3.5,2.9,2.4,1.9,1.4,.8
2410 DATA .3,-.3,-1.3,-2.7,-4.5,-6.7,-9.1,-12.0,-15.1,-18.6,-22.4,-26.5,-29.9
2420 DATA -31.0,-31.9,-32.6,-33.2,-33.6,-34.1,-34.5,-34.8,-35.1,-35.4,-35.7
2430 DATA -35.9,-36.0,-36.1
2500 DATA -25.0,-24.8,-24.7,-24.6,-24.5,-24.4,-24.4,-24.2,-24.1,-24.0,-23.8
2510 DATA -23.6,-23.5,-24.6,-25.2,-25.5,-25.7,-25.7,-25.8,-25.8,-25.7,-25.7
2520 DATA -25.6,-25.5,-25.4,-25.3,-25.1,-24.9,-24.6,-24.0,-23.1,-21.8,-20.1
2530 DATA -18.0,-15.6,-12.8,-9.7,-6.2,-2.3,0,0,0,0,0,0,0,0,0,0,0,0,0,0,0,.2
2540 DATA .9,1.9,3.2,5.0,7.0,9.5,12.2,15.3,18.7,22.5,24.4,25.1,25.4,25.6,25.7
2550 DATA 25.8,25.8,25.8,25.7,25.7,25.6,25.5,25.3,25.2,25.0,24.8,24.7,24.6,24.6
2560 DATA 24.5,24.4,24.3,24.1,24.0,23.8,23.6,23.6,24.7,25.2,25.5,25.7,25.8,25.8
2570 DATA 25.8,25.8,25.7,25.6,25.6,25.4,25.3,25.1,24.9,24.7,24.1,23.1,21.8,20.1
2580 DATA 18.0,15.6,12.8,9.7,6.2,2.3,0,0,0,0,0,0,0,0,0,0,0,0,0,0,-.2,-.9
2590 DATA -1.9,-3.3,-5.0,-7.1,-9.5,-12.3,-15.4,-18.8,-22.5,-24.4,-25.1,-25.5
2600 DATA -25.7,-25.8,-25.8,-25.8,-25.8,-25.8,-25.7,-25.6,-25.5,-25.4,-25.2
2610 DATA -25.0
2700 DATA -35.9,-36.5,-37.4,-38.8,-40.4,-42.5,-44.8,-47.4,-50.3,-53.6,-57.1
2710 DATA -60.9,-64.9,-68.2,-70.0,-71.2,-71.9,-72.5,-73.0,-73.3,-73.5,-73.7
2720 DATA -73.8,-73.8,-73.7,-73.6,-73.3,-73.0,-72.6,-71.9,-71.0,-69.6,-67.8
2730 DATA -65.6,-63.0,-60.0,-56.5,-52.7,-48.4,-43.9,-44.9,-45.1,-45.1,-44.8
2740 DATA -44.4,-43.9,-43.3,-42.7,-42.0,-41.2,-40.4,-39.6,-38.7,-37.8,-36.8
2750 DATA -35.6,-33.6,-30.9,-27.5,-23.4,-18.5,-12.9,-6.6,.4,8.1,16.5,22.1,23.9
2760 DATA 25.2,26.4,27.4,28.4,29.3,30.2,31.0,31.8,32.6,33.4,34.1,34.8,35.4
2770 DATA 36.0,37.0,38.3,40.0,42.0,44.4,47.0,49.9,53.2,56.7,60.5,64.5,67.8,69.6
2780 DATA 70.7,71.5,72.1,72.5,72.9,73.1,73.3,73.4,73.4,73.3,73.1,72.9,72.6,72.2
2790 DATA 71.5,70.5,69.2,67.4,65.2,62.5,59.5,56.1,52.2,47.9,43.4,44.4,44.7,44.6
2800 DATA 44.3,43.9,43.4,42.8,42.2,41.5,40.8,40.0,39.1,38.3,37.3,36.4,35.1,33.1
2810 DATA 30.4,27.0,22.9,18.0,12.4,6.1,-1.0,-8.6,-17.0,-22.6,-24.4,-25.8,-26.9
2820 DATA -27.9,-28.9,-29.8,-30.7,-31.5,-32.3,-33.1,-33.9,-34.6,-35.3,-35.9
3500 REM Printer On/Tabular Output/Plotted Output Subroutine
3510 CLS:INPUT"          Your printer must be on to run this program.  Is it on (
Y or N)";C$
3530 IF C$="n" OR C$="N" THEN PRINT"For operation without a printer go to DOS. T
ype INFORMATION": END
```

Appendix B

```
3560 CLS:PRINT"                    PROGRAM IS IN PROGRESS
                                   ------- -- -- --------"
3570 RETURN
4000 REM Volt-Ampere / Power Subroutine
4010 IF A$="P" OR A$="p" THEN GOTO 4050
4012 PRINT"                   Volt-Ampere & Power Analysis"
4014 PRINT"                   ----------------------------":PRINT
4020 PRINT TAB(10);" Harmonic";TAB(20)"  Voltage";TAB(30)" V. Angle"TAB(40);"
    Amperes";TAB(50);" A. Angle";TAB(60)"   Power"
4025 PRINT TAB(10);" --------";TAB(20)"  --------";TAB(30)" --------"TAB(40);"
    --------";TAB(50);" --------";TAB(60)"   -----"
4030 FOR I=N7 TO N8
4035 PRINT TAB(14);:PRINT USING"###";I;
4036 PRINT TAB(22);:PRINT USING"#####.##";V(I);
4037 PRINT TAB(32);:PRINT USING"#####.##";AV(I);
4038 PRINT TAB(42);:PRINT USING"#####.##";E(I);
4039 PRINT TAB(52);:PRINT USING"#####.##";A2(I);
4040 P=V(I)*E(I)*COS((AV(I)-A2(I))*PI/180)*.5
4041 PRINT TAB(61);:PRINT USING"######.##";P
4042 VR=VR+V(I)^2/2:IR=IR+E(I)^2/2:PS=PS+P
4043 NEXT I:VR=SQR(V0^2+VR):IR=SQR(A0^2+IR):VAS=VR*IR:PS=V0*A0+PS
4044 PRINT:PRINT:PRINT"           RMS Voltage =";VR
4045 PRINT"           RMS Current =";IR
4046 PRINT"           Volt Amperes =";VAS
4047 PRINT"              Power =";PS
4048 PRINT"           Power Factor =";PS/VAS:VR=0:IR=0:PS=0:VAS=0
4049 RETURN
4050 LPRINT"                   Volt-Ampere & Power Analysis"
4052 LPRINT"                   ----------------------------":LPRINT
4054 LPRINT TAB(10);" Harmonic";TAB(20)"  Voltage";TAB(30)" V. Angle"TAB(40);
    " Amperes";TAB(50);" A. Angle";TAB(60)"   Power"
4055 LPRINT TAB(10);" --------";TAB(20)"  --------";TAB(30)" --------"TAB(40);
    " --------";TAB(50);" --------";TAB(60)"   -----"
4060 FOR I=N7 TO N8
4065 LPRINT TAB(14);:LPRINT USING"###";I;
4066 LPRINT TAB(22);:LPRINT USING"#####.##";V(I);
4067 LPRINT TAB(32);:LPRINT USING"#####.##";AV(I);
4068 LPRINT TAB(42);:LPRINT USING"#####.##";E(I);
4069 LPRINT TAB(52);:LPRINT USING"#####.##";A2(I);
4070 P=V(I)*E(I)*COS((AV(I)-A2(I))*PI/180)*.5
4071 LPRINT TAB(61);:LPRINT USING"######.##";P
4072 VR=VR+V(I)^2/2:IR=IR+E(I)^2/2:PS=PS+P
4073 NEXT I:VR=SQR(V0^2+VR):IR=SQR(A0^2+IR):VAS=VR*IR:PS=V0*A0+PS
4074 LPRINT:LPRINT:LPRINT"           RMS Voltage =";VR
4075 LPRINT"           RMS Current =";IR
4076 LPRINT"           Volt Amperes =";VAS
4077 LPRINT"              Power =";PS
4078 LPRINT"           Power Factor =";PS/VAS:VR=0:IR=0:PS=0:VAS=0
4079 RETURN



10 REM APPENDIX.9 -- Bessel Function Of The First Kind --
20 CLS:PRINT"                Bessel Functions Of The First Kind
                            ------ --------- -- --- ----- ----":PRINT
30 PRINT" 1. This program calculates and prints out the values of Bessel functio
ns of      the first kind if your computer is an IBM compatible AT or XT syste
m with     a version of BASIC 2.0 or higher.":PRINT
```

```
40 INPUT" 2. What is the maximum order of the Bessel function (up to 10)";NMAX:P
RINT
50 INPUT" 3. What is the argument (mp or mf) of the function (up to 12)";X:PRINT
60 INPUT" 4. Do you want a printer copy or a screen display (P or S)";A$:IF A$="
P" OR A$="p" THEN GOTO 80
70 IF A$="S" OR A$="s" THEN GOTO 100 ELSE PRINT"      --- Your symbol is in error.
   Please repeat. ---":GOTO 60
80 PRINT"     Make certain that your printer is turned on.  Then press F5 key.":E
ND
90 LPRINT CHR$(27);"@";CHR$(15);CHR$(27);"G";CHR$(27);"M";CHR$(10);:WIDTH "LPT1:
",135
100 DEFDBL D,S,J:PRINT
110 DIM D(50):DIM N(50):DIM X(50)
120 IF A$="S" OR A$="s"THEN PRINT"             Bessel Functions Of First
 Kind":PRINT:GOTO 150
130 LPRINT"                                    Bessel Functions Of Fir
st Kind"
140 LPRINT"                                    ------ --------- -- ---
-- ----":LPRINT
150 REM -- N = Order of Bessel function --
160 FOR N=0 TO NMAX
170 NF=N*NF:IF N<=1 THEN NF=1
180 REM -- T = Number of terms in evaluation series --
190 FOR T=1 TO 14
200 IF T>1 THEN GOTO 220
210 D(T)=-2*(2*N+2):GOTO 230
220 D(T)=D(T-1)*2*T*(2*N+2*T)*(-1)
230 NEXT T
240 REM -- X = Argument of the Bessel function --
250 FOR T=1 TO 14
260 SUM=SUM+X^(2*T)/D(T)
270 NEXT T
280 J=(X^N/((2^N)*NF))*(1+SUM)
290 SUM=0
300 IF A$="P" OR A$="p" THEN GOTO 330
310 PRINT"J";N;"(";:PRINT USING"##.##";X;:PRINT") =";:PRINT USING"#.####";J;:PRI
NT":    ";:COUNT=COUNT+1:IF COUNT=3 THEN PRINT"":COUNT=0:PRINT
320 GOTO 350
330 LPRINT"J";CHR$(27);"S";CHR$(1);N;CHR$(27);"T";"(";:LPRINT USING"##.##";X;:LP
RINT") =";:LPRINT " ";:LPRINT USING"#.####";J;:LPRINT":    ";:COUNT=COUNT+1:IF CO
UNT=5 THEN LPRINT"":COUNT=0
340 GOTO 350
350 NEXT N
360 PRINT"---------------------------------------------------------------------":
IF A$="S" OR A$="s" THEN GOTO 380
370 LPRINT"---------------------------------------------------------------------
--------------------------------------------------------":LPRINT:LPRINT
380 COUNT=0:END



10 REM APPENDIX.11 -- DC Motor Acceleration Analysis Program --
20 REM To run program with C1 and R6, change line 80 to N=4 and delete line 550
30 CLS
40 DIM X(49)
50 DIM D(49)
60 DIM E(49)
70 DIM F(49)
```

Appendix B

```
80 N=3
90 T1=.005
100 T2=.05
110 T3=2
120 T=0
130 P=0
140 R1=25
150 R2=50
160 R3=90
170 R4=10
180 R5=.05
190 R6=10
200 L=50
210 C=4
220 C1=.03
230 K6=90
240 V=250
250 PRINT"   Integration Step =";T1;"Seconds"
260 PRINT
270 GOSUB 510
280 GOSUB 580
290 REM INTEGRATION ROUTINE
300 GOSUB 510
310 FOR A = 1 TO N
320 E(A)=X(A)
330 F(A)=D(A)
340 X(A)=X(A)+T1*D(A)
350 NEXT A
360 T=T+T1
370 GOSUB 510
380 FOR A = 1 TO N
390 X(A)=E(A)+T1/2*(F(A)+D(A))
400 NEXT A
410 P=P+T1
420 IF P+T1/5>T2 THEN GOTO 440
430 GOTO 300
440 P=0
450 REM PRINTOUT ROUTINE
460 GOSUB 610
470 IF T+T1/5>T3 THEN GOTO 490
480 GOTO 300
490 PRINT"---------------------------------"
500 END
510 REM EQUATION SUBROUTINE
520 X(2)=(V-R1*X(1)+R3*C1*D(4))/(R1+R3+R4)
530 D(1)=(V-(R1+R2)*X(1)-R1*X(2))/L
540 D(3)=(K6*X(1)-X(3))/(R5*C)
550 RETURN
560 D(4)=(R3*X(2)-X(4))/((R3+R6)*C1)
570 RETURN
580 REM PRINT SUBROUTINE
590 PRINT"Transient Numerical Integration Program":PRINT
600 PRINT TAB(11);"Seconds";TAB(22);"Current":PRINT
610 PRINT TAB(11);INT(100*T+.5)/100; TAB(22);INT(1000*C*D(3)+.5)/1000
620 RETURN
```

```
10 REM APPENDIX.12 -- Transient Numerical Integration Program --
11 CLS
12 DIM X(49)
14 DIM D(49)
16 DIM E(49)
18 DIM F(49)
20 N=1
32 T1=.1
34 T2=.1
36 T3=3
38 T=0
40 P=0
50 R=1
60 L=1
70 V=1
80 PRINT"   Integration Step =";T1;"Seconds"
82 PRINT
86 GOSUB 1000
90 GOSUB 2000
100 REM INTEGRATION ROUTINE
110 GOSUB 1000
120 FOR A = 1 TO N
130 E(A)=X(A)
140 F(A)=D(A)
150 X(A)=X(A)+T1*D(A)
160 NEXT A
165 T=T+T1
170 GOSUB 1000
180 FOR A = 1 TO N
190 X(A)=E(A)+T1/2*(F(A)+D(A))
200 NEXT A
220 P=P+T1
230 IF P+T1/5>T2 THEN GOTO 250
240 GOTO 110
250 P=0
260 REM PRINTOUT ROUTINE
270 GOSUB 2010
280 IF T+T1/5>T3 THEN GOTO 295
290 GOTO 110
295 PRINT"--------------------------------"
300 END
1000 REM EQUATION SUBROUTINE
1010 D(1)=(V-X(1)*R)/L
1020 RETURN
2000 REM PRINT SUBROUTINE
2002 PRINT"Transient Numerical Integration Program":PRINT
2005 PRINT TAB(11);"Seconds";TAB(22);"Current":PRINT
2010 PRINT TAB(11);INT(10*T+.5)/10; TAB(22);INT(1000*X(1)+.5)/1000
2020 RETURN


10 REM APPENDIX.13 -- Simplified Power Supply Transient Analysis --
11 GOSUB 3500
13 DIM X(49)
14 DIM D(49)
16 DIM E(49)
18 DIM F(49)
```

Appendix B

```
22 LPRINT CHR$(27);"@";CHR$(15);CHR$(27);"A";CHR$(8);CHR$(27);"Q";CHR$(120);CHR$
(27);"6";CHR$(27);"1";CHR$(10);
30 N=2
32 T1=.001
34 T2=.005
36 T3=.35
38 T=0
40 P=0
50 X(1)=60
60 X(2)=10000
70 VI=10060
72 L=2
74 C=.00027
76 R=1
78 R1=166.7
80 PRINT"      INTEGRATION STEP = ";T1;" SECONDS"
81 PRINT
86 GOSUB 1000
90 GOSUB 2000
100 REM INTEGRATION ROUTINE
110 GOSUB 1000
112 IF T>.02 THEN R1=1000000!
114 IF T>.075 THEN R1=166.7
120 FOR A = 1 TO N
130 E(A)=X(A)
140 F(A)=D(A)
150 X(A)=X(A)+T1*D(A)
152 IF X(1)<0 THEN X(1)=0
160 NEXT A
165 T=T+T1
170 GOSUB 1000
180 FOR A = 1 TO N
190 X(A)=E(A)+T1/2*(F(A)+D(A))
192 IF X(1)<0 THEN X(1)=0
200 NEXT A
220 P=P+T1
230 IF P+T1/5>T2 THEN GOTO 250
240 GOTO 110
250 P=0
260 REM PRINTOUT ROUTINE
270 GOSUB 2010
280 IF T+T1/5>T3 THEN GOTO 292
290 GOTO 110
291 PRINT
292 IF A$="P" OR A$="p" THEN LPRINT"!....+....!....+....!....+....!....+....!...
.+....!....+....!....+....!....+....!":END
294 LPRINT"--------------------------------"
300 END
1000 REM EQUATION SUBROUTINE
1005 D(1)=-(R/L)*X(1)-(1/L)*X(2)+VI/L
1010 D(2)=(1/C)*X(1)-(1/(R1*C))*X(2)
1020 RETURN
2000 REM PRINT SUBROUTINE
2002 IF A$="P" OR A$="p" THEN LPRINT CHR$(27);"6";CHR$(27);"A";CHR$(8);:WIDTH "L
PT1:",(120):GOTO 2012 THE RESPONSE THEN ENTER P PRINT P
2003 LPRINT"--DC Power Supply Response--":LPRINT
2004 LPRINT TAB(1);"MILLISEC.";TAB(12);"AMPERES";TAB(23);"VOLTS"
2005 LPRINT TAB(1);"    T   ";TAB(12);" X(1)" ;TAB(23);" X(2)":LPRINT
```

```
2010 IF A$="P" OR A$="p" THEN GOTO 2012
2011 GOTO 2015
2012 IF T=0 THEN GOSUB 3000
2014 GOSUB 3040:GOTO 2030
2015 LPRINT TAB(4);INT((1000#T)+.5);:LPRINT TAB(12);:LPRINT USING "###.#";X(1);:
LPRINT TAB(23);:LPRINT USING "#####";X(2)
2030 RETURN
3000 REM Plotting Subroutine
3005 LPRINT"-- Integration Step =";T1;"Seconds --":LPRINT
3010 LPRINT"                         Transient Response Of D-C Power Supply"
3015 LPRINT"                         --------- -------- -- --- ----- ------":LPRIN
T
3020 LPRINT"0         2000      4000      6000      8000      10000     12000
14000     16000"
3030 LPRINT"!....+....!....+....!....+....!....+....!....+....!....+....!....+..
..!....+....!   Volts"
3032 SC=-2
3040 SC=SC+1:IF SC=-1 THEN RETURN
3060 PP=1+X(2)/200
3100 IF SC=0 OR SC=5 THEN LPRINT TAB(1);"--T=";:LPRINT USING ".###";T;:SC=0:GOTO
 3210
3205 LPRINT TAB(1);"-";
3210 LPRINT TAB(PP);"+";TAB(85);"";:LPRINT USING"#####";X(2);:RETURN
3500 REM Printer On/Tabular Output/Plotted Output Subroutine
3510 CLS:PRINT"Your printer must be on to run this program.  Is it on (Y or N)?"
3520 A$=INKEY$:IF A$="" THEN 3520
3530 IF A$="n" OR A$="N" THEN PRINT"For operation without a printer go to DOS.
Type INFORMATION.":END
3540 PRINT"For a plotted response enter P.  For a tabular response enter any oth
er letter."
3550 A$=INKEY$:IF A$="" THEN 3550
3560 CLS:PRINT"                        PROGRAM IS IN PROGRESS
                                       ------- -- -- --------"
3570 RETURN


10 REM APPENDIX.14 -- Operational Amplifier Transient Analysis --
11 GOSUB 3500
14 DIM X(49)
16 DIM D(49)
18 DIM E(49)
20 DIM F(49)
22 LPRINT CHR$(27);"@";CHR$(15);CHR$(27);"A";CHR$(8);CHR$(27);"Q";CHR$(92);CHR$(
27);"6";CHR$(27);"1";CHR$(10);
30 N=4
32 T1=2.5E-10
34 T2=2.5E-08
36 T3=.000001
38 T=0
40 P=0
50 R1=665000!
52 R2=5000000!
54 R3=150
56 C1=1.6E-08
58 C2=1.6E-14
60 C4=1E-10
62 L3=.001
```

```
64 GM=467
66 VI=1
80 LPRINT"     INTEGRATION STEP = ";T1;" SECONDS"
81 LPRINT
86 GOSUB 1000
90 GOSUB 2000
100 REM INTEGRATION ROUTINE
110 GOSUB 1000
120 FOR A = 1 TO N
130 E(A)=X(A)
140 F(A)=D(A)
150 X(A)=X(A)+T1*D(A)
160 NEXT A
165 T=T+T1
170 GOSUB 1000
180 FOR A = 1 TO N
190 X(A)=E(A)+T1/2*(F(A)+D(A))
200 NEXT A
220 P=P+T1
230 IF P+T1/5>T2 THEN GOTO 250
240 GOTO 110
250 P=0
260 REM PRINTOUT ROUTINE
270 GOSUB 2010
280 IF T+T1/5>T3 THEN GOTO 292
290 GOTO 110
291 LPRINT
292 IF A$="P" OR A$="p" THEN GOSUB 3030:GOTO 296
294 LPRINT"--------------------------------"
296 LPRINT
300 END
1000 REM EQUATION SUBROUTINE
1005 D(1)=((VI-X(1)-X(4))/R1-X(3))/C1
1010 D(2)=X(3)/C2
1015 D(3)=(X(1)-X(2)-R2*X(3))/L3
1020 D(4)=(GM*(L3*D(3)+X(2))+X(3)+C1*D(1)-X(4)/R3)/C4
1025 RETURN
2000 REM PRINT SUBROUTINE
2001 LPRINT"Operational Amplifier Transient Response":LPRINT
2002 IF A$="P" OR A$="p" THEN GOTO 2012
2004 LPRINT TAB(1);"Nanosec. ";TAB(12);"Volts Out":PRINT
2010 IF A$="P" OR A$="p" THEN GOTO 2012
2011 GOTO 2015
2012 IF T=0 THEN GOSUB 3000:GOSUB 3100:GOTO 2030
2014 GOSUB 3040:GOTO 2030
2015 LPRINT TAB(1);INT((1E+09*T)+.5);TAB(12);INT(1000*X(4)+.5)/1000
2030 RETURN
3000 REM Plotting Subroutine
3005 WIDTH"LPT1:",132:LPRINT CHR$(15);CHR$(27);"A";CHR$(8);CHR$(27);"Q";CHR$(135
);CHR$(27);"G";
3007 LPRINT"--Integration Time =";T1;"Seconds--":LPRINT
3010 LPRINT"                        Transient Response Of uA741 Operation
al Amplifier"
3012 LPRINT"             --------- -------- -- ----- ---------
-- ---------":LPRINT
3020 LPRINT"0         .1        .2        .3       .4       .5       .6
   .70      .8       .9      1.0      1.1"
3030 LPRINT"!....+....!....+....!....+....!....+....!....+....!....+.."
```

Appendix B

```
                ..!....+....!....+....!....+....!....+....! Volts"
3032 RETURN
3040 SC=SC+1
3044 PP=(1+100*X(4))
3060 IF SC<5 THEN LPRINT TAB(1);"-";:GOTO 3200
3100 LPRINT TAB(1);"--T=";INT(1E+09*T+.5);"nSec.";:SC=0:IF PP<16 THEN GOTO 3220
3200 IF PP<1.51 THEN GOTO 3220
3210 LPRINT TAB(PP);"+";TAB(112);"";:LPRINT USING"##.###";X(4):RETURN
3220 LPRINT TAB(112);"";:LPRINT USING"##.###";X(4):RETURN
3500 REM Printer On/Tabular Output/Plotted Output Subroutine
3510 CLS:PRINT"Your printer must be on to run this program.  Is it on (Y or N)?"
3520 A$=INKEY$:IF A$="" THEN 3520
3530 IF A$="n" OR A$="N" THEN PRINT"For operation without a printer go to DOS.
Type INFORMATION.":END
3540 PRINT"For a plotted response enter P.  For a tabular response enter any oth
er letter."
3550 A$=INKEY$:IF A$="" THEN 3550
3560 CLS:PRINT"                   PROGRAM IS IN PROGRESS
                                   ------- -- -- --------"
3570 RETURN


10 REM APPENDIX.15 -- Band Pass Amplifier Transient Analysis --
11 GOSUB 3500
14 DIM X(49)
16 DIM D(49)
18 DIM E(49)
20 DIM F(49)
22 LPRINT CHR$(27);"@";CHR$(15);CHR$(27);"A";CHR$(8);CHR$(27);"Q";CHR$(92);CHR$(
27);"6";CHR$(27);"1";CHR$(10);
30 N=10
32 T1=1E-08
34 T2=2E-08
36 T3=.000005
38 T=0
40 P=0
50 R1=1600
52 C1=4.3E-10
54 L1=1.05E-06
56 L2=7.56E-07
58 L3=5.94E-06
60 C2=1.14E-10
62 C3=2.39E-09
64 L4=3.26E-07
66 L5=5.64E-06
68 C4=1.3E-10
70 C5=1.317E-09
72 L6=5.940001E-07
74 R2=50
76 W=2*3.141572654#*5000000!
80 LPRINT"    INTEGRATION STEP = ";T1;" SECONDS"
81 LPRINT
86 GOSUB 1000
90 GOSUB 2000
100 REM INTEGRATION ROUTINE
110 GOSUB 1000
120 FOR A = 1 TO N
130 E(A)=X(A)
```

Appendix B

```
140 F(A)=D(A)
150 X(A)=X(A)+T1*D(A)
160 NEXT A
165 T=T+T1
170 GOSUB 1000
180 FOR A = 1 TO N
190 X(A)=E(A)+T1/2*(F(A)+D(A))
200 NEXT A
220 P=P+T1
230 IF P+T1/5>T2 THEN GOTO 250
240 GOTO 110
250 P=0
260 REM PRINTOUT ROUTINE
270 GOSUB 2010
280 IF T+T1/5>T3 THEN GOTO 292
290 GOTO 110
291 LPRINT
292 IF A$="P" OR A$="p" THEN GOSUB 3030:GOTO 296
294 LPRINT"--------------------------------"
296 LPRINT
300 END
1000 REM EQUATION SUBROUTINE
1005 D(1)=(SIN(W*T)-X(1)/R1-X(2))/C1
1010 D(2)=((1+L3/L2)*X(1)-X(4)-X(5))/(L1+L1*L3/L2+L3)
1015 D(3)=((L1+L2)*D(2)-X(1))/L2
1020 D(4)=X(3)/C2
1025 D(5)=(X(3)-X(6)-X(7))/C3
1030 D(6)=X(5)/L4
1035 D(7)=(X(5)-X(8)-X(9))/L5
1040 D(8)=X(7)/C4
1045 D(9)=(X(7)-X(9)/R2-X(10))/C5
1050 D(10)=X(9)/L6
1055 RETURN
2000 REM PRINT SUBROUTINE
2001 LPRINT"Band Pass Amplifier Transient Analysis":LPRINT
2002 IF A$="P" OR A$="p" THEN GOTO 2012
2004 LPRINT TAB(1);"Nanosec. ";TAB(12);"Volts Out":PRINT
2010 IF A$="P" OR A$="p" THEN GOTO 2012
2011 GOTO 2015
2012 IF T=0 THEN GOSUB 3000:GOSUB 3100:GOTO 2030
2014 GOSUB 3040:GOTO 2030
2015 LPRINT TAB(1);INT((1E+09*T)+.5);TAB(12);:LPRINT USING "####.#";X(9)
2030 RETURN
3000 REM Plotting Subroutine
3005 WIDTH"LPT1:",140:LPRINT CHR$(15);CHR$(27);"A";CHR$(8);CHR$(27);"Q";CHR$(135
);CHR$(27);"6";
3007 LPRINT"--Integration Time =";T1;"Seconds--":LPRINT
3010 LPRINT"                                        Transient Response Of Band
pass Amplifier"
3012 LPRINT"                            --------- -------- -- ----
---- ---------":LPRINT
3020 LPRINT"-120      -100      -80       -60       -40       -20        0
 20        40        60        80       100       120"
3030 LPRINT"|....+....|....+....|....+....|....+....|....+....|....+....|....+..
..|....+....|....+....|....+....|....+....|....+....| Volts"
3032 RETURN
3040 SC=SC+1
3044 PP=61+X(9)/2
```

```
3060 IF SC<5 THEN LPRINT TAB(1);"-";:GOTO 3210
3100 LPRINT TAB(1);"--T=";INT(1E+09#T+.5);"nSec.";:SC=0:IF PP<16 THEN GOTO 3220
3210 LPRINT TAB(PP);"+";TAB(120);"";:LPRINT USING"####.#";X(9):RETURN
3220 LPRINT TAB(120);"";:LPRINT USING"####.#";X(9):RETURN
3500 REM Printer On/Tabular Output/Plotted Output Subroutine
3510 CLS:PRINT"Your printer must be on to run this program.  Is it on (Y or N)?"
3520 A$=INKEY$:IF A$="" THEN 3520
3530 IF A$="n" OR A$="N" THEN PRINT"For operation without a printer go to DOS.  Type INFORMATION.":END
3540 PRINT"For a plotted response enter P.  For a tabular response enter any other letter."
3550 A$=INKEY$:IF A$="" THEN 3550
3560 CLS:PRINT"                    PROGRAM IS IN PROGRESS
                                    ------- -- -- --------"
3570 RETURN


10 REM APPENDIX.16   -- DC Power Supply Transient Analysis - 1 Ph. C.T. --
11 GOSUB 3500
13 DIM X(49)
14 DIM D(49)
16 DIM E(49)
18 DIM F(49)
19 PI=3.141592654#:SC=-1
22 LPRINT CHR$(27);"@";CHR$(15);CHR$(27);"A";CHR$(8);CHR$(27);"G";CHR$(27);"M";CHR$(6);:WIDTH "LPT1:",140
30 N=3
32 T1=.000025
34 T2=.002
36 T3=.32
38 T=0
40 P=0
50 X(1)=0
60 X(2)=0
70 X(3)=0
71 W=2*PI#62.5
72 VP=15708
73 LL=.1414
74 L=2
75 C=.00005
76 R=500
77 A=L-LL
78 B=L+LL
80 LPRINT"  -- 1 Phase Center Tap Rectifier Integration Time =";T1;"Seconds --":LPRINT
86 GOSUB 1000
90 GOSUB 2000
100 REM INTEGRATION ROUTINE
110 GOSUB 1000
120 FOR A = 1 TO N
130 E(A)=X(A)
140 F(A)=D(A)
150 X(A)=X(A)+T1#D(A)
152 IF X(1)<0 THEN X(1)=0
154 IF X(2)<0 THEN X(2)=0
160 NEXT A
165 T=T+T1
```

```
170 GOSUB 1000
180 FOR A = 1 TO N
190 X(A)=E(A)+T1/2*(F(A)+D(A))
192 IF X(1)<0 THEN X(1)=0
194 IF X(2)<0 THEN X(2)=0
200 NEXT A
202 CO=CO+1:SUM=SUM+X(3):IF CO=INT(8.000001E-03/T1+.5) THEN AVG = SUM/INT(8.0000
01E-03/T1+.5):CO=0:SUM=0
220 P=P+T1
230 IF P+T1/5>T2 THEN GOTO 250
240 GOTO 110
250 P=0
260 REM PRINTOUT ROUTINE
270 GOSUB 2010
280 IF T+T1/5>T3 THEN GOTO 292
290 GOTO 110
292 IF A$="P" OR A$="p"THEN LPRINT"          |....+....|....+....|....+....|....+.
...|....+....|....+....|....+....|....+....|....+....|....+....|....+.
...|":END
294 LPRINT"-------------------------------------------------------------
-----------":END
1000 REM EQUATION SUBROUTINE
1005 V=VP*SIN (W*T)
1010 IF X(2)>0 THEN GOTO 1035
1015 IF X(1)>0 AND L*D(1)+X(3)>-V THEN GOSUB 1125:RETURN
1020 GOSUB 1205:RETURN
1035 IF X(1)>0 THEN GOSUB 1205:RETURN
1040 IF L*D(2)+X(3)>V THEN GOSUB 1165:RETURN
1045 GOSUB 1205:RETURN
1100 REM Equation Subroutine
1125 D(1)=(V-X(3))/B
1130 D(2)=(-V-X(3))/B
1135 D(3)=(X(1)-X(3)/R)/C
1136 IF A$="p" OR A$="P" THEN RETURN
1140 RETURN
1165 D(1)=(V-X(3))/B
1170 D(2)=(-V-X(3))/B
1175 D(3)=(X(2)-X(3)/R)/C
1176 IF A$="p" OR A$="P" THEN RETURN
1180 RETURN
1205 D(1)=(V+2*L*V/LL-X(3))/(2*L+LL)
1210 D(2)=D(1)-2*V/LL
1215 D(3)=(X(1)+X(2)-X(3)/R)/C
1216 IF A$="p" OR A$="P" THEN RETURN
1220 RETURN
2000 REM PRINT SUBROUTINE
2002 IF A$="P" OR A$="p" THEN LPRINT CHR$(27);"G";WIDTH "LPT1:",140:GOTO 2012
2004 LPRINT TAB(1);"MILLISEC.";TAB(10);"   X(1)";TAB(20);"   X(2)";TAB(30);"   X
(3)";TAB(40);"    Vac":LPRINT
2010 IF A$="P" OR A$="p" THEN GOTO 2012
2011 GOTO 2015
2012 IF T=0 THEN GOSUB 3000
2014 GOSUB 3040:RETURN
2015 LPRINT TAB(1);:LPRINT USING"###.###";T*1000;
2025 LPRINT TAB(10);:LPRINT USING"###.#####";X(1);
2030 LPRINT TAB(20);:LPRINT USING"###.#####";X(2);
2035 LPRINT TAB(30);:LPRINT USING"#####.#";X(3);
2040 LPRINT TAB(40);:LPRINT USING"######.#";V;
```

```
2045 IF SUM=0 THEN LPRINT" ---Avg. X(3) =";AVG:RETURN
2050 LPRINT"":RETURN
3000 REM PLOTTING SUBROUTINE
3010 LPRINT"                         Transient Response In kV For Single Phase Center Tap DC Power Supply"
3015 LPRINT"                         --------- ------- -- -- --- ------ --- ------ --- -- ----- ------"
3020 LPRINT"        0         3         4         5         6         7         8         9        10        11        12        13        14"
3030 LPRINT"        !....+....!....+....!....+....!....+....!....+....!....+....!....+....!....+....!....+....!....+....!....+....!....+....!....+....!....+...."
3032 LPRINT TAB(1);"X(3) kV";TAB(22);"X(1) A.";TAB(30);"X(2) A.":RETURN
3040 IF T<.1201 THEN GOTO 3043
3041 IF T>.2999 THEN GOTO 3043
3042 IF T>.12 THEN SC=-1:RETURN
3043 SC=SC+1
3044 PP=(-11+X(3)/100)
3060 IF SC < 5 THEN LPRINT USING "##.###";X(3)/1000;:LPRINT TAB(9);"-";
3062 IF SC < 5 THEN GOTO 3150
3100 LPRINT USING "##.###";X(3)/1000;:LPRINT TAB(9);"--T=";:LPRINT USING "#.###";T;:LPRINT " SEC.";:SC=0
3150 LPRINT TAB(23);:LPRINT USING"##.##";X(1);:LPRINT TAB(31);:LPRINT USING"##.##";X(2);:IF PP<37 THEN LPRINT"":RETURN
3160 IF PP>135 THEN LPRINT TAB(131)"+->":RETURN
3200 LPRINT TAB(PP);"+":RETURN
3500 REM Printer On/Tabular Output/Plotted Output Subroutine
3510 CLS:PRINT"Your printer must be on to run this program.  Is it on (Y or N)?"
3520 A$=INKEY$:IF A$="" THEN 3520
3530 IF A$="n" OR A$="N" THEN PRINT"For operation without a printer go to DOS. Type INFORMATION.":END
3540 PRINT"For a plotted response enter P.  For a tabular response enter any other letter."
3550 A$=INKEY$:IF A$="" THEN 3550
3560 CLS:PRINT"                         PROGRAM IS IN PROGRESS
                                        ------- -- -- --------"
3570 RETURN


10 REM APPENDIX.17  -- DC Power Supply Transient Analysis - 1 Ph. Bridge --
11 GOSUB 3500
13 DIM X(49)
14 DIM D(49)
16 DIM E(49)
18 DIM F(49)
19 PI=3.141592654#:SC=-1
22 LPRINT CHR$(27);"@";CHR$(15);CHR$(27);"A";CHR$(8);CHR$(27);"Q";CHR$(145);CHR$(27);"G";CHR$(27);"M";CHR$(6);
30 N=3
32 T1=.000025
34 T2=.002
36 T3=.32
38 T=0
40 P=0
50 X(1)=0
60 X(2)=0
70 X(3)=0
71 W=2*PI*62.5
```

```
72 VP=15708
73 LL=.1414
74 L=2
75 C=.00005
76 R=500
77 A=L-LL
78 B=L+LL
80 LPRINT" -- 1 Phase Bridge Rectifier Integration Time =";T1;"Seconds --"
86 GOSUB 1000
90 GOSUB 2000
100 REM INTEGRATION ROUTINE
110 GOSUB 1000
111 IF T=>.12 AND T<.12+T1 THEN LPRINT:LPRINT
120 FOR A = 1 TO N
130 E(A)=X(A)
140 F(A)=D(A)
150 X(A)=X(A)+T1*D(A)
152 IF X(1)<0 THEN X(1)=0
154 IF X(2)<0 THEN X(2)=0
160 NEXT A
165 T=T+T1
170 GOSUB 1000
180 FOR A = 1 TO N
190 X(A)=E(A)+T1/2*(F(A)+D(A))
192 IF X(1)<0 THEN X(1)=0
194 IF X(2)<0 THEN X(2)=0
200 NEXT A
202 CO=CO+1:SUM=SUM+X(3):IF CO=INT(8.000001E-03/T1+.5) THEN AVG=SUM/INT(8.000001
E-03/T1+.5):CO=0:SUM=0
220 P=P+T1
230 IF P+T1/5>T2 THEN GOTO 250
240 GOTO 110
250 P=0
260 REM PRINTOUT ROUTINE
270 GOSUB 2010
280 IF T+T1/5>T3 THEN GOTO 292
290 GOTO 110
291 LPRINT
292 IF A$="P" OR A$="p"THEN LPRINT"          !....+....!....+....!....+....!....+.
...!....+....!....+....!....+....!....+....!....+....!....+....!....+....!....+.
...!":END
294 LPRINT"-----------------------------------------------------------------------
----------"
296 LPRINT
300 END
1000 REM EQUATION SUBROUTINE
1005 V=VP*SIN (W*T)
1010 IF X(2)>0 THEN GOTO 1035
1015 IF X(1)>0 THEN GOTO 1025
1025 IF L*D(1)+X(3)>0 THEN GOSUB 1125:RETURN
1030 GOSUB 1105:RETURN
1035 IF X(1)>0 THEN GOSUB 1185:RETURN
1040 IF L*D(2)+X(3)>0 THEN GOSUB 1165:RETURN
1045 GOSUB 1145
1100 REM Equation Subroutine
1105 D(1)=V/(2*LL)-X(3)/(2*L)
1110 D(2)=-V/(2*LL)-X(3)/(2*L)
1115 D(3)=(X(1)-X(3)/R)/C
```

```
1116 IF A$="p" OR A$="P" THEN RETURN
1117 RETURN
1120 LPRINT"a";:RETURN
1125 D(1)=(V-X(3))/B
1130 D(2)=(-V-X(3))/B
1135 D(3)=(X(1)-X(3)/R)/C
1136 IF A$="p" OR A$="P" THEN RETURN
1137 RETURN
1140 LPRINT"b";:RETURN
1145 D(1)=V/(2*LL)-X(3)/(2*L)
1150 D(2)=-V/(2*LL)-X(3)/(2*L)
1155 D(3)=(X(2)-X(3)/R)/C
1156 IF A$="p" OR A$="P" THEN RETURN
1157 RETURN
1160 LPRINT"c";:RETURN
1165 D(1)=(V-X(3))/B
1170 D(2)=(-V-X(3))/B
1175 D(3)=(X(2)-X(3)/R)/C
1176 IF A$="p" OR A$="P" THEN RETURN
1177 RETURN
1180 LPRINT"d";:RETURN
1185 D(1)=V/(2*LL)-X(3)/(2*L)
1190 D(2)=-V/(2*LL)-X(3)/(2*L)
1195 D(3)=(X(1)+X(2)-X(3)/R)/C
1196 IF A$="p" OR A$="P" THEN RETURN
1197 RETURN
1200 LPRINT"e";:RETURN
2000 REM PRINT SUBROUTINE
2002 IF A$="P" OR A$="p" THEN LPRINT CHR$(27);"G";WIDTH "LPT1:",140:GOTO 2012
2004 LPRINT TAB(1);"MILLISEC.";TAB(10);"   X(1)";TAB(20);"   X(2)";TAB(30);"   X(3)";TAB(40);"    Vac":LPRINT
2010 IF A$="P" OR A$="p" THEN GOTO 2012
2011 GOTO 2015
2012 IF T=0 THEN GOSUB 3000
2014 GOSUB 3040:GOTO 2050
2015 LPRINT TAB(2);:LPRINT USING"###.###";T*1000;
2025 LPRINT TAB(10);:LPRINT USING"###.###";X(1);
2030 LPRINT TAB(20);:LPRINT USING"###.###";X(2);
2035 LPRINT TAB(30);:LPRINT USING"#####.#";X(3);
2040 LPRINT TAB(40);:LPRINT USING"######.#";V;
2045 IF SUM=0 THEN LPRINT"  ---Avg. X(3) =";AVG:RETURN
2050 RETURN
3000 REM Plotting Subroutine
3010 LPRINT"                           Transient Response In kV For Single Phase Bridge DC Power Supply"
3015 LPRINT"                     --------- -------- -- -- --- ------ -- --- ------ -- ----- ------"
3020 LPRINT"        0          3     4     5     6      7      8      10     11    12    13    14"
3030 LPRINT"        |....+.  .|....+.|....+.|....+.|....+.|....+....|....+....|....+....|....+....|....+....|....+....|"
3032 LPRINT TAB(1);"X(3) kV";TAB(22);"X(1) A.";TAB(30);"X(2) A.":RETURN
3040 IF T<.1201 THEN GOTO 3043
3041 IF T>.2999 THEN GOTO 3043
3042 IF T>.12 THEN SC=-1:RETURN
3043 SC=SC+1
3044 PP=(-11+X(3)/100)
3060 IF SC < 5 THEN LPRINT USING "##.###";X(3)/1000;:LPRINT TAB(9);"-";
```

```
3062 IF SC < 5 THEN GOTO 3150
3100 LPRINT USING "##.###";X(3)/1000;:LPRINT TAB(9);"--T=";:LPRINT USING "#.###";T;:LPRINT "SEC.";:SC=0
3150 LPRINT TAB(23);:LPRINT USING"##.##";X(1);:LPRINT TAB(31);:LPRINT USING"##.##";X(2);:IF PP<37 THEN LPRINT"":RETURN
3160 IF PP>130 THEN LPRINT TAB(129)"+->":RETURN
3200 LPRINT TAB(PP);"+":RETURN
3500 REM Printer On/Tabular Output/Plotted Output Subroutine
3510 CLS:PRINT"Your printer must be on to run this program.  Is it on (Y or N)?"
3520 A$=INKEY$:IF A$="" THEN 3520
3530 IF A$="n" OR A$="N" THEN PRINT"For operation without a printer go to DOS. Type INFORMATION.":END
3540 PRINT"For a plotted response enter P.  For a tabular response enter any other letter."
3550 A$=INKEY$:IF A$="" THEN 3550
3560 CLS:PRINT"                    PROGRAM IS IN PROGRESS
                                ------- -- -- --------"
3570 RETURN


10 REM APPENDIX.18  -- DC Power Supply Transient Analysis - 3 Ph. Bridge --
12 GOSUB 3500
13 PI=3.141592654#:SC=-1
14 F=62.5:A1=PI/6:A2=5*PI/6:A3=3*PI/2
16 W=2*PI*F
21 LPRINT CHR$(27);"@";CHR$(15);CHR$(27);"A";CHR$(8);CHR$(27);"G";CHR$(27);"M";CHR$(6);
22 DIM X(49)
24 DIM D(49)
26 DIM E(49)
28 DIM F(49)
30 N=4
32 T1=.000025
34 T2=.002
36 T3=.32
38 T=0
40 P=0
50 X(1)=0
51 X(2)=0:X(3)=0
52 X(4)=0
53 V1=3023
54 V2=3023
55 V3=-6046
58 L=2
60 LL=.0667
62 C=.00005
64 R=500
66 VP=6046
68 K2=LL
70 K3=L+LL
72 K4=L+2*LL
80 LPRINT"     --- 3 Phase Bridge Rectifier Integration Time =";T1;" Seconds ---":LPRINT
84 GOSUB 1000
86 GOSUB 2000
100 REM Integration Routine
110 GOSUB 400
```

```
120 FOR A = 1 TO N
130 E(A)=X(A)
140 F(A)=D(A)
150 X(A)=X(A)+T1*D(A)
152 IF X(A)<0 THEN X(A)=0
160 NEXT A
165 T=T+T1
170 GOSUB 400
180 FOR A = 1 TO N
190 X(A)=E(A)+T1/2*(F(A)+D(A))
192 IF X(A)<0 THEN X(A)=0
200 NEXT A
220 P=P+T1
230 IF P+T1/5>T2 THEN GOTO 250
240 GOTO 110
250 P=0
260 REM PRINTOUT ROUTINE
270 GOSUB 2013
280 IF T+T1/5>T3 THEN GOTO 291
290 GOTO 110
291 IF A$="P" OR A$="p" THEN GOSUB 3030:END
292 LPRINT
294 LPRINT"----------------------------------------------------------------
--------"
300 END
400 REM Commutation Subroutine
401 K=W*T:V1=VP*SIN(A1+K)
402 V2=VP*SIN(A2+K)
403 V3=VP*SIN(A3+K)
404 IF (V1+LL*D(1)-V3)>(V2+LL*D(2)-V3) AND (V1+LL*D(2)-V2)<(V1+LL*D(3)-V3) THEN
GOSUB 1010:RETURN
405 IF (V1+LL*D(2)-V2)>(V1+LL*D(3)-V3) AND (V3+LL*D(3)-V2)<(V1+LL*D(1)-V2) THEN
GOSUB 1030:RETURN
406 IF (V3+LL*D(3)-V2)>(V1+LL*D(1)-V2) AND (V3+LL*D(1)-V1)<(V3+LL*D(2)-V2) THEN
GOSUB 1050:RETURN
407 IF (V3+LL*D(1)-V1)>(V3+LL*D(2)-V2) AND (V2+LL*D(2)-V1)<(V3+LL*D(3)-V1) THEN
GOSUB 1070:RETURN
408 IF (V2+LL*D(2)-V1)>(V3+LL*D(3)-V1) AND (V2+LL*D(3)-V3)<(V2+LL*D(1)-V1) THEN
GOSUB 1090:RETURN
409 IF (V2+LL*D(3)-V3)>(V2+LL*D(1)-V1) AND (V1+LL*D(1)-V3)<(V2+LL*D(2)-V3) THEN
GOSUB 1110:RETURN
1000 REM Equation Subroutine
1010 REM Forward paths 1 & 2.  Return path 6.
1012 IF X(2)=0 THEN GOTO 1020
1014 D(1)=((V3-V2+X(4))*K3-(V3-V1+X(4))*K4)/(K4^2-K3^2)
1016 D(2)=((V3-V2+X(4))*K4-(V3-V1+X(4))*K3)/(K3^2-K4^2)
1017 D(3)=D(1)+D(2):VL=L*(D(1)+D(2))
1018 D(4)=(X(1)+X(2)-X(4)/R)/C:TS=1:RETURN
1020 REM Forward path 1.  Return path 4.
1022 D(1)=(V1-V3-X(4))/K4
1023 D(3)=D(1):D(2)=0:X(2)=0::VL=L*D(1)
1024 D(4)=(X(1)-X(4)/R)/C:TS=2:RETURN
1030 REM Forward path 1.  Return paths 5 & 6.
1032 IF X(3)=0 THEN GOTO 1040
1034 D(2)=((V3-V1+X(4))*K3-(V2-V1+X(4))*K4)/(K4^2-K3^2)
1036 D(3)=((V3-V1+X(4))*K4-(V2-V1+X(4))*K3)/(K3^2-K4^2)
1037 D(1)=D(2)+D(3):VL=L*(D(2)+D(3))
1038 D(4)=(X(2)+X(3)-X(4)/R)/C:TS=3:RETURN
```

```
1040 REM Forward path 1.  Return path 5.
1042 D(2)=(V1-V2-X(4))/K4
1043 D(1)=D(2):D(3)=0:X(3)=0:VL=L*D(2)
1044 D(4)=(X(2)-X(4)/R)/C:TS=4:RETURN
1050 REM Forward paths 1 & 3.  Return paths 5.
1052 IF X(1)=0 THEN GOTO 1060
1054 D(3)=((V2-V1+X(4))*K3-(V2-V3+X(4))*K4)/(K4^2-K3^2)
1056 D(1)=((V2-V1+X(4))*K4-(V2-V3+X(4))*K3)/(K3^2-K4^2)
1057 D(2)=D(1)+D(3):VL=L*(D(1)+D(3))
1058 D(4)=(X(1)+X(3)-X(4)/R)/C:TS=5:RETURN
1060 REM Forward path 3.  Return path 5.
1062 D(3)=(V3-V2-X(4))/K4
1063 D(2)=D(3):D(1)=0:X(1)=0:VL=L*D(3)
1064 D(4)=(X(3)-X(4)/R)/C:TS=6:RETURN
1070 REM Forward paths 3.  Return paths 4 & 5.
1072 IF X(2)=0 THEN GOTO 1080
1074 D(1)=((V2-V3+X(4))*K3-(V1-V3+X(4))*K4)/(K4^2-K3^2)
1076 D(2)=((V2-V3+X(4))*K4-(V1-V3+X(4))*K3)/(K3^2-K4^2)
1077 D(3)=D(1)+D(2):VL=L*(D(1)+D(2))
1078 D(4)=(X(1)+X(2)-X(4)/R)/C:TS=7:RETURN
1080 REM Forward path 3.  Return path 4.
1082 D(1)=(V3-V1-X(4))/K4
1083 D(3)=D(1):D(2)=0:X(2)=0:VL=L*D(1)
1084 D(4)=(X(1)-X(4)/R)/C:TS=8:RETURN
1090 REM Forward paths 2 & 3.  Return path 4.
1092 IF X(3)=0 THEN GOTO 1100
1094 D(2)=((V1-V3+X(4))*K3-(V1-V2+X(4))*K4)/(K4^2-K3^2)
1096 D(3)=((V1-V3+X(4))*K4-(V1-V2+X(4))*K3)/(K3^2-K4^2)
1097 D(1)=D(2)+D(3):VL=L*(D(2)+D(3))
1098 D(4)=(X(2)+X(3)-X(4)/R)/C:TS=9:RETURN
1100 REM Forward path 2.  Return path 4.
1102 D(2)=(V2-V1-X(4))/K4
1103 D(1)=D(2):D(3)=0:X(3)=0:VL=L*D(2)
1104 D(4)=(X(2)-X(4)/R)/C:TS=10:RETURN
1110 REM Forward path 2.  Return paths 4 & 6.
1112 IF X(1)=0 THEN GOTO 1120
1114 D(3)=((V1-V2+X(4))*K3-(V3-V2+X(4))*K4)/(K4^2-K3^2)
1116 D(1)=((V1-V2+X(4))*K4-(V3-V2+X(4))*K3)/(K3^2-K4^2)
1117 D(2)=D(1)+D(3):VL=L*(D(1)+D(3))
1118 D(4)=(X(1)+X(3)-X(4)/R)/C:TS=11:RETURN
1120 REM Forward path 2.  Return path 6.
1122 D(3)=(V2-V3-X(4))/K4
1123 D(2)=D(3):D(1)=0:X(1)=0:VL=L*D(3)
1124 D(4)=(X(3)-X(4)/R)/C:TS=12:RETURN
2000 REM Print Subroutine
2002 IF A$="P" OR A$="p" THEN LPRINT CHR$(27);"G";:WIDTH "LPT1:",140:GOTO 2013
2004 LPRINT TAB(1);"Sec.   TS";TAB(14);"V1";TAB(21);"V2";TAB(28);"V3";TAB(33);"X(
1)";TAB(40);"X(2)";TAB(47);"X(3)";TAB(54);"X(4)";TAB(61);"X(5)";TAB(68);"X(6)";T
AB(75);"X(7)":LPRINT
2013 FOR Z=1 TO 7:I(Z)=0:NEXT Z
2014 IF TS=1 THEN I(1)=X(1):I(2)=X(2):I(6)=X(1)+X(2):I(7)=X(4):GOTO 2026
2015 IF TS=2 THEN I(1)=X(1):I(2)=X(2):I(6)=X(1):I(7)=X(4):GOTO 2026
2016 IF TS=3 THEN I(1)=X(2)+X(3):I(5)=X(2):I(6)=X(3):I(7)=X(4):GOTO 2026
2017 IF TS=4 THEN I(1)=X(2):I(5)=X(2):I(6)=X(3):I(7)=X(4):GOTO 2026
2018 IF TS=5 THEN I(1)=X(1):I(3)=X(3):I(5)=X(1)+X(3):I(7)=X(4):GOTO 2026
2019 IF TS=6 THEN I(1)=X(3):I(3)=X(3):I(5)=X(3):I(7)=X(4):GOTO 2026
2020 IF TS=7 THEN I(3)=X(1)+X(2):I(4)=X(1):I(5)=X(2):I(7)=X(4):GOTO 2026
2021 IF TS=8 THEN I(3)=X(1):I(4)=X(1):I(5)=X(2):I(7)=X(4):GOTO 2026
```

```
2022 IF TS=9 THEN I(2)=X(2):I(3)=X(3):I(4)=X(2)+X(3):I(7)=X(4):GOTO 2026
2023 IF TS=10 THEN I(2)=X(2):I(3)=X(3):I(4)=X(2):I(7)=X(4):GOTO 2026
2024 IF TS=11 THEN I(2)=X(1)+X(3):I(4)=X(1):I(6)=X(3):I(7)=X(4):GOTO 2026
2025 IF TS=12 THEN I(2)=X(3):I(4)=X(1):I(6)=X(3):I(7)=X(4):GOTO 2026
2026 IF A$="P" OR A$="p" THEN GOTO 2028
2027 GOTO 2040
2028 IF T=0 THEN GOSUB 3000:RETURN
2029 GOSUB 3040:RETURN
2040 LPRINT TAB(1);:LPRINT USING".#####";T;:LPRINT;TS;
2041 LPRINT TAB(11);:LPRINT USING"######";V1;
2042 LPRINT TAB(18);:LPRINT USING"######";V2;
2043 LPRINT TAB(25);:LPRINT USING"######";V3;
2044 LPRINT TAB(31);:LPRINT USING"###.#";I(1);
2045 LPRINT TAB(38);:LPRINT USING"###.#";I(2);
2046 LPRINT TAB(45);:LPRINT USING"###.#";I(3);
2047 LPRINT TAB(52);:LPRINT USING"###.#";I(4);
2048 LPRINT TAB(59);:LPRINT USING"###.#";I(5);
2049 LPRINT TAB(66);:LPRINT USING"###.#";I(6);
2050 LPRINT TAB(73);:LPRINT USING"######";I(7)
2051 RETURN
3000 REM PLOTTING SUBROUTINE
3010 LPRINT"                         Transient Response In kV For A Three Pha
se Bridge DC Power Supply"
3015 LPRINT"                         --------- ------- -- -- --- - ----- ---
-- ------ -- ----- ------"
3020 LPRINT"        0         3         4         5         6         7
8         9        10        11        12        13        14"
3030 LPRINT"        !....+.  .!....+....!....+....!....+....!....+....!....+....
!....+....!....+....!....+....!....+....!....+....!....+....!":IF T>0 THEN RETUR
N
3032 LPRINT TAB(1);"X(4)kV";TAB(23);"I(1)";TAB(29);"I(2)";TAB(35);"I(3)";TAB(41)
;"I(4)";TAB(47);"I(5)";TAB(53);"I(6)"
3040 IF T<.1201 THEN GOTO 3043
3041 IF T>.2999 THEN GOTO 3043
3042 IF T>.12 THEN SC=-1: RETURN
3043 SC=SC+1
3044 PP=(-11+X(4)/100)
3060 IF SC<5 THEN LPRINT USING "##.###";X(4)/1000;:LPRINT TAB(9);"-";
3062 IF SC<5 THEN GOTO 3072
3070 LPRINT USING "##.###";X(4)/1000;:LPRINT TAB(9);"--T=";:LPRINT USING"#.###";
T;:LPRINT" Sec.";:SC=0
3072 LPRINT TAB(23);:LPRINT USING"##.#";I(1);:LPRINT TAB(29);:LPRINT USING"##.#"
;I(2);:LPRINT TAB(35);:LPRINT USING"##.#";I(3);:LPRINT TAB(41);:LPRINT USING"##.
#";I(4);:LPRINT TAB(47);:LPRINT USING"##.#";I(5);:LPRINT TAB(53);:LPRINT USING"#
#.#";I(6);
3074 IF PP<58 THEN LPRINT"":RETURN
3100 IF PP<10 THEN LPRINT"":RETURN
3110 IF PP>130 THEN LPRINT TAB(129)"+->":RETURN
3200 LPRINT TAB(PP);"+":RETURN
3500 REM Printer On/Tabular Output/Plotted Output Subroutine
3510 CLS:PRINT"Your printer must be on to run this program.  Is it on (Y or N)?"
3520 A$=INKEY$:IF A$="" THEN 3520
3530 IF A$="n" OR A$="N" THEN PRINT"For operation without a printer go to DOS. T
ype INFORMATION": END
3540 PRINT"For a plotted response enter P.  For a tabular response enter any oth
er letter."
3550 A$=INKEY$:IF A$="" THEN 3550
3560 CLS:PRINT"                         PROGRAM IS IN PROGRESS
                         ------- -- -- --------"
```

Appendix B

```
3570 RETURN


10 REM APPENDIX.18C   3 Phase F.W.P.S., RC Snub., Rev.Comm.Rout.,& Expanded Plot
12 GOSUB 3500
13 PI=3.141592654#:SC=-1
14 F=62.5:A1=PI/6:A2=5*PI/6:A3=3*PI/2
16 W=2*PI*F
21 LPRINT CHR$(27);"@";CHR$(15);CHR$(27);"A";CHR$(8);CHR$(27);"G";CHR$(27);"M";C
HR$(6);:WIDTH "LPT1:",140
22 DIM X(49)
24 DIM D(49)
26 DIM E(49)
28 DIM F(49)
30 N=6
32 T1=.0001:TP=T1
34 T2=.001
36 T3=.08
38 T=0
40 P=0
50 X(1)=0
51 X(2)=18.842:X(3)=18.842
52 X(4)=9665.344:X(5)=19.209:X(6)=9525.03
53 V1=3023
54 V2=3023
55 V3=-6046
58 L=2
60 LL=.0667
62 C=.00005:CI=.000002
64 R=500:RI=500
66 VP=6046
80 LPRINT"     --- 3 Phase Bridge Rectifier Integration Time =";T1;" Seconds ---"
:LPRINT
84 GOSUB 1000
86 GOSUB 2000
100 REM Integration Subroutine
110 GOSUB 400
120 FOR A = 1 TO N
130 E(A)=X(A)
140 F(A)=D(A)
150 X(A)=X(A)+T1*D(A)
160 NEXT A
165 T=T+T1:K=W*T:V1=VP*SIN(A1+K):V2=VP*SIN(A2+K):V3=VP*SIN(A3+K)
170 IF TS=<2 THEN GOSUB 1010 ELSE IF TS=<4 THEN GOSUB 1030 ELSE IF TS=<6 THEN GO
SUB 1050 ELSE IF TS=<8 THEN GOSUB 1070 ELSE IF TS=<10 THEN GOSUB 1090 ELSE IF TS
=< 12 THEN GOSUB 1110
180 FOR A = 1 TO N
190 X(A)=E(A)+T1/2*(F(A)+D(A))
200 NEXT A
202 IF X(1)=>0 AND X(2)=>0 AND X(3)=>0 THEN GOTO 220
204 IF X(1)<0 THEN B=1 ELSE IF X(2)<0 THEN B=2 ELSE IF X(3)<0 THEN B=3
206 T=T-T1:TC=E(B)*T1/(E(B)-X(B)):FOR A=1 TO N:X(A)=E(A)+TC*F(A):NEXT A:T=T+TC:X
(B)=0:D(B)=0:K=W*T:V1=VP*SIN(A1+K):V2=VP*SIN(A2+K):V3=VP*SIN(A3+K)
208 IF TS=<2 THEN GOSUB 1010 ELSE IF TS=<4 THEN GOSUB 1030 ELSE IF TS=<6 THEN GO
SUB 1050 ELSE IF TS=<8 THEN GOSUB 1070 ELSE IF TS=<10 THEN GOSUB 1090 ELSE IF TS
=< 12 THEN GOSUB 1110
210 TX=TP-TC:FOR A=1 TO N:X(A)=X(A)+TX*D(A):NEXT A:T=T+TX:B=0
```

```
220 P=P+T1
230 IF P+T1/5>T2 THEN GOTO 250
240 GOTO 110
250 P=0
260 REM PRINTOUT ROUTINE
270 GOSUB 2013
280 IF T+T1/5>T3 THEN GOTO 291
290 GOTO 110
291 IF A$="P" OR A$="p" THEN GOSUB 3030:CLS:END
292 LPRINT
294 LPRINT"----------------------------------------------------------
----------":END
300 CLS:END
400 REM Commutation Subroutine
404 IF V1=>(V2-LL*D(2)) AND -V2<(-LL*D(3)-V3) THEN GOSUB 1010:RETURN
405 IF -V2=>(-LL*D(3)-V3) AND V3<(V1-LL*D(1)) THEN GOSUB 1030:RETURN
406 IF V3=>(V1-LL*D(1)) AND -V1<(-LL*D(2)-V2) THEN GOSUB 1050:RETURN
407 IF -V1=>(-LL*D(2)-V2) AND V2<(V3-LL*D(3)) THEN GOSUB 1070:RETURN
408 IF V2=>(V3-LL*D(3)) AND -V3<(-LL*D(1)-V1) THEN GOSUB 1090:RETURN
409 IF -V3=>(-LL*D(1)-V1) AND V1<(V2-LL*D(2)) THEN GOSUB 1110:RETURN
410 RETURN
1000 REM Equation Subroutine
1010 REM Forward paths 1 & 2.  Return path 6.
1011 IF X(2)=0 THEN GOTO 1020
1012 D(4)=(X(1)+X(2)-X(5))/CI
1013 D(5)=(X(4)+CI*RI*D(4)-X(6))/L
1014 D(1)=(2*V1-V2-V3-CI*RI*D(4)-X(4))/(3*LL)
1015 D(2)=(2*V2-V1-V3-CI*RI*D(4)-X(4))/(3*LL)
1016 D(3)=D(1)+D(2)
1017 D(6)=(X(5)-X(6)/R)/C:TS=1:RETURN
1020 REM Forward path 1.  Return path 4.
1022 D(4)=(X(1)-X(5))/CI
1023 D(5)=(X(4)+CI*RI*D(4)-X(6))/L
1024 D(1)=(V1-V3-CI*RI*D(4)-X(4))/(2*LL)
1025 D(3)=D(1):D(2)=0:X(2)=0:X(3)=X(1)
1026 D(6)=(X(5)-X(6)/R)/C:TS=2:RETURN
1030 REM Forward path 1.  Return paths 5 & 6.
1031 IF X(3)=0 THEN GOTO 1040
1032 D(4)=(X(2)+X(3)-X(5))/CI
1033 D(5)=(X(4)+CI*RI*D(4)-X(6))/L
1034 D(2)=(V1-2*V2+V3-CI*RI*D(4)-X(4))/(3*LL)
1035 D(3)=(V1-2*V3+V2-CI*RI*D(4)-X(4))/(3*LL)
1036 D(1)=D(2)+D(3)
1037 D(6)=(X(5)-X(6)/R)/C:TS=3:RETURN
1040 REM Forward path 1.  Return path 5.
1042 D(4)=(X(2)-X(5))/CI
1043 D(5)=(X(4)+CI*RI*D(4)-X(6))/L
1044 D(2)=(V1-V2-CI*RI*D(4)-X(4))/(2*LL)
1045 D(1)=D(2):D(3)=0:X(3)=0:X(1)=X(2)
1046 D(6)=(X(5)-X(6)/R)/C:TS=4:RETURN
1050 REM Forward paths 1 & 3.  Return paths 5.
1051 IF X(1)=0 THEN GOTO 1060
1052 D(4)=(X(1)+X(3)-X(5))/CI
1053 D(5)=(X(4)+RI*CI*D(4)-X(6))/L
1054 D(1)=(2*V1-V2-V3-RI*CI*D(4)-X(4))/(3*LL)
1055 D(3)=(2*V3-V2-V1-RI*CI*D(4)-X(4))/(3*LL)
1056 D(2)=D(3)+D(1)
```

```
1057 D(6)=(X(5)-X(6)/R)/C:TS=5:RETURN
1060 REM Forward path 3.  Return path 5.
1062 D(4)=(X(3)-X(5))/CI
1063 D(5)=(X(4)+RI*CI*D(4)-X(6))/L
1064 D(3)=(V3-V2-RI*CI*D(4)-X(4))/(2*LL)
1065 D(2)=D(3):D(1)=0:X(1)=0:X(2)=X(3)
1066 D(6)=(X(5)-X(6)/R)/C:TS=6:RETURN
1070 REM Forward path 1. Return paths 2 and 3.
1071 IF X(2)=0 THEN GOTO 1080
1072 D(4)=(X(1)+X(2)-X(5))/CI
1073 D(5)=(X(4)+CI*RI*D(4)-X(6))/L
1074 D(1)=(V3-2*V1+V2-CI*RI*D(4)-X(4))/(3*LL)
1075 D(2)=(V3-2*V2+V1-CI*RI*D(4)-X(4))/(3*LL)
1076 D(3)=D(1)+D(2)
1077 D(6)=(X(5)-X(6)/R)/C:TS=7:RETURN
1080 REM Forward path 3.  Return path 4.
1082 D(4)=(X(1)-X(5))/CI
1083 D(5)=(X(4)+CI*RI*D(4)-X(6))/L
1084 D(1)=(V3-V1-CI*RI*D(4)-X(4))/(2*LL)
1085 D(3)=D(1):D(2)=0:X(2)=0:X(3)=X(1)
1086 D(6)=(X(5)-X(6)/R)/C:TS=8:RETURN
1090 REM Forward paths 2 & 3.  Return path 4.
1091 IF X(3)=0 THEN GOTO 1100
1092 D(4)=(X(2)+X(3)-X(5))/CI
1093 D(5)=(X(4)+CI*RI*D(4)-X(6))/L
1094 D(2)=(2*V2-V3-V1-CI*RI*D(4)-X(4))/(3*LL)
1095 D(3)=(2*V3-V1-V2-CI*RI*D(4)-X(4))/(3*LL)
1096 D(1)=D(2)+D(3)
1097 D(6)=(X(5)-X(6)/R)/C:TS=9:RETURN
1100 REM Forward path 2.  Return path 4.
1102 D(4)=(X(2)-X(5))/CI
1103 D(5)=(X(4)+RI*CI*D(4)-X(6))/L
1104 D(2)=(V2-V1-CI*RI*D(4)-X(4))/(2*LL)
1105 D(1)=D(2):D(3)=0:X(3)=0:X(1)=X(2)
1106 D(6)=(X(5)-X(6)/R)/C:TS=10:RETURN
1110 REM Forward path 2.  Return paths 4 & 6.
1111 IF X(1)=0 THEN GOTO 1120
1112 D(4)=(X(1)+X(3)-X(5))/CI
1113 D(5)=(X(4)+RI*CI*D(4)-X(6))/L
1114 D(1)=(V2-2*V1+V3-RI*CI*D(4)-X(4))/(3*LL)
1115 D(3)=(V2-2*V3+V1-RI*CI*D(4)-X(4))/(3*LL)
1116 D(2)=D(1)+D(3)
1117 D(6)=(X(5)-X(6)/R)/C:TS=11:RETURN
1120 REM Forward path 2.  Return path 6.
1122 D(4)=(X(3)-X(5))/CI
1123 D(5)=(X(4)+RI*CI*D(4)-X(6))/L
1124 D(3)=(V2-V3-RI*CI*D(4)-X(4))/(2*LL)
1125 D(2)=D(3):D(1)=0:X(1)=0:X(2)=X(3)
1126 D(6)=(X(5)-X(6)/R)/C:TS=12:RETURN
2000 REM Print Subroutine
2002 IF A$="P" OR A$="p" THEN LPRINT CHR$(27);"G";:WIDTH "LPT1:",140:GOTO 2013
2004 LPRINT TAB(1);"Sec.    TS";TAB(14);"V1";TAB(20);"V2";TAB(26);"V3";TAB(31);"X
(1)";TAB(36);"X(2)";TAB(41);"X(3)";TAB(46);"X(4)";TAB(51);"X(5)";TAB(56);"X(6)";
TAB(62);"X(7)";
2005 LPRINT TAB(68)"X(5)";TAB(73);"X(4)":LPRINT
2013 FOR Z=1 TO 7:I(Z)=0:NEXT Z
2014 IF TS=1 THEN I(1)=X(1):I(2)=X(2):I(6)=X(1)+X(2):I(7)=X(6):GOTO 2026
2015 IF TS=2 THEN I(1)=X(1):I(2)=X(2):I(6)=X(1):I(7)=X(6):GOTO 2026
```

```
2016 IF TS=3 THEN I(1)=X(2)+X(3):I(5)=X(2):I(6)=X(3):I(7)=X(6):GOTO 2026
2017 IF TS=4 THEN I(1)=X(2):I(5)=X(2):I(6)=X(3):I(7)=X(6):GOTO 2026
2018 IF TS=5 THEN I(1)=X(1):I(3)=X(3):I(5)=X(1)+X(3):I(7)=X(6):GOTO 2026
2019 IF TS=6 THEN I(1)=X(1):I(3)=X(3):I(5)=X(3):I(7)=X(6):GOTO 2026
2020 IF TS=7 THEN I(3)=X(1)+X(2):I(4)=X(1):I(5)=X(2):I(7)=X(6):GOTO 2026
2021 IF TS=8 THEN I(3)=X(1):I(4)=X(1):I(5)=X(2):I(7)=X(6):GOTO 2026
2022 IF TS=9 THEN I(2)=X(2):I(3)=X(3):I(4)=X(2)+X(3):I(7)=X(6):GOTO 2026
2023 IF TS=10 THEN I(2)=X(2):I(3)=X(3):I(4)=X(2):I(7)=X(6):GOTO 2026
2024 IF TS=11 THEN I(2)=X(1)+X(3):I(4)=X(1):I(6)=X(3):I(7)=X(6):GOTO 2026
2025 IF TS=12 THEN I(2)=X(3):I(4)=X(1):I(6)=X(3):I(7)=X(6):GOTO 2026
2026 IF B=>1 THEN GOTO 2040
2027 IF A$="P" OR A$="p" THEN GOTO 2029
2028 GOTO 2040
2029 IF T=0 THEN GOSUB 3000:RETURN
2030 GOSUB 3040:RETURN
2040 LPRINT TAB(1);:LPRINT USING".#####";T;:LPRINT;TS;
2041 LPRINT TAB(12);:LPRINT USING"#####";V1;
2042 LPRINT TAB(18);:LPRINT USING"#####";V2;
2043 LPRINT TAB(24);:LPRINT USING"#####";V3;
2044 LPRINT TAB(30);:LPRINT USING"###.#";I(1);
2045 LPRINT TAB(35);:LPRINT USING"###.#";I(2);
2046 LPRINT TAB(40);:LPRINT USING"###.#";I(3);
2047 LPRINT TAB(45);:LPRINT USING"###.#";I(4);
2048 LPRINT TAB(50);:LPRINT USING"###.#";I(5);
2049 LPRINT TAB(55);:LPRINT USING"###.#";I(6);
2050 LPRINT TAB(60);:LPRINT USING"######";I(7);
2051 LPRINT TAB(66);:LPRINT USING"###.#";X(5);
2052 LPRINT TAB(71);:LPRINT USING"######";RI*CI*D(4)+X(4)
2053 RETURN
3000 REM Plotting Subroutine
3010 LPRINT"               Transient Response In kV For A Three Pha
se Bridge DC Power Supply"
3015 LPRINT"       --------- -------- -- -- --- - ----- ---
-- ------ -- ----- ------"
3020 LPRINT"        9430    9440    9450    9460    9470    9480    94
90    9500    9510    9520    9530    9540    9550"
3030 LPRINT"        |....+....|....+....|....+....|....+....|....+....
|....+....|....+....|....+....|....+....|....+....|....+....|";:IF T>0 THEN RETUR
N
3032 LPRINT TAB(1);" X(6) V";TAB(23);"I(1)";TAB(29);"I(2)";TAB(35);"I(3)";TAB(41
);"I(4)";TAB(47);"I(5)";TAB(53);"I(6)"
3040 SC=SC+1
3044 PP=X(6)-9421
3060 IF SC<5 THEN LPRINT USING "#####.#";X(6);:LPRINT TAB(9);"-";
3062 IF SC<5 THEN GOTO 3072
3070 LPRINT USING "#####.#";X(6);:LPRINT TAB(9);"--T=";:LPRINT USING"#.###";T;:L
PRINT" Sec.";:SC=0
3072 LPRINT TAB(23);:LPRINT USING"##.#";I(1);:LPRINT TAB(29);:LPRINT USING"##.#"
;I(2);:LPRINT TAB(35);:LPRINT USING"##.#";I(3);:LPRINT TAB(41);:LPRINT USING"##.
#";I(4);:LPRINT TAB(47);:LPRINT USING"##.#";I(5);:LPRINT TAB(53);:LPRINT USING"#
#.#";I(6);
3074 IF PP<58 THEN LPRINT"":RETURN
3100 IF PP<10 THEN LPRINT"":RETURN
3110 IF PP>130 THEN LPRINT TAB(129)"+->":RETURN
3200 LPRINT TAB(PP);"+":RETURN
3500 REM Printer On/Tabular Output/Plotted Output Subroutine
3510 CLS:PRINT"       Your printer must be on to run this program.  Is it on (
Y or N)?"
```

```
3520 A$=INKEY$:IF A$="" THEN 3520
3530 IF A$="n" OR A$="N" THEN PRINT"For operation without a printer go to DOS. T
ype INFORMATION": END
3540 PRINT"For a plotted response enter P.  For a tabular response enter any oth
er letter."
3550 A$=INKEY$:IF A$="" THEN 3550
3560 CLS:PRINT"                       PROGRAM IS IN PROGRESS
                                      ------- -- -- --------"
3570 RETURN



10 REM APPENDIX.18D --Controlled Three Phase Bridge Converter For Shunt Motor--
11 CLEAR ,49152!,3000
12 GOSUB 3500
13 PI=3.141592654#:SC=-1:COMU=1:RIN=80:CP=.016:TF=.0034
14 F=62.5:A1=PI/6:A2=5*PI/6:A3=3*PI/2
16 W=2*PI*F
21 LPRINT CHR$(27);"@";CHR$(15);CHR$(27);"A";CHR$(8);CHR$(27);"M";CHR$(1);CHR$(2
7);"6";CHR$(27);"N";CHR$(11);:WIDTH "LPT1:",136
22 DIM X(49)
24 DIM D(49)
26 DIM E(49)
28 DIM F(49)
30 N=6
32 T1=.0001:TP=T1
34 T2=.016
36 T3=1
38 T=0
40 P=0
50 X(1)=0
51 X(2)=1:X(3)=1
52 X(4)=1:X(5)=1:X(6)=0
58 L=.001
60 LL=.0001467
62 C=1:CI=.0002
64 R=83.33:RI=2.42:RA=.05
66 VP=169.7:K=W*T:V3=VP*SIN(3*PI/2+K):V2=VP*SIN(5*PI/6+K):V1=VP*SIN(PI/6+K)
80 LPRINT"      --- 3 Phase Bridge Rectifier Integration Time =";T1;" Seconds ---"
:LPRINT
84 TS=1:GOSUB 400
86 GOSUB 2000
100 REM Integration Routine
110 IF T=>(CCY+1)*CP THEN CCY=CCY+1:TCY=CCY*CP
111 IF T<TF THEN GOTO 117
112 IF X(5)>350 THEN TF=TF+T1 ELSE GOTO 115
113 IF TF=>.004 THEN TF=.004
114 GOTO 117
115 IF X(5)<325 THEN TF=TF-T1
116 IF TF=<0 THEN TF=0
117 CLS:PRINT"             TF =";TF:GOSUB 400
118 IF T>.8 THEN R=.83333
120 FOR A = 1 TO N
130 E(A)=X(A)
140 F(A)=D(A)
150 X(A)=X(A)+T1*D(A)
160 NEXT A
165 T=T+T1:K=W*T:V3=VP*SIN(3*PI/2+K):V2=VP*SIN(5*PI/6+K):V1=VP*SIN(PI/6+K)
```

```
170 IF TS=1 THEN GOSUB 1012 ELSE IF TS=2 THEN GOSUB 1020 ELSE IF TS=3 THEN GOSUB
 1032 ELSE IF TS=4 THEN GOSUB 1040 ELSE IF TS=5 THEN GOSUB 1052 ELSE IF TS=6 THE
N GOSUB 1060 ELSE IF TS=7 THEN GOSUB 1072 ELSE IF TS=8 THEN GOSUB 1080
171 IF TS=9 THEN GOSUB 1092 ELSE IF TS=10 THEN GOSUB 1100 ELSE IF TS=11 THEN GOS
UB 1112 ELSE IF TS=12 THEN GOSUB 1120
180 FOR A=1 TO N
190 X(A)=E(A)+T1/2*(F(A)+D(A))
200 NEXT A
202 IF X(1)=>0 AND X(2)=>0 AND X(3)=>0 THEN GOTO 212
204 IF X(1)<0 THEN B=1 ELSE IF X(2)<0 THEN B=2 ELSE IF X(3)<0 THEN B=3
206 T=T-T1:TC=E(B)*T1/(E(B)-X(B)):FOR A=1 TO N:X(A)=E(A)+TC*F(A):NEXT A:T=T+TC:X
(B)=0:D(B)=0:K=W*T:V3=VP*SIN(3*PI/2+K):V2=VP*SIN(5*PI/6+K):V1=VP*SIN(PI/6+K)
208 IF TS=<2 THEN GOSUB 1012 ELSE IF TS=<4 THEN GOSUB 1032 ELSE IF TS=<6 THEN GO
SUB 1052 ELSE IF TS=<8 THEN GOSUB 1072 ELSE IF TS=<10 THEN GOSUB 1092 ELSE IF TS
=< 12 THEN GOSUB 1112
210 TX=TP-TC:FOR A=1 TO N:X(A)=X(A)+TX*D(A):NEXT A:T=T+TX:B=0
212 IF X(1)<0 THEN X(1)=0 AND D(1)=0
214 IF X(2)<0 THEN X(2)=0 AND D(2)=0
216 IF X(3)<0 THEN X(3)=0 AND D(3)=0
220 P=P+T1
230 IF P+T1/5>T2 THEN GOTO 250
240 GOTO 110
250 P=0
260 REM PRINTOUT ROUTINE
270 GOSUB 2013
280 IF T+T1/5>T3 THEN GOTO 291
290 GOTO 110
291 IF A$="P" OR A$="p" THEN GOSUB 3030:CLS:END
292 LPRINT:FOR N=1 TO 18:LPRINT"X(";N;") =";X(N):NEXT N:LPRINT"T =";T
294 LPRINT"----------------------------------------------------------------
------------------------------------------------------------":END
300 CLS:END
400 IF BL1=1 THEN GOSUB 1010 ELSE IF BL2=1 THEN GOSUB 1030 ELSE IF BL3=1 THEN GO
SUB 1050 ELSE IF BL4=1 THEN GOSUB 1070 ELSE IF BL5=1 THEN GOSUB 1090 ELSE IF BL6
=1 THEN GOSUB 1110
401 IF BL1=1 OR BL2=1 OR BL3=1 OR BL4=1 OR BL5=1 OR BL6=1 THEN RETURN
404 IF V1=>V2 AND -V2<-V3 THEN GOSUB 1010:RETURN
405 IF -V2=>-V3 AND V3<V1 THEN GOSUB 1030:RETURN
406 IF V3=>V1 AND -V1<-V2 THEN GOSUB 1050:RETURN
407 IF -V1=>-V2 AND V2<V3 THEN GOSUB 1070:RETURN
408 IF V2=>V3 AND -V3<-V1 THEN GOSUB 1090:RETURN
409 IF -V3=>-V1 AND V1<V2 THEN GOSUB 1110:RETURN
410 STOP
1000 REM Equation Subroutine
1010 REM Forward paths 1 & 2. Return path 3
1011 IF T-TCY>=0 AND T-TCY=<TF THEN BL1=1 ELSE BL1=0
1012 IF BL1=1 THEN GOTO 1112
1013 IF X(2)=0 THEN GOTO 1020 ELSE D(4)=(X(1)+X(2)-X(5))/CI
1014 D(5)=(X(4)+CI*RI*D(4)-RA*X(5)-X(6))/L
1015 D(1)=(2*V1-V2-V3-CI*RI*D(4)-X(4))/(3*LL)
1016 D(2)=(2*V2-V1-V3-CI*RI*D(4)-X(4))/(3*LL)
1017 D(3)=D(1)+D(2)
1018 D(6)=(X(5)-X(6)/R)/C:TS=1:RETURN
1020 REM Forward path 1.  Return path 4.
1022 D(4)=(X(1)-X(5))/CI
1023 D(5)=(X(4)+CI*RI*D(4)-RA*X(5)-X(6))/L
1024 D(1)=(V1-V3-CI*RI*D(4)-X(4))/(2*LL)
1025 D(3)=D(1):D(2)=0:X(2)=0:X(3)=X(1)
```

Appendix B

```
1026 D(6)=(X(5)-X(6)/R)/C:TS=2:RETURN
1030 REM Forward path 1.  Return paths 5 & 6.
1031 IF T-TCY>=CP/6 AND T-TCY=<CP/6+TF THEN BL2=1 ELSE BL2=0
1032 IF BL2=1 THEN GOTO 1012
1033 IF X(3)=0 THEN GOTO 1040 ELSE D(4)=(X(2)+X(3)-X(5))/CI
1034 D(5)=(X(4)+CI*RI*D(4)-RA*X(5)-X(6))/L
1035 D(2)=(V1-2*V2+V3-CI*RI*D(4)-X(4))/(3*LL)
1036 D(3)=(V1-2*V3+V2-CI*RI*D(4)-X(4))/(3*LL)
1037 D(1)=D(2)+D(3)
1038 D(6)=(X(5)-X(6)/R)/C:TS=3:RETURN
1040 REM Forward path 1.  Return path 5.
1042 D(4)=(X(2)-X(5))/CI
1043 D(5)=(X(4)+CI*RI*D(4)-RA*X(5)-X(6))/L
1044 D(2)=(V1-V2-CI*RI*D(4)-X(4))/(2*LL)
1045 D(1)=D(2):D(3)=0:X(3)=0:X(1)=X(2)
1046 D(6)=(X(5)-X(6)/R)/C:TS=4:RETURN
1050 REM Forward paths 1 & 3.  Return paths 5.
1051 IF T-TCY>=CP/3 AND T-TCY=<CP/3+TF THEN BL3=1 ELSE BL3=0
1052 IF BL3=1 THEN GOTO 1032
1053 IF X(1)=0 THEN GOTO 1060 ELSE D(4)=(X(1)+X(3)-X(5))/CI
1054 D(5)=(X(4)+RI*CI*D(4)-RA*X(5)-X(6))/L
1055 D(1)=(2*V1-V2-V3-RI*CI*D(4)-X(4))/(3*LL)
1056 D(3)=(2*V3-V2-V1-RI*CI*D(4)-X(4))/(3*LL)
1057 D(2)=D(3)+D(1)
1058 D(6)=(X(5)-X(6)/R)/C:TS=5:RETURN
1060 REM Forward path 3.  Return path 5.
1062 D(4)=(X(3)-X(5))/CI
1063 D(5)=(X(4)+RI*CI*D(4)-RA*X(5)-X(6))/L
1064 D(3)=(V3-V2-RI*CI*D(4)-X(4))/(2*LL)
1065 D(2)=D(3):D(1)=0:X(1)=0:X(2)=X(3)
1066 D(6)=(X(5)-X(6)/R)/C:TS=6:RETURN
1070 REM Forward path 1. Return paths 2 and 3.
1071 IF T-TCY>=CP/2 AND T-TCY=<CP/2+TF THEN BL4=1 ELSE BL4=0
1072 IF BL4=1 THEN GOTO 1052
1073 IF X(2)=0 THEN GOTO 1080 ELSE D(4)=(X(1)+X(2)-X(5))/CI
1074 D(5)=(X(4)+CI*RI*D(4)-RA*X(5)-X(6))/L
1075 D(1)=(V3-2*V1+V2-CI*RI*D(4)-X(4))/(3*LL)
1076 D(2)=(V3-2*V2+V1-CI*RI*D(4)-X(4))/(3*LL)
1077 D(3)=D(1)+D(2)
1078 D(6)=(X(5)-X(6)/R)/C:TS=7:RETURN
1080 REM Forward path 3.  Return path 4.
1082 D(4)=(X(1)-X(5))/CI
1083 D(5)=(X(4)+CI*RI*D(4)-RA*X(5)-X(6))/L
1084 D(1)=(V3-V1-CI*RI*D(4)-X(4))/(2*LL)
1085 D(3)=D(1):D(2)=0:X(2)=0:X(3)=X(1)
1086 D(6)=(X(5)-X(6)/R)/C:TS=8:RETURN
1090 REM Forward paths 2 & 3.  Return path 4.
1091 IF T-TCY>=2*CP/3 AND T-TCY=<2*CP/3+TF THEN BL5=1 ELSE BL5=0
1092 IF BL5=1 THEN GOTO 1072
1093 IF X(3)=0 THEN GOTO 1100 ELSE D(4)=(X(2)+X(3)-X(5))/CI
1094 D(5)=(X(4)+CI*RI*D(4)-RA*X(5)-X(6))/L
1095 D(2)=(2*V2-V3-V1-CI*RI*D(4)-X(4))/(3*LL)
1096 D(3)=(2*V3-V1-V2-CI*RI*D(4)-X(4))/(3*LL)
1097 D(1)=D(2)+D(3)
1098 D(6)=(X(5)-X(6)/R)/C:TS=9:RETURN
1100 REM Forward path 2.  Return path 4.
1102 D(4)=(X(2)-X(5))/CI
1103 D(5)=(X(4)+RI*CI*D(4)-RA*X(5)-X(6))/L
```

```
1104 D(2)=(V2-V1-CI*RI*D(4)-X(4))/(2*LL)
1105 D(1)=D(2):D(3)=0:X(3)=0:X(1)=X(2)
1106 D(6)=(X(5)-X(6)/R)/C:TS=10:RETURN
1110 REM Forward path 2.  Return paths 4 & 6.
1111 IF T-TCY>=5*CP/6 AND T-TCY=<5*CP/6+TF THEN BL6=1 ELSE BL6=0
1112 IF BL6=1 THEN GOTO 1092
1113 IF X(1)=0 THEN GOTO 1120 ELSE D(4)=(X(1)+X(3)-X(5))/CI
1114 D(5)=(X(4)+RI*CI*D(4)-RA*X(5)-X(6))/L
1115 D(1)=(V2-2*V1+V3-RI*CI*D(4)-X(4))/(3*LL)
1116 D(3)=(V2-2*V3+V1-RI*CI*D(4)-X(4))/(3*LL)
1117 D(2)=D(1)+D(3)
1118 D(6)=(X(5)-X(6)/R)/C:TS=11:RETURN
1120 REM Forward path 2.  Return path 6.
1122 D(4)=(X(3)-X(5))/CI
1123 D(5)=(X(4)+RI*CI*D(4)-RA*X(5)-X(6))/L
1124 D(3)=(V2-V3-RI*CI*D(4)-X(4))/(2*LL)
1125 D(2)=D(3):D(1)=0:X(1)=0:X(2)=X(3)
1126 D(6)=(X(5)-X(6)/R)/C:TS=12:RETURN
2000 REM Print Subroutine
2002 IF A$="P" OR A$="p" THEN LPRINT CHR$(27);"G";:WIDTH "LPT1:",140:GOTO 2013
2013 FOR Z=1 TO 7:I(Z)=0:NEXT Z
2014 IF TS=1 THEN I(1)=X(1):I(2)=X(2):I(6)=X(1)+X(2):I(7)=X(6):GOTO 2026
2015 IF TS=2 THEN I(1)=X(1):I(2)=X(2):I(6)=X(1):I(7)=X(6):GOTO 2026
2016 IF TS=3 THEN I(1)=X(2)+X(3):I(5)=X(2):I(6)=X(3):I(7)=X(6):GOTO 2026
2017 IF TS=4 THEN I(1)=X(2):I(5)=X(2):I(6)=X(3):I(7)=X(6):GOTO 2026
2018 IF TS=5 THEN I(1)=X(1):I(3)=X(3):I(5)=X(1)+X(3):I(7)=X(6):GOTO 2026
2019 IF TS=6 THEN I(1)=X(1):I(3)=X(3):I(5)=X(3):I(7)=X(6):GOTO 2026
2020 IF TS=7 THEN I(3)=X(1)+X(2):I(4)=X(1):I(5)=X(2):I(7)=X(6):GOTO 2026
2021 IF TS=8 THEN I(3)=X(1):I(4)=X(1):I(5)=X(2):I(7)=X(6):GOTO 2026
2022 IF TS=9 THEN I(2)=X(2):I(3)=X(3):I(4)=X(2)+X(3):I(7)=X(6):GOTO 2026
2023 IF TS=10 THEN I(2)=X(2):I(3)=X(3):I(4)=X(2):I(7)=X(6):GOTO 2026
2024 IF TS=11 THEN I(2)=X(1)+X(3):I(4)=X(1):I(6)=X(3):I(7)=X(6):GOTO 2026
2025 IF TS=12 THEN I(2)=X(3):I(4)=X(1):I(6)=X(3):I(7)=X(6):GOTO 2026
2026 IF A$="P" OR A$="p" THEN GOTO 2029
2027 IF A$="P" OR A$="p" THEN GOTO 2029
2028 GOTO 2036
2029 IF T=0 THEN GOSUB 3000:RETURN
2030 GOSUB 3040:RETURN
2036 IF CO=0 THEN GOTO 2038
2037 GOTO 2040
2038 CY=CY+1
2039 IF HEAD=0 THEN GOSUB 4300
2040 LPRINT TAB(1);:LPRINT USING".#####";T;:LPRINT;TS;
2041 LPRINT TAB(12);:LPRINT USING"#####";V1;
2042 LPRINT TAB(18);:LPRINT USING"#####";V2;
2043 LPRINT TAB(24);:LPRINT USING"#####";V3;
2044 LPRINT TAB(30);:LPRINT USING"#####";I(1);
2045 LPRINT TAB(35);:LPRINT USING"#####";I(2);
2046 LPRINT TAB(40);:LPRINT USING"#####";I(3);
2047 LPRINT TAB(45);:LPRINT USING"#####";I(4);
2048 LPRINT TAB(50);:LPRINT USING"#####";I(5);
2049 LPRINT TAB(55);:LPRINT USING"#####";I(6);
2050 LPRINT TAB(61);:LPRINT USING"###.#";I(7);
2051 LPRINT TAB(66);:LPRINT USING"#####";X(5);
2052 LPRINT TAB(71);:LPRINT USING"#####";RI*CI*D(4)+X(4);
2053 LPRINT TAB(76);:LPRINT USING"####.#";X(7);
2054 LPRINT TAB(83);:LPRINT USING"####.#";X(8);
2055 LPRINT TAB(90);:LPRINT USING"####.#";X(9);
```

```
2056 LPRINT TAB(96);:LPRINT USING"###.#";DEL1;
2057 LPRINT TAB(102);:LPRINT USING"###.#";DEL2;
2058 LPRINT TAB(108);:LPRINT USING"###.#";DEL3;
2059 LPRINT TAB(113);:LPRINT USING"###";I1;
2069 LPRINT TAB(116);:LPRINT USING"###";I2;
2070 LPRINT TAB(119);:LPRINT USING"###";I3;
2071 LPRINT TAB(122);:LPRINT USING"###.#";X(10);
2072 LPRINT TAB(127);:LPRINT USING"###.#";X(11);
2073 LPRINT TAB(132);:LPRINT USING"###.#";X(12)
2100 REM RMS Subroutine
2110 CO=CO+1
2120 RMS1=RMS1+V1^2:RMS2=RMS2+V2^2:RMS3=RMS3+V3^2
2122 RMS4=RMS4+X(7)^2:RMS5=RMS5+X(8)^2:RMS6=RMS6+X(9)^2
2124 RMSI1=RMSI1+I1^2:RMSI2=RMSI2+I2^2:RMSI3=RMSI3+I3^2
2125 RMSX10=RMSX10+X(10)^2:RMSX11=RMSX11+X(11)^2:RMSX12=RMSX12+X(12)^2
2126 RMSX1=RMSX1+X(1)^2:RMSX2=RMSX2+X(2)^2:RMSX3=RMSX3+X(3)^2:RMSRI=RMSRI+(CI*D(4))^2
2130 IF CO<RIN THEN RETURN
2140 RMS1=SQR(RMS1/RIN):RMS2=SQR(RMS2/RIN):RMS3=SQR(RMS3/RIN)
2142 RMS4=SQR(RMS4/RIN):RMS5=SQR(RMS5/RIN):RMS6=SQR(RMS6/RIN)
2150 LPRINT:LPRINT" Cycle";CY;" Line Voltage RMS Values Are: V1 =";RMS1;"  V2 = ";RMS2;"  V3 =";RMS3;"  VRMSavg =";(RMS1+RMS2+RMS3)/3
2152 LPRINT" Cycle";CY;" Line Current RMS Values Are: X(7)=";RMS4;"  X(8)=";RMS5;" X(9)=";RMS6;" IRMSavg =";(RMS4+RMS5+RMS6)/3;" >>> Vdc =";X(6);"<<<"
2155 LPRINT:RMS1=0:RMS2=0:RMS3=0:RMS4=0:RMS5=0:RMS6=0:CO=0:RETURN
3000 REM Plotting Subroutine
3010 LPRINT"                              DC Output Voltage Of A Controlled Three Phase Bridge DC Power System"
3015 LPRINT"                              -- ------ ------- -- - ---------- ----- ----- ------ -- ----- ------"
3020 LPRINT"                      0        25       50       75      100       125      150     175     200     225     250     275"
3022 LPRINT"      X(5)    X(6)   !....+....!....+....!....+....!....+....!....+....!....+....!....+....!....+....!....+....!....+....!....+....!....+....!":IF T>0 THEN RETURN
3026 GOTO 3040
3030 LPRINT"                             !....+....!....+....!....+....!....+....!....+....!....+....!....+....!....+....!....+....!....+....!....+....!....+....!":RETURN
3040 SC=SC+1
3044 PP=X(6)/2.5+19
3060 IF SC<5 THEN LPRINT TAB(5);:LPRINT USING"#####";X(5);:LPRINT TAB(12);:LPRINT USING"###.#";X(6);:LPRINT TAB(19);"-";
3062 IF SC<5 THEN GOTO 3100
3070 LPRINT TAB(5);:LPRINT USING"#####";X(5);:LPRINT TAB(12);:LPRINT USING"###.#";X(6);:LPRINT TAB(19);"--";:LPRINT TAB(21);:LPRINT USING"#.##";T;:SC=0:GOTO 3100
3100 IF X(6)>300 THEN RETURN
3110 IF T=0 THEN RETURN ELSE LPRINT TAB(PP);"+":RETURN
3500 REM Printer On/Tabular Output/Plotted Output Subroutine
3510 CLS:PRINT"           Your printer must be on to run this program.  Is it on (Y or N)?"
3520 A$=INKEY$:IF A$="" THEN 3520
3530 IF A$="n" OR A$="N" THEN PRINT"For operation without a printer go to DOS.  Type INFORMATION": END
3540 PRINT"For a plotted response enter P.  For a tabular response enter any other letter."
3550 A$=INKEY$:IF A$="" THEN 3550
3560 CLS:PRINT"                        PROGRAM IS IN PROGRESS
                                       ------- -- -- --------"
```

```
3570 RETURN
4300 REM Tabular Heading
4310 HEAD=HEAD+1
4320 LPRINT TAB(123);"X";TAB(128);"X";TAB(133);"X"
4330 LPRINT TAB(1);"Sec.    TS";TAB(14);"V1";TAB(20);"V2";TAB(26);"V3";TAB(31);"I
(1)";TAB(36);"I(2)";TAB(41);"I(3)";TAB(46);"I(4)";TAB(51);"I(5)";TAB(56);"I(6)";
TAB(63);"X(6)";
4340 LPRINT TAB(68);"X(5)";TAB(73);"Vin";TAB(77);"X(7)";TAB(84);"x(8)";TAB(91);"
X(9)";TAB(96);"Del 1";TAB(102);"Del 2";TAB(108);"Del 3";TAB(114);"I1";TAB(117);"
I2";TAB(120);"I3";TAB(123);"(10)";TAB(128);"(11)";TAB(133);"(12)":LPRINT
4350 RETURN


10 REM APPENDIX.18E -Controlled Three Phase Bridge Inverter For Shunt Generator-
11 CLEAR ,49152!,3000
12 GOSUB 3500
13 PI=3.141592654#:SC=-1:CP=.016:TF=0
14 F=62.5:A1=PI/6:A2=5*PI/6:A3=3*PI/2
16 W=2*PI*F
21 LPRINT CHR$(27);"@";CHR$(15);CHR$(27);"A";CHR$(8);CHR$(27);"M";CHR$(1);CHR$(2
7);"G";CHR$(27);"N";CHR$(8);:WIDTH "LPT1:",136
22 DIM X(49)
24 DIM D(49)
26 DIM E(49)
28 DIM F(49)
29 PRINT:INPUT"Do you also want to have a conservation of energy analysis perfor
med (Y or N)";C$
30 N=6
31 LPRINT CHR$(27);"S";CHR$(0);CHR$(27);"A";CHR$(5);
32 T1=.00005:TP=T1
34 T2=.0005:RIN=CP/T2:RIN1=CP/(4*T2)
36 T3=.016
38 T=0
40 P=0
50 X(1)=0
51 X(2)=300:X(3)=300
52 X(4)=-250:X(5)=300:X(6)=-250
58 L=.001
60 LL=.0001467
62 C=1000000!:CI=.0002
64 R=1000000!:RI=2.42:RA=.05:RL=.576
66 VP=169.7:K=W*T:V3=VP*SIN(3*PI/2+K):V2=VP*SIN(5*PI/6+K):V1=VP*SIN(PI/6+K)
80 LPRINT"    --- 3 Phase Bridge Rectifier Integration Time =";T1;" Seconds ---"
:LPRINT
84 TS=1:GOSUB 400
86 GOSUB 2000
100 REM Integration Routine
110 IF T=>(CCY+1)*CP THEN CCY=CCY+1:TCY=CCY*CP
115 GOSUB 400
120 FOR A = 1 TO N
130 E(A)=X(A):VP1=V1:VP2=V2:VP3=V3
140 F(A)=D(A)
150 X(A)=X(A)+T1*D(A)
160 NEXT A
165 T=T+T1:K=W*T:V3=VP*SIN(3*PI/2+K):V2=VP*SIN(5*PI/6+K):V1=VP*SIN(PI/6+K)
170 IF TS=1 THEN GOSUB 1012 ELSE IF TS=2 THEN GOSUB 1020 ELSE IF TS=3 THEN GOSUB
  1032 ELSE IF TS=4 THEN GOSUB 1040 ELSE IF TS=5 THEN GOSUB 1052 ELSE IF TS=6 THE
N GOSUB 1060 ELSE IF TS=7 THEN GOSUB 1072 ELSE IF TS=8 THEN GOSUB 1080
```

```
171 IF TS=9 THEN GOSUB 1092 ELSE IF TS=10 THEN GOSUB 1100 ELSE IF TS=11 THEN GOS
UB 1112 ELSE IF TS=12 THEN GOSUB 1120
180 FOR A=1 TO N
190 X(A)=E(A)+T1/2*(F(A)+D(A))
200 NEXT A
202 IF X(1)=>0 AND X(2)=>0 AND X(3)=>0 THEN GOTO 212
204 IF X(1)<0 THEN B=1 ELSE IF X(2)<0 THEN B=2 ELSE IF X(3)<0 THEN B=3
206 T=T-T1:TC=E(B)*T1/(E(B)-X(B)):FOR A=1 TO N:X(A)=E(A)+TC*F(A):NEXT A:T=T+TC:X
(B)=0:D(B)=0:K=W*T:V3=VP*SIN(3*PI/2+K):V2=VP*SIN(5*PI/6+K):V1=VP*SIN(PI/6+K)
208 IF TS=<2 THEN GOSUB 1012 ELSE IF TS=<4 THEN GOSUB 1032 ELSE IF TS=<6 THEN GO
SUB 1052 ELSE IF TS=<8 THEN GOSUB 1072 ELSE IF TS=<10 THEN GOSUB 1092 ELSE IF TS
=< 12 THEN GOSUB 1112
210 TX=TP-TC:FOR A=1 TO N:X(A)=X(A)+TX*D(A):NEXT A:T=T+TX:B=0
212 IF X(1)<0 THEN X(1)=0 AND D(1)=0
214 IF X(2)<0 THEN X(2)=0 AND D(2)=0
216 IF X(3)<0 THEN X(3)=0 AND D(3)=0
222 P=P+T1
230 IF P+T1/5>T2 THEN GOTO 250
240 GOTO 110
250 P=0
260 REM PRINTOUT ROUTINE
270 GOSUB 2013
280 IF T+T1/5>T3 THEN GOTO 291
290 GOTO 110
291 IF A$="P" OR A$="p" THEN GOSUB 3030:CLS:END
294 LPRINT"------------------------------------------------------------------
--------------------------------------------------------":END
300 CLS:END
400 IF BL1=1 THEN GOSUB 1010 ELSE IF BL2=1 THEN GOSUB 1030 ELSE IF BL3=1 THEN GO
SUB 1050 ELSE IF BL4=1 THEN GOSUB 1070 ELSE IF BL5=1 THEN GOSUB 1090 ELSE IF BL6
=1 THEN GOSUB 1110
401 IF BL1=1 OR BL2=1 OR BL3=1 OR BL4=1 OR BL5=1 OR BL6=1 THEN RETURN
404 IF V1=>V2 AND -V2<-V3 THEN GOSUB 1010:RETURN
405 IF -V2=>-V3 AND V3<V1 THEN GOSUB 1030:RETURN
406 IF V3=>V1 AND -V1<-V2 THEN GOSUB 1050:RETURN
407 IF -V1=>-V2 AND V2<V3 THEN GOSUB 1070:RETURN
408 IF V2=>V3 AND -V3<-V1 THEN GOSUB 1090:RETURN
409 IF -V3=>-V1 AND V1<V2 THEN GOSUB 1110:RETURN
410 STOP
1000 REM Equation Subroutine
1010 REM Forward paths 1 & 2. Return path 3
1011 IF T-TCY>=0 AND T-TCY=<TF THEN BL1=1 ELSE BL1=0
1012 IF BL1=1 THEN GOTO 1112
1013 IF X(2)=0 THEN GOTO 1020 ELSE D(4)=(-X(1)-X(2)+X(5))/CI
1014 D(5)=(-X(4)-CI*RI*D(4)-RA*X(5)+X(6))/L
1015 D(1)=(2*V1-V2-V3+CI*RI*D(4)+X(4))/(3*LL)
1016 D(2)=(2*V2-V1-V3+CI*RI*D(4)+X(4))/(3*LL)
1017 D(3)=D(1)+D(2)
1018 D(6)=(X(5)+X(6)/R)/C:TS=1:RETURN
1020 REM Forward path 1.  Return path 4.
1022 D(4)=(-X(1)+X(5))/CI
1023 D(5)=(-X(4)-CI*RI*D(4)-RA*X(5)+X(6))/L
1024 D(1)=(V1-V3+CI*RI*D(4)+X(4))/(2*LL)
1025 D(3)=D(1):D(2)=0:X(2)=0:X(3)=X(1)
1026 D(6)=(X(5)+X(6)/R)/C:TS=2:RETURN
1030 REM Forward path 1.  Return paths 5 & 6.
1031 IF T-TCY>=CP/6 AND T-TCY=<CP/6+TF THEN BL2=1 ELSE BL2=0
1032 IF BL2=1 THEN GOTO 1012
```

```
1033 IF X(3)=0 THEN GOTO 1040 ELSE D(4)=(-X(3)-X(2)+X(5))/CI
1034 D(5)=(-X(4)-CI*RI*D(4)-RA*X(5)+X(6))/L
1035 D(2)=(V1-2*V2+V3+CI*RI*D(4)+X(4))/(3*LL)
1036 D(3)=(V1-2*V3+V2+CI*RI*D(4)+X(4))/(3*LL)
1037 D(1)=D(2)+D(3)
1038 D(6)=(X(5)+X(6)/R)/C:TS=3:RETURN
1040 REM Forward path 1.  Return path 5.
1042 D(4)=(-X(2)+X(5))/CI
1043 D(5)=(-X(4)-CI*RI*D(4)-RA*X(5)+X(6))/L
1044 D(2)=(V1-V2+CI*RI*D(4)+X(4))/(2*LL)
1045 D(1)=D(2):D(3)=0:X(3)=0:X(1)=X(2)
1046 D(6)=(X(5)+X(6)/R)/C:TS=4:RETURN
1050 REM Forward paths 1 & 3.  Return paths 5.
1051 IF T-TCY>=CP/3 AND T-TCY=<CP/3+TF THEN BL3=1 ELSE BL3=0
1052 IF BL3=1 THEN GOTO 1032
1053 IF X(1)=0 THEN GOTO 1060 ELSE D(4)=(-X(1)-X(3)+X(5))/CI
1054 D(5)=(-X(4)-RI*CI*D(4)-RA*X(5)+X(6))/L
1055 D(1)=(2*V1-V2-V3+RI*CI*D(4)+X(4))/(3*LL)
1056 D(3)=(2*V3-V2-V1+RI*CI*D(4)+X(4))/(3*LL)
1057 D(2)=D(3)+D(1)
1058 D(6)=(X(5)+X(6)/R)/C:TS=5:RETURN
1060 REM Forward path 3.  Return path 5.
1062 D(4)=(-X(3)+X(5))/CI
1063 D(5)=(-X(4)-RI*CI*D(4)-RA*X(5)+X(6))/L
1064 D(3)=(V3-V2+RI*CI*D(4)+X(4))/(2*LL)
1065 D(2)=D(3):D(1)=0:X(1)=0:X(2)=X(3)
1066 D(6)=(X(5)+X(6)/R)/C:TS=6:RETURN
1070 REM Forward path 1. Return paths 2 and 3.
1071 IF T-TCY>=CP/2 AND T-TCY=<CP/2+TF THEN BL4=1 ELSE BL4=0
1072 IF BL4=1 THEN GOTO 1052
1073 IF X(2)=0 THEN GOTO 1080 ELSE D(4)=(-X(1)-X(2)+X(5))/CI
1074 D(5)=(-X(4)-CI*RI*D(4)-RA*X(5)+X(6))/L
1075 D(1)=(V3-2*V1+V2+CI*RI*D(4)+X(4))/(3*LL)
1076 D(2)=(V3-2*V2+V1+CI*RI*D(4)+X(4))/(3*LL)
1077 D(3)=D(1)+D(2)
1078 D(6)=(X(5)+X(6)/R)/C:TS=7:RETURN
1080 REM Forward path 3.  Return path 4.
1082 D(4)=(-X(1)+X(5))/CI
1083 D(5)=(-X(4)-CI*RI*D(4)-RA*X(5)+X(6))/L
1084 D(1)=(V3-V1+CI*RI*D(4)+X(4))/(2*LL)
1085 D(3)=D(1):D(2)=0:X(2)=0:X(3)=X(1)
1086 D(6)=(X(5)+X(6)/R)/C:TS=8:RETURN
1090 REM Forward paths 2 & 3.  Return path 4.
1091 IF T-TCY>=2*CP/3 AND T-TCY=<2*CP/3+TF THEN BL5=1 ELSE BL5=0
1092 IF BL5=1 THEN GOTO 1072
1093 IF X(3)=0 THEN GOTO 1100 ELSE D(4)=(-X(2)-X(3)+X(5))/CI
1094 D(5)=(-X(4)-CI*RI*D(4)-RA*X(5)+X(6))/L
1095 D(2)=(2*V2-V3-V1+CI*RI*D(4)+X(4))/(3*LL)
1096 D(3)=(2*V3-V1-V2+CI*RI*D(4)+X(4))/(3*LL)
1097 D(1)=D(2)+D(3)
1098 D(6)=(X(5)+X(6)/R)/C:TS=9:RETURN
1100 REM Forward path 2.  Return path 4.
1102 D(4)=(-X(2)+X(5))/CI
1103 D(5)=(-X(4)-RI*CI*D(4)-RA*X(5)+X(6))/L
1104 D(2)=(V2-V1+CI*RI*D(4)+X(4))/(2*LL)
1105 D(1)=D(2):D(3)=0:X(3)=0:X(1)=X(2)
1106 D(6)=(X(5)+X(6)/R)/C:TS=10:RETURN
1110 REM Forward path 2.  Return paths 4 & 6.
```

```
1111 IF T-TCY>=5*CP/6 AND T-TCY=<5*CP/6+TF THEN BL6=1 ELSE BL6=0
1112 IF BL6=1 THEN GOTO 1092
1113 IF X(1)=0 THEN GOTO 1120 ELSE D(4)=(-X(1)-X(3)+X(5))/CI
1114 D(5)=(-X(4)-RI*CI*D(4)-RA*X(5)+X(6))/L
1115 D(1)=(V2-2*V1+V3+RI*CI*D(4)+X(4))/(3*LL)
1116 D(3)=(V2-2*V3+V1+RI*CI*D(4)+X(4))/(3*LL)
1117 D(2)=D(1)+D(3)
1118 D(6)=(X(5)+X(6)/R)/C:TS=11:RETURN
1120 REM Forward path 2.  Return path 6.
1122 D(4)=(-X(3)+X(5))/CI
1123 D(5)=(-X(4)-RI*CI*D(4)-RA*X(5)+X(6))/L
1124 D(3)=(V2-V3+RI*CI*D(4)+X(4))/(2*LL)
1125 D(2)=D(3):D(1)=0:X(1)=0:X(2)=X(3)
1126 D(6)=(X(5)+X(6)/R)/C:TS=12:RETURN
2000 REM Print Subroutine
2002 IF A$="P" OR A$="p" THEN LPRINT CHR$(27);"G";:WIDTH "LPT1:",140:GOTO 2013
2013 FOR Z=1 TO 7:I(Z)=0:NEXT Z
2014 IF TS=1 THEN I(1)=X(1):I(2)=X(2):I(6)=X(1)+X(2):I(7)=X(6):GOTO 2026
2015 IF TS=2 THEN I(1)=X(1):I(2)=X(2):I(6)=X(1):I(7)=X(6):GOTO 2026
2016 IF TS=3 THEN I(1)=X(2)+X(3):I(5)=X(2):I(6)=X(3):I(7)=X(6):GOTO 2026
2017 IF TS=4 THEN I(1)=X(2):I(5)=X(2):I(6)=X(3):I(7)=X(6):GOTO 2026
2018 IF TS=5 THEN I(1)=X(1):I(3)=X(3):I(5)=X(1)+X(3):I(7)=X(6):GOTO 2026
2019 IF TS=6 THEN I(1)=X(1):I(3)=X(3):I(5)=X(3):I(7)=X(6):GOTO 2026
2020 IF TS=7 THEN I(3)=X(1)+X(2):I(4)=X(1):I(5)=X(2):I(7)=X(6):GOTO 2026
2021 IF TS=8 THEN I(3)=X(1):I(4)=X(1):I(5)=X(2):I(7)=X(6):GOTO 2026
2022 IF TS=9 THEN I(2)=X(2):I(3)=X(3):I(4)=X(2)+X(3):I(7)=X(6):GOTO 2026
2023 IF TS=10 THEN I(2)=X(2):I(3)=X(3):I(4)=X(2):I(7)=X(6):GOTO 2026
2024 IF TS=11 THEN I(2)=X(1)+X(3):I(4)=X(1):I(6)=X(3):I(7)=X(6):GOTO 2026
2025 IF TS=12 THEN I(2)=X(3):I(4)=X(1):I(6)=X(3):I(7)=X(6):GOTO 2026
2026 IF A$="P" OR A$="p" THEN GOTO 2029
2027 IF A$="P" OR A$="p" THEN GOTO 2029
2028 GOTO 2036
2029 IF T=.32 THEN GOSUB 3000:RETURN
2030 GOSUB 3040:RETURN
2036 IF CO=0 THEN GOTO 2038
2037 GOTO 2040
2038 CY=CY+1
2039 IF HEAD=0 THEN GOSUB 4300
2040 LPRINT TAB(1);:LPRINT USING".#####";T;:LPRINT;TS;
2041 LPRINT TAB(12);:LPRINT USING"#####";V1;
2042 LPRINT TAB(18);:LPRINT USING"#####";V2;
2043 LPRINT TAB(24);:LPRINT USING"#####";V3;
2044 LPRINT TAB(30);:LPRINT USING"#####";I(1);
2045 LPRINT TAB(35);:LPRINT USING"#####";I(2);
2046 LPRINT TAB(40);:LPRINT USING"#####";I(3);
2047 LPRINT TAB(45);:LPRINT USING"#####";I(4);
2048 LPRINT TAB(50);:LPRINT USING"#####";I(5);
2049 LPRINT TAB(55);:LPRINT USING"#####";I(6);
2050 LPRINT TAB(61);:LPRINT USING"#####";I(7);
2051 LPRINT TAB(66);:LPRINT USING"#####";X(5);
2052 LPRINT TAB(71);:LPRINT USING"#####";RI*CI*D(4)+X(4);
2053 GOSUB 4400:LPRINT TAB(76);:LPRINT USING"#####";X(1);
2054 LPRINT TAB(83);:LPRINT USING"#####";X(2);
2055 LPRINT TAB(90);:LPRINT USING"#####";X(3);
2056 LPRINT TAB(96);:LPRINT USING"#####";X(1)-X(2);
2057 LPRINT TAB(102);:LPRINT USING"#####";X(2)-X(3);
2058 LPRINT TAB(108);:LPRINT USING"#####";X(3)-X(1);
2059 LPRINT TAB(113);:LPRINT USING"####";V1/RL;
```

```
2069 LPRINT TAB(117);:LPRINT USING"####";V2/RL;
2070 LPRINT TAB(121);:LPRINT USING"####";V3/RL;
2071 LPRINT TAB(125);:IG1=V1/RL-X(1):LPRINT USING"####";IG1;
2072 LPRINT TAB(129);:IG2=V2/RL-X(2):LPRINT USING"####";IG2;
2073 LPRINT TAB(133);:IG3=V3/RL-X(3):LPRINT USING"####";IG3:GOSUB 4400:PCT=PCT+1
2078 IF C$="Y" OR C$="y" THEN GOTO 2079 ELSE GOTO 2100
2079 IF PCT=>RIN1 THEN LPRINT:LPRINT"For period T =";:LPRINT USING".#####";T-T1;
:LPRINT" to T =";:LPRINT USING".#####";T;:LPRINT" Sec.":LPRINT:LPRINT"  STORAGE
ELEMENTS" ELSE GOTO 2100
2080 WLL1=.5*LL*(X(1)^2-E(1)^2):LPRINT"    WLL1 =";:LPRINT USING"######.##";WLL1
2081 WLL2=.5*LL*(X(2)^2-E(2)^2):LPRINT"    WLL2 =";:LPRINT USING"######.##";WLL2
2082 WLL3=.5*LL*(X(3)^2-E(3)^2):LPRINT"    WLL3 =";:LPRINT USING"######.##";WLL3
2083  WCI=.5*CI*(X(4)^2-E(4)^2):LPRINT"     WCI =";:LPRINT USING"######.##";WCI
2084 WL=.5*L*(X(5)^2-E(5)^2):LPRINT"      WL =";:LPRINT USING"######.##";WL
2085 PSSUM=(WLL1+WLL2+WLL3+WCI+WL)/T1:LPRINT"   PSSUM =";:LPRINT USING"######.##";
PSSUM:LPRINT
2086 LPRINT"  DISSIPATIVE ELEMENTS"
2087 PRI=.5*RI*CI^2*(F(4)^2+D(4)^2):LPRINT"     PRI =";:LPRINT USING"######.##";PR
I
2088 PRA=.5*RA*(E(5)^2+X(5)^2):LPRINT"     PRA =";:LPRINT USING"######.##";PRA
2089 PRL1=(V1^2+VP1^2)/(2*RL):PRL2=(V2^2+VP2^2)/(2*RL):PRL3=(V3^2+VP3^2)/(2*RL)
2090 PDSUM=PRI+PRA:LPRINT"   PDSUM =";:LPRINT USING"######.##";PDSUM:LPRINT
2091 LPRINT"  GENERATORS"
2092 PDC=-.5*(E(6)*E(5)+X(6)*X(5)):LPRINT"     PDC =";:LPRINT USING"######.##";PDC
2093 GOSUB 4400:PAC1=.5*(V1*IG1+VP1*(VP1/RL-E(1))):LPRINT"    PAC1 =";:LPRINT USIN
G"######.##";PAC1
2094 PAC2=.5*(V2*IG2+VP2*(VP2/RL-E(2))):LPRINT"    PAC2 =";:LPRINT USING"######.##
";PAC2
2095 PAC3=.5*(V3*IG3+VP3*(VP3/RL-E(3))):LPRINT"    PAC3 =";:LPRINT USING"######.##
";PAC3:GOSUB 4400
2096 PGSUM=PDC+PAC1+PAC2+PAC3:LPRINT"   PGSUM =";:LPRINT USING"######.##";PGSUM:LP
RINT" PSSUM+PDSUM+PGSUM =";PSSUM+PDSUM+PGSUM:LPRINT:PCT=0
2097 LPRINT"  AC LOAD ELEMENTS":LPRINT"    PRL1 =";:LPRINT USING"######.##";PRL1:L
PRINT"    PRL2 =";:LPRINT USING"######.##";PRL2:LPRINT"    PRL3 =";:LPRINT USING"###
###.##";PRL3
2098 LPRINT"    PSUM =";:LPRINT USING"######.##";PRL1+PRL2+PRL3:LPRINT
2099 STOP
2100 REM RMS Subroutine
2110 CO=CO+1
2120 RMS1=RMS1+V1^2:RMS2=RMS2+V2^2:RMS3=RMS3+V3^2
2122 RMS4=RMS4+X(1)^2:RMS5=RMS5+X(2)^2:RMS6=RMS6+X(3)^2
2124 RMSI1=RMSI1+I1^2:RMSI2=RMSI2+I2^2:RMSI3=RMSI3+I3^2
2130 IF CO<RIN THEN RETURN
2140 RMS1=SQR(RMS1/RIN):RMS2=SQR(RMS2/RIN):RMS3=SQR(RMS3/RIN)
2142 RMS4=SQR(RMS4/RIN):RMS5=SQR(RMS5/RIN):RMS6=SQR(RMS6/RIN)
2150 LPRINT" Cycle";CY;" Line Voltage RMS Values Are: V1 =";RMS1;"  V2 =";RMS2;
" V3 =";RMS3;" VRMSavg =";(RMS1+RMS2+RMS3)/3
2152 LPRINT" Cycle";CY;" Line Current RMS Values Are: X(7)=";RMS4;"  X(8)=";RMS
5;" X(9)=";RMS6;" IRMSavg =";(RMS4+RMS5+RMS6)/3;"  >>> Vdc =";X(6);"<<<"
2155 LPRINT:RMS1=0:RMS2=0:RMS3=0:RMS4=0:RMS5=0:RMS6=0:CO=0:RETURN
3000 REM Plotting Subroutine
3500 REM Printer On/Tabular Output/Plotted Output Subroutine
3510 CLS:PRINT"         Your printer must be on to run this program.  Is it on (
Y or N)?"
3520 A$=INKEY$:IF A$="" THEN 3520
3530 IF A$="n" OR A$="N" THEN PRINT"For operation without a printer go to DOS. T
ype INFORMATION": END
3560 CLS:PRINT"                       PROGRAM IS IN PROGRESS
                               ------- -- -- --------"
```

```
3570 RETURN
4300 REM Tabular Heading
4310 HEAD=HEAD+1
4330 LPRINT TAB(1);"Sec.    TS";TAB(14);"V1";TAB(20);"V2";TAB(26);"V3";TAB(31);"I
(1)";TAB(36);"I(2)";TAB(41);"I(3)";TAB(46);"I(4)";TAB(51);"I(5)";TAB(56);"I(6)";
TAB(63);"X(6)";
4340 LPRINT TAB(68);"X(5)";TAB(73);"Vin";TAB(77);"X(1)";TAB(84);"x(2)";TAB(91);"
X(3)";TAB(96);"Del 1";TAB(102);"Del 2";TAB(108);"Del 3";TAB(114);"IL1";TAB(118);
"IL2";TAB(122);"IL3";TAB(126);"IG1";TAB(130);"IG2";TAB(134);"IG3":LPRINT
4350 RETURN
4400 REM Line Current Conversion Subroutine
4410 IF TS=<2 THEN X(3)=-X(3):E(3)=-E(3):RETURN
4420 IF TS=<4 THEN X(2)=-X(2):X(3)=-X(3):E(2)=-E(2):E(3)=-E(3):RETURN
4430 IF TS=<6 THEN X(2)=-X(2):E(2)=-E(2):RETURN
4440 IF TS=<8 THEN X(1)=-X(1):X(2)=-X(2):E(1)=-E(1):E(2)=-E(2):RETURN
4450 IF TS=<10 THEN X(1)=-X(1):E(1)=-E(1):RETURN
4460 IF TS=<12 THEN X(3)=-X(3):X(1)=-X(1):E(3)=-E(3):E(1)=-E(1):RETURN


10 REM APPENDIX.18G -- 12 Pulse F.W.P.S. --
20 DEFDBL B-O,Q-X
30 GOSUB 1850
40 PI=3.141592654#:SC=-1:SF=5
50 F=62.5:A1=PI/6:A2=5*PI/6:A3=3*PI/2:A4=2*PI/3:A5=4*PI/3
60 W=2*PI*F
70 LPRINT CHR$(27);"@";CHR$(15);CHR$(27);"A";CHR$(8);CHR$(27);"M";CHR$(1);CHR$(2
7);"N";CHR$(8);CHR$(27);"G":WIDTH "LPT1:",136
80 DIM X(49)
90 DIM D(49)
100 DIM E(49)
110 DIM F(49)
120 DIM I(49)
130 N=10
140 T1=.00005#:TP=T1
150 T2=.0002#
160 T3=.024#
170 T=0#
180 P=0
190 X(1)=0:D(1)=0
200 X(2)=19.23595:D(2)=-3239.642:X(3)=19.23595:D(3)=-3239.642
210 X(4)=9890.909:D(4)=-490917.4:X(5)=20.22405:D(5)=161.8497:X(6)=19243.89:D(6)=
-688.9628:X(7)=-7.345432:D(7)=969.169:X(8)=7.345432:D(8)=939.169:X(9)=14.56383:D
(9)=1938.099:X(10)=9309.999:D(10)=860190.4:TS=12:TSD=14
220 K=W*T:VP=6046:V1=VP*SIN(A1+K):V7=V1*SQR(3)
230 V2=VP*SIN(A2+K):V8=V2*SQR(3)
240 V3=VP*SIN(A3+K):V9=V3*SQR(3)
250 L=2
260 LL=.2094/4:LLD=3*LL
270 C=.00005:CI=.000002:CID=CI
280 R=950:RI=500:RID=RI
290 VPD=VP*SQR(3)
300 LPRINT"      --- 12 Pulse Rectifier Integration Time =";T1;" Seconds ---":LPRI
NT
310 GOSUB 940
320 REM Y Integration Subroutine
330 GOSUB 660
340 FOR A = 1 TO 4
350 E(A)=X(A)
```

```
360 F(A)=D(A)
370 X(A)=X(A)+T1*D(A)
380 NEXT A
390 T=T+T1:K=W*T:V1=VP*SIN(A1+K):V2=VP*SIN(A2+K):V3=VP*SIN(A3+K)
400 IF TS=<2 THEN GOSUB 750 ELSE IF TS=<4 THEN GOSUB 760 ELSE IF TS=<6 THEN GOSU
B 770 ELSE IF TS=<8 THEN GOSUB 780 ELSE IF TS=<10 THEN GOSUB 790 ELSE IF TS=< 12
 THEN GOSUB 800
410 FOR A = 1 TO 4
420 X(A)=E(A)+T1/2*(F(A)+D(A))
430 NEXT A
440 IF X(I1)=>0 THEN GOTO 510
450 IF X(1)<0 THEN B=1 ELSE IF X(2)<0 THEN B=2 ELSE IF X(3)<0 THEN B=3
460 T=T-T1:TC=E(B)*T1/(E(B)-X(B))
470 FOR A=1 TO 4:X(A)=E(A)+TC*(X(A)-E(A))/T1:NEXT A
480 T=T+TC:X(B)=0:D(B)=0:K=W*T:V1=VP*SIN(A1+K):V2=VP*SIN(A2+K):V3=VP*SIN(A3+K)
490 IF TS=<2 THEN GOSUB 750 ELSE IF TS=<4 THEN GOSUB 760 ELSE IF TS=<6 THEN GOSU
B 770 ELSE IF TS=<8 THEN GOSUB 780 ELSE IF TS=<10 THEN GOSUB 790 ELSE IF TS=< 12
 THEN GOSUB 800
500 TX=TP-TC:FOR A=1 TO 4:X(A)=X(A)+TX*D(A):NEXT A:T=T+TX:B=0
510 IF X(1)<0 THEN X(1)=0
520 IF X(2)<0 THEN X(2)=0
530 IF X(3)<0 THEN X(3)=0
540 T=T-T1:GOSUB 1930:P=P+T1
550 IF P+T1/5>T2 THEN GOTO 570
560 GOTO 330
570 P=0
580 REM PRINTOUT ROUTINE
590 GOSUB 980
600 IF T+T1/5>T3 THEN GOTO 620
610 GOTO 330
620 IF A$="P" OR A$="p" THEN GOSUB 1600:CLS:END
630 LPRINT
640 LPRINT"--------------------------------------------------------------
----------":END
650 CLS:END
660 REM Commutation Subroutine
670 IF V1=>V2-LL*D(2) AND -V2<-V3-LL*D(3) THEN GOSUB 750:RETURN
680 IF -V2=>-V3-LL*D(3) AND V3<V1-LL*D(1) THEN GOSUB 760:RETURN
690 IF V3=>V1-LL*D(1) AND -V1<-V2-LL*D(2) THEN GOSUB 770:RETURN
700 IF -V1=>-V2-LL*D(2) AND V2<V3-LL*D(3) THEN GOSUB 780:RETURN
710 IF V2=>V3-LL*D(3) AND -V3<-V1-LL*D(1) THEN GOSUB 790:RETURN
720 IF -V3=>-V1-LL*D(1) AND V1<V2-LL*D(2) THEN GOSUB 800:RETURN
730 STOP
740 REM Equation Subroutine
750 I1=2:I2=1:I3=3:E1=V1:E2=V2:E3=V3:TS=1:GOTO 820
760 I1=3:I2=2:I3=1:E1=-V2:E2=-V3:E3=-V1:TS=3:GOTO 820
770 I1=1:I2=3:I3=2:E1=V3:E2=V1:E3=V2:TS=5:GOTO 820
780 I1=2:I2=1:I3=3:E1=-V1:E2=-V2:E3=-V3:TS=7:GOTO 820
790 I1=3:I2=2:I3=1:E1=V2:E2=V3:E3=V1:TS=9:GOTO 820
800 I1=1:I2=3:I3=2:E1=-V3:E2=-V1:E3=-V2:TS=11:GOTO 820
810 STOP
820 REM Dual Rectification Path Equations
830 IF X(I1)=0 THEN GOTO 890
840 IF X(4)=>0 THEN D(4)=(X(I2)+X(I1)-X(5))/CI ELSE D(4)=0
850 D(I2)=(2*E1-E2-E3-CI*RI*D(4)-X(4))/(3*LL)
860 D(I1)=(2*E2-E1-E3-CI*RI*D(4)-X(4))/(3*LL)
870 D(I3)=D(I2)+D(I1)
880 RETURN
```

```
890 REM Single Rectification Path Equations
900 IF X(4)=>0 THEN D(4)=(X(I2)-X(5))/CI ELSE D(4)=0
910 D(I2)=(E1-E3-CI*RI*D(4)-X(4))/(2*LL)
920 D(I3)=D(I2):D(I1)=0:X(I1)=0:X(I3)=X(I2)
930 TS=TS+1:RETURN
940 REM Print Subroutine
950 IF A$="P" OR A$="p" THEN WIDTH "LPT1:",140:GOTO 980
960 LPRINT TAB(1);"Sec.    TS";TAB(14);"V1";TAB(20);"V2";TAB(26);"V3";TAB(31);"I(
1)";TAB(36);"I(2)";TAB(41);"I(3)";TAB(46);"I(4)";TAB(51);"I(5)";TAB(56);"I(6)";T
AB(61);"I(7)";
970 LPRINT TAB(66);"I(8)";TAB(71);"I(9)";TAB(75);"I(10)";TAB(80);"I(11)";TAB(85)
;"I(12)";TAB(92);"X(5)";TAB(100);"X(6)";TAB(106);"X(4)";TAB(111);"X(10)";TAB(120
);"IL1";TAB(126);"IL2";TAB(132);"IL3":LPRINT
980 FOR Z=1 TO 12:I(Z)=0:NEXT Z
990 IF TS=1 THEN I(1)=X(1):I(2)=X(2):I(6)=X(1)+X(2):GOTO 1110
1000 IF TS=2 THEN I(1)=X(1):I(2)=X(2):I(6)=X(1):GOTO 1110
1010 IF TS=3 THEN I(1)=X(2)+X(3):I(5)=X(2):I(6)=X(3):GOTO 1110
1020 IF TS=4 THEN I(1)=X(2):I(5)=X(2):I(6)=X(3):GOTO 1110
1030 IF TS=5 THEN I(1)=X(1):I(3)=X(3):I(5)=X(1)+X(3):GOTO 1110
1040 IF TS=6 THEN I(1)=X(1):I(3)=X(3):I(5)=X(3):GOTO 1110
1050 IF TS=7 THEN I(3)=X(1)+X(2):I(4)=X(1):I(5)=X(2):GOTO 1110
1060 IF TS=8 THEN I(3)=X(1):I(4)=X(1):I(5)=X(2):GOTO 1110
1070 IF TS=9 THEN I(2)=X(2):I(3)=X(3):I(4)=X(2)+X(3):GOTO 1110
1080 IF TS=10 THEN I(2)=X(2):I(3)=X(3):I(4)=X(2):GOTO 1110
1090 IF TS=11 THEN I(2)=X(1)+X(3):I(4)=X(1):I(6)=X(3):GOTO 1110
1100 IF TS=12 THEN I(2)=X(3):I(4)=X(1):I(6)=X(3)
1110 IF TSD=13 THEN I(8)=X(8)+X(9):I(10)=X(8)-X(7):I(12)=X(7)+X(9):GOTO 1230
1120 IF TSD=14 THEN I(8)=X(8)+X(9):I(12)=X(7)+X(9):GOTO 1230
1130 IF TSD=15 THEN I(7)=X(7)+X(8):I(8)=X(9)-X(8):I(12)=X(7)+X(9):GOTO 1230
1140 IF TSD=16 THEN I(7)=X(7)+X(8):I(12)=X(7)+X(9):GOTO 1230
1150 IF TSD=17 THEN I(7)=X(7)+X(8):I(11)=X(8)+X(9):I(12)=X(7)-X(9):GOTO 1230
1160 IF TSD=18 THEN I(7)=X(7)+X(8):I(11)=X(8)+X(9):GOTO 1230
1170 IF TSD=19 THEN I(7)=X(8)-X(7):I(9)=X(7)+X(9):I(11)=X(8)+X(9):GOTO 1230
1180 IF TSD=20 THEN I(9)=X(7)+X(9):I(11)=X(8)+X(9):GOTO 1230
1190 IF TSD=21 THEN I(9)=X(7)+X(9):I(10)=X(7)+X(8):I(11)=X(9)-X(8):GOTO 1230
1200 IF TSD=22 THEN I(9)=X(7)+X(9):I(10)=X(7)+X(8):GOTO 1230
1210 IF TSD=23 THEN I(8)=X(8)+X(9):I(9)=X(7)-X(9):I(10)=X(7)+X(8):GOTO 1230
1220 IF TSD=24 THEN I(8)=X(8)+X(9):I(10)=X(7)+X(8)
1230 IF B=>1 THEN GOTO 1320
1240 IF A$="P" OR A$="p" THEN GOTO 1260
1250 GOTO 1320
1260 IF T=0# THEN GOTO 1270 ELSE GOTO 1300
1270 INPUT"Do you want a dc output voltage or a line current plot (Enter V or L)
";P$
1280 IF P$="L" OR P$="l" THEN GOSUB 2560 ELSE GOSUB 1560
1290 RETURN
1300 IF P$="L" OR P$="l" THEN GOSUB 2610 ELSE GOSUB 1620
1310 RETURN
1320 LPRINT TAB(1);:LPRINT USING".####";T;:LPRINT TAB(6);:LPRINT USING"##";TS;:L
PRINT TAB(9);:LPRINT USING"##";TSD;
1330 LPRINT TAB(12);:LPRINT USING"#####";V1;
1340 LPRINT TAB(18);:LPRINT USING"#####";V2;
1350 LPRINT TAB(24);:LPRINT USING"#####";V3;
1360 LPRINT TAB(30);:LPRINT USING"###.#";I(1);
1370 LPRINT TAB(35);:LPRINT USING"###.#";I(2);
1380 LPRINT TAB(40);:LPRINT USING"###.#";I(3);
1390 LPRINT TAB(45);:LPRINT USING"###.#";I(4);
1400 LPRINT TAB(50);:LPRINT USING"###.#";I(5);
```

```
1410 LPRINT TAB(55);:LPRINT USING"###.#";I(6);
1420 LPRINT TAB(60);:LPRINT USING"###.#";I(7);
1430 LPRINT TAB(65);:LPRINT USING"###.#";I(8);
1440 LPRINT TAB(70);:LPRINT USING"###.#";I(9);
1450 LPRINT TAB(75);:LPRINT USING"###.#";I(10);
1460 LPRINT TAB(80);:LPRINT USING"###.#";I(11);
1470 LPRINT TAB(85);:LPRINT USING"###.#";I(12);
1480 LPRINT TAB(91);:LPRINT USING"###.#";X(5);
1490 LPRINT TAB(97);:LPRINT USING"#####.#";X(6);
1500 LPRINT TAB(105);:LPRINT USING"#####";X(4);
1510 LPRINT TAB(111);:LPRINT USING"#####";X(10);
1520 GOSUB 2430:LPRINT TAB(117);:LPRINT USING"####.#";X(2)+X(8)-X(1)-X(7);
1530 LPRINT TAB(123);:LPRINT USING"####.#";X(3)+X(9)-X(2)-X(8);
1540 LPRINT TAB(129);:LPRINT USING"####.#";X(1)+X(7)-X(3)-X(9):GOSUB 2430
1550 RETURN
1560 REM Plotting Subroutine
1570 LPRINT"                          Transient Response In kV For A Three Pha
se 12 Pulse DC Power Supply"
1580 LPRINT"                          --------- -------- -- -- --- - ----- ---
-- -- ----- -- ----- ------"
1590 LPRINT"       0          1          2          3          4          5
6          7          8          9         10         11         12"
1600 LPRINT"       |....+....|....+....|....+....|....+....|....+....|....+....
|....+....|....+....|....+....|....+....|....+....|....+....|":IF T>0 THEN RETUR
N
1610 LPRINT TAB(1);"Vdc Out    A./Rect.-->";TAB(23);"I(1)";TAB(28);"I(2)";TAB(33
);"I(3)";TAB(38);"I(4)";TAB(43);"I(5)";TAB(48);"I(6)";TAB(53);"I(7)";TAB(58);"I(
8)";TAB(63);"I(9)";TAB(68);"I(10)";TAB(74);"I(11)";TAB(80);"I(12)"
1620 SC=SC+1
1630 IF T>.016025 AND T<.016225 THEN GOTO 1640 ELSE GOTO 1660
1640 LPRINT"        |....+....|....+....|....+....|....+....|....+....|....+....
|....+....|....+....|....+....|....+....|....+....|....+....|":LPRINT:LPRINT
1642 LPRINT"     19243.0    19243.2    19243.4    19243.6    19243.8    19244.0    192
44.2    19244.4    19244.6    19244.8    19245.0    19245.2    19245.4":T2=.0001
1650 LPRINT"        |....+....|....+....|....+....|....+....|....+....|....+....
|....+....|....+....|....+....|....+....|....+....|....+....|"
1660 IF T>.016025 THEN PP=X(6)#50-962141!:PP1=0:PP2=0:GOTO 1690
1670 PP1=(9+(X(4)+RI#CI#D(4))/100)
1680 PP2=(9+(X(10)+RID#CID#D(10))/100)
1690 IF SC<5 THEN LPRINT USING "#####.##";X(6);:LPRINT TAB(9);"-";
1700 IF SC<5 AND T>.016025 THEN GOTO 1770
1710 IF SC<5 THEN GOTO 1730
1720 LPRINT USING "#####.##";X(6);:LPRINT TAB(9);"--T=";:LPRINT USING".####";T;:
LPRINT" Sec.";:SC=0:IF T>.016025 THEN GOTO 1780
1730 LPRINT TAB(23);:LPRINT USING"##.#";I(1);:LPRINT TAB(28);:LPRINT USING"##.#"
;I(2);:LPRINT TAB(33);:LPRINT USING"##.#";I(3);:LPRINT TAB(38);:LPRINT USING"##.
#";I(4);:LPRINT TAB(43);:LPRINT USING"##.#";I(5);:LPRINT TAB(48);:LPRINT USING"#
#.#";I(6);
1740 LPRINT TAB(53);:LPRINT USING"##.#";I(7);:LPRINT TAB(58);:LPRINT USING"##.#"
;I(8);:LPRINT TAB(63);:LPRINT USING"##.#";I(9);:LPRINT TAB(69);:LPRINT USING"##.
#";I(10);:LPRINT TAB(75);:LPRINT USING"##.#";I(11);:LPRINT TAB(81);:LPRINT USING
"##.#";I(12);
1750 IF PP1 OR PP2 >85 THEN GOTO 1810 ELSE IF PP>85 THEN GOTO 1770
1760 LPRINT"":RETURN
1770 IF PP>10 THEN GOTO 1790 ELSE LPRINT"":RETURN
1780 IF PP<23 THEN LPRINT"":RETURN
1790 IF PP>130 THEN LPRINT TAB(129)"+->":RETURN
1800 LPRINT TAB(PP);"+":RETURN
```

```
1810 IF CINT(PP1)=CINT(PP2) THEN LPRINT TAB(PP1);"o":RETURN
1820 IF CINT(PP1)<CINT(PP2) THEN LPRINT TAB(PP1);"x";TAB(PP2);"+":RETURN
1830 IF CINT(PP2)<CINT(PP1) THEN LPRINT TAB(PP2);"+";TAB(PP1);"x":RETURN
1840 LPRINT TAB(PP);"+":RETURN
1850 REM Printer On/Tabular Output/Plotted Output Subroutine
1860 CLS:PRINT"            Your printer must be on to run this program.  Is it on (
Y or N)?"
1870 A$=INKEY$:IF A$="" THEN 1870
1880 IF A$="n" OR A$="N" THEN PRINT"For operation without a printer go to DOS. T
ype INFORMATION": END
1890 PRINT"For a plotted response enter P.  For a tabular response enter any oth
er letter."
1900 A$=INKEY$:IF A$="" THEN 1900
1910 CLS:PRINT"                    PROGRAM IS IN PROGRESS
                                   ------- -- -- --------"
1920 RETURN
1930 REM D Integration Subroutine
1940 GOSUB 2120
1950 FOR A = 5 TO N
1960 E(A)=X(A)
1970 F(A)=D(A)
1980 X(A)=X(A)+T1*D(A)
1990 NEXT A
2000 T=T+T1:K=W*T:V7=VPD*SIN(A1+K):V8=VPD*SIN(A2+K):V9=VPD*SIN(A3+K)
2010 IF TSD=<14 THEN GOSUB 2210 ELSE IF TSD=<16 THEN GOSUB 2220 ELSE IF TSD=<18
THEN GOSUB 2230 ELSE IF TSD=<20 THEN GOSUB 2240 ELSE IF TSD=<22 THEN GOSUB 2250
ELSE IF TSD=<24 THEN GOSUB 2260
2020 FOR A = 5 TO N
2030 X(A)=E(A)+T1/2*(F(A)+D(A))
2040 NEXT A
2050 IF X(I8)=<X(I7) THEN GOTO 2110
2060 DEN=((E(I7)-E(I8))-(X(I7)-X(I8)))
2070 T=T-T1:TC=(E(I7)-E(I8))*T1/DEN:IF TC>T1 OR TC<.000001 THEN TC=T1/10
2080 FOR A=5 TO N:X(A)=E(A)+TC*(X(A)-E(A))/T1:NEXT A:T=T+TC:X(I8)=X(I7):D(I8)=D(
I7):K=W*T
2090 V7=VPD*SIN(A1+K):V8=VPD*SIN(A2+K):V9=VPD*SIN(A3+K):IF TSD=<14 THEN GOSUB 22
10 ELSE IF TSD=<16 THEN GOSUB 2220 ELSE IF TSD=<18 THEN GOSUB 2230 ELSE IF TSD=<
20 THEN GOSUB 2240 ELSE IF TSD=<22 THEN GOSUB 2250 ELSE IF TSD=<24 THEN GOSUB 22
60
2100 TX=TP-TC:FOR A=5 TO N:X(A)=X(A)+TX*D(A):NEXT A:T=T+TX
2110 RETURN
2120 REM Commutation Subroutine
2130 IF -V9-LLD*D(9)+1=>V8-LLD*D(8) AND V8-LLD*D(8)>0 AND -V9>0 THEN GOSUB 2210:
RETURN
2140 IF V7-LLD*D(7)+1=>-V9-LLD*D(9) AND -V9-LLD*D(9)>0 AND V7>0 THEN GOSUB 2220:
RETURN
2150 IF -V8-LLD*D(8)+1=>V7-LLD*D(7) AND V7-LLD*D(7)>0 AND -V8>0 THEN GOSUB 2230:
RETURN
2160 IF V9-LLD*D(9)+1=>-V8-LLD*D(8) AND -V8-LLD*D(8)>0 AND V9>0 THEN GOSUB 2240:
RETURN
2170 IF -V7-LLD*D(7)+1=>V9-LLD*D(9) AND V9-LLD*D(9)>0 AND -V7>0 THEN GOSUB 2250:
RETURN
2180 IF V8-LLD*D(8)+1=>-V7-LLD*D(7) AND -V7-LLD*D(7)>0 AND V8>0 THEN GOSUB 2260:
RETURN
2190 STOP
2200 REM Delta Equation Subroutine
2210 I7=8:I8=7:I9=9:E7= V7:E8= V8:E9= V9:TSD=13:GOTO 2270
2220 I7=9:I8=8:I9=7:E7=-V8:E8=-V9:E9=-V7:TSD=15:GOTO 2270
```

```
2230 I7=7:I8=9:I9=8:E7= V9:E8= V7:E9= V8:TSD=17:GOTO 2270
2240 I7=8:I8=7:I9=9:E7=-V7:E8=-V8:E9=-V9:TSD=19:GOTO 2270
2250 I7=9:I8=8:I9=7:E7= V8:E8= V9:E9= V7:TSD=21:GOTO 2270
2260 I7=7:I8=9:I9=8:E7=-V9:E8=-V7:E9=-V8:TSD=23
2270 REM Dual Rectification Path Equations
2280 IF TSD=TSDP THEN GOTO 2290 ELSE X(I8)=-X(I8):TSDP=TSD
2290 D(10)=(X(I7)+X(I9)-X(5))/CID
2300 D(5)=(X(4)+RI#CI#D(4)+X(10)+RID#CID#D(10)-X(6))/L
2310 D(6)=(X(5)-X(6)/R)/C
2320 IF X(5)<0 AND X(4)+RI#CI#D(4)+X(10)+RID#CID#D(10)>E1-E3-E9 THEN X(7)=0:X(8)
=0:X(9)=0:D(7)=0:D(8)=0:D(9)=0:RETURN
2330 IF X(7)=0 AND X(8)=0 AND X(9)=0 AND X(4)+RI#CI#D4+X(10)+RID#CID#D(10)>E1-E3
-E9 THEN X(7)=0:X(8)=0:X(9)=0:D(7)=0:D(8)=0:D(9)=0:RETURN
2340 IF X(I8)=>X(I7) THEN GOTO 2380
2350 D(I7)=(E8-X(10)-RID#CID#D(10))/LLD
2360 D(I9)=(-E9-X(10)-RID#CID#D(10))/LLD:D(I8)=E7/LLD
2370 RETURN
2380 REM Single Rectification Path Equations
2390 D(I9)=(-E9-X(10)-RID#CID#D(10))/LLD
2400 D(I8)=(E7+E8-X(10)-RID#CID#D(10))/(2#LLD):D(I7)=D(I8)
2410 TSD=TSD+1
2420 RETURN
2430 REM Line Current Conversion Subroutine
2440 IF TS=<2 THEN X(3)=-X(3):GOTO 2500
2450 IF TS=<4 THEN X(2)=-X(2):X(3)=-X(3):GOTO 2500
2460 IF TS=<6 THEN X(2)=-X(2):GOTO 2500
2470 IF TS=<8 THEN X(1)=-X(1):X(2)=-X(2):GOTO 2500
2480 IF TS=<10 THEN X(1)=-X(1):GOTO 2500
2490 IF TS=<12 THEN X(3)=-X(3):X(1)=-X(1)
2500 IF TSD=<14 THEN X(9)=-X(9):RETURN
2510 IF TSD=<16 THEN X(8)=-X(8):X(9)=-X(9):RETURN
2520 IF TSD=<18 THEN X(8)=-X(8):RETURN
2530 IF TSD=<20 THEN X(7)=-X(7):X(8)=-X(8):RETURN
2540 IF TSD=<22 THEN X(7)=-X(7):RETURN
2550 IF TSD=<24 THEN X(9)=-X(9):X(7)=-X(7):RETURN
2560 REM Line Current Plotting Subroutine
2570 LPRINT"                           Transient Response In Amperes For Three
Phase 12 Pulse Power Supply IL1"
2580 LPRINT"                 --------- -------- -- ------- --- -----
----- -- ----- ----- ------ ---"
2590 LPRINT"     -120     -100      -80      -60      -40      -20
0       20       40       60       80       100      120"
2600 LPRINT"      !....+....!....+....!....+....!....+....!....+....!....+....
!....+....!....+....!....+....!....+....!....+....!....+....!":SF=2:IF T>0 THEN
RETURN
2610 SC=SC+1
2650 GOSUB 2430:IL1=X(2)+X(8)-X(1)-X(7):GOSUB 2430:PP=69+IL1/SF
2660 IF SC<5 THEN LPRINT USING "####.##";IL1;:LPRINT TAB(9);"-";
2670 IF SC<5 THEN GOTO 2690
2680 LPRINT USING "####.##";IL1;:LPRINT TAB(9);"--T=";:LPRINT USING".####";T;:LP
RINT" Sec.";:SC=0:IF PP<23 THEN LPRINT"":RETURN
2690 IF PP<10 THEN LPRINT"":RETURN
2700 IF PP>130 THEN LPRINT TAB(129)"+->":RETURN
2710 LPRINT TAB(PP);"+":RETURN



10 REM APPENDIX.18U -- Non-sync. 3 Ph. Y-Delta UPS Driving A Resistance Load --
11 CLEAR ,49152!,3000
```

```
12 GOSUB 3500
13 PI=3.141592654#:SC=-1:CP=.016:TF=0
21 LPRINT CHR$(27);"@";CHR$(15);CHR$(27);"A";CHR$(8);CHR$(27);"M";CHR$(10);CHR$(
27);"G";CHR$(27);"N";CHR$(1);:WIDTH "LPT1:",136
22 DIM X(49)
24 DIM D(49)
26 DIM E(49)
28 DIM F(49)
30 N=3
32 T1=.00005:TP=T1
34 T2=.0002:RIN=CP/T2:RIN1=CP/(4#T2)
36 T3=.0322
38 T=0
40 P=0
50 X(1)=0
51 X(2)=281:X(3)=281:VDC=176
58 L=0
60 LL=.0001467
64 RA=.05:RL=.576
66 V1=RL#X(1):V2=RL#X(2):V3=-V1-V2
80 LPRINT"       --- 3 Phase Bridge Rectifier Integration Time =";T1;" Seconds ---"
:LPRINT
84 TS=1
86 GOSUB 2000
100 REM Integration Routine
110 IF T=>(CCY+1)#CP THEN CCY=CCY+1:TCY=CCY#CP
115 GOSUB 404
120 FOR A = 1 TO N
130 E(A)=X(A):VP1=V1:VP2=V2:VP3=V3
140 F(A)=D(A)
150 X(A)=X(A)+T1#D(A)
160 NEXT A
165 T=T+T1
170 IF TS=1 THEN GOSUB 1013 ELSE IF TS=2 THEN GOSUB 1020 ELSE IF TS=3 THEN GOSUB
 1033 ELSE IF TS=4 THEN GOSUB 1040 ELSE IF TS=5 THEN GOSUB 1053 ELSE IF TS=6 THE
N GOSUB 1060 ELSE IF TS=7 THEN GOSUB 1073 ELSE IF TS=8 THEN GOSUB 1080
171 IF TS=9 THEN GOSUB 1093 ELSE IF TS=10 THEN GOSUB 1100 ELSE IF TS=11 THEN GOS
UB 1113 ELSE IF TS=12 THEN GOSUB 1120
180 FOR A=1 TO N
190 X(A)=E(A)+T1/2#(F(A)+D(A))
200 NEXT A
202 IF X(1)=>0 AND X(2)=>0 AND X(3)=>0 THEN GOTO 212
204 IF X(1)<0 THEN B=1 ELSE IF X(2)<0 THEN B=2 ELSE IF X(3)<0 THEN B=3
206 T=T-T1:TC=E(B)#T1/(E(B)-X(B)):FOR A=1 TO N:X(A)=E(A)+TC#F(A):NEXT A:T=T+TC:X
(B)=0:D(B)=0
208 IF TS=<2 THEN GOSUB 1013 ELSE IF TS=<4 THEN GOSUB 1033 ELSE IF TS=<6 THEN GO
SUB 1053 ELSE IF TS=<8 THEN GOSUB 1073 ELSE IF TS=<10 THEN GOSUB 1093 ELSE IF TS
=< 12 THEN GOSUB 1113
210 TX=TP-TC:FOR A=1 TO N:X(A)=X(A)+TX#D(A):NEXT A:T=T+TX:B=0
212 IF X(1)<0 THEN X(1)=0 AND D(1)=0
214 IF X(2)<0 THEN X(2)=0 AND D(2)=0
216 IF X(3)<0 THEN X(3)=0 AND D(3)=0
218 GOSUB 4400:V1=RL#X(1):V2=RL#X(2):V3=RL#X(3):GOSUB 4400
222 P=P+T1
230 IF P+T1/5>T2 THEN GOTO 250
240 GOTO 110
250 P=0
260 REM PRINTOUT ROUTINE
```

```
270 GOSUB 2013
280 IF T+T1/5>T3 THEN GOTO 291
290 GOTO 110
291 IF A$="P" OR A$="p" THEN GOSUB 3030:CLS:END
294 LPRINT"----------------------------------------------------------------
------------------------------------------------------------------":END
300 CLS:END
400 REM Switching Subroutine
404 IF T-TCY=>0 AND T-TCY<CP/6 THEN GOSUB 1010:RETURN
405 IF T-TCY=>CP/6 AND T-TCY<CP/3 THEN GOSUB 1030:RETURN
406 IF T-TCY=>CP/3 AND T-TCY<CP/2 THEN GOSUB 1050:RETURN
407 IF T-TCY=>CP/2 AND T-TCY<2*CP/3 THEN GOSUB 1070:RETURN
408 IF T-TCY=>2*CP/3 AND T-TCY<5*CP/6 THEN GOSUB 1090:RETURN
409 IF T-TCY=>5*CP/6 AND T-TCY<CP THEN GOSUB 1110:RETURN
410 STOP
1000 REM Equation Subroutine
1010 REM Forward paths 1 & 2. Return path 3
1011 IF T-TCY>0 AND T-TCY=<TF THEN GOTO 1012 ELSE GOTO 1013
1012 D(1)=0:D(2)=-RL*X(2)/LL:D(3)=-RL*X(3)/LL:RETURN
1013 D(2)=-RL*X(2)/LL
1015 D(1)=(VDC-(RL+RA)*X(1))/(L+LL)
1016 D(3)=(RL*(X(1)-X(3))+LL*D(1))/LL
1017 IF X(2)=0 THEN TS=2 ELSE TS=1:RETURN
1030 REM Forward path 1.  Return paths 5 & 6.
1031 IF T-TCY>CP/6 AND T-TCY=<CP/6+TF THEN GOTO 1032 ELSE GOTO 1033
1032 D(2)=0:D(3)=-RL*X(3)/LL:D(1)=-RL*X(1)/LL:RETURN
1033 D(3)=-RL*X(3)/LL
1035 D(2)=(VDC-(RL+RA)*X(2))/(L+LL)
1036 D(1)=(RL*(X(2)-X(1))+LL*D(2))/LL
1037 IF X(3)=0 THEN TS=4 ELSE TS=3:RETURN
1050 REM Forward paths 1 & 3.  Return path 5.
1051 IF T-TCY>CP/3 AND T-TCY=<CP/3+TF THEN GOTO 1052 ELSE GOTO 1053
1052 D(3)=0:D(1)=-RL*X(1)/LL:D(2)=-RL*X(2)/LL:RETURN
1053 D(1)=-RL*X(1)/LL
1055 D(3)=(VDC-(RL+RA)*X(3))/(L+LL)
1056 D(2)=(RL*(X(3)-X(2))+LL*D(3))/LL
1057 IF X(1)=0 THEN TS=6 ELSE TS=5:RETURN
1070 REM Forward path 1. Return paths 2 and 3.
1071 IF T-TCY>CP/2 AND T-TCY=<CP/2+TF THEN GOTO 1072 ELSE GOTO 1073
1072 D(1)=0:D(2)=-RL*X(2)/LL:D(3)=-RL*X(3)/LL:RETURN
1073 D(2)=-RL*X(2)/LL
1075 D(1)=(VDC-(RL+RA)*X(1))/(L+LL)
1076 D(3)=(RL*(X(1)-X(3))+LL*D(1))/LL
1077 IF X(2)=0 THEN TS=8 ELSE TS=7:RETURN
1090 REM Forward paths 2 & 3.  Return path 4.
1091 IF T-TCY>2*CP/3 AND T-TCY=<2*CP/3+TF THEN GOTO 1092 ELSE GOTO 1093
1092 D(2)=0:D(3)=-RL*X(3)/LL:D(1)=-RL*X(1)/LL:RETURN
1093 D(3)=-RL*X(3)/LL
1095 D(2)=(VDC-(RL+RA)*X(2))/(L+LL)
1096 D(1)=(RL*(X(2)-X(1))+LL*D(2))/LL
1097 IF X(3)=0 THEN TS=10 ELSE TS=9:RETURN
1110 REM Forward path 2.  Return paths 4 & 6.
1111 IF T-TCY>5*CP/6 AND T-TCY=<5*CP/6+TF THEN GOTO 1112 ELSE GOTO 1113
1112 D(3)=0:D(1)=-RL*X(1)/LL:D(2)=-RL*X(2)/LL:RETURN
1113 D(1)=-RL*X(1)/LL
1115 D(3)=(VDC-(RL+RA)*X(3))/(L+LL)
1116 D(2)=(RL*(X(3)-X(2))+LL*D(3))/LL
1117 IF X(1)=0 THEN TS=12 ELSE TS=11:RETURN
```

```
2000 REM Print Subroutine
2002 IF A$="P" OR A$="p" THEN LPRINT CHR$(27);"G";:WIDTH "LPT1:",140:GOTO 2013
2013 FOR Z=1 TO 7:I(Z)=0:NEXT Z
2014 IF TS=1 THEN I(1)=X(1):I(2)=X(3):I(6)=X(1)+X(3):GOTO 2026
2015 IF TS=2 THEN I(1)=X(1):I(2)=X(3):I(6)=X(1)+X(3):GOTO 2026
2016 IF TS=3 THEN I(1)=X(1)+X(2):I(5)=X(2):I(6)=X(1):GOTO 2026
2017 IF TS=4 THEN I(1)=X(1)+X(2):I(5)=X(2):I(6)=X(1):GOTO 2026
2018 IF TS=5 THEN I(1)=X(2):I(3)=X(3):I(5)=X(2)+X(3):GOTO 2026
2019 IF TS=6 THEN I(1)=X(2):I(3)=X(3):I(5)=X(2)+X(3):GOTO 2026
2020 IF TS=7 THEN I(3)=X(1)+X(3):I(4)=X(1):I(5)=X(3):GOTO 2026
2021 IF TS=8 THEN I(3)=X(1)+X(3):I(4)=X(1):I(5)=X(3):GOTO 2026
2022 IF TS=9 THEN I(2)=X(2):I(3)=X(1):I(4)=X(1)+X(2):GOTO 2026
2023 IF TS=10 THEN I(2)=X(2):I(3)=X(1):I(4)=X(1)+X(2):GOTO 2026
2024 IF TS=11 THEN I(2)=X(2)+X(3):I(4)=X(2):I(6)=X(3):GOTO 2026
2025 IF TS=12 THEN I(2)=X(2)+X(3):I(4)=X(2):I(6)=X(3)
2026 IF T=>.016+T2-T1/2 THEN GOTO 2029 ELSE GOTO 2036
2028 GOTO 2036
2029 IF T>.016+T2-T1/2 AND T<.016+T2+T1/2 THEN LPRINT:LPRINT:LPRINT:LPRINT:LPRIN
T:LPRINT:LPRINT:LPRINT:LPRINT:LPRINT:LPRINT:LPRINT:LPRINT:LPRINT:LPRINT:A$="P":G
OSUB 3000:RETURN
2030 GOSUB 3040:RETURN
2036 IF CO=0 THEN GOTO 2038
2037 GOTO 2040
2038 CY=CY+1
2039 IF HEAD=0 THEN GOSUB 4300
2040 LPRINT TAB(1);:LPRINT USING".#####";T;:LPRINT;TS;
2041 LPRINT TAB(12);:LPRINT USING"#####";V1;
2042 LPRINT TAB(18);:LPRINT USING"#####";V2;
2043 LPRINT TAB(24);:LPRINT USING"#####";V3;
2044 LPRINT TAB(30);:LPRINT USING"#####";I(1);
2045 LPRINT TAB(35);:LPRINT USING"#####";I(2);
2046 LPRINT TAB(40);:LPRINT USING"#####";I(3);
2047 LPRINT TAB(45);:LPRINT USING"#####";I(4);
2048 LPRINT TAB(50);:LPRINT USING"#####";I(5);
2049 LPRINT TAB(55);:LPRINT USING"#####";I(6);
2050 LPRINT TAB(61);:LPRINT USING"#####";VDC;
2051 LPRINT TAB(66);:LPRINT USING"#####";IDC;
2052 LPRINT TAB(71);:LPRINT USING"#####";RI#CI#D(4)+X(4);
2053 GOSUB 4400:LPRINT TAB(76);:LPRINT USING"#####";X(1);
2054 LPRINT TAB(83);:LPRINT USING"#####";X(2);
2055 LPRINT TAB(90);:LPRINT USING"#####";X(3):GOSUB 4400
2100 REM RMS Subroutine
2110 CO=CO+1
2120 RMS1=RMS1+V1^2:RMS2=RMS2+V2^2:RMS3=RMS3+V3^2
2122 RMS4=RMS4+(V1-V3)^2:RMS5=RMS5+(V2-V1)^2:RMS6=RMS6+(V3-V2)^2
2130 IF CO<RIN THEN RETURN
2140 RMS1=SQR(RMS1/RIN):RMS2=SQR(RMS2/RIN):RMS3=SQR(RMS3/RIN)
2142 RMS4=SQR(RMS4/RIN):RMS5=SQR(RMS5/RIN):RMS6=SQR(RMS6/RIN)
2150 LPRINT" Cycle";CY;" Line To Neutral RMS Values Are: V1 =";RMS1;" V2 =";RM
S2;" V3 =";RMS3;" VLNavg =";(RMS1+RMS2+RMS3)/3
2152 LPRINT" Cycle";CY;" Line To Line RMS Values Are: V13=";RMS4;" V21=";RMS5;
" V32=";RMS6;" VLLavg =";(RMS4+RMS5+RMS6)/3
2155 LPRINT:RMS1=0:RMS2=0:RMS3=0:RMS4=0:RMS5=0:RMS6=0:CO=0:RETURN
3000 REM Plotting Subroutine
3010 LPRINT"                        AC Line To Line Voltage Generated By An
Uninterruptable 6 Pulse System
3015 LPRINT"                                  -- ---- -- ---- ------- --------- -- --
---------------- - ----- ------"
```

```
3020 LPRINT"                    -500     -400     -300     -200     -100     
0     100       200      300      400      500"
3022 LPRINT" X(1)      V1-V3    !....+....!....+....!....+....!....+....!....+....
!....+....!....+....!....+....!....+....!....+....!":IF T>.016+T2+T1/2 THEN RETU
RN
3026 GOTO 3040
3030 LPRINT"                             !....+....!....+....!....+....!....+....
!....+....!....+....!....+....!....+....!....+....!....+....!":RETURN
3040 SC=SC+1
3044 PP=69+(V1-V3)/10
3060 IF SC<5 THEN LPRINT TAB(1);:LPRINT USING"####.#";X(1);:LPRINT TAB(10);:LPRI
NT USING"####.#";V1-V3;:LPRINT TAB(19);"-";
3062 IF SC<5 THEN GOTO 3100
3070 LPRINT TAB(1);:LPRINT USING"####.#";X(1);:LPRINT TAB(10);:LPRINT USING"####
.#";V1-V3;:LPRINT TAB(19);"--";:LPRINT TAB(21);:LPRINT USING"###.#";T*1000;:SC=0
:GOTO 3100
3100 LPRINT TAB(PP);"+":RETURN
3500 REM Printer On/Tabular Output/Plotted Output Subroutine
3510 CLS:PRINT"         Your printer must be on to run this program.  Is it on (
Y or N)?"
3520 A$=INKEY$:IF A$="" THEN 3520
3530 IF A$="n" OR A$="N" THEN PRINT"For operation without a printer go to DOS. T
ype INFORMATION": END
3540 PRINT"For a plotted response enter P.  For a tabular response enter any oth
er letter."
3550 A$=INKEY$:IF A$="" THEN 3550
3560 CLS:PRINT"                    PROGRAM IS IN PROGRESS
                                    ------- -- -- --------"
3570 RETURN
4300 REM Tabular Heading
4310 HEAD=HEAD+1
4330 LPRINT TAB(1);"Sec.    TS";TAB(14);"V1";TAB(20);"V2";TAB(26);"V3";TAB(31);"I
(1)";TAB(36);"I(2)";TAB(41);"I(3)";TAB(46);"I(4)";TAB(51);"I(5)";TAB(56);"I(6)";
TAB(63);"VDC";
4340 LPRINT TAB(68);"";TAB(73);"";TAB(77);"X(1)";TAB(84);"x(2)";TAB(91);"X(3)":L
PRINT
4350 RETURN
4400 REM Line Current Conversion Subroutine
4410 IF TS=<2 THEN X(3)=-X(3):E(3)=-E(3):RETURN
4420 IF TS=<4 THEN X(2)=-X(2):X(3)=-X(3):E(2)=-E(2):E(3)=-E(3):RETURN
4430 IF TS=<6 THEN X(2)=-X(2):E(2)=-E(2):RETURN
4440 IF TS=<8 THEN X(1)=-X(1):X(2)=-X(2):E(1)=-E(1):E(2)=-E(2):RETURN
4450 IF TS=<10 THEN X(1)=-X(1):E(1)=-E(1):RETURN
4460 IF TS=<12 THEN X(3)=-X(3):X(1)=-X(1):E(3)=-E(3):E(1)=-E(1):RETURN


10 REM APPENDIX.19 -- DC-DC Converter Power Supply Transient Analysis --
11 GOSUB 3500
13 DIM X(49)
14 DIM D(49)
16 DIM E(49)
18 DIM F(49)
19 SC=-1
22 LPRINT CHR$(27);"@";CHR$(15);CHR$(27);"M";CHR$(3);CHR$(27);"Q";CHR$(128):WIDT
H "LPT1:",128
30 N=5
32 T1=1E-09
34 T2=.000001
```

```
36 T3=.00008
38 T=0
40 P=0
50 X(1)=10
52 X(2)=0
54 X(3)=0
56 X(4)=10
58 X(5)=0
60 LS=.005:LP=.005
62 M=.99975*LP
64 C=.000001
66 R=50
68 V=500
70 R5=1000000!:R4=1:RC=1
80 IF A$="P" OR A$="p" THEN GOTO 86:LPRINT"            -- DC-DC Converter Integrati
on Time =";T1;"Seconds --":LPRINT
86 SR=2:GOSUB 1000
90 GOSUB 2000
100 REM INTEGRATION ROUTINE
110 REM Equation Subroutine Selection
111 TR=TR+T1:IF TR>.00002 THEN TR=T1
112 IF TR=<.00001 THEN R5=RC:R4=1000000!:GOTO 115
113 IF TR=<.00002 THEN R4=RC:R5=1000000!
115 IF R4=RC AND X(2)>0 THEN SR=1:GOTO 119
116 IF R4=RC AND X(1)>=0 THEN SR=2:GOTO 119
117 IF R5=RC AND X(1)>0 THEN SR=3:GOTO 119
118 IF R5=RC AND X(2)>=0 THEN SR=4
119 GOSUB 1000
120 FOR A = 1 TO N
130 E(A)=X(A)
150 X(A)=X(A)+T1*D(A)
152 IF X(1)<0 THEN X(1)=0
154 IF X(2)<0 THEN X(2)=0
160 NEXT A
169 T=T+T1
170 GOSUB 1000
180 FOR A = 1 TO N
190 X(A)=E(A)+T1/2*(F(A)+D(A))
192 IF X(1)<0 THEN X(1)=0
194 IF X(2)<0 THEN X(2)=0
200 NEXT A
220 P=P+T1
230 IF P+T1/5>T2 THEN GOTO 250
240 GOTO 110
250 P=0
260 REM PRINTOUT ROUTINE
270 GOSUB 2010
280 IF T+T1/5>T3 THEN GOTO 292
290 GOTO 110
292 IF A$="P" OR A$="p" THEN GOSUB 3030:END
294 LPRINT"-----------------------------------------------------------
----------"
300 END
1000 REM Equation Subroutine
1005 IF SR=1 THEN GOSUB 1100:RETURN
1010 IF SR=2 THEN GOSUB 1110:RETURN
1015 IF SR=3 THEN GOSUB 1120:RETURN
1020 IF SR=4 THEN GOSUB 1130:RETURN
```

```
1100 REM --- If X(2)>0 and R4=RC ---
1101 X(4)=-X(5):T1=1E-09
1102 D(5)=0:X(5)=0:D(4)=(LS*V-LS*R4*X(4)+M*X(3))/(LS*LP-M^2)
1104 D(2)=(-X(3)-M*D(4))/LS
1106 D(3)=(X(2)-X(3)/R)/C:S$="a":RETURN
1110 REM --- If X(1)>0 and R4=RC ---
1111 T1=.0000001
1112 D(5)=0:X(5)=0:D(4)=(LS*V-LS*R4*X(4)-M*X(3))/(LS*LP-M^2)
1114 D(1)=(-X(3)+M*D(4))/LS
1116 D(3)=(X(1)-X(3)/R)/C:S$="b":RETURN
1120 REM --- If X(1)>0 and R5=RC ---
1121 X(5)=-X(4):T1=1E-09
1122 D(4)=0:X(4)=0:D(5)=(LS*V-LS*R5*X(5)+M*X(3))/(LS*LP-M^2)
1124 D(1)=(-X(3)-M*D(5))/LS
1126 D(3)=(X(1)-X(3)/R)/C:S$="c":RETURN
1130 REM --- If X(2)>0 and R5=RC ---
1131 T1=.0000001
1132 D(4)=0:X(4)=0:D(5)=(LS*V-LS*R5*X(5)-M*X(3))/(LS*LP-M^2)
1134 D(2)=(-X(3)+M*D(5))/LS
1136 D(3)=(X(2)-X(3)/R)/C:S$="d":RETURN
2000 REM PRINT SUBROUTINE
2001 LPRINT"DC-DC Power Supply Transient Analysis":LPRINT
2002 IF A$="P" OR A$="p" THEN LPRINT CHR$(27);"G":WIDTH "LPT1:",140:GOTO 2012
2004 LPRINT TAB(1);"MICROSEC.";TAB(11);"X(1)";TAB(18);"X(2)";TAB(25);"X(3)";TAB(
32);"X(4)";TAB(39);"X(5)"
2010 IF A$="P" OR A$="p" THEN GOTO 2012
2011 GOTO 2015
2012 IF T=0 THEN GOSUB 3000
2014 GOSUB 3040:RETURN
2015 LPRINT TAB(2);:LPRINT USING"####.###";T*1000000!;
2025 LPRINT TAB(11);:LPRINT USING"###.##";X(1);
2030 LPRINT TAB(18);:LPRINT USING"###.##";X(2);
2040 LPRINT TAB(25);:LPRINT USING"###.##";X(3);
2042 LPRINT TAB(32);:LPRINT USING"###.##";X(4);
2044 LPRINT TAB(39);:LPRINT USING"####.#";X(5);:LPRINT"   R4=";R4;": R5=";R5;": S
$=";S$
2050 RETURN
3000 REM Plotting Subroutine
3010 LPRINT"                              Transient Response For A Single P
hase DC-DC Converter Power Supply"
3015 LPRINT"                              --------- -------- --- - ------ -
---- ----- --------- ----- ------"
3020 LPRINT"                 0        100       200       300       400
  500       600       700       800       900      1000"
3030 LPRINT"                 |....+....|....+....|....+....|....+....|....+.
...|....+....|....+....|....+....|....+....|....+....|"
3032 LPRINT TAB(1);"X(3)V.";TAB(8);"X(1)A.";TAB(15);"X(2)A.":RETURN
3040 SC=SC+1
3044 PP=(22+X(3)/10)
3060 IF SC<5 THEN LPRINT USING"####.#";X(3);:LPRINT TAB(8);:LPRINT USING"###.##"
;X(1);:LPRINT TAB(15);:LPRINT USING"###.##";X(2);:LPRINT TAB(22);"-";
3062 IF SC<5 THEN GOTO 3150
3100 LPRINT USING"####.#";X(3);:LPRINT TAB(8);:LPRINT USING"###.##";X(1);:LPRINT
 TAB(15);:LPRINT USING"###.##";X(2);:LPRINT TAB(22);"--";:LPRINT USING" ##";T*10
00000!;:LPRINT" uS.";:SC=0:IF PP<33 THEN LPRINT"":RETURN
3150 IF PP<23 THEN LPRINT"":RETURN
3200 LPRINT TAB(PP);"+":RETURN
3500 REM Printer On/Tabular Output/Plotted Output Subroutine
```

```
3510 CLS:PRINT"Your printer must be on to run this program.  Is it on (Y or N)?"
3520 A$=INKEY$:IF A$="" THEN 3520
3530 IF A$="n" OR A$="N" THEN PRINT"For operation without a printer go to DOS. T
ype INFORMATION.":END
3540 PRINT"For a plotted response enter P.  For a tabular response enter any oth
er letter."
3550 A$=INKEY$:IF A$="" THEN 3550
3560 CLS:PRINT"                        PROGRAM IS IN PROGRESS
                                       ------- -- -- --------"
3570 RETURN


10 REM APPENDIX.21 -- Phased Array Pulsing System Analysis --
11 GOSUB 3500
13 DIM X(49)
14 DIM D(49)
16 DIM E(49)
18 DIM F(49)
19 SC=-1
22 LPRINT CHR$(27);"@";CHR$(15);CHR$(27);"A";CHR$(8);CHR$(27);"M";CHR$(7);CHR$(2
7);"Q";CHR$(128):WIDTH "LPT1:",128
30 N=7
32 T1=1E-09
34 T2=.000001
36 T3=.00052
38 T=0
40 P=0
44 TP=.00001
48 TI=.0001
50 X(1)=148.9
52 X(2)=0
54 X(3)=55.9
56 X(4)=19.7
58 X(5)=0
59 X(6)=55.1:X(7)=113.2
60 LS=7.8125E-05:LP=.005:LT=.000001
62 M=.9995*SQR(LS*LP)
64 C=.0001:CS=.02
66 R=100
68 V=460
70 R5=1000000!:R4=1:RC=1:RS=.006
80 LPRINT"          -- Minimum Integration Time =";T1;"Seconds --":LPRINT
86 SR=2:GOSUB 1000
90 GOSUB 2000
100 REM INTEGRATION ROUTINE
110 REM Equation Subroutine Selection
111 TR=TR+T1:IF TR>.00002 THEN TR=T1
112 IF TR=<.00001 THEN R5=RC:R4=1000000!:GOTO 115
113 IF TR=<.00002 THEN R4=RC:R5=1000000!
115 IF R4=RC AND X(2)>0 THEN SR=1:GOTO 119
116 IF R4=RC AND X(1)>=0 THEN SR=2:GOTO 119
117 IF R5=RC AND X(1)>0 THEN SR=3:GOTO 119
118 IF R5=RC AND X(2)>=0 THEN SR=4
119 TT=TT+T1:IF TT>TI THEN TT=T1
120 IF TT=<.00001 THEN R=.05:GOTO 128
121 IF TT=<TI THEN R=100
128 GOSUB 1000
```

```
129 FOR A = 1 TO N
130 E(A)=X(A)
140 F(A)=D(A)
150 X(A)=X(A)+T1*D(A)
152 IF X(1)<0 THEN X(1)=0
154 IF X(2)<0 THEN X(2)=0
160 NEXT A
169 T=T+T1
170 GOSUB 1000
180 FOR A = 1 TO N
190 X(A)=E(A)+T1/2*(F(A)+D(A))
192 IF X(1)<0 THEN X(1)=0
194 IF X(2)<0 THEN X(2)=0
200 NEXT A
220 P=P+T1
230 IF P+T1/5>T2 THEN GOTO 250
240 GOTO 110
250 P=0
260 REM PRINTOUT ROUTINE
270 GOSUB 2010
280 IF T+T1/5>T3 THEN GOTO 292
290 GOTO 110
291 PRINT
292 IF A$="P" OR A$="p" THEN GOSUB 3030:END
294 LPRINT"--------------------------------------------------------------------
----------"
300 END
1000 REM Equation Subroutine
1005 IF SR=1 THEN GOSUB 1100:RETURN
1010 IF SR=2 THEN GOSUB 1110:RETURN
1015 IF SR=3 THEN GOSUB 1120:RETURN
1020 IF SR=4 THEN GOSUB 1130:RETURN
1100 REM--- If x(2)>0 and R4=rc
1101 X(4)=-X(5):T1=1E-09
1102 D(5)=0:X(5)=0:D(4)=(LS*V-LS*R4*X(4)+M*X(3))/(LS*LP-M^2)
1104 D(2)=(-X(3)-M*D(4))/LS
1106 D(3)=(X(2)-X(7))/C:D(6)=(R*X(7)-X(6))/((R+RS)*CS):D(7)=(X(3)-X(6)+RS*CS*D(6
))/LT:S$="a":RETURN
1110 REM --- If X(1)>0 and R4=RC ---
1111 T1=.0000001
1112 D(5)=0:X(5)=0:D(4)=(LS*V-LS*R4*X(4)-M*X(3))/(LS*LP-M^2)
1114 D(1)=(-X(3)+M*D(4))/LS
1116 D(3)=(X(1)-X(7))/C:D(6)=(R*X(7)-X(6))/((R+RS)*CS):D(7)=(X(3)-X(6)+RS*CS*D(6
))/LT:S$="b":RETURN
1120 REM --- If X(1)>0 and R5=RC ---
1121 X(5)=-X(4):T1=1E-09
1122 D(4)=0:X(4)=0:D(5)=(LS*V-LS*R5*X(5)+M*X(3))/(LS*LP-M^2)
1124 D(1)=(-X(3)-M*D(5))/LS
1126 D(3)=(X(1)-X(7))/C:D(6)=(R*X(7)-X(6))/((R+RS)*CS):D(7)=(X(3)-X(6)+RS*CS*D(6
))/LT:S$="c":RETURN
1130 REM --- If X(2)>0 and R5=RC ---
1131 T1=.0000001
1132 D(4)=0:X(4)=0:D(5)=(LS*V-LS*R5*X(5)-M*X(3))/(LS*LP-M^2)
1134 D(2)=(-X(3)+M*D(5))/LS
1136 D(3)=(X(2)-X(7))/C:D(6)=(R*X(7)-X(6))/((R+RS)*CS):D(7)=(X(3)-X(6)+RS*CS*D(6
))/LT:S$="d":RETURN
2000 REM PRINT SUBROUTINE
2002 IF A$="P" OR A$="p" THEN LPRINT CHR$(27);"G":WIDTH "LPT1:",140:GOTO 2012
```

```
2004 LPRINT:LPRINT"      -- Response Of Phased Array Low Level Modulation System
--":LPRINT
2006 LPRINT TAB(1)"Microsec.";TAB(11)"X(1)";TAB(18);"X(2)";TAB(25);"X(3)";TAB(32
);"X(4)";TAB(39);"X(5)";TAB(46);"X(6)";TAB(53);"X(7)";TAB(60);" VO";TAB(67);" CS
#D(6)":LPRINT
2010 IO=(RS#X(7)+X(6))/(RS+R):IF A$="P" OR A$="p" THEN GOTO 2012
2011 GOTO 2015
2012 IO=(RS#X(7)+X(6))/(RS+R):IF T=0 THEN GOSUB 3000
2013 GOSUB 3040:RETURN
2015 LPRINT TAB(2);:LPRINT USING"####.###";T#1000000!;
2025 LPRINT TAB(11);:LPRINT USING"###.##";X(1);
2030 LPRINT TAB(18);:LPRINT USING"###.##";X(2);
2040 LPRINT TAB(25);:LPRINT USING"###.##";X(3);
2042 LPRINT TAB(32);:LPRINT USING"###.##";X(4);
2044 LPRINT TAB(39);:LPRINT USING"###.##";X(5);
2046 LPRINT TAB(46);:LPRINT USING"###.##";X(6);
2048 LPRINT TAB(53);:LPRINT USING"###.##";X(7);
2050 LPRINT TAB(60);:LPRINT USING"###.##";R#IO;
2052 LPRINT TAB(67);:LPRINT USING"####.##";CS#D(6)
2058 RETURN
3000 REM  Plotting Subroutine
3010 LPRINT"                              Output Voltage Response For Ph
ased Array Low Level Modulation System"
3015 LPRINT"                              ------ ------- -------- --- --
---- ----- --- ----- ---------- ------"
3020 LPRINT"                 0        10        20        30        40
  50        60        70        80        90       100"
3030 LPRINT"                 !....+....!....+....!....+....!....+....!..
..!....+....!....+....!....+....!....+....!....+....!"
3032 LPRINT TAB(1);"VO  V.";TAB(8);"X(1)A.";TAB(15);"X(2)A.";TAB(36);"CS#D(6)":R
ETURN
3040 SC=SC+1
3044 PP=(22+IO#R)
3060 IF SC < 5 THEN LPRINT USING"###.##";IO#R;:LPRINT TAB(8);:LPRINT USING"###.#
#";X(1);:LPRINT TAB(15);:LPRINT USING"###.##";X(2);:LPRINT TAB(22);"-";TAB(36);:
LPRINT USING"####.##";CS#D(6);
3062 IF SC < 5 THEN GOTO 3150
3100 LPRINT USING "###.##";IO#R;:LPRINT TAB(8);:LPRINT USING"###.##";X(1);:LPRIN
T TAB(15);:LPRINT USING"###.##";X(2);:LPRINT TAB(22);"--T=";:LPRINT USING" ##.#"
;T#1000000!;:LPRINT" uS.";TAB(36);:LPRINT USING"####.##";CS#D(6);:SC=0
3150 IF PP<23 THEN LPRINT"":RETURN
3200 LPRINT TAB(PP);"+":RETURN
3500 REM Printer On/Tabular Output/Plotted Output Subroutine
3510 CLS:PRINT"Your printer must be on to run this program.  Is it on (Y or N)?"
3520 A$=INKEY$:IF A$="" THEN 3520
3530 IF A$="n" OR A$="N" THEN PRINT"For operation without a printer go to DOS. T
ype INFORMATION.":END
3540 PRINT"For a plotted response enter P.  For a tabular response enter any oth
er letter."
3550 A$=INKEY$:IF A$="" THEN 3550
3560 CLS:PRINT"                        PROGRAM IS IN PROGRESS
                                       ------- -- -- --------"
3570 RETURN


10 REM APPENDIX.22 -- Phased Array Pulsing System Analysis (modified) --
11 GOSUB 3500
13 DIM X(49)
```

```
14 DIM D(49)
16 DIM E(49)
18 DIM F(49)
19 SC=-1
22 LPRINT CHR$(27);"@";CHR$(15);CHR$(27);"A";CHR$(8);CHR$(27);"M";CHR$(7);CHR$(2
7);"Q";CHR$(128):WIDTH "LPT1:",128
30 N=9
32 T1=1E-09
34 T2=2E-08
36 T3=.000012
38 T=0
40 P=0
44 TP=.00001
48 TI=.0001
50 X(1)=159.7
52 X(2)=0
54 X(3)=55.5
56 X(4)=20.9
58 X(5)=0
59 X(6)=55.1:X(7)=114.4:X(8)=.5575:X(9)=55.75
60 LS=7.8125E-05:LP=.005:LT=.000001:LC=1E-09
62 M=.9995*SQR(LS*LP)
64 C=.0001:CS=.01:CB=.000001
66 R=100
68 V=460
70 R5=1000000!:R4=1:RC=1:RS=.006
80 LPRINT"          -- Minimum Integration Time =";T1;"Seconds --":LPRINT
86 SR=2:GOSUB 1000
90 GOSUB 2000
100 REM INTEGRATION ROUTINE
110 REM Equation Subroutine Selection
111 TR=TR+T1:IF TR>.00002 THEN TR=T1
112 IF TR=<.00001 THEN R5=RC:R4=1000000!:GOTO 115
113 IF TR=<.00002 THEN R4=RC:R5=1000000!
115 IF R4=RC AND X(2)>0 THEN SR=1:GOTO 119
116 IF R4=RC AND X(1)>=0 THEN SR=2:GOTO 119
117 IF R5=RC AND X(1)>0 THEN SR=3:GOTO 119
118 IF R5=RC AND X(2)>=0 THEN SR=4
119 TT=TT+T1:IF TT>TI THEN TT=T1
120 IF TT=<.00001 THEN R=.05:GOTO 128
121 IF TT=<TI THEN R=100
128 GOSUB 1000
129 FOR A = 1 TO N
130 E(A)=X(A)
140 F(A)=D(A)
150 X(A)=X(A)+T1*D(A)
152 IF X(1)<0 THEN X(1)=0
154 IF X(2)<0 THEN X(2)=0
160 NEXT A
169 T=T+T1
170 GOSUB 1000
180 FOR A = 1 TO N
190 X(A)=E(A)+T1/2*(F(A)+D(A))
192 IF X(1)<0 THEN X(1)=0
194 IF X(2)<0 THEN X(2)=0
200 NEXT A
220 P=P+T1
230 IF P+T1/5>T2 THEN GOTO 250
```

Appendix B

```
240 GOTO 110
250 P=0
260 REM PRINTOUT ROUTINE
270 GOSUB 2010
280 IF T+T1/5>T3 THEN GOTO 292
290 GOTO 110
291 PRINT
292 IF A$="P" OR A$="p" THEN GOSUB 3030:END
294 LPRINT"-----------------------------------------------------------------------
----------"
300 END
1000 REM Equation Subroutine
1005 IF SR=1 THEN GOSUB 1100:RETURN
1010 IF SR=2 THEN GOSUB 1110:RETURN
1015 IF SR=3 THEN GOSUB 1120:RETURN
1020 IF SR=4 THEN GOSUB 1130:RETURN
1100 REM--- If x(2)>0 and R4=RC
1101 X(4)=-X(5):T1=1E-09:T2=2E-08
1102 D(5)=0:X(5)=0:D(4)=(LS*V-LS*R4*X(4)+M*X(3))/(LS*LP-M^2)
1104 D(2)=(-X(3)-M*D(4))/LS
1106 D(3)=(X(2)-X(7))/C:D(6)=(X(7)-X(8))/CS:D(7)=(X(3)-X(9))/LT:D(8)=(X(6)+LC*D(
7)+RS*X(7)-RS*X(8)-X(9))/LC:D(9)=(X(8)-X(9))/R/CB:S$="a":RETURN
1110 REM --- If X(1)>0 and R4=RC ---
1111 T1=2E-08:T2=2E-08
1112 D(5)=0:X(5)=0:D(4)=(LS*V-LS*R4*X(4)-M*X(3))/(LS*LP-M^2)
1114 D(1)=(-X(3)+M*D(4))/LS
1116 D(3)=(X(1)-X(7))/C:D(6)=(X(7)-X(8))/CS:D(7)=(X(3)-X(9))/LT:D(8)=(X(6)+LC*D(
7)+RS*X(7)-RS*X(8)-X(9))/LC:D(9)=(X(8)-X(9))/R/CB:S$="b":RETURN
1120 REM --- If X(1)>0 and R5=RC ---
1121 X(5)=-X(4):T1=1E-09:T2=2E-08
1122 D(4)=0:X(4)=0:D(5)=(LS*V-LS*R5*X(5)+M*X(3))/(LS*LP-M^2)
1124 D(1)=(-X(3)-M*D(5))/LS
1126 D(3)=(X(1)-X(7))/C:D(6)=(X(7)-X(8))/CS:D(7)=(X(3)-X(9))/LT:D(8)=(X(6)+LC*D(
7)+RS*X(7)-RS*X(8)-X(9))/LC:D(9)=(X(8)-X(9))/R/CB:S$="c":RETURN
1130 REM --- If X(2)>0 and R5=RC ---
1131 T1=2E-08:T2=2E-08
1132 D(4)=0:X(4)=0:D(5)=(LS*V-LS*R5*X(5)-M*X(3))/(LS*LP-M^2)
1134 D(2)=(-X(3)+M*D(5))/LS
1136 D(3)=(X(2)-X(7))/C:D(6)=(X(7)-X(8))/CS:D(7)=(X(3)-X(9))/LT:D(8)=(X(6)+LC*D(
7)+RS*X(7)-RS*X(8)-X(9))/LC:D(9)=(X(8)-X(9))/R/CB:S$="d":RETURN
2000 REM PRINT SUBROUTINE
2002 IF A$="P" OR A$="p" THEN LPRINT CHR$(27);"G":WIDTH "LPT1:",140:GOTO 2012
2004 LPRINT:LPRINT"       -- Response Of Phased Array Low Level Modulation System
--":LPRINT
2006 LPRINT TAB(1)"Microsec.";TAB(11)"X(1)";TAB(18);"X(2)";TAB(25);"X(3)";TAB(32
);"X(4)";TAB(39);"X(5)";TAB(46);"X(6)";TAB(53);"X(7)";TAB(60);"X(8)";TAB(67);"X(
9)";TAB(74);"  R";TAB(81);"  CS*D(6)":LPRINT
2010 IF A$="P" OR A$="p" THEN GOTO 2012
2011 GOTO 2015
2012 IO=(RS*X(7)+X(6))/(RS+R):IF T=0 THEN GOSUB 3000
2013 GOSUB 3040:RETURN
2015 LPRINT TAB(2);:LPRINT USING"####.###";T*1000000!;
2025 LPRINT TAB(11);:LPRINT USING"###.##";X(1);
2030 LPRINT TAB(18);:LPRINT USING"###.##";X(2);
2040 LPRINT TAB(25);:LPRINT USING"###.##";X(3);
2042 LPRINT TAB(32);:LPRINT USING"###.##";X(4);
2044 LPRINT TAB(39);:LPRINT USING"###.##";X(5);
2046 LPRINT TAB(46);:LPRINT USING"###.##";X(6);
```

```
2048 LPRINT TAB(53);:LPRINT USING"###.##";X(7);
2049 LPRINT TAB(60);:LPRINT USING"####.##";X(8);
2050 LPRINT TAB(67);:LPRINT USING"###.##";X(9);
2051 LPRINT TAB(74);:LPRINT USING"###.##";R;
2052 LPRINT TAB(82);:LPRINT USING"#####.##";CS*D(6)
2058 RETURN
3000 REM  Plotting Subroutine
3010 LPRINT"                              Output Voltage Response For Ph
ased Array Low Level Modulation System"
3015 LPRINT"                              ------ ------- -------- --- --
---- ----- --- ----- ---------- ------"
3020 LPRINT"           0        10       20       30       40
 50       60       70       80       90      100"
3030 LPRINT"           !....+....!....+....!....+....!....+....!....+.
...!....+....!....+....!....+....!....+....!....+....!"
3032 LPRINT TAB(1);"VO   V.";TAB(8);"X(1)A.";TAB(15);"X(2)A.":RETURN
3040 SC=SC+1
3044 PP=(22+X(9))
3060 IF SC < 5 THEN LPRINT USING"###.##";X(9);:LPRINT TAB(8):LPRINT USING"###.#
#";X(1);:LPRINT TAB(15);:LPRINT USING"###.##";X(2);:LPRINT TAB(22);"-";
3062 IF SC < 5 THEN GOTO 3150
3100 LPRINT USING "###.##";X(9);:LPRINT TAB(8);:LPRINT USING"###.##";X(1);:LPRIN
T TAB(15);:LPRINT USING"###.##";X(2);:LPRINT TAB(22);"--T=";:LPRINT USING" ##.#"
;T*1000000!;:LPRINT" uS.";:SC=0:IF PP<33 THEN LPRINT"":RETURN
3150 IF PP<23 THEN LPRINT"":RETURN
3200 LPRINT TAB(PP);"+":RETURN
3500 REM Printer On/Tabular Output/Plotted Output Subroutine
3510 CLS:PRINT"Your printer must be on to run this program.  Is it on (Y or N)?"
3520 A$=INKEY$:IF A$="" THEN 3520
3530 IF A$="n" OR A$="N" THEN PRINT"For operation without a printer go to DOS. T
ype INFORMATION.":END
3540 PRINT"For a plotted response enter P.  For a tabular response enter any oth
er letter."
3550 A$=INKEY$:IF A$="" THEN 3550
3560 CLS:PRINT"                      PROGRAM IS IN PROGRESS
                      ------- -- -- --------"
3570 RETURN


10 REM APPENDIX.24 -- Grid Pulser Analysis Program --
11 GOSUB 3500
13 DIM X(49)
14 DIM D(49)
16 DIM E(49)
18 DIM F(49)
20 N=8
24 LPRINT CHR$(27);"@";CHR$(15);CHR$(27);"A";CHR$(8);CHR$(27);"M";CHR$(7);CHR$(2
7);"Q";CHR$(128):WIDTH "LPT1:",128
32 T1=2.5E-10
34 T2=1E-08
36 T3=5.5E-07
38 T=0
40 P=0
51 CI=4E-09
52 CF=7E-11
53 CO=2.5E-10
54 CG=1E-10
55 RF1=1000000!:RF2=RF1
```

```
56 RD=2.5
57 RP=150
58 RG=100000!
59 LG=.0000005
60 VD2=5:VD1=0
61 V=500
62 G=3.5
70 X(1)=0
71 X(2)=1000
72 X(3)=1000
73 X(4)=-.0049925
74 X(5)=-499.25
75 X(6)=5
76 X(7)=-5
77 X(8)=0
80 LPRINT" TWT Grid Pulser Integration Time =";T1*1000000!;"Microseconds --":LP
RINT
86 GOSUB 1000
90 GOSUB 2000
100 REM INTEGRATION ROUTINE
110 IF T<.0000004 AND VD1=>5 THEN GOTO 120
111 IF T<.0000004 AND VD1<5 THEN GOTO 114
112 IF T=>.0000004 AND VD2=>5 THEN GOTO 120
113 IF T=>.0000004 AND VD2<5 THEN GOTO 116
114 VD1=VD1+.0625:VD2=VD2-.0625:GOTO 120
116 VD2=VD2+.0625:VD1=VD1-.0625
120 FOR A = 1 TO N
130 E(A)=X(A)
140 F(A)=D(A)
150 X(A)=X(A)+T1*D(A)
152 GM1=G:IF X(3)<10 THEN GM1=GM1*X(3)/10
153 IF X(3)<0 THEN RF1=.25
154 IF X(1)<0 THEN GM1=0
156 GM2=G:IF X(8)<10 THEN GM2=GM2*X(8)/10
157 IF X(8)<0 THEN RF2=.25
158 IF X(6)<0 THEN GM2=0
160 NEXT A
165 T=T+T1
170 GOSUB 1000
180 FOR A = 1 TO N
190 X(A)=E(A)+T1/2*(F(A)+D(A))
192 GM1=G:IF X(3)<10 THEN GM1=GM1*X(3)/10
193 IF X(3)<0 THEN RF1=.25
194 IF X(1)<0 THEN GM1=0
196 GM2=G:IF X(8)<10 THEN GM2=GM2*X(8)/10
197 IF X(8)<0 THEN RF2=.25
198 IF X(6)<0 THEN GM2=0
200 NEXT A
220 P=P+T1
230 IF P+T1/5>T2 THEN GOTO 250
240 GOTO 110
250 P=0
260 REM PRINTOUT ROUTINE
270 GOSUB 2010
280 IF T+T1/5>T3 THEN GOTO 292
290 GOTO 110
291 PRINT
292 IF A$="P" OR A$="p" THEN GOSUB 3030:END
```

```
294 LPRINT"------------------------------------------------------------------
----------"
300 END
1000 REM EQUATION SUBROUTINE
1005 II1=(VD1-X(1))/RD
1010 II2=(VD2-X(6))/RD
1015 D(3)=(X(4)+(CF/(CI+CF))*(II1-II2)-X(3)/RF1-GM1*X(1)+X(8)/RF2+GM2*X(6))/(2*C
O+2*CI*CF/(CI+CF))
1020 D(2)=(CI*D(3)-II1)/(CI+CF)
1030 D(1)=D(3)-D(2)
1040 D(4)=(V-X(3)-X(5)-RP*X(4))/LG
1050 D(5)=(X(4)-X(5))/CG
1070 D(7)=-(CI*D(3)+II2)/(CI+CF)
1080 D(6)=-(D(3)+D(7))
1090 D(8)=-D(3)
1100 RETURN
2000 REM PRINT SUBROUTINE
2002 IF A$="P" OR A$="p" THEN GOSUB 3000:RETURN
2005 LPRINT TAB(1)"Microsec.";TAB(10)"X(1)";TAB(20);"X(2)";TAB(30);"X(3)";TAB(40
);"X(4)";TAB(50);"X(5)";TAB(60);"X(6)";TAB(70);"X(7)";TAB(80);"X(8)";TAB(90);"FE
T1 Cur.";TAB(100);"FET2 Cur.":LPRINT:GOTO 2015
2010 IF A$="P" OR A$="p" THEN GOSUB 3040:RETURN
2015 LPRINT TAB(1);:LPRINT USING"##.###";T*1000000!;
2025 LPRINT TAB(10);:LPRINT USING"####.###";X(1);
2030 LPRINT TAB(20);:LPRINT USING"####.###";X(2);
2040 LPRINT TAB(30);:LPRINT USING"####.###";X(3);
2042 LPRINT TAB(40);:LPRINT USING"####.###";X(4);
2044 LPRINT TAB(50);:LPRINT USING"####.###";X(5);
2046 LPRINT TAB(60);:LPRINT USING"####.###";X(6);
2048 LPRINT TAB(70);:LPRINT USING"####.###";X(7);
2050 LPRINT TAB(80);:LPRINT USING"####.###";X(8);
2052 LPRINT TAB(90);:LPRINT USING"####.###";GM1*X(1);
2054 LPRINT TAB(100);:LPRINT USING"####.###";GM2*X(6)
2058 RETURN
3000 REM  Plotting Subroutine
3010 LPRINT"                                                  Transient Response Of
 TWT Grid Pulser"
3015 LPRINT"                                                  --------- -------- --
--- ---- ------"
3020 LPRINT"              -500     -400     -300     -200     -100      0
     100      200      300      400      500"
3030 LPRINT"             |....+....|....+....|....+....|....+....|....+...|..
..+....|....+....|....+....|....+....|....+....|"
3032 LPRINT"Grid Volts":LPRINT USING" ####.###";X(5);:LPRINT TAB(16);"--":RETURN
3040 SC=SC+1
3044 PP=(66+X(5)/10)
3060 IF SC < 5 THEN LPRINT USING" ####.###";X(5);:LPRINT TAB(16);"-";
3062 IF SC < 5 THEN GOTO 3150
3100 LPRINT USING" ####.###";X(5);:LPRINT TAB(16);"--T=";:LPRINT USING".###";T*1
000000!;:LPRINT" uS.";:SC=0:IF PP>30 THEN GOTO 3200
3110 LPRINT"":RETURN
3150 IF PP<17 THEN LPRINT"":RETURN
3200 LPRINT TAB(PP);"+":RETURN
3500 REM Printer On/Tabular Output/Plotted Output Subroutine
3510 CLS:PRINT"Your printer must be on to run this program.  Is it on (Y or N)?"
3520 A$=INKEY$:IF A$="" THEN 3520
3530 IF A$="n" OR A$="N" THEN PRINT"For operation without a printer go to DOS. T
ype INFORMATION.":END
```

```
3540 PRINT"For a plotted response enter P.  For a tabular response enter any oth
er letter."
3550 A$=INKEY$:IF A$="" THEN 3550
3560 CLS:PRINT"                        PROGRAM IS IN PROGRESS
                                        ------- -- -- --------"
3570 RETURN



10 REM APPENDIX.25 -- Hard Switched Analysis Program --
11 GOSUB 3500
13 DIM X(49)
14 DIM D(49)
16 DIM E(49)
18 DIM F(49)
20 N=7:NS=20
24 LPRINT CHR$(27);"@";CHR$(15);CHR$(27);"A";CHR$(8);CHR$(27);"M";CHR$(10);CHR$(
27);"Q";CHR$(128):WIDTH "LPT1:",128
32 T1=2.5E-10
34 T2=1E-08
36 T3=.000001
38 T=0
40 P=0
50 CI=4E-09/NS
51 CF=7E-11/NS
52 CO=2.5E-10/NS
53 CS=2E-08
54 C=1E-10
55 RR=1000000!*NS
56 RD=2.5*NS
57 RC=10000
58 R=.25*NS
59 LC=.06
60 L=.0015
61 VD=0
62 V=15000
63 G=3.5/NS
70 X(1)=0
71 X(2)=15000
72 X(3)=15000
73 X(4)=.00075
74 X(5)=15000
75 X(6)=0
76 X(7)=0
80 LPRINT"Integration Step =";T1*1000000!;"Microseconds --":LPRINT
86 GOSUB 1000
90 GOSUB 2000
100 REM INTEGRATION ROUTINE
110 IF T<.0000004 AND VD=>5*NS THEN GOTO 114
112 IF T<.0000004 AND VD<5*NS THEN VD=VD+.0625*NS:GOTO 114
113 VD=VD-.0625*NS:IF VD<0 THEN VD=0
114 IF X(6)<0 THEN R=.25*NS:GOTO 120
116 IF X(6)>0 THEN R=100000!
118 IF X(6)>14000 THEN R=1000-(X(6)-14000)/10
120 FOR A = 1 TO N
130 E(A)=X(A)
140 F(A)=D(A)
150 X(A)=X(A)+T1*D(A)
152 GM=G:IF X(3)<10*NS THEN GM=GM*X(3)/(10*NS)
```

```
153 IF X(3)<0 THEN RR=.25*NS
154 IF X(1)<0 THEN GM=0
160 NEXT A
165 T=T+T1
170 GOSUB 1000
180 FOR A = 1 TO N
190 X(A)=E(A)+T1/2*(F(A)+D(A))
192 GM=G:IF X(3)<10*NS THEN GM=GM*X(3)/(10*NS)
193 IF X(3)<0 THEN RR=.25*NS
194 IF X(1)<0 THEN GM=0
200 NEXT A
220 P=P+T1
230 IF P+T1/5>T2 THEN GOTO 250
240 GOTO 110
250 P=0
260 REM PRINTOUT ROUTINE
270 GOSUB 2010
280 IF T+T1/5>T3 THEN GOTO 292
290 GOTO 110
292 IF A$="P" OR A$="p" THEN GOSUB 3030:END
294 LPRINT"---------------------------------------------------------------
----------"
300 END
1000 REM EQUATION SUBROUTINE
1005 II=(VD-X(1))/RD
1010 D(4)=(V-X(3)-RC*X(4))/LC
1015 D(3)=(X(4)-GM*X(1)-X(3)/RR+CF*II/(CF+CI)+CS*(X(6)/R+X(7))/(CS+C))/(CO+CF*CI
/(CF+CI)+C*CS/(CS+C))
1020 D(2)=(CI*D(3)-II)/(CI+CF)
1025 D(1)=D(3)-D(2)
1030 D(5)=(C*D(3)-X(7)-X(6)/R)/(CS+C)
1035 D(6)=D(5)-D(3)
1040 D(7)=X(6)/L
1100 RETURN
2000 REM PRINT SUBROUTINE
2002 IF A$="P" OR A$="p" THEN GOSUB 3000:RETURN
2003 LPRINT"                    Response Of A Hard Switched Type Modulator For A
 Magnetron"
2004 LPRINT"       ---------------------------------------------------
----------"
2005 LPRINT TAB(1)"Microsec.";TAB(10)"  X(1)";TAB(20);"  X(2)";TAB(30);"  X(3)";
TAB(40);" X(4)";TAB(50);"  X(5)";TAB(60);"  X(6)";TAB(70);"  X(7)";TAB(80);"FET
 Cur.";TAB(90);"Mag.Cur.":LPRINT:GOTO 2015
2010 IF A$="P" OR A$="p" THEN GOSUB 3040:RETURN
2015 LPRINT TAB(1);:LPRINT USING"##.###";T*1000000!;
2025 LPRINT TAB(10);:LPRINT USING"####.###";X(1);
2030 LPRINT TAB(20);:LPRINT USING"######.#";X(2);
2040 LPRINT TAB(30);:LPRINT USING"######.#";X(3);
2042 LPRINT TAB(40);:LPRINT USING"###.####";X(4);
2044 LPRINT TAB(50);:LPRINT USING"#####.##";X(5);
2046 LPRINT TAB(60);:LPRINT USING"######.#";X(6);
2048 LPRINT TAB(70);:LPRINT USING"###.####";X(7);
2050 LPRINT TAB(80);:LPRINT USING"###.####";GM*X(1);
2052 LPRINT TAB(90);:LPRINT USING"###.####";X(6)/R
2058 RETURN
3000 REM  Plotting Subroutine
3010 LPRINT"                                    Transient Response Of Mag
netron Voltage - kV"
```

```
3015 LPRINT"-                                        --------- -------- -- ---
------ ------- - --"
3020 LPRINT"              -18         -16         -14         -12         -10         -8
       -6         -4         -2          0          2"
3030 LPRINT"              |....+....|....+....|....+....|....+....|....+....|....+....|..
..+....|....+....|....+....|....+....|":LPRINT" Magnetron kV"
3032 LPRINT USING"    ###.###";-X(6)/1000;:LPRINT TAB(16);"--";:PP=(106-X(6)/200)
:GOTO 3200
3040 SC=SC+1
3044 PP=(106-X(6)/200)
3060 IF SC < 5 THEN LPRINT USING"    ###.###";-X(6)/1000;:LPRINT TAB(16);"-";
3062 IF SC < 5 THEN GOTO 3150
3100 LPRINT USING"    ###.###";-X(6)/1000;:LPRINT TAB(16);"--T=";:LPRINT USING"#.
##";T#1000000!;:LPRINT" uS.";:SC=0:IF PP>29 THEN GOTO 3200
3110 LPRINT"":RETURN
3150 IF PP<17 THEN LPRINT"":RETURN
3200 LPRINT TAB(PP);"+":RETURN
3500 REM Printer On/Tabular Output/Plotted Output Subroutine
3510 CLS:PRINT"Your printer must be on to run this program.  Is it on (Y or N)?"
3520 A$=INKEY$:IF A$="" THEN 3520
3530 IF A$="n" OR A$="N" THEN PRINT"For operation without a printer go to DOS.
Type INFORMATION.":END
3540 PRINT"For a plotted response enter P.  For a tabular response enter any oth
er letter."
3550 A$=INKEY$:IF A$="" THEN 3550
3560 CLS:PRINT"                    PROGRAM IS IN PROGRESS
                                ------- -- -- --------"
3570 RETURN


10 REM APPENDIX.26 -- Pulse Forming Network Analysis Program --
12 DIM X(49)
14 DIM D(49)
16 DIM E(49)
18 DIM F(49)
20 N=10
22 GOSUB 3500
25 LPRINT CHR$(27);"@";CHR$(15);CHR$(27);"A";CHR$(8);CHR$(27);"G";CHR$(27);"Q";C
HR$(128);CHR$(27);"M";CHR$(10):WIDTH "LPT1:",128
32 T1=.005
34 T2=.02
36 T3=1.5
38 T=0
40 P=0
50 L1=.0805
51 L2=.0711
52 L3=.0668
53 L4=.0674
54 L5=.0801
55 M12=.0077
56 M23=.0118
57 M34=.0122
58 M45=.0113
59 C6=.091
60 C7=C6:C8=C6:C9=C6:C10=C6
61 X(6)=2
62 X(7)=X(6):X(8)=X(6):X(9)=X(6):X(10)=X(6)
63 R=1
```

```
64 K1=M45^2-L5*L4
65 K2=-L5*M34
66 K4=M34*K2-K1*L3
67 K5=-K1*M23
68 K7=M23*K5-K4*L2
69 K8=-K4*M12
80 LPRINT"   Integration Step =";T1;"Seconds":LPRINT
86 GOSUB 1000
90 GOSUB 2000
100 REM INTEGRATION ROUTINE
110 GOSUB 1000
120 FOR A = 1 TO N
130 E(A)=X(A)
140 F(A)=D(A)
150 X(A)=X(A)+T1*D(A)
160 NEXT A
165 T=T+T1
170 GOSUB 1000
180 FOR A = 1 TO N
190 X(A)=E(A)+T1/2*(F(A)+D(A))
200 NEXT A
220 P=P+T1
230 IF P+T1/5>T2 THEN GOTO 250
240 GOTO 110
250 P=0
260 REM PRINTOUT ROUTINE
270 GOSUB 2008
280 IF T+T1/5>T3 THEN GOTO 292
290 GOTO 110
292 IF A$="P" OR A$="p" THEN GOSUB 3030:END
295 LPRINT"----------------------------------------------------------------
--------"
300 END
1000 REM EQUATION SUBROUTINE
1005 K3=M45*(X(10)-R*X(5))-L5*(X(9)-X(10))
1010 K6=M34*K3-K1*(X(8)-X(9))
1015 K9=M23*K6-K4*(X(7)-X(8))
1020 D(1)=(M12*K9-K7*(X(6)-X(7)))/(M12*K8-K7*L1)
1025 D(2)=(X(6)-X(7)-L1*D(1))/M12
1030 D(3)=(X(7)-X(8)-M12*D(1)-L2*D(2))/M23
1035 D(4)=(X(8)-X(9)-M23*D(2)-L3*D(3))/M34
1040 D(5)=(X(9)-X(10)-M34*D(3)-L4*D(4))/M45
1045 D(6)=-X(1)/C6
1050 D(7)=(X(1)-X(2))/C7
1055 D(8)=(X(2)-X(3))/C8
1060 D(9)=(X(3)-X(4))/C9
1065 D(10)=(X(4)-X(5))/C10
1070 RETURN
2000 REM PRINT SUBROUTINE
2002 IF A$="P" OR A$="p" THEN GOSUB 3000:RETURN
2004 LPRINT"              Response Of Type E Five Section PFN"
2005 LPRINT"              -------- -- ---- - ---- ------- ---"
2006 LPRINT TAB(2);"Sec.";TAB(8);"x(5)";TAB(15);"x(4)";TAB(22);"x(3)";TAB(29);"x
(2)";TAB(36);"x(1)";TAB(43);"x(10)";TAB(49);"x(9)";TAB(55);"x(8)";TAB(61);"x(7)"
;TAB(67);"x(6)";TAB(73);"Vout":LPRINT
2008 IF A$="P" OR A$="p" THEN GOSUB 3040:RETURN
2010 LPRINT TAB(2);:LPRINT USING"#.##";T;
2012 LPRINT TAB(7);:LPRINT USING"##.###";X(5);
```

```
2014 LPRINT TAB(14);:LPRINT USING"##.###";X(4);
2016 LPRINT TAB(21);:LPRINT USING"##.###";X(3);
2018 LPRINT TAB(28);:LPRINT USING"##.###";X(2);
2020 LPRINT TAB(35);:LPRINT USING"##.###";X(1);
2022 LPRINT TAB(42);:LPRINT USING"##.##";X(10);
2024 LPRINT TAB(48);:LPRINT USING"##.##";X(9);
2026 LPRINT TAB(54);:LPRINT USING"##.##";X(8);
2028 LPRINT TAB(60);:LPRINT USING"##.##";X(7);
2030 LPRINT TAB(66);:LPRINT USING"##.##";X(6);
2032 LPRINT TAB(72);:LPRINT USING"##.###";X(5)*R
2040 RETURN
3000 REM PLOTTING SUBROUTINE
3010 LPRINT"                          Voltage Response Of A Type E Fi
ve Section PFN"
3015 LPRINT"                          ------- -------- -- - ---- - --
-- ------- ---"
3020 LPRINT"       -.6     -.4     -.2      0      .2      .4      .6
    .8     1.0     1.2     1.4    1.6"
3030 LPRINT"       !....+....!....+....!....+....!....+....!....+....!
....+....!....+....!....+....!....+....!....+....!....+....!":IF T>0 THEN RETURN
3032 LPRINT" Vout":LPRINT USING"##.###";X(5)*R;:LPRINT TAB(8);"--":RETURN
3040 SC=SC+1
3044 PP=(38+X(5)*R*50)
3060 IF SC<5 THEN LPRINT USING"##.###";X(5)*R;:LPRINT TAB(8)"-";
3062 IF SC<5 THEN GOTO 3150
3100 LPRINT USING"##.###";X(5)*R;:LPRINT TAB(8);"--T=";:LPRINT USING"#.##";T;:LP
RINT"Sec.";:SC=0:IF PP>20 THEN GOTO 3200
3110 LPRINT"":RETURN
3150 IF PP<9 THEN LPRINT"":RETURN
3200 LPRINT TAB(PP);"+":RETURN
3500 REM Printer On/Tabular Output/Plotted Output Subroutine
3510 CLS:PRINT"Your printer must be on to run this program.  Is it on (Y or N)?"
3520 A$=INKEY$:IF A$="" THEN 3520
3530 IF A$="n" OR A$="N" THEN PRINT"For operation without a printer go to DOS. T
ype INFORMATION.":END
3540 PRINT"For a plotted response enter P.  For a tabular response enter any oth
er letter."
3550 A$=INKEY$:IF A$="" THEN 3550
3560 CLS:PRINT"                  PROGRAM IS IN PROGRESS
                                  ------- -- -- --------"
3570 RETURN


10 REM APPENDIX.27 -- Line Type Pulser Analysis Program --
12 DIM X(49)
14 DIM D(49)
16 DIM E(49)
18 DIM F(49)
20 N=12
22 GOSUB 3500
25 LPRINT CHR$(27);"@";CHR$(15);CHR$(27);"A";CHR$(8);CHR$(27);"G";CHR$(27);"Q";C
HR$(128);CHR$(27);"M";CHR$(10):WIDTH "LPT1:",128
32 T1=.005
34 T2=.02
36 T3=1.5
38 T=0
40 P=0
50 L1=.0805
```

```
51 L2=.0711
52 L3=.0668
53 L4=.0674
54 L5=.0801
55 M12=.0077
56 M23=.0118
57 M34=.0122
58 M45=.0113
59 C6=.091
60 C7=C6:C8=C6:C9=C6:C10=C6
61 X(6)=2
62 X(7)=X(6):X(8)=X(6):X(9)=X(6):X(10)=X(6)
64 LP=10:LS=10
65 K=.999
66 M=K*SQR(LP*LS)
67 Z0=1
68 LL=LS*(1-K^2)
69 C=LL/(Z0*LS/LP)^2
70 PV=1
71 S1=LS*(LP+L5)-M^2
72 S2=LS*M45
73 K1=M45*S2-S1*L4
74 K2=-S1*M34
75 K4=M34*K2-K1*L3
76 K5=-K1*M23
77 K7=M23*K5-K4*L2
78 K8=-K4*M12
80 LPRINT"  Integration Step =";T1;"Seconds":LPRINT
86 GOSUB 1000
90 GOSUB 2000
100 REM INTEGRATION ROUTINE
110 GOSUB 1000
120 FOR A = 1 TO N
130 E(A)=X(A)
140 F(A)=D(A)
150 X(A)=X(A)+T1*D(A)
160 NEXT A
165 T=T+T1
170 GOSUB 1000
180 FOR A = 1 TO N
190 X(A)=E(A)+T1/2*(F(A)+D(A))
200 NEXT A
220 P=P+T1
230 IF P+T1/5>T2 THEN GOTO 250
240 GOTO 110
250 P=0
260 REM PRINTOUT ROUTINE
270 GOSUB 2008
280 IF T+T1/5>T3 THEN GOTO 292
290 GOTO 110
292 IF A$="P" OR A$="p" THEN GOSUB 3030:END
295 LPRINT"-----------------------------------------------------------------------"
300 END
1000 REM EQUATION SUBROUTINE
1005 K3=M45*S3-S1*(X(9)-X(10))
1010 K6=M34*K3-K1*(X(8)-X(9))
1015 K9=M23*K6-K4*(X(7)-X(8))
```

```
1017 S3=-M*X(12)+LS*X(10)
1020 D(1)=(M12*K9-K7*(X(6)-X(7)))/(M12*K8-K7*L1)
1025 D(2)=(X(6)-X(7)-L1*D(1))/M12
1030 D(3)=(X(7)-X(8)-M12*D(1)-L2*D(2))/M23
1035 D(4)=(X(8)-X(9)-M23*D(2)-L3*D(3))/M34
1040 D(5)=(X(9)-X(10)-M34*D(3)-L4*D(4))/M45
1045 D(6)=-X(1)/C6
1050 D(7)=(X(1)-X(2))/C7
1055 D(8)=(X(2)-X(3))/C8
1060 D(9)=(X(3)-X(4))/C9
1065 D(10)=(X(4)-X(5))/C10
1070 D(11)=(-X(12)+M*D(5))/LS
1075 IF X(12)<0 THEN IL=0:GOTO 1085
1080 IL=PV*X(12)^1.5
1085 D(12)=(X(11)-IL)/C
1090 RETURN
2000 REM PRINT SUBROUTINE
2002 IF A$="P" OR A$="p" THEN GOSUB 3000:RETURN
2004 LPRINT"      Output Response Of 5 Section PFN, Pulse Transformer, And Non-l
inear Load"
2005 LPRINT"      ------ -------- -- - ------- ---- ----- ------------ --- -----
----- ----"
2006 LPRINT TAB(2);"Sec.";TAB(8);"x(5)";TAB(15);"x(4)";TAB(22);"x(3)";TAB(29);"x
(2)";TAB(36);"x(1)";TAB(43);"x(10)";TAB(49);"x(9)";TAB(55);"x(8)";TAB(61);"x(7)"
;TAB(67);"x(6)";TAB(73);"x(11)";TAB(80);"x(12)":LPRINT
2008 IF A$="P" OR A$="p" THEN GOSUB 3040:RETURN
2010 LPRINT TAB(2);:LPRINT USING"#.##";T;
2012 LPRINT TAB(7);:LPRINT USING"##.###";X(5);
2014 LPRINT TAB(14);:LPRINT USING"##.###";X(4);
2016 LPRINT TAB(21);:LPRINT USING"##.###";X(3);
2018 LPRINT TAB(28);:LPRINT USING"##.###";X(2);
2020 LPRINT TAB(35);:LPRINT USING"##.###";X(1);
2022 LPRINT TAB(42);:LPRINT USING"##.##";X(10);
2024 LPRINT TAB(48);:LPRINT USING"##.##";X(9);
2026 LPRINT TAB(54);:LPRINT USING"##.##";X(8);
2028 LPRINT TAB(60);:LPRINT USING"##.##";X(7);
2030 LPRINT TAB(66);:LPRINT USING"##.##";X(6);
2032 LPRINT TAB(72);:LPRINT USING"##.##";X(11);
2034 LPRINT TAB(79);:LPRINT USING"##.###";X(12)
2040 RETURN
3000 REM PLOTTING SUBROUTINE
3010 LPRINT"                              Output Response Of 5 Section PFN, Pulse Tr
ansformer, And Non-linear Load"
3015 LPRINT"                              ------ -------- -- - ------- ---- ----- --
---------- --- ---------- ----"
3020 LPRINT"      -.6      -.4      -.2       0       .2       .4       .6
     .8      1.0      1.2      1.4      1.6"
3030 LPRINT"      !....+....!....+....!....+....!....+....!....+....!....+....!
....+....!....+....!....+....!....+....!....+....!":IF T>0 THEN RETURN
3032 GOTO 3044
3040 SC=SC+1
3044 PP=(38+X(12)*50)
3060 IF SC<5 THEN LPRINT USING"##.###";X(12);:LPRINT TAB(8)"-";
3062 IF SC<5 THEN GOTO 3150
3100 LPRINT USING"##.###";X(12);:LPRINT TAB(8);"--T=";:LPRINT USING"#.##";T;:LPR
INT"Sec.";:SC=0:IF PP>20 THEN GOTO 3200
3110 LPRINT"":RETURN
3150 IF PP<9 THEN LPRINT"":RETURN
```

```
3200 LPRINT TAB(PP);"+":RETURN
3500 REM Printer On/Tabular Output/Plotted Output Subroutine
3510 CLS:PRINT"Your printer must be on to run this program.  Is it on (Y or N)?"
3520 A$=INKEY$:IF A$="" THEN 3520
3530 IF A$="n" OR A$="N" THEN PRINT"For operation without a printer go to DOS. T
ype INFORMATION.":END
3540 PRINT"For a plotted response enter P.  For a tabular response enter any oth
er letter."
3550 A$=INKEY$:IF A$="" THEN 3550
3560 CLS:PRINT"                         PROGRAM IS IN PROGRESS
                                        ------- -- -- --------"
3570 RETURN


10 REM APPENDIX.28 - D-C Resonant Charging / Inverse Diode Analysis -
12 DIM X(49)
14 DIM D(49)
16 DIM E(49)
18 DIM F(49)
20 N=2
22 GOSUB 3500
25 LPRINT CHR$(27);"@";CHR$(15);CHR$(27);"A";CHR$(8);CHR$(27);"G";CHR$(27);"Q";C
HR$(128);CHR$(27);"M";CHR$(10):WIDTH "LPT1:",128
32 T1=10
34 T2=20
36 T3=1100
38 T=0
40 P=0
45 REM TP = 1 Sec.: PRR = 1 / 1000  Pulses / Sec.
50 L= 222684!
52 C=.455
54 R=35
56 RI=44
58 V=1
60 X(2)=-.2
80 LPRINT" Integration Step =";T1;"Seconds":LPRINT
86 GOSUB 1000
90 GOSUB 2000
100 REM INTEGRATION ROUTINE
110 GOSUB 1000
120 FOR A = 1 TO N
130 E(A)=X(A)
140 F(A)=D(A)
150 X(A)=X(A)+T1*D(A)
154 IF X(1)<0 THEN X(1)=0
160 NEXT A
165 T=T+T1
170 GOSUB 1000
180 FOR A = 1 TO N
190 X(A)=E(A)+T1/2*(F(A)+D(A))
194 IF X(1)<0 THEN X(1)=0
200 NEXT A
220 P=P+T1
230 IF P+T1/5>T2 THEN GOTO 250
240 GOTO 110
250 P=0
260 REM PRINTOUT ROUTINE
270 GOSUB 2008
```

```
280 IF T+T1/5>T3 THEN GOTO 292
290 GOTO 110
292 IF A$="P" OR A$="p" THEN GOSUB 3030:END
294 LPRINT"---------------------------------------------"
300 END
888 DIM X(49)
1000 REM EQUATION SUBROUTINE
1010 D(1)=(V-R*X(1)-X(2))/L
1015 II=X(2)/RI:IF II>0 THEN II=0
1020 D(2)=(X(1)-II)/C
1030 RETURN
2000 REM PRINT SUBROUTINE
2002 IF A$="P" OR A$="p" THEN GOSUB 3000:RETURN
2004 LPRINT"D-C Resonant Charging And Inverse Diode Circuit"
2005 LPRINT"---------------------------------------------":LPRINT
2006 LPRINT TAB(10);"Sec.";TAB(20);"x(1)";TAB(30);"x(2)":LPRINT
2008 IF A$="P" OR A$="p" THEN GOSUB 3040:RETURN
2010 LPRINT TAB(10);:LPRINT USING"####";T;
2012 LPRINT TAB(20);:LPRINT USING".#####";X(1);
2014 LPRINT TAB(30);:LPRINT USING"##.###";X(2)
2020 RETURN
3000 REM  Plotting Subroutine
3010 LPRINT"                        D-C Resonant Charging And Inverse Diode Circuit"
3015 LPRINT"                       ---------------------------------------------"
3020 LPRINT"     -.2     0     .2     .4     .6     .8     1.0     1.2     1.4     1.6     1.8     2.0"
3030 LPRINT"     !....+....!....+....!....+....!....+....!....+....!....+....!....+....!....+....!....+....!....+....!....+....!":IF T>0 THEN RETURN
3032 GOTO 3044
3040 SC=SC+1
3044 PP=(18+X(2)*50):IF T=0 THEN GOTO 3100
3060 IF SC < 5 THEN LPRINT USING"##.###";X(2);:LPRINT TAB(8);"-";
3062 IF SC < 5 THEN GOTO 3150
3100 LPRINT USING "##.###";X(2);:LPRINT TAB(8);"--T=";:LPRINT USING"####";T;:LPRINT" Sec.";:SC=0:IF PP>20 THEN GOTO 3200
3110 LPRINT"":RETURN
3150 IF PP<9 THEN LPRINT"":RETURN
3200 LPRINT TAB(PP);"+":RETURN
3500 REM Printer On/Tabular Output/Plotted Output Subroutine
3510 CLS:PRINT"Your printer must be on to run this program.  Is it on (Y or N)?"
3520 A$=INKEY$:IF A$="" THEN 3520
3530 IF A$="n" OR A$="N" THEN PRINT"For operation without a printer go to DOS. Type INFORMATION.":END
3540 PRINT"For a plotted response enter P.  For a tabular response enter any other letter."
3550 A$=INKEY$:IF A$="" THEN 3550
3560 CLS:PRINT"                    PROGRAM IS IN PROGRESS
                                    ------- -- -- --------"
3570 RETURN


10 REM APPENDIX.29 -Three Phase F.W.P.S. With DC Res. Charging & Inverse Diode-
12 GOSUB 3500
13 PI=3.141592654#:SC=-1:COMU=1
14 F=62.5:A1=PI/6:A2=5*PI/6:A3=3*PI/2
```

```
16 W=2*PI*F
21 LPRINT CHR$(27);"@";CHR$(15);CHR$(27);"A";CHR$(8);CHR$(27);"G";CHR$(27);"M";C
HR$(6);:WIDTH "LPT1:",140
22 DIM X(49)
24 DIM D(49)
26 DIM E(49)
28 DIM F(49)
30 N=8
32 T1=.00001:TP=T1
34 T2=.001
36 T3=.08
38 T=0
40 P=0
50 X(1)=0
51 X(2)=18.84197:X(3)=18.84197
52 X(4)=9665.344:X(5)=19.20942:X(6)=9525.029:X(8)=-925:II=-46.25
53 V1=3023:VS1=V1
54 V2=3023:VS2=V2
55 V3=-6046:VS3=V3
58 L=2
60 LL=.0667
62 C=.00005:CI=.000002
64 RI=1000
66 VP=6046
70 LCH=.10132
72 CN=.000001
74 RCH=16
76 RIN=20
80 LPRINT"       --- 3 Phase Bridge Rectifier Integration Time =";T1;" Seconds ---"
:LPRINT
84 GOSUB 1000
86 GOSUB 2000
100 REM Integration Subroutine
110 TT=TT+T1
112 IF TR>.001-T1/2 THEN VCH=X(8):TR=0:X(8)=-.05*VCH
114 GOSUB 400
120 FOR A = 1 TO N
130 E(A)=X(A)
140 F(A)=D(A)
150 X(A)=X(A)+T1*D(A)
152 IF X(7)<0 THEN X(7)=0
160 NEXT A
165 T=T+T1:K=W*T:V1=VP*SIN(A1+K):V2=VP*SIN(A2+K):V3=VP*SIN(A3+K):TR=TR+T1
170 IF TS=<2 THEN GOSUB 1010 ELSE IF TS=<4 THEN GOSUB 1030 ELSE IF TS=<6 THEN GO
SUB 1050 ELSE IF TS=<8 THEN GOSUB 1070 ELSE IF TS=<10 THEN GOSUB 1090 ELSE IF TS
=< 12 THEN GOSUB 1110
180 FOR A = 1 TO N
190 X(A)=E(A)+T1/2*(F(A)+D(A))
192 IF X(7)<0 THEN X(7)=0
200 NEXT A
202 IF X(1)=>0 AND X(2)=>0 AND X(3)=>0 THEN GOTO 220
204 IF X(1)<0 THEN B=1 ELSE IF X(2)<0 THEN B=2 ELSE IF X(3)<0 THEN B=3
206 T=T-T1:TC=E(B)*T1/(E(B)-X(B)):FOR A=1 TO N:X(A)=E(A)+TC*F(A):NEXT A:T=T+TC:X
(B)=0:D(B)=0:K=W*T:V1=VP*SIN(A1+K):V2=VP*SIN(A2+K):V3=VP*SIN(A3+K)
208 IF TS=<2 THEN GOSUB 1010 ELSE IF TS=<4 THEN GOSUB 1030 ELSE IF TS=<6 THEN GO
SUB 1050 ELSE IF TS=<8 THEN GOSUB 1070 ELSE IF TS=<10 THEN GOSUB 1090 ELSE IF TS
=< 12 THEN GOSUB 1110
210 TX=TP-TC:FOR A=1 TO N:X(A)=X(A)+TX*D(A):NEXT A:T=T+TX:B=0
```

```
220 P=P+T1
230 IF P+T1/5>T2 THEN GOTO 250
240 GOTO 110
250 P=0
260 REM PRINTOUT ROUTINE
270 GOSUB 2013
280 IF T+T1/5>T3 THEN GOTO 291
290 GOTO 110
291 IF A$="P" OR A$="p" THEN GOSUB 3030:CLS:END
292 LPRINT
294 LPRINT"----------------------------------------------------------------
----------":END
300 CLS:END
400 REM Commutation Subroutine
401 D(7)=(X(6)-RCH*X(7)-X(8))/LCH:II=X(8)/RIN:IF II>0 THEN II=0
402 D(8)=(X(7)-II)/CN
404 IF V1=>(V2-LL*D(2)) AND -V2<(-LL*D(3)-V3) THEN GOSUB 1010:RETURN
405 IF -V2=>(-LL*D(3)-V3) AND V3<(V1-LL*D(1)) THEN GOSUB 1030:RETURN
406 IF V3=>(V1-LL*D(1)) AND -V1<(-LL*D(2)-V2) THEN GOSUB 1050:RETURN
407 IF -V1=>(-LL*D(2)-V2) AND V2<(V3-LL*D(3)) THEN GOSUB 1070:RETURN
408 IF V2=>(V3-LL*D(3)) AND -V3<(-LL*D(1)-V1) THEN GOSUB 1090:RETURN
409 IF -V3=>(-LL*D(1)-V1) AND V1<(V2-LL*D(2)) THEN GOSUB 1110:RETURN
410 RETURN
1000 REM Equation Subroutine
1010 REM Forward paths 1 & 2.  Return path 6.
1011 IF X(2)=0 THEN GOTO 1020
1012 D(4)=(X(1)+X(2)-X(5))/CI
1013 D(5)=(X(4)+CI*RI*D(4)-X(6))/L
1014 D(1)=(2*V1-V2-V3-CI*RI*D(4)-X(4))/(3*LL)
1015 D(2)=(2*V2-V1-V3-CI*RI*D(4)-X(4))/(3*LL)
1016 D(3)=D(1)+D(2)
1017 D(6)=(X(5)-X(7))/C:TS=1:RETURN
1020 REM Forward path 1.  Return path 4.
1022 D(4)=(X(1)-X(5))/CI
1023 D(5)=(X(4)+CI*RI*D(4)-X(6))/L
1024 D(1)=(V1-V3-CI*RI*D(4)-X(4))/(2*LL)
1025 D(3)=D(1):D(2)=0:X(2)=0:X(3)=X(1)
1026 D(6)=(X(5)-X(7))/C:TS=2:RETURN
1030 REM Forward path 1.  Return paths 5 & 6.
1031 IF X(3)=0 THEN GOTO 1040
1032 D(4)=(X(2)+X(3)-X(5))/CI
1033 D(5)=(X(4)+CI*RI*D(4)-X(6))/L
1034 D(2)=(V1-2*V2+V3-CI*RI*D(4)-X(4))/(3*LL)
1035 D(3)=(V1-2*V3+V2-CI*RI*D(4)-X(4))/(3*LL)
1036 D(1)=D(2)+D(3)
1037 D(6)=(X(5)-X(7))/C:TS=3:RETURN
1040 REM Forward path 1.  Return path 5.
1042 D(4)=(X(2)-X(5))/CI
1043 D(5)=(X(4)+CI*RI*D(4)-X(6))/L
1044 D(2)=(V1-V2-CI*RI*D(4)-X(4))/(2*LL)
1045 D(1)=D(2):D(3)=0:X(3)=0:X(1)=X(2)
1046 D(6)=(X(5)-X(7))/C:TS=4:RETURN
1050 REM Forward paths 1 & 3.  Return paths 5.
1051 IF X(1)=0 THEN GOTO 1060
1052 D(4)=(X(1)+X(3)-X(5))/CI
1053 D(5)=(X(4)+RI*CI*D(4)-X(6))/L
1054 D(1)=(2*V1-V2-V3-RI*CI*D(4)-X(4))/(3*LL)
1055 D(3)=(2*V3-V2-V1-RI*CI*D(4)-X(4))/(3*LL)
```

```
1056 D(2)=D(3)+D(1)
1057 D(6)=(X(5)-X(7))/C:TS=5:RETURN
1060 REM Forward path 3.  Return path 5.
1062 D(4)=(X(3)-X(5))/CI
1063 D(5)=(X(4)+RI*CI*D(4)-X(6))/L
1064 D(3)=(V3-V2-RI*CI*D(4)-X(4))/(2*LL)
1065 D(2)=D(3):D(1)=0:X(1)=0:X(2)=X(3)
1066 D(6)=(X(5)-X(7))/C:TS=6:RETURN
1070 REM Forward path 1. Return paths 2 and 3.
1071 IF X(2)=0 THEN GOTO 1080
1072 D(4)=(X(1)+X(2)-X(5))/CI
1073 D(5)=(X(4)+CI*RI*D(4)-X(6))/L
1074 D(1)=(V3-2*V1+V2-CI*RI*D(4)-X(4))/(3*LL)
1075 D(2)=(V3-2*V2+V1-CI*RI*D(4)-X(4))/(3*LL)
1076 D(3)=D(1)+D(2)
1077 D(6)=(X(5)-X(7))/C:TS=7:RETURN
1080 REM Forward path 3.  Return path 4.
1082 D(4)=(X(1)-X(5))/CI
1083 D(5)=(X(4)+CI*RI*D(4)-X(6))/L
1084 D(1)=(V3-V1-CI*RI*D(4)-X(4))/(2*LL)
1085 D(3)=D(1):D(2)=0:X(2)=0:X(3)=X(1)
1086 D(6)=(X(5)-X(7))/C:TS=8:RETURN
1090 REM Forward paths 2 & 3.  Return path 4.
1091 IF X(3)=0 THEN GOTO 1100
1092 D(4)=(X(2)+X(3)-X(5))/CI
1093 D(5)=(X(4)+CI*RI*D(4)-X(6))/L
1094 D(2)=(2*V2-V3-V1-CI*RI*D(4)-X(4))/(3*LL)
1095 D(3)=(2*V3-V1-V2-CI*RI*D(4)-X(4))/(3*LL)
1096 D(1)=D(2)+D(3)
1097 D(6)=(X(5)-X(7))/C:TS=9:RETURN
1100 REM Forward path 2.  Return path 4.
1102 D(4)=(X(2)-X(5))/CI
1103 D(5)=(X(4)+RI*CI*D(4)-X(6))/L
1104 D(2)=(V2-V1-CI*RI*D(4)-X(4))/(2*LL)
1105 D(1)=D(2):D(3)=0:X(3)=0:X(1)=X(2)
1106 D(6)=(X(5)-X(7))/C:TS=10:RETURN
1110 REM Forward path 2.  Return paths 4 & 6.
1111 IF X(1)=0 THEN GOTO 1120
1112 D(4)=(X(1)+X(3)-X(5))/CI
1113 D(5)=(X(4)+RI*CI*D(4)-X(6))/L
1114 D(1)=(V2-2*V1+V3-RI*CI*D(4)-X(4))/(3*LL)
1115 D(3)=(V2-2*V3+V1-RI*CI*D(4)-X(4))/(3*LL)
1116 D(2)=D(1)+D(3)
1117 D(6)=(X(5)-X(7))/C:TS=11:RETURN
1120 REM Forward path 2.  Return path 6.
1122 D(4)=(X(3)-X(5))/CI
1123 D(5)=(X(4)+RI*CI*D(4)-X(6))/L
1124 D(3)=(V2-V3-RI*CI*D(4)-X(4))/(2*LL)
1125 D(2)=D(3):D(1)=0:X(1)=0:X(2)=X(3)
1126 D(6)=(X(5)-X(7))/C:TS=12:RETURN
2000 REM Print Subroutine
2002 IF A$="P" OR A$="p" THEN LPRINT CHR$(27);"G";:WIDTH "LPT1:",140:GOTO 2013
2004 LPRINT TAB(1);"Sec.   TS";TAB(14);"V1";TAB(20);"V2";TAB(26);"V3";TAB(31);"X
(1)";TAB(36);"X(2)";TAB(41);"X(3)";TAB(46);"X(4)";TAB(51);"X(5)";TAB(56);"X(6)";
TAB(62);"X(7)";
2005 LPRINT TAB(68)"X(5)";TAB(73);"X(7)";TAB(80);"Vch";TAB(87);"II":LPRINT
2013 FOR Z=1 TO 7:I(Z)=0:NEXT Z
2014 IF TS=1 THEN I(1)=X(1):I(2)=X(2):I(6)=X(1)+X(2):I(7)=X(6):GOTO 2026
```

```
2015 IF TS=2 THEN I(1)=X(1):I(2)=X(2):I(6)=X(1):I(7)=X(6):GOTO 2026
2016 IF TS=3 THEN I(1)=X(2)+X(3):I(5)=X(2):I(6)=X(3):I(7)=X(6):GOTO 2026
2017 IF TS=4 THEN I(1)=X(2):I(5)=X(2):I(6)=X(3):I(7)=X(6):GOTO 2026
2018 IF TS=5 THEN I(1)=X(1):I(3)=X(3):I(5)=X(1)+X(3):I(7)=X(6):GOTO 2026
2019 IF TS=6 THEN I(1)=X(1):I(3)=X(3):I(5)=X(3):I(7)=X(6):GOTO 2026
2020 IF TS=7 THEN I(3)=X(1)+X(2):I(4)=X(1):I(5)=X(2):I(7)=X(6):GOTO 2026
2021 IF TS=8 THEN I(3)=X(1):I(4)=X(1):I(6)=X(3):I(7)=X(6):GOTO 2026
2022 IF TS=9 THEN I(2)=X(2):I(3)=X(3):I(4)=X(2)+X(3):I(7)=X(6):GOTO 2026
2023 IF TS=10 THEN I(2)=X(2):I(3)=X(3):I(4)=X(2):I(7)=X(6):GOTO 2026
2024 IF TS=11 THEN I(2)=X(1)+X(3):I(4)=X(1):I(6)=X(3):I(7)=X(6):GOTO 2026
2025 IF TS=12 THEN I(2)=X(3):I(4)=X(1):I(6)=X(3):I(7)=X(6):GOTO 2026
2026 IF B=>1 THEN GOTO 2040
2027 IF A$="P" OR A$="p" THEN GOTO 2029
2028 GOTO 2040
2029 IF T=0 THEN GOSUB 3000:RETURN
2030 GOSUB 3040:RETURN
2040 LPRINT TAB(1);:LPRINT USING".#####";T;:LPRINT;TS;
2041 LPRINT TAB(12);:LPRINT USING"#####";V1;
2042 LPRINT TAB(18);:LPRINT USING"#####";V2;
2043 LPRINT TAB(24);:LPRINT USING"#####";V3;
2044 LPRINT TAB(30);:LPRINT USING"###.#";I(1);
2045 LPRINT TAB(35);:LPRINT USING"###.#";I(2);
2046 LPRINT TAB(40);:LPRINT USING"###.#";I(3);
2047 LPRINT TAB(45);:LPRINT USING"###.#";I(4);
2048 LPRINT TAB(50);:LPRINT USING"###.#";I(5);
2049 LPRINT TAB(55);:LPRINT USING"###.#";I(6);
2050 LPRINT TAB(60);:LPRINT USING"######";I(7);
2051 LPRINT TAB(67);:LPRINT USING"###.#";X(5);
2052 LPRINT TAB(72);:LPRINT USING"###.#";X(7);
2053 LPRINT TAB(77);:LPRINT USING"######";X(8);
2054 LPRINT TAB(84);:LPRINT USING"###.#";II
2055 RETURN
2100 REM RMS Subroutine
2110 CO=CO+1
2120 RMS1=RMS1+VS1^2:RMS2=RMS2+VS2^2:RMS3=RMS3+VS3^2
2130 IF CO<40 GOTO 2160
2140 RMS1=SQR(RMS1/40):RMS2=SQR(RMS2/40):RMS3=SQR(RMS3/40)
2150 LPRINT:LPRINT" ----- For The Last Cycle The RMS Values Are: VS1 =";RMS1;"
 VS2 =";RMS2;" VS3 =";RMS3;" -----":LPRINT:RMS1=0:RMS2=0:RMS3=0:CO=0
2160 RETURN
2200 REM RMS Subroutine
3000 REM Plotting Subroutine
3010 LPRINT"                              DC Resonant Charging From A Three Phase
 Bridge DC Power Supply"
3015 LPRINT"                       -- -------- -------- ---- - ----- -----
------ -- ----- ------"
3020 LPRINT"         18.0      18.1      18.2      18.3      18.4      18.5      18
.6      18.7      18.8      18.9      19.0      19.1      19.2"
3030 LPRINT"         !....+....!....+....!....+....!....+....!....+....!....+....!....+....!....+....!....+....!....+....!....+....!....+....!":IF T>0 THEN RETUR
N
3032 LPRINT TAB(1);"X(6) kV";TAB(23);"I(1)";TAB(29);"I(2)";TAB(35);"I(3)";TAB(41
);"I(4)";TAB(47);"I(5)";TAB(53);"I(6)"
3040 SC=SC+1
3044 PP=X(8)/10-1791
3060 IF SC<5 THEN LPRINT USING "#####.#";X(8);:LPRINT TAB(9);"-";
3062 IF SC<5 THEN GOTO 3072
3070 LPRINT USING "#####.#";X(8);:LPRINT TAB(9);"--T=";:LPRINT USING"#.###";T;:L
```

```
PRINT" Sec.";:SC=0
3072 LPRINT TAB(23);:LPRINT USING"##.#";I(1);:LPRINT TAB(29);:LPRINT USING"##.#"
;I(2);:LPRINT TAB(35);:LPRINT USING"##.#";I(3);:LPRINT TAB(41);:LPRINT USING"##.
#";I(4);:LPRINT TAB(47);:LPRINT USING"##.#";I(5);:LPRINT TAB(53);:LPRINT USING"#
#.#";I(6);
3074 IF PP<58 THEN LPRINT"":RETURN
3100 IF PP<10 THEN LPRINT"":RETURN
3110 IF PP>130 THEN LPRINT TAB(129)"+->":RETURN
3200 LPRINT TAB(PP);"+":RETURN
3500 REM Printer On/Tabular Output/Plotted Output Subroutine
3510 CLS:PRINT"          Your printer must be on to run this program.  Is it on (
Y or N)?"
3520 A$=INKEY$:IF A$="" THEN 3520
3530 IF A$="n" OR A$="N" THEN PRINT"For operation without a printer go to DOS. T
ype INFORMATION": END
3540 PRINT"For a plotted response enter P.  For a tabular response enter any oth
er letter."
3550 A$=INKEY$:IF A$="" THEN 3550
3560 CLS:PRINT"                     PROGRAM IS IN PROGRESS
                                    ------- -- -- --------"
3570 RETURN


10 REM APPENDIX.31 - MTI Stability Of TWT RF Power Generator (Voltage Source)
12 DIM X(49)
14 DIM D(49)
16 DIM E(49)
18 DIM F(49)
20 N=16
21 GOSUB 3500
24 LPRINT CHR$(27);"@";CHR$(15);CHR$(27);"A";CHR$(8);CHR$(27);"G";CHR$(27);"Q";C
HR$(128);CHR$(27);"M";CHR$(4):WIDTH "LPT1:",128
32 T1=1E-09:TR=T1
34 T2=.0000002
36 T3=.000015
38 T=0
40 P=0
51 CI=4E-09
52 CF=7E-11
53 CO=2.5E-10
54 C=3.13E-11
55 CS=.0000001
56 LP=.004
57 LS=3.6
58 M=.11994
59 RB=33333!
60 RD=2.5:RF1=1000000!:RF2=RF1:RF3=RF1:RF4=RF1
61 V=1000
62 G=3.5
70 X(1)=0:X(2)=925:X(3)=925
71 X(4)=0:X(5)=925:X(6)=925
72 X(7)=0:X(8)=75:X(9)=75
73 X(10)=0:X(11)=75:X(12)=75
74 X(13)=-25000
75 X(14)=25000.01
76 X(15)=-17.5
77 X(16)=-.583
```

```
79 K1=CF/(CI+CF):K2=(2*C0+2*CI*K1):K3=CI+CF
80 LPRINT"Minimum Regulator Integration Step = ";:LPRINT USING".####";T1*1000000
!;:LPRINT" Microseconds":LPRINT
84 GOSUB 1000
86 GOSUB 2000
100 REM Integration Routine
110 GOSUB 1000
120 FOR A = 1 TO N
130 E(A)=X(A)
140 F(A)=D(A)
150 X(A)=X(A)+T1*D(A)
160 NEXT A
165 T=T+T1
170 GOSUB 1000
180 FOR A = 1 TO N
190 X(A)=E(A)+T1/2*(F(A)+D(A))
200 NEXT A
220 P=P+T1
230 IF P+T1/5>T2 THEN GOTO 250
240 GOTO 110
250 P=0
260 REM PRINTOUT ROUTINE
270 GOSUB 2010
280 IF T+T1/5>T3 THEN GOTO 295
290 GOTO 110
295 IF A$="P" OR A$="p" THEN LPRINT"          +....!....+....!....+....!....+...
.!....+....!....+....!....+....!....+....!....+....!....+....!....+....!....+":E
ND
300 LPRINT"------------------------------------------------------------------":END
1000 REM Equation Subroutine
1001 GOSUB 1100
1002 D(15)=(LS*(V-X(3)-X(6))-M*X(13))/(LP*LS-M^2)
1004 D(16)=(M*D(15)-X(13))/LS
1006 II1=(VD1-X(1))/RD
1008 II2=(VD2-X(4))/RD
1010 II3=(VD3-X(7))/RD
1012 II4=(VD4-X(10))/RD
1014 D(3)=(X(15)+X(12)/RF4-X(3)/RF1+GM4*X(10)-GM1*X(1)+K1*(II1-II4))/K2
1016 D(2)=(CI*D(3)-II1)/K3
1018 D(1)=D(3)-D(2)
1020 D(12)=-D(3)
1022 D(11)=-(CI*D(3)+II4)/K3
1024 D(10)=-(D(3)+D(11))
1026 D(9)=(-X(15)+X(6)/RF2-X(9)/RF3+GM2*X(4)-GM3*X(7)+K1*(II3-II2))/K2
1028 D(8)=(CI*D(9)-II3)/K3
1030 D(7)=D(9)-D(8)
1032 D(6)=-D(9)
1034 D(5)=-(CI*D(9)+II2)/K3
1036 D(4)=-(D(9)+D(5))
1038 D(13)=X(16)/C:D(14)=-X(14)/(RB*CS):IF ABS(X(13))=>X(14) AND SGN(X(13))=SGN(
X(16)) THEN GOTO 1044
1040 IF DR=0 THEN GOSUB 1330:RETURN
1042 GOSUB 1300:RETURN
1044 GOSUB 1330
1046 X(13)=SGN(X(13))*X(14):D(14)=(ABS(X(16))-X(14)/RB)/(C+CS):D(13)=SGN(D(13))*
D(14):RETURN
1100 REM Switching Subroutine
```

```
1102 IF T>.000001 THEN RB=5E+07
1104 IF T>.000005 THEN RB=33333!
1106 IF T>.000006 THEN RB=5E+07
1108 IF T>.00001 THEN RB=33333!
1110 IF T>.000011 THEN RB=5E+07
1112 IF X(14)>25000 THEN VD1=0:VD2=0:VD3=0:VD4=0:DR=0:GOTO 1170
1114 DR=1
1116 IF RB=33333! THEN VD1=0:VD2=0:VD3=0:VD4=0:DR=0:GOTO 1170
1118 IF T=>.0000104 AND VD3=>5 THEN VD3=5:VD4=5:VD1=0:VD2=0:GOTO 1170
1120 IF T=>.0000104 AND VD3<5 THEN GOTO 1160
1122 IF T=>.0000004 AND VD1=>5 THEN VD1=5:VD2=5:VD3=0:VD4=0:GOTO 1170
1124 IF T=>.0000004 AND VD1<5 THEN GOTO 1140
1126 IF T=<.0000004 AND VD3=>5 THEN VD3=5:VD4=5:VD1=0:VD2=0:GOTO 1170
1130 IF T=<.0000004 AND VD3<5 THEN GOTO 1160
1140 VD1=VD1+.0625:VD2=VD1:VD3=VD3-.0625:VD4=VD3:GOTO 1170
1160 VD3=VD3+.0625:VD4=VD3:VD1=VD1-.0625:VD2=VD1
1170 RF1=1000000!:RF2=1000000!:RF3=1000000!:RF4=1000000!:GM1=0:GM2=0:GM3=0:GM4=0
1180 IF X(3)<0 OR X(6)<0 THEN RF1=1:RF2=1:GOTO 1220
1190 IF X(1)<0 OR X(4)<0 THEN GOTO 1220
1200 IF X(3)<15*X(1) OR X(6)<15*X(4) THEN GM1=X(3)*G/(15*X(1)):GM2=GM1:GOTO 1220
1210 GM1=G:GM2=G
1220 IF X(9)<0 OR X(12)<0 THEN RF3=1:RF4=1:RETURN
1230 IF X(7)<0 OR X(10)<0 THEN RETURN
1240 IF X(9)<15*X(7) OR X(12)<15*X(10) THEN GM3=X(9)*G/(15*X(7)):GM4=GM3:RETURN
1250 GM3=G:GM4=G:RETURN
1300 REM Integration Period Change
1310 IF P<>0 THEN RETURN
1320 T1=TR:RETURN
1330 IF P<>0 THEN RETURN
1340 T1=5*TR:RETURN
2000 REM Print Subroutine
2002 IF A$="P" OR A$="p" THEN GOSUB 3000:RETURN
2003 LPRINT:LPRINT"                            Tabulated Characteristics Of A
 D-C To D-C Regulator Using A Voltage Source":LPRINT
2004 LPRINT TAB(1);"Microsec.";TAB(10);"   X(1)";TAB(17);"   X(2)";TAB(24);"   X(3)
";TAB(31);"   X(4)";TAB(38);"   X(5)";TAB(45);"   X(6)";TAB(52);"   X(7)";TAB(59);"
   X(8)";TAB(66);"   X(9)";TAB(72);"   X(10)";TAB(79);"   X(11)";TAB(86);"   X(12)";
2006 LPRINT TAB(94);"    X(13)";TAB(103);"     X(14)";TAB(112);"   X(15)";TAB(119)
;"   X(16)"
2008 LPRINT:GOTO 2012
2010 IF A$="P" OR A$="p" THEN GOSUB 3040:RETURN
2012 LPRINT TAB(1);:LPRINT USING"##.###";T*1000000!;:LPRINT TAB(10);:LPRINT USIN
G"####.#";X(1);:LPRINT TAB(17);:LPRINT USING"####.#";X(2);:LPRINT TAB(24);:LPRIN
T USING"####.#";X(3);:LPRINT TAB(31);:LPRINT USING"####.#";X(4);
2014 LPRINT TAB(38);:LPRINT USING"####.#";X(5);:LPRINT TAB(45);:LPRINT USING"###
#.#";X(6);:LPRINT TAB(52);:LPRINT USING"####.#";X(7);:LPRINT TAB(59);:LPRINT USI
NG"####.#";X(8);:LPRINT TAB(66);:LPRINT USING"####.#";X(9);
2016 LPRINT TAB(73);:LPRINT USING"####.#";X(10);:LPRINT TAB(80);:LPRINT USING"##
##.#";X(11);:LPRINT TAB(87);:LPRINT USING"####.#";X(12);:LPRINT TAB(94);:LPRINT
USING"######.##";X(13);:LPRINT TAB(103);:LPRINT USING"######.##";X(14);
2018 LPRINT TAB(113);:LPRINT USING"###.##";X(15);:LPRINT TAB(120);:LPRINT USING"
##.###";X(16):RETURN
3000 REM Plotting Subroutine
3010 LPRINT"                                              Transient Response Of
Switching Voltage"
3015 LPRINT"                                            --------- -------- --
--------- -------"
3020 LPRINT"        -25000    -20000    -15000    -10000    -5000        0
```

```
                   5000      10000      15000      20000      25000"
3030 LPRINT"               +....!....+....!....+....!....+....!....+....!....+....!..
..+....!....+....!....+....!....+....!....+....!....+"
3032 LPRINT"Beam Volts":LPRINT USING"######.##";X(14);:LPRINT TAB(11);"--";:GOTO
 3044
3040 SC=SC+1
3044 PP=(66+ X(13)/500):IF T=0 THEN GOTO 3200
3060 IF SC<5 THEN LPRINT USING "######.##";X(14);:LPRINT TAB(11);"-";
3062 IF SC<5 THEN GOTO 3150
3100 LPRINT USING "######.##";X(14);:LPRINT TAB(11);"--T=";:LPRINT USING"##.##";
T*1000000!;:LPRINT" uS.";:SC=0:IF PP>25 THEN GOTO 3200
3110 LPRINT"":RETURN
3150 IF PP<12 THEN LPRINT"":RETURN
3200 LPRINT TAB(PP);"+":RETURN
3500 REM Printer On/Tabular Output/Plotted Output Subroutine
3510 CLS:PRINT"Your printer must be on to run this program.  Is it on (Y or N)?"
3520 A$=INKEY$:IF A$="" THEN 3520
3530 IF A$="n" OR A$="N" THEN PRINT"For operation without a printer go to DOS. T
ype INFORMATION.":END
3540 PRINT"For a plotted response enter P.  For a tabular response enter any oth
er letter."
3550 A$=INKEY$:IF A$="" THEN 3550
3560 CLS:PRINT"                        PROGRAM IS IN PROGRESS
                                        ------- -- -- --------"
3570 RETURN


10 REM APPENDIX.33 - MTI Stability Of TWT RF Pwr.Gen.(Cur.Source/ Voltage Lim.)
12 DIM X(49)
14 DIM D(49)
16 DIM E(49)
18 DIM F(49)
20 N=16
21 GOSUB 3500
24 LPRINT CHR$(27);"@";CHR$(15);CHR$(27);"A";CHR$(6);CHR$(27);"G";:WIDTH "LPT1:"
,128
32 T1=2E-09
34 T2=.0000001
36 T3=.0000075
38 T=0:Q=T
40 P=0
51 CI=4E-09
52 CF=7E-11
53 CO=2.5E-10
54 C=3.13E-11
55 CS=.0000001
56 LP=.004
57 LS=3.6
58 M=.11994
59 RB=5E+07
60 RD=2.5:RF1=1000000!:RF2=RF1:RF3=RF1:RF4=RF1
61 V=1000
62 G=3.5
70 X(1)=0:X(2)=900:X(3)=900
71 X(4)=0:X(5)=900:X(6)=900
72 X(7)=0:X(8)=100:X(9)=100
73 X(10)=0:X(11)=100:X(12)=100
74 X(13)=-21300
```

```
75 X(14)=25000.3
76 X(15)=0
77 X(16)=0
78 IS=16.5:M$="I"
79 K1=CF/(CI+CF):K2=(2*CO+2*CI*K1):K3=CI+CF
80 LPRINT"Regulator Integration Step = ";:LPRINT USING".####";T1*1000000!;:LPRIN
T" Microseconds":LPRINT
84 GOSUB 1000
86 GOSUB 2000
100 REM Integration Routine
110 GOSUB 1000
120 FOR A = 1 TO N
130 E(A)=X(A)
140 F(A)=D(A)
150 X(A)=X(A)+T1*D(A)
160 NEXT A
165 T=T+T1
170 GOSUB 1000
180 FOR A = 1 TO N
190 X(A)=E(A)+T1/2*(F(A)+D(A))
200 NEXT A
220 P=P+T1
230 IF P+T1/5>T2 THEN GOTO 250
240 GOTO 110
250 P=0
260 REM PRINTOUT ROUTINE
270 GOSUB 2010
280 IF T+T1/5>T3 THEN GOTO 295
290 GOTO 110
295 IF A$="P" OR A$="p" THEN LPRINT"              |....+....|....+....|....+...
.|....+....|....+....|....+....|....+....|....+....|....+....|....+....|....+...
.|":END
300 LPRINT" ---------------------------------------------------------------
--------------------------------------------------":END
1000 REM Equation Subroutine
1002 GOSUB 1100
1004 II1=(VD1-X(1))/RD
1006 II2=(VD2-X(4))/RD
1008 II3=(VD3-X(7))/RD
1010 II4=(VD4-X(10))/RD
1012 IF M$="I" THEN GOTO 1050
1014 REM Voltage Source
1016 D(3)=(X(15)+X(12)/RF4-X(3)/RF1+GM4*X(10)-GM1*X(1)+K1*(II1-II4))/K2:D(2)=(CI
*D(3)-II1)/K3:D(1)=D(3)-D(2):D(12)=-D(3):D(11)=-(CI*D(3)+II4)/K3:D(10)=-(D(3)+D(
11))
1018 D(9)=(-X(15)+X(6)/RF2-X(9)/RF3+GM2*X(4)-GM3*X(7)+K1*(II3-II2))/K2:D(8)=(CI*
D(9)-II3)/K3:D(7)=D(9)-D(8):D(6)=-D(9):D(5)=-(CI*D(9)+II2)/K3:D(4)=-(D(9)+D(5))
1020 I1=CO*D(3)+CF*D(2)+GM1*X(1)+X(3)/RF1
1022 I2=CO*D(9)+CF*D(8)+GM3*X(7)+X(9)/RF3
1024 IF ABS(X(15))<IS AND (I1+I2)>IS THEN M$="I"
1026 IF ABS(X(15))>IS AND (I1-I2)<IS THEN M$="I"
1028 D(15)=(LS*(V-X(3)-X(6))-M*X(13))/(LP*LS-M^2)
1030 D(16)=(M*D(15)-X(13))/LS
1032 D(13)=X(16)/C:D(14)=-X(14)/(RB*CS):IF ABS(X(13))=>X(14) AND SGN(X(13))=SGN(
X(16)) THEN GOTO 1036
1034 RETURN
1036 X(13)=SGN(X(13))*X(14):D(14)=(ABS(X(16))-X(14)/RB)/(C+CS):D(13)=SGN(D(13))*
D(14):RETURN
```

```
1050 REM Current Source
1052 D(3)=((IS+X(15))/2-X(3)/RF1-GM1*X(1)+K1*II1)/(CO+CI*K1):D(2)=(CI*D(3)-II1)/
K3:D(1)=D(3)-D(2):D(6)=D(3):D(5)=D(2):D(4)=D(1)
1054 D(9)=((IS-X(15))/2-X(9)/RF3-GM3*X(7)+K1*II3)/(CO+CI*K1):D(8)=(CI*D(9)-II3)/
K3:D(7)=D(9)-D(8):D(12)=D(9):D(11)=D(8):D(10)=D(7)
1060 IF (X(3)+X(12))=>V THEN M$="V":GOTO 1014
1062 D(15)=(LS*(X(12)-X(3))-M*X(13))/(LP*LS-M^2)
1064 D(16)=(M*D(15)-X(13))/LS
1068 D(13)=X(16)/C:D(14)=-X(14)/(RB*CS):IF ABS(X(13))=>X(14) AND SGN(X(13))=SGN(
X(16)) THEN GOTO 1072
1070 RETURN
1072 X(13)=SGN(X(13))*X(14):D(14)=(ABS(X(16))-X(14)/RB)/(C+CS):D(13)=SGN(D(13))*
D(14):RETURN
1100 REM Switching Subroutine
1101 IF T>.000001 THEN RB=33333!
1102 IF T>.000002 THEN RB=5E+07
1103 IF T>.000006 THEN RB=33333!
1104 IF T>.000007 THEN RB=5E+07
1105 IF T>.000011 THEN RB=33333!
1106 IF T>.000012 THEN RB=5E+07
1107 IF T>.000016 THEN RB=33333!
1108 IF T>.000017 THEN RB=5E+07
1109 IF T>.000021 THEN RB=33333!
1110 IF T>.000022 THEN RB=5E+07
1111 IF T>.000026 THEN RB=33333!
1112 IF T>.000027 THEN RB=5E+07
1114 IF X(14)>25000 THEN VD1=0:VD2=0:VD3=0:VD4=0:GOTO 1170
1115 IF RB=33333! AND ABS(X(13))> 20000 THEN VD1=0:VD2=0:VD3=0:VD4=0:GOTO 1170
1116 IF T=>.0000204 AND VD1=>5 THEN VD1=5:VD2=5:VD3=0:VD4=0:GOTO 1170
1117 IF T=>.0000204 AND VD1<5 THEN GOTO 1140
1118 IF T=>.0000104 AND VD3=>5 THEN VD3=5:VD4=5:VD1=0:VD2=0:GOTO 1170
1119 IF T=>.0000104 AND VD3<5 THEN GOTO 1160
1120 IF T=>.0000004 AND VD1=>5 THEN VD1=5:VD2=5:VD3=0:VD4=0:GOTO 1170
1122 IF T=>.0000004 AND VD1<5 THEN GOTO 1140
1126 IF T=<.0000004 AND VD3=>5 THEN VD3=5:VD4=5:VD1=0:VD2=0:GOTO 1170
1130 IF T=<.0000004 AND VD3<5 THEN GOTO 1160
1140 VD1=VD1+.0625:VD2=VD1:VD3=VD3-.0625:VD4=VD3:GOTO 1170
1160 VD3=VD3+.0625:VD4=VD3:VD1=VD1-.0625:VD2=VD1
1170 RF1=1000000!:RF2=1000000!:RF3=1000000!:RF4=1000000!:GM1=0:GM2=0:GM3=0:GM4=0
1180 IF X(3)<0 OR X(6)<0 THEN RF1=1:RF2=1
1190 IF X(9)<0 OR X(12)<0 THEN RF3=1:RF4=1:RETURN
1200 IF X(1)<0 OR X(4)<0 THEN GOTO 1230
1210 IF X(3)<15*X(1) OR X(6)<15*X(4) THEN GM1=X(3)*G/(15*X(1)):GM2=GM1:GOTO 1230
1220 GM1=G:GM2=G
1230 IF X(7)<0 OR X(10)<0 THEN RETURN
1240 IF X(9)<15*X(7) OR X(12)<15*X(10) THEN GM3=X(9)*G/(15*X(7)):GM4=GM3:RETURN
1250 GM3=G:GM4=G:RETURN
2000 REM Print Subroutine
2002 IF A$="P" OR A$="p" THEN GOSUB 3000:RETURN
2003 LPRINT:LPRINT"              Tabulated Characteristics Of A D-C To D-C Reg
ulator With A Voltage Limited Current Source":LPRINT
2004 LPRINT TAB(1);"uSec.";TAB(7);"Mode";TAB(12);"X(1)";TAB(17);"  X(2)";TAB(24)
;"  X(3)";TAB(31);"  X(4)";TAB(38);"  X(5)";TAB(45);"  X(6)";TAB(52);"  X(7)";T
AB(59);"  X(8)";TAB(66);"  X(9)";TAB(72);"  X(10)";TAB(79);"  X(11)";TAB(86);"
X(12)";
2006 LPRINT TAB(94);"   X(13)";TAB(103);"   X(14)";TAB(112);" X(15)";TAB(119)
;"  X(16)"
2008 LPRINT:GOTO 2012
```

```
2010 IF A$="P" OR A$="p" THEN GOSUB 3040:RETURN
2012 LPRINT TAB(1);:LPRINT USING"##.###";T#1000000!;:LPRINT TAB(8);M$;TAB(10);:L
PRINT USING"####.#";X(1);:LPRINT TAB(17);:LPRINT USING"####.#";X(2);:LPRINT TAB(
24);:LPRINT USING"####.#";X(3);:LPRINT TAB(31);:LPRINT USING"####.#";X(4);
2014 LPRINT TAB(38);:LPRINT USING"####.#";X(5);:LPRINT TAB(45);:LPRINT USING"###
#.#";X(6);:LPRINT TAB(52);:LPRINT USING"####.#";X(7);:LPRINT TAB(59);:LPRINT USI
NG"####.#";X(8);:LPRINT TAB(66);:LPRINT USING"####.#";X(9);
2016 LPRINT TAB(73);:LPRINT USING"####.#";X(10);:LPRINT TAB(80);:LPRINT USING"##
##.#";X(11);:LPRINT TAB(87);:LPRINT USING"####.#";X(12);:LPRINT TAB(94);:LPRINT
USING"######.##";X(13);:LPRINT TAB(103);:LPRINT USING"######.##";X(14);
2018 LPRINT TAB(113);:LPRINT USING"###.##";X(15);:LPRINT TAB(120);:LPRINT USING"
##.###";X(16):RETURN
3000 REM Plotting Subroutine
3010 LPRINT"                                              Transient Response Of
Switching Voltage"
3015 LPRINT"
--------- ------- "
3020 LPRINT"                        -25000    -20000    -15000    -10000    -5000
          0      5000     10000     15000     20000     25000"
3030 LPRINT"Beam Volts    |....+....|....+....|....+....|....+....|....+....|..
..+....|....+....|....+....|....+....|....+....|"
3032 LPRINT USING"######.##";X(14);:LPRINT TAB(11);"--T=";:LPRINT USING"##.##";Q
#1000000!;:LPRINT" uS.";:GOTO 3044
3040 SC=SC+1
3044 PP=(76+ X(13)/500):IF T=Q THEN GOTO 3200
3060 IF SC<5 THEN LPRINT USING "######.##";X(14);:LPRINT TAB(11);"-";
3062 IF SC<5 THEN GOTO 3150
3100 LPRINT USING "######.##";X(14);:LPRINT TAB(11);"--T=";:LPRINT USING"##.##";
T#1000000!;:LPRINT" uS.";:SC=0:IF PP>25 THEN GOTO 3200
3110 LPRINT"":RETURN
3150 IF PP<12 THEN LPRINT"":RETURN
3200 LPRINT TAB(PP);"+":RETURN
3500 REM Printer On/Tabular Output/Plotted Output Subroutine
3510 CLS:PRINT"Your printer must be on to run this program.  Is it on (Y or N)?"
3520 A$=INKEY$:IF A$="" THEN 3520
3530 IF A$="n" OR A$="N" THEN PRINT"For operation without a printer go to DOS. T
ype INFORMATION.":END
3540 PRINT"For a plotted response enter P.  For a tabular response enter any oth
er letter."
3550 A$=INKEY$:IF A$="" THEN 3550
3560 CLS:PRINT"                        PROGRAM IS IN PROGRESS
                        ------- -- -- --------"
3570 RETURN


/* FOURIER.C  This is a graphical fourier analysis program */
#include <stdio.h>
#include <math.h>
#define Y 500
int n,n7,n8,i,n3,x,j; float t,a,t3,a0,C,D,E,A2,t1,v0,p,vr,ir,ps,vas;
float f[Y];float c[Y];float d[Y];float e[Y];float a2[Y];float v[Y];float av[Y];
void load_calculate(void);void printout(void);

main()
{
 /* Enter analysis parameters */
 printf("\n\n      Enter the following information from the keyboard:\n");
```

```
  printf("\n1. Number ordinates (number must be odd);<Enter>\n");scanf("%d",&n);
  printf("    Number = %d",n);
  printf("\n2. Lowest harmonic;space;highest harmonic,<Enter>.\n");scanf("%d%d",&n7,&n8);
  printf("    Lowest harmonic = %d, highest harmonic = %d",n7,n8);
  printf("\n3. Period of waveform in seconds.\n");scanf("%f",&t);
  printf("    Period = %f\n",t);
  load_calculate();
  printout();
}
  void load_calculate(void)
  {printf("\n  Type waveform ordinate;<Enter>\n");
   for (i=1;i<n+1;i++)
    {scanf("%f",&f[i]);printf("        Ordinate %d = %f\n",i-1,f[i]);}
   n3=n-1;a=2*PI/t;t3=t/n3;a0=0;
   for (i=1;i<n3+1;i+=2) {a0=a0+f[i]+4*f[i+1]+f[i+2];}
   a0=a0*t3/(3*t);if (fabs(a0)<.001) a0=0;
   printf("\n\n        --- Calculations In Progress ---\n");
   for (x=n7;x<n8+1;x++)
    {C=0;D=0;E=0;A2=0;t1=-t3;
     for (j=1;j<n3+1;j+=2)
      {t1=t1+2*t3;
       C=C+(f[j]*cos(x*a*(t1-t3)))+(4*f[j+1]*cos(x*a*t1))+(f[j+2]*cos(x*a*(t1+t3)));
       D=D+(f[j]*sin(x*a*(t1-t3)))+(4*f[j+1]*sin(x*a*t1))+(f[j+2]*sin(x*a*(t1+t3)));}
     C=2*C*t3/(3*t);D=2*D*t3/(3*t);if (fabs(C)<.0001) C=0;if (fabs(D)<.0001) D=0;
     E=sqrt(C*C+D*D);A2=atan2(D,C)*180/PI;
     printf("     Harmonic %d = %f @ %f deg.\n",x,E,A2);
     c[x]=C;d[x]=D;e[x]=E;a2[x]=A2;v[x]=e[x];av[x]=a2[x];v0=a0;}}

  void printout(void)
  {printf("\n\n Harmonic    Voltage     V.Angle     Amperes     A.Angle      Power\n");
   for (i=n7;i<n8+1;i++)
    {p=v[i]*e[i]*cos((av[i]-a2[i])*PI/180)*.5;
     printf("\n%10d%10f%15f%10f%15f%15f",i,v[i],av[i],e[i],a2[i],p);
     vr=vr+v[i]*v[i]/2;ir=ir+e[i]*e[i]/2;ps=ps+p;}
   vr=sqrt(v0*v0+vr);ir=sqrt(a0*a0+ir);vas=vr*ir;ps=v0*a0+ps;
   printf("\n\n        Rms voltage = %f",vr);
   printf("\n        Rms current = %f",ir);
   printf("\n        Volt-amperes = %f",vas);
   printf("\n             Power = %f",ps);
   printf("\n        Power factor = %f\n\n",ps/vas);}


/* TRANS_LI.C  Steady state transmission line analysis program. */
#include <stdio.h>
#include <math.h>
float f,vl,rl,xl,zl,la,rdc,z0,db,pc,l,lf,a,lr,chr,chi,ch,cha,shr,shi,sh,sha;
float vsr,vsi,vs,vsa,il,isr,isi,is,isa,rc,vswr,n,d,na,da,rca;
void hyperbolic(void);void printout(void);void printtab(void);

main()
{
 /* Enter line parameters */
  printf("\n\n             Enter the following values from the keyboard:\n");
  printf("\n1. Freq.in MHz;space;load voltage;space;load resis.;space;load react.;
  <Enter>.\n");
  scanf("    %f%f%f%f",&f,&vl,&rl,&xl);
  printf("    f = %f, vl = %f, rl = %f, xl = %f.\n",f,vl,rl,xl);zl=sqrt(rl*rl+xl*xl);
```

```c
  la=atan(x1/r1);rdc=180/PI;
  printf("\n2. Line Z0;space,dB loss/100 ft;space;prop.const.;space;length in ft;
  <Enter>.\n");
  scanf("%f%f%f%f",&z0,&db,&pc,&l);lf=1;
  printf("   z0 = %f, db = %f, pc = %f, l = %f.\n",z0,db,pc,l);
  /* Solve for v and i at ends of line and print results */
  hyperbolic();printout();
  printf("\n---------------------------------------------------------");
  /* Solve for v, i, and z status along the line and print table of results */
  printf("\nFt To Load     Voltage     Current    Impedance   Angle deg.\n");
  for (lf=0;lf<l+1;lf +=5) {hyperbolic();printtab();}
  printf("\n---------------------------------------------------------\n\n");
}
void hyperbolic(void)
{a=lf*.1151*db/100;lr=2*lf*PI*f/(984*pc);chr=cosh(a)*cos(lr);
chi=sinh(a)*sin(lr);ch=sqrt(chr*chr+chi*chi);cha=atan2(chi,chr);
shr=sinh(a)*cos(lr);shi=cosh(a)*sin(lr);sh=sqrt(shr*shr+shi*shi);
sha=atan2(shi,shr);vsr=vl*(chr+(z0/zl)*sh*cos(sha-la));
vsi=vl*(chi+(z0/zl)*sh*sin(sha-la));vs=sqrt(vsr*vsr+vsi*vsi);
vsa=atan2(vsi,vsr);il=vl/zl;isr=il*(chr+(zl/z0)*sh*cos(sha+la));
isi=il*(chi+(zl/z0)*sh*sin(sha+la));is=sqrt(isr*isr+isi*isi);
isa=atan2(isi,isr)-la;}

void printout(void)
{n=sqrt(x1*x1+(r1-z0)*(r1-z0));na=atan2(x1,(r1-z0));
d=sqrt(x1*x1+(r1+z0)*(r1+z0));da=atan2(x1,(r1+z0));rc=n/d;rca=na-da;
printf("\n  Load impedance = %f @ %f deg.",zl,la*rdc);
printf("\nLoad refl. coef. = %f @ %f deg.",rc,rca*rdc);vswr=(1+rc)/(1-rc);
printf("\n        Load vswr = %f",vswr);
printf("\n     Load voltage = %f @ 0 deg.",vl);
printf("\n     Load current = %f @ %f deg.",il,-la*rdc);
printf("\n                  = %f ",il*cos(la*rdc));
if (il*sin(-la)>0) printf("+ j %f",il*sin(-la));else printf("- j %f",fabs(il*sin(-la)));
printf("\n   Source voltage = %f @ %f deg.",vs,vsa*rdc);
printf("\n                  = %f ",vs*cos(vsa));
if (vs*sin(vsa)>0) printf("+ j %f",vs*sin(vsa));else printf("- j %f",fabs(vs*sin(vsa)));
printf("\n   Source current = %f @ %f deg.",is,isa*rdc);
printf("\n                  = %f ",is*cos(isa));
if (is*sin(isa)>0) printf("+ j %f\n",is*sin(isa));else printf("- j %f\n",
fabs(isa*sin(isa)));}
void printtab(void)
{printf("\n%10f  %10f  %10f  %10f  %10f",lf,vs,is,vs/is,(vsa-isa)*rdc);}


/* SSANAL.C  Steady state analysis program for use with R L C and FET elements */
#include <stdio.h>
#include <math.h>
#define Y 10
  float a[Y][Y];float b[Y][Y];float p[Y][Y];float q[Y][Y];float b1[Y][Y];
  float q1[Y][Y]; char el[5];
  int  i,j,n,fn1,e,f,k,l,i1,n1,j2;
  float vs,g,h,d,f1,f2,w,u,u2,b1,u3,du,u1,d1,d2,s,s1,t,v;
  void elements(void);void load_r(void);void load_c(void);void load_l(void);
  void load_n(void);void load_f(void);void load_p_q_q1(void);
  void i_o_sweep(void);void solutions(void);void print(void);
  void determinant(void);
```

```
main()
{
/* Define elements from keyboard */
 printf("\n          Enter system elements from keyboard as follows.");
 printf("\n1. Type:<source voltage value><Enter>\n");scanf("%f",&vs);
 printf("   Source voltage = %f.\n",vs);
 printf("2. Type element symbol:<R L C or F><Enter>");
 printf("\n3. For R L or C type:<first node#><space><second node#><space><value>
 <Enter>");
 printf("\n   Value is in ohms, hy., or uF. After final element entry type:<E>
 <Enter>");
 printf("\n4. Type F input similarly except nodes are for G S D and value is GM.\n");
/* Load element arrays */
 for(;;)
  {elements(); if(fnl>0) break;}
/* Load system arrays and obtain solutions */
 load_p_q_ql();
 i_o_sweep();
}
 void elements(void)
 {scanf("%s",el);if ((el[0]==101)||(el[0]==69)) {printf("End of list\n");fnl=1;
 el[4]='\0';return;}
 if ((el[0]==102)||(el[0]==70)) scanf("%d%d%d%f",&k,&j,&i,&v);else scanf
 ("%d%d%f",&i,&j,&v);
 if ((el[0]==114)||(el[0]==82)) {printf("%s %d %d %f ohms.\n",el,i,j,v);el[4]='\0';
 v=1/v;load_r();return;}
 if ((el[0]==99)||(el[0]==67)) {printf("%s %d %d %f uF.\n",el,i,j,v);el[4]='\0';
 v=v*1e-6;load_c();return;}
 if ((el[0]==108)||(el[0]==76)) {printf("%s %d %d %f hy.\n",el,i,j,v);el[4]='\0';
 v=-1/v;load_l();return;}
 if ((el[0]==102)||(el[0]==70)) {printf("%s g=%d s=%d d=%d gm=%f mhos.\n",el,k,j,
 i,v);el[4]='\0';l=j;load_f();}}

 void load_r(void)
 {if (i==0) {a[j][j]=a[j][j]+v;load_n();return;}
  a[i][i]=a[i][i]+v;if (j==0) {load_n();return;}
  a[i][j]=a[i][j]-v;a[j][i]=a[j][i]-v;a[j][j]=a[j][j]+v;load_n();}

 void load_c(void)
 {if (i==0) {b[j][j]=b[j][j]+v;load_n();return;}
  b[i][i]=b[i][i]+v;if (j==0) {load_n();return;}
  b[i][j]=b[i][j]-v;b[j][i]=b[j][i]-v;b[j][j]=b[j][j]+v;load_n();}

 void load_l(void)
 {if (i==0) {bl[j][j]=bl[j][j]+v;load_n();return;}
  bl[i][i]=bl[i][i]+v;if (j==0) {load_n();return;}
  bl[i][j]=bl[i][j]-v;bl[j][i]=bl[j][i]-v;bl[j][j]=bl[j][j]+v;load_n();}

 void load_f(void)
 {if((i!=0)&&(k!=0)) a[i][k]=a[i][k]+v;if((j!=0)&&(l!=0)) a[j][l]=a[j][l]+v;
  if((j!=0)&&(k!=0)) a[j][k]=a[j][k]-v;if((i!=0)&&(l!=0)) a[i][l]=a[i][l]-v;if(k>n)
  n=k;if(l>n) n=l;load_n();}

 void load_n(void)
 {if (i>n) n=i;if (j>n) n=j;}

 void load_p_q_ql(void)
 {for(i=0;i<n+1;i++)
```

414   Appendix B

```
   {for(j=0;j<n+1;j++) {p[i][j]=a[i][j];q1[i][j]=b1[i][j];q[i][j]=b[i][j];}}}

void i_o_sweep(void)
{printf("\nType analysis nodes:<input node><space><output node><Enter>\n");
 scanf("%d%d",&e,&f);printf("Input node = %d and output node = %d.\n",e,f);
 printf("Type sweep values in hz:<low><space><high><space><delta><Enter>\n");
 scanf("%f%f%f",&g,&h,&d);printf("Lowest = %f Hz, highest = %f Hz, increment = %f Hz.\n",g,h,d);
 printf("\n  Freq=Mhz    V=volts      Phase=deg.     Real V         Imag. V    Gain=dB");
 if (d<0) f2=-d;else f2=1+(h-g)/d;
 if (d<0) d=-pow((h/g),(1/(-d-1)));f1=g;
 for(i1=1;i1<f2+1;i1++)
  {w=2*PI*f1;d1=e;d2=f;solutions();v=B1;u=d2;
  if(pow((-1),(e+f))>0) {d1=e;d2=e;} else {d1=e;d2=e;u=u-180;}
  solutions();
  if ((v==0)||(B1==0)) u=u-d2;
  else {v=v/B1;u=u-d2;}
  if (u>180) u=u-360;
  if (u<-180) u=u+360;
  print();du=u1-u;u1=u;
  if (d<0) f1=-u3*d; else f1=u3+d;}
 printf("\n\nFor an analysis at another circuit node enter Y. \n");
 scanf("%s",el);if ((el[0]==89)||(el[0]==121)) {printf("%s\n",el);el[4]='\0';
 i_o_sweep();}}

void solutions(void)
 {n1=n;n=n-1;i=0;
  {for(k=1;k<n+1;k++)
   {if (k!=d1)  j=0; else {i=1;j=0;}
    {for(l=1;l<n+1;l++)
     {if (l!=d2) a[k][l]=p[k+i][l+j]; else {j=1;a[k][l]=p[k+i][l+j];}
      b[k][l]=w*q[k+i][l+j]+q1[k+i][l+j]/w;}}}
  determinant();n=n1;B1=sqrt((d1*d1)+(d2*d2));
  if ((d1==0)&&(d2==0)) return;
  if ((d1<0)&&(d2==0)) {d2=180;return;}
  d2=(360/PI)*atan(d2/(B1+d1));}

void print(void)
 {u3=f1;printf("\n%10f    %10f%15f%15f",f1*1e-6,v*vs,u,v*vs*cos(u*PI/180));
        printf("%15f    %10f",v*vs*sin(u*PI/180),20*log10(v));}

void determinant(void)
 {if (n>1) {d1=1;d2=0;k=1;} else {d1=a[n][n];d2=b[n][n];return;}
  for(;;)
   {l=k;s=fabs(a[k][k])+fabs(b[k][k]);
    for(i=k;i<n+1;i++)
     {t=fabs(a[i][k])+fabs(b[i][k]);if (s<t) {l=i;s=t;}}
    if (l!=k)
     for(j=1;j<n+1;j++)
      {s=-a[k][j];a[k][j]=a[l][j];a[l][j]=s;s1=-b[k][j];
       b[k][j]=b[l][j];b[l][j]=s1;}
    l=k+1;
    for (i=l;i<n+1;i++)
     {s1=a[k][k]*a[k][k]+b[k][k]*b[k][k];s=(a[i][k]*a[k][k]+b[i][k]*b[k][k])/s1;
      b[i][k]=(a[k][k]*b[i][k]-a[i][k]*b[k][k])/s1;a[i][k]=s;}j2=k-1;
    if (j2!=0)
     {for (j=1;j<n+1;j++)
```

```c
       {for (i=1;i<j2+1;i++)
         {a[k][j]=a[k][j]-a[k][i]*a[i][j]+b[k][i]*b[i][j];
          b[k][j]=b[k][j]-b[k][i]*a[i][j]-a[k][i]*b[i][j];}}}
      j2=k;k=k+1;
      {for (i=k;i<n+1;i++)
        {for (j=1;j<j2+1;j++)
          {a[i][k]=a[i][k]-a[i][j]*a[j][k]+b[i][j]*b[j][k];
           b[i][k]=b[i][k]-b[i][j]*a[j][k]-a[i][j]*b[j][k];}}}
      if (k==n) break;}
  l=1;j2=n/2;if (n!=2*j2) {l=0;d1=a[n][n];d2=b[n][n];}
  for (i=1;i<j2+1;i++)
    {j=n-i+l;s=a[i][i]*a[j][j]-b[i][i]*b[j][j];
     s1=a[i][i]*b[j][j]+a[j][j]*b[i][i];t=d1*s-d2*s1;d2=d2*s+d1*s1;d1=t;}}


/* SSANAL2.C Steady state bandpass amplifier analysis. Load Z entered from keyboard */
#include <stdio.h>
#include <math.h>
#define Y 11
float a[Y][Y];float b[Y][Y];float p[Y][Y];float q[Y][Y];float b1[Y][Y];
float q1[Y][Y];
int i,j,n,fn1,e,f,k,l,i1,n1,j2;char el;char lo[2];
float vs,g,h,d,f1,f2,w,u,u2,b1,u3,du,u1,d1,d2,s,s1,t,v,fr,eq;
void elements(void);void load_r(void);void load_c(void);void load_l(void);
void load_n(void);void load_f(void);void load_p_q_q1(void);
void i_o_sweep(void);void solutions(void);void print(void);
void determinant(void);void fixed_el(void);

main()
{
/* Load fixed system elements*/
  fixed_el();
/* Define variables from keyboard */
  printf("\n    Enter variables from the keyboard as follows.");
  printf("\n1. Type:<source voltage value><Enter>\n");scanf("%f",&vs);
  printf("   Source voltage = %f.\n",vs);
  printf("2. Type element symbol:<C L R or X>;<space>;<Enter>. Then type:
  <first node#>;");
  printf("\n   <space><second node#><space><value><Enter>. Note: value = uf.,hy. or
  ohms.");
  printf("\n3. After final element entry type:<E><Enter>\n");
/* Load element arrays */
  for(;;)
    {scanf("%s",lo);if ((lo[0]==101)||lo[0]==69) {printf("End of list.\n");lo[1]='\0';
    break;}
     scanf("%d%d%f",&i,&j,&v);if ((lo[0]==88)||(lo[0]==120))
      {printf("%s %d %d %f. Enter reactance measurement freq.in Hz.\n",lo,i,j,v);
       scanf("%f",&fr);printf("At %f Hz, equivalent element = ",fr);
       if (v>0) {el='L';v=v/(2*PI*fr);} else {el='C';v=-1E6/(2*PI*fr*v);};elements();}}
/* Load system arrays and obtain solutions */
  load_p_q_q1();
  i_o_sweep();
}
  void fixed_el(void)
   {printf("\n\n The fixed circuit element are:\n\n");
    el='F';k=1;j=0;i=2;v=1;elements();
    el='R';i=2;j=0;v=1600;elements();
```

Appendix B

```
el='C';i=2;j=0;v=.00043;elements();
el='L';i=2;j=3;v=1.05e-6;elements();
el='L';i=3;j=0;v=7.56e-7;elements();
el='C';i=3;j=0;v=.0000057;elements();
el='L';i=3;j=8;v=2.97e-6;elements();
el='C';i=8;j=0;v=.0000101;elements();
el='L';i=8;j=4;v=2.97e-6;elements();
el='C';i=4;j=0;v=.0000057;elements();
el='L';i=5;j=0;v=3.26e-7;elements();
el='C';i=4;j=5;v=.000114;elements();
el='C';i=5;j=0,v=.00239;elements();
el='C';i=5;j=6;v=.00013;elements();
el='C';i=6;j=0;v=.0000057;elements();
el='L';i=6;j=9;v=2.82e-6;elements();
el='C';i=9;j=0;v=9.62e-6;elements();
el='L';i=9;j=7;v=2.82e-6;elements();
el='L';i=7;j=0;v=5.94e-7;elements();
el='C';i=7;j=0;v=.001317;elements();}

void elements(void)
{if ((el=='R')||(el=='r'))
  {printf("%c %d %d %f ohms.\n",el,i,j,v);v=1/v;load_r();el='\0';return;}
 if ((lo[0]==114)||(lo[0]==82))
  {printf("%s %d %d %f ohms.\n",lo,i,j,v);v=1/v;load_r();lo[1]='\0';return;}
 if ((el=='C')||(el=='c'))
  {printf("%c %d %d %f uF.\n",el,i,j,v);v=v*1e-6;load_c();el='\0';return;}
 if ((lo[0]==99)||(lo[0]==67))
  {printf("%s %d %d %f uF.\n",lo,i,j,v);v=v*1e-6;load_c();lo[1]='\0';return;}
 if ((el=='L')||(el=='l'))
  {printf("%c %d %d %f hy.\n",el,i,j,v);v=-1/v;load_l();el='\0';return;}
 if ((lo[0]==108)||(lo[0]==76))
  {printf("%s %d %d %f hy.\n",lo,i,j,v);v=-1/v;load_l();lo[1]='\0';return;}
 if ((el=='F')||(el=='f'))
  {printf("%c g=%d s=%d d=%d gm=%f mhos.\n",el,k,j,i,v);l=j;load_f();el='\0';}}

void load_r(void)
{if (i==0) {a[j][j]=a[j][j]+v;load_n();return;}
 a[i][i]=a[i][i]+v;if (j==0) {load_n();return;}
 a[i][j]=a[i][j]-v;a[j][i]=a[j][i]-v;a[j][j]=a[j][j]+v;load_n();}

void load_c(void)
{if (i==0) {b[j][j]=b[j][j]+v;load_n();return;}
 b[i][i]=b[i][i]+v;if (j==0) {load_n();return;}
 b[i][j]=b[i][j]-v;b[j][i]=b[j][i]-v;b[j][j]=b[j][j]+v;load_n();}

void load_l(void)
{if (i==0) {b1[j][j]=b1[j][j]+v;load_n();return;}
 b1[i][i]=b1[i][i]+v;if (j==0) {load_n();return;}
 b1[i][j]=b1[i][j]-v;b1[j][i]=b1[j][i]-v;b1[j][j]=b1[j][j]+v;load_n();}

void load_f(void)
{if((i!=0)&&(k!=0)) a[i][k]=a[i][k]+v;if((j!=0)&&(l!=0)) a[j][l]=a[j][l]+v;
 if((j!=0)&&(k!=0)) a[j][k]=a[j][k]-v;if((i!=0)&&(l!=0)) a[i][l]=a[i][l]-v;
 if(k>n) n=k;if(l>n) n=l;load_n();}

void load_n(void)
{if (i>n) n=i;if (j>n) n=j;}
```

Appendix B

```
void load_p_q_q1(void)
{for(i=0;i<n+1;i++)
 {for(j=0;j<n+1;j++) {p[i][j]=a[i][j];q1[i][j]=b1[i][j];q[i][j]=b[i][j];}}}

void i_o_sweep(void)
{printf("\nType analysis nodes:<input node><space><output node><Enter>\n");
 scanf("%d%d",&e,&f);printf("Input node = %d and output node = %d.\n",e,f);
 printf("Type sweep values in hz:<low><space><high><space><delta><Enter>\n");
 scanf("%f%f%f",&g,&h,&d);printf("Lowest = %f Hz, highest = %f Hz, increment = %f Hz.\n",g,h,d);
 printf("\n Freq=Mhz      V=volts       Phase=deg.       Real V        Imag. V     Gain=dB");
 if (d<0) f2=-d;else f2=1+(h-g)/d;
 if (d<0) d=-pow((h/g),(1/(-d-1)));f1=g;
 for(i1=1;i1<f2+1;i1++)
  {w=2*PI*f1;d1=e;d2=f;solutions();v=B1;u=d2;
   if(pow((-1),(e+f))>0) {d1=e;d2=e;} else {d1=e;d2=e;u=u-180;}
   solutions();
   if ((v==0)||(B1==0)) u=u-d2;
   else {v=v/B1;u=u-d2;}
   if (u>180) u=u-360;
   if (u<-180) u=u+360;
   print();du=u1-u;u1=u;
   if (d<0) f1=-u3*d; else f1=u3+d;}
 printf("\n\nFor an analysis at another circuit node enter Y. \n");
 scanf("%s",lo);if ((lo[0]==89)||(lo[0]==121)) {printf("%s\n",lo);i_o_sweep();}}

void solutions(void)
{n1=n;n=n-1;i=0;
 {for(k=1;k<n+1;k++)
   {if (k!=d1)  j=0; else {i=1;j=0;}
    {for(l=1;l<n+1;l++)
     {if (l!=d2) a[k][l]=p[k+i][l+j]; else {j=1;a[k][l]=p[k+i][l+j];}
      b[k][l]=w*q[k+i][l+j]+q1[k+i][l+j]/w;}}}
  determinant();n=n1;B1=sqrt((d1*d1)+(d2*d2));
  if ((d1==0)&&(d2==0)) return;
  if ((d1<0)&&(d2==0)) {d2=180;return;}
  d2=(360/PI)*atan(d2/(B1+d1));}

void print(void)
{u3=f1;printf("\n%10f   %10f%15f%15f",f1*1e-6,v*vs,u,v*vs*cos(u*PI/180));
        printf("%15f   %10f",v*vs*sin(u*PI/180),20*log10(v));}

void determinant(void)
{if (n>1) {d1=1;d2=0;k=1;} else {d1=a[n][n];d2=b[n][n];return;}
 for(;;)
  {l=k;s=fabs(a[k][k])+fabs(b[k][k]);
   for(i=k;i<n+1;i++)
    {t=fabs(a[i][k])+fabs(b[i][k]);if (s<t) {l=i;s=t;}}
   if (l!=k)
    for(j=1;j<n+1;j++)
     {s=-a[k][j];a[k][j]=a[l][j];a[l][j]=s;s1=-b[k][j];
      b[k][j]=b[l][j];b[l][j]=s1;}
   l=k+1;
   for (i=l;i<n+1;i++)
    {s1=a[k][k]*a[k][k]+b[k][k]*b[k][k];s=(a[i][k]*a[k][k]+b[i][k]*b[k][k])/s1;
     b[i][k]=(a[k][k]*b[i][k]-a[i][k]*b[k][k])/s1;a[i][k]=s;}j2=k-1;
   if (j2!=0)
```

Appendix B

```
    {for (j=1;j<n+1;j++)
      {for (i=1;i<j2+1;i++)
        {a[k][j]=a[k][j]-a[k][i]*a[i][j]+b[k][i]*b[i][j];
         b[k][j]=b[k][j]-b[k][i]*a[i][j]-a[k][i]*b[i][j];}}}
       j2=k;k=k+1;
      {for (i=k;i<n+1;i++)
        {for (j=1;j<j2+1;j++)
          {a[i][k]=a[i][k]-a[i][j]*a[j][k]+b[i][j]*b[j][k];
           b[i][k]=b[i][k]-b[i][j]*a[j][k]-a[i][j]*b[j][k];}}}
     if (k==n) break;}
   l=1;j2=n/2;if (n!=2*j2) {l=0;d1=a[n][n];d2=b[n][n];}
   for (i=1;i<j2+1;i++)
     {j=n-i+l;s=a[i][i]*a[j][j]-b[i][i]*b[j][j];
      s1=a[i][i]*b[j][j]+a[j][j]*b[i][i];t=d1*s-d2*s1;d2=d2*s+d1*s1;d1=t;}}


/* OPAMP.C  Operational amplifier tabular transient response program */
#include <stdio.h>
#define MAX 5
float t1=2.5e-10, t2=5e-8, t3=1e-6,t=0,po=0,vi=1,r1=6.65e5,r2=5e6;
float r3=150,c1=1.6e-8,c2=1.6e-14,c4=1e-10,l3=.001,gm=467;int a;
float d[MAX];float e[MAX];float f[MAX];float x[MAX];
void equations(void);void solutions(void);
main()
{
/* Print out headings and initial values */
   printf("\nIntegration Step = %f uS.\n",t1*1e6);
   equations();
   printf(" Time=uS    x[1]=uV     x[2]=uV     x[3]=nA     x[4]=V");
   solutions();
   /* Calculate and print out solution values */
 for(;;)
 {
   equations();
   for(a=1;a<MAX;a++)
   {
   e[a]=x[a];
   f[a]=d[a];
   x[a]=x[a]+t1*d[a];
   }
   t=t+t1;
   equations();
   for(a=1;a<MAX;a++)
   {
   x[a]=e[a]+t1/2*(f[a]+d[a]);
   }
   po=po+t1;
   if((po+t1/5)>t2)
   {
   solutions();po=0;
   }
   if((t+t1/5)>t3)
   break;
 }
printf("\n");
}
      void equations(void)
      {
```

Appendix B

```c
        d[1]=((vi-x[1]-x[4])/r1-x[3])/c1;
        d[2]=x[3]/c2;
        d[3]=(x[1]-x[2]-r2*x[3])/l3;
        d[4]=(gm*(l3*d[3]+x[2])+x[3]+c1*d[1]-x[4]/r3)/c4;
    }
    void solutions(void)
    {
        printf("\n%10f %10f %10f %10f %10f",t*1e6,x[1]*1e6,x[2]*1e6,x[3]*1e9,x[4]);
    }


/* OPAMPP.C  Operational amplifier transient response plotting program */
#include <stdio.h>
#include <stdlib.h>
#define MAX 5
float t1=2.5e-10, t2=5e-8, t3=1e-6,t=0,po=0,vi=1,r1=6.65e5,r2=5e6;
float r3=150,c1=1.6e-8,c2=1.6e-14,c4=1e-10,l3=.001,gm=467;char count,*ptr,*p;int a,
sc,ts=-1;
float d[MAX];float e[MAX] = {0,0,0,0,0};float f[MAX];float x[MAX];
void equations(void);void solutions(void);void plot(void);void vscale(void);
main()
{
/* Print out headings and initial values */
    printf("\nIntegration Step = %f uS.\n",t1*1e6);
    equations();
    printf("         0       .2       .4       .6       .8      1.0      1.2
 Volts\n");
    vscale();
    plot();

    /* Calculate and print out solution values */
 for(;;)
 {
    equations();
    for(a=1;a<MAX;a++)
    {
    e[a]=x[a];
    f[a]=d[a];
    x[a]=x[a]+t1*d[a];
    }
    t=t+t1;
    equations();
    for(a=1;a<MAX;a++)
    {
    x[a]=e[a]+t1/2*(f[a]+d[a]);
    }
    po=po+t1;
    if((po+t1/5)>t2)
    {
    plot();po=0;
    }
    if((t+t1/5)>t3)
    break;
 }
vscale();printf("\n");
}
        void equations(void)
```

```c
    {
    d[1]=((vi-x[1]-x[4])/r1-x[3])/c1;
    d[2]=x[3]/c2;
    d[3]=(x[1]-x[2]-r2*x[3])/l3;
    d[4]=(gm*(l3*d[3]+x[2])+x[3]+c1*d[1]-x[4])/r3)/c4;
    }

    void solutions(void)
    {
    printf("\t%f \t%f \t%f \t%f \t%f", t*1e6,x[1]*1e6,x[2]*1e6,x[3]*1e9,x[4]);
    }

    void plot(void)
    {
    /* Plot string allocation and update */
    ptr=malloc(80);p=ptr;ts=ts+1;
    /* Time scale plot */
        if(ts==5){sc=sc+250;if(sc<1000)printf(" %d nS--",sc);
                  else printf("%d nS--",sc);ts=0;}
        else
          if(t==0)printf("    %d nS--",t); else
          printf("          -");
    /* Function value plot */
        for(count=10;count<((x[4]+.2)*50);count++) *p++=32;
        *p++=43;
        *p='\0';
        puts(ptr);
    }

    void vscale(void)
    {
    printf("          !....+....!....+....!....+....!....+....!....+....!....+....!
    ....+....!\n");
    }


/* 3PHFWPS.C   Three phase full wave power supply transient analysis. */
#include <stdio.h>
#include <stdlib.h>
#include <math.h>
#define MAX 7
float x[7]={0,0,.1,.1,.1,.1,0}; float d[MAX];
float e[MAX]; float f[MAX]; float i[MAX]; float v[4]={0,3023,3023,-6046};
float t1=.0001, t2=.001, t3=.32, t=0, po=0, co=50e-6, ci=2e-6;
float l1=.0667, vp=6046, w=2*PI*62.5, a1=PI/6, a2=5*PI/6, a3=3*PI/2;
float tc, tx, k,;
int a, b, c, A, B, C, D, ts, l=2, r=500, ri=500, si;
void equations(void); void heading(void); void timing(void);
void printout(void); void integration(void);
main()
{
/* Printout heading and initial conditions */
 heading(); timing(); equations();printout();
/* Analysis / solutions */
 for (;;)
   {integration(); po=po+t1; if (po+t1/5)>t2) {po=0;printout();}
   if(t+t1/5>t3) break;}
```

Appendix B   421

```c
}
void timing(void)
{if ((v[1]>=v[2]-11*d[2])&&(-v[2]<-v[3]-11*d[3]))
    {A=1;B=2;C=3;si=1;ts=1;return;}
 if ((-v[2])>=-v[3]-11*d[3])&&(v[3]<v[1]-11*d[1]))
    {A=2;B=3;C=1;si=-1;ts=2;return;}
 if ((v[3]>=v[1]-11*d[1])&&(-v[1]<-v[2]-11*d[2]))
    {A=3;B=1;C=2;si=1;ts=3;return;}
 if ((-v[1])>=-v[2]-11*d[2])&&(v[2]<v[3]-11*d[3]))
    {A=1;B=2;C=3;si=-1;ts=4;return;}
 if ((v[2]>=v[3]-11*d[3])&&(-v[3]<-v[1]-11*d[1]))
    {A=2;B=3;C=1;si=1;ts=5;return;}
 if ((-v[3])>=-v[1]-11*d[1])&&(v[1]<v[2]-11*d[2]))
    {A=3;B=1;C=2;si=-1;ts=6;return;}}

void equations(void)
{if (x[B]==0) {d[4]=(x[A]-x[5])/ci; d[5]=(x[4]+ci*ri*d[4]-x[6])/l;
   d[A]=(si*(v[A]-v[C])-ci*ri*d[4]-x[4])/(2*11); d[C]=d[A]; d[B]=0; x[B]=0;
   x[C]=x[A]; d[6]=(x[5]-x[6]/r)/co; return;}
 else {d[4]=(x[A]+x[B]-x[5])/ci; d[5]=(x[4]+ci*ri*d[4]-x[6])/l;
   d[A]=(si*(2*v[A]-v[B]-v[C])-ci*ri*d[4]-x[4])/(3*11);
   d[B]=(si*(-v[A]+2*v[B]-v[C])-ci*ri*d[4]-x[4])/(3*11); d[C]=d[A]+d[B];
   d[6]=(x[5]-x[6]/r)/co; return;}}

void heading(void)
{printf("\n  Time ts i[1] i[2] i[3] i[4] i[5] i[6] x[5]      vin     vout  v[1]  v[2]  v[3]\n");}

void printout(void)
{if (ts==1) {i[1]=x[1];i[2]=x[2];i[6]=x[1]+x[2];}
 if (ts==2) {i[1]=x[2]+x[3];i[5]=x[2];i[6]=x[3];}
 if (ts==3) {i[1]=x[1];i[3]=x[3];i[5]=x[1]+x[3];}
 if (ts==4) {i[3]=x[1]+x[2];i[4]=x[1];i[5]=x[2];}
 if (ts==5) {i[2]=x[2];i[3]=x[3];i[4]=x[2]+x[3];}
 if (ts==6) {i[2]=x[1]+x[3];i[4]=x[1];i[6]=x[3];}
 printf("%7.4f%2d%5.1f%5.1f%5.1f%5.1f%5.1f",t,ts,i[1],i[2],i[3],i[4],i[5]);
 printf("%5.1f%5.1f%8.1f%8.1f%6.f%6.f%6.f\n",i[6],x[5],ri*ci*d[4]+x[4],x[6],v[1],
   v[2],v[3]);}

void integration(void)
{timing(); equations();
 for (a=1;a<MAX;a++) {e[a]=x[a]; f[a]=d[a]; x[a]=x[a]+t1*d[a];}
 t=t+t1;k=w*t;v[1]=vp*sin(a1+k);v[2]=vp*sin(a2+k);v[3]=vp*sin(a3+k);
 equations();
 for (a=1;a<MAX;a++) x[a]=e[a]+t1/2*(f[a]+d[a]);
 if ((x[1]>=0)&&(x[2]>=0)&&(x[3]>=0)) return;
 if(x[1]<0)D=1;if(x[2]<0)D=2;if(x[3]<0)D=3; t=t-t1; tc=e[D]*t1/(e[D]-x[D]);
 for (a=1;a<MAX;a++) x[a]=e[a]+tc*f[a]; t=t+tc; x[D]=0; d[D]=0; k=w*t;
 equations(); tx=t1-tc;
 for (a=1;a<MAX;a++) x[a]=x[a]+tx*d[a]; t=t+tx; D=0;}


/* 12PULSE.C  Twelve pulse dc power supply transient analysis. */
#include <stdio.h>
#include <stdlib.h>
#include <math.h>
#define MAX 11
float x[11]={0,0,19.23595,19.23595,9890.909,20.22405,19243.89,-7.345432,7.345432,
```

```c
14.56383,9309.999};
float d[11]={0,0,-3239.642,-3239.642,-490917.4,161.8497,-688.9628,969.169,939.169,
1938.099,860190.4};
float v[10]={0,3023,3023,-6046,0,0,0,5236,5236,-10472};
float e[MAX]; float f[MAX]; float i[MAX];
float t1=.00005, t2=.0002, t3=.016, t=0, po=0, co=50e-6, ci=2e-6, cid=2e-6;
float ll=.05235, lld=.15705, vp=6046, vpd=10472, w=2*PI*62.5, a1=PI/6, den;
float a2=5*PI/6, a3=3*PI/2, tc, tx, k;
int a, b, A, B, C, D, E, F, G, ts=6, tsd=8, tsdp; l=2, r=950, ri=500, rid=500, si, sid;
void equations(void); void equations_d(void); void heading(void);
void timing(void); void timing_d(void); void printout(void); void liney(void);
void lined(void); void integration(void); void integration_d(void); void reset(void);
main()
{
/* Printout heading and initial conditions */
 heading(); printout();
/* Analysis / solutions */
 for (;;)
  {integration(); if(x[1]<0) x[1]=0; if(x[2]<0) x[2]=0; if(x[3]<0) x[3]=0;
   integration_d(); po=po+t1; if(po+t1/5>t2){po=0;printout();}
  if(t+t1/5>t3) break;}
}
 void timing(void)
 {if ((v[1]>=v[2]-ll*d[2])&&(-v[2]<-v[3]-ll*d[3]))
     {A=1;B=2;C=3;si=1;ts=1;return;}
  if ((-v[2]>=-v[3]-ll*d[3])&&(v[3]<v[1]-ll*d[1]))
     {A=2;B=3;C=1;si=-1;ts=2;return;}
  if ((v[3]>=v[1]-ll*d[1])&&(-v[1]<-v[2]-ll*d[2]))
     {A=3;B=1;C=2;si=1;ts=3;return;}
  if ((-v[1]>=-v[2]-ll*d[2])&&(v[2]<v[3]-ll*d[3]))
     {A=1;B=2;C=3;si=-1;ts=4;return;}
  if ((v[2]>=v[3]-ll*d[3])&&(-v[3]<-v[1]-ll*d[1]))
     {A=2;B=3;C=1;si=1;ts=5;return;}
  if ((-v[3]>=-v[1]-ll*d[1])&&(v[1]<v[2]-ll*d[2]))
     {A=3;B=1;C=2;si=-1;ts=6;return;}}

 void timing_d(void)
 {if ((-v[9]-lld*d[9]+1>=v[8]-lld*d[8])&&(v[8]-lld*d[8]>0)&&-v[9]>0){tsd=7;return;}
  if ((v[7]-lld*d[7]+1>=-v[9]-lld*d[9])&&(-v[9]-lld*d[9]>0)&&v[7]>0){tsd=9;return;}
  if ((-v[8]-lld*d[8]+1>=v[7]-lld*d[7])&&(v[7]-lld*d[7]>0)&&-v[8]>0){tsd=11;return;}
  if ((v[9]-lld*d[9]+1>=-v[8]-lld*d[8])&&(-v[8]-lld*d[8]>0)&&v[9]>0){tsd=13;return;}
  if ((-v[7]-lld*d[7]+1>=v[9]-lld*d[9])&&(v[9]-lld*d[9]>0)&&-v[7]>0){tsd=15;return;}
  if ((v[8]-lld*d[8]+1>=-v[7]-lld*d[7])&&(-v[7]-lld*d[7]>0)&&v[8]>0){tsd=17;return;}}

 void equations(void)
 {if (x[B]==0) {if (x[4]<0) d[4]=0; else d[4]=(x[A]-x[5])/ci;
     d[A]=(si*(v[A]-v[C])-ci*ri*d[4]-x[4])/(2*ll); d[C]=d[A]; d[B]=0; x[B]=0;
     x[C]=x[A]; return;}
  else {if (x[4]<0) d[4]=0; else d[4]=(x[A]+x[B]-x[5])/ci;
       d[A]=(si*(2*v[A]-v[B]-v[C])-ci*ri*d[4]-x[4])/(3*ll);
       d[B]=(si*(2*v[B]-v[A]-v[C])-ci*ri*d[4]-x[4])/(3*ll);
       d[C]=d[A]+d[B]; return;}}

 void equations_d(void)
 {if (tsd!=tsdp) {x[E]=-x[E];tsdp=tsd;}; d[10]=(x[F]+x[G]-x[5])/cid;
  d[5]=(x[4]+ri*ci*d[4]+x[10]+rid*cid*d[10]-x[6])/1; d[6]=(x[5]-x[6]/r)/co;
  if (x[5]<0&&x[4]+ri*ci*d[4]+x[10]+rid*cid*d[10]>si*(v[A]-v[C])-sid*v[G])
```

Appendix B

```
    {x[7]=0;x[8]=0;x[9]=0;d[7]=0;d[8]=0;d[9]=0;return;}
 if (x[7]==0&&x[8]==0&&x[9]==0&&x[4]+ri*ci*d[4]+x[10]+rid*cid*d[10])-sid*v[G]
    +si*(v[A]-v[C])) {x[7]=0; x[8]=0; x[9]=0; d[7]=0; d[8]=0; d[9]=0; return;}
 if (x[E]>=x[F]) {d[G]=(-sid*v[G]-x[10]-rid*cid*d[10])/lld;
     d[E]=(sid*(v[E]+v[F])-x[10]-rid*cid*d[10])/(2*lld);
     d[F]=d[E]; tsd=tsd+1; return;}
  else d[F]=(sid*v[F]-x[10]-rid*cid*d[10])/lld;
     d[G]=(-sid*v[G]-x[10]-rid*cid*d[10])/lld; d[E]=sid*v[E]/lld;return;}

void heading(void)
{printf("\n          y d  i  i  i  i  i  i  i  i  i  i  i  i  x        dc");
 printf("\n  Time  ts ts [1] [2] [3] [4] [5] [6] [7] [8] [9][10][11][12] [5]
 vout\n");}

void printout(void)
{for (b=1;b<12;b++) i[b]=0;
 if (ts==1) {i[1]=x[1];i[2]=x[2];i[6]=x[1]+x[2];}
 if (ts==2) {i[1]=x[2]+x[3];i[5]=x[2];i[6]=x[3];}
 if (ts==3) {i[1]=x[1];i[3]=x[3];i[5]=x[1]+x[3];}
 if (ts==4) {i[3]=x[1]+x[2];i[4]=x[1];i[5]=x[2];}
 if (ts==5) {i[2]=x[2];i[3]=x[3];i[4]=x[2]+x[3];}
 if (ts==6) {i[2]=x[1]+x[3];i[4]=x[1];i[6]=x[3];}
 if (tsd==7) {i[8]=x[8]+x[9];i[10]=x[8]-x[7];i[12]=x[7]+x[9];}
 if (tsd==8) {i[8]=x[8]+x[9];i[12]=x[7]+x[9];}
 if (tsd==9) {i[7]=x[7]+x[8];i[8]=x[9]-x[8];i[12]=x[7]+x[9];}
 if (tsd==10){i[7]=x[7]+x[8];i[12]=x[7]+x[9];}
 if (tsd==11){i[7]=x[7]+x[8];i[11]=x[8]+x[9];i[12]=x[7]-x[9];}
 if (tsd==12){i[7]=x[7]+x[8];i[11]=x[8]+x[9];}
 if (tsd==13){i[7]=x[8]-x[7];i[9]=x[7]+x[9];i[11]=x[8]+x[9];}
 if (tsd==14){i[9]=x[7]+x[9];i[11]=x[8]+x[9];}
 if (tsd==15){i[9]=x[7]+x[9];i[10]=x[7]+x[8];i[11]=x[9]-x[8];}
 if (tsd==16){i[9]=x[7]+x[9];i[10]=x[7]+x[8];}
 if (tsd==17){i[8]=x[8]+x[9];i[9]=x[7]-x[9];i[10]=x[7]+x[8];}
 if (tsd==18){i[8]=x[8]+x[9];i[10]=x[7]+x[8];}
printf("%7.4f%3d%3d%4.0f%4.0f%4.0f%4.0f%4.0f%4.0f",t,ts,tsd,i[1],i[2],i[3],i[4],
i[5],i[6]);
printf("%4.0f%4.0f%4.0f%4.0f%4.0f%4.0f%4.0f%9.2f\n",i[7],i[8],i[9],i[10],i[11],
i[12],x[5],x[6]);
}

void liney(void)
{if (ts==1) x[3]=-x[3]; return; if (ts==2) x[2]=-x[2]; x[3]=-x[3]; return;
 if (ts==3) x[2]=-x[2]; return; if (ts==4) x[1]=-x[1]; x[2]=-x[2]; return;
 if (ts==5) x[1]=-x[1]; return; if (ts==6) x[3]=-x[3]; x[1]=-x[1];}

void lined(void)
{if (ts<=7)  x[9]=-x[9]; return; if (ts<=9) x[8]=-x[8]; x[9]=-x[9]; return;
 if (ts<=11) x[8]=-x[8]; return; if (ts<=13)x[7]=-x[7]; x[8]=-x[8]; return;
 if (ts<=15) x[7]=-x[7]; return; if (ts<=17)x[9]=-x[9]; x[7]=-x[7];}

void integration(void)
{timing();equations();
 for (a=1;a<5;a++) {e[a]=x[a]; f[a]=d[a]; x[a]=x[a]+t1*d[a];}
 t=t+t1;k=w*t;v[1]=vp*sin(a1+k);v[2]=vp*sin(a2+k);v[3]=vp*sin(a3+k);
 equations();
 for (a=1;a<5;a++) x[a]=e[a]+t1/2*(f[a]+d[a]);if (x[B]>=0) {t=t-t1; return;}
 if(x[1]<0)D=1;if(x[2]<0)D=2;if(x[3]<0)D=3; t=t-t1; tc=e[D]*t1/(e[D]-x[D]);
 for (a=1;a<5;a++) x[a]=e[a]+tc*(x[a]-e[a])/t1; t=t+tc; x[D]=0; d[D]=0;
```

```
equations(); tx=t1-tc;
for (a=1;a<5;a++) x[a]=x[a]+tx*d[a]; t=t+tx; D=0; t=t-t1;}

void integration_d(void)
{timing_d(); reset(); equations_d();
 for (a=5;a<11;a++) {e[a]=x[a]; f[a]=d[a]; x[a]=x[a]+t1*d[a];}
 t=t+t1; k=w*t; v[7]=vpd*sin(a1+k); v[8]=vpd*sin(a2+k); v[9]=vpd*sin(a3+k);
 reset(); equations_d();
 for (a=5;a<11;a++) x[a]=e[a]+t1/2*(f[a]+d[a]);
 if (x[E]<=x[F]) return;
 den=((e[F]-e[E])-(x[F]-x[E])); t=t-t1; tc=(e[F]-e[E])*t1/den;
 if (tc>t1||tc<1e-6) tc=t1/10;
 for (a=5;a<11;a++) x[a]=e[a]+tc*(x[a]-e[a])/t1; t=t+tc;x[E]=x[F];d[E]=d[F];
 k=w*t; v[7]=vpd*sin(a1+k); v[8]=vpd*sin(a2+k); v[9]=vpd*sin(a3+k);
 reset(); equations_d(); tx=t1-tc; for (a=5;a<11;a++) x[a]=x[a]+tx*d[a];
 t=t+tx; return;}

void reset(void)
{if(tsd<= 8){tsd=7;E=7;F=8;G=9;sid=1;return;}
 if(tsd<=10){tsd=9;E=8;F=9;G=7;sid=-1;return;}
 if(tsd<=12){tsd=11;E=9;F=7;G=8;sid=1;return;}
 if(tsd<=14){tsd=13;E=7;F=8;G=9;sid=-1;return;}
 if(tsd<=16){tsd=15;E=8;F=9;G=7;sid=1;return;}
 if(tsd<=18){tsd=17;E=9;F=7;G=8;sid=-1;return;}}
```

Index

A
Amplitude modulation, (definition), 102

B
Backup power source, (definition), 21
Bandpass amplifier, analysis, 144
BASIC
   disk procedures, 302
   programming language capabilities, 282
Bessel functions, application, 103
Bipolar power transistors, capabilities, 115

C
C, ANSI standard
   BASIC to C conversion, 283
   disk procedures, 304
   programming language capabilities, 282
Capacitance, (definition), 131
Charging, dc resonant, characteristics, 251
Circuit breaker
   definition, 9
   fault interruption current, 13
Circulator, RF capabilities, 75
Clutter, (definition), 259
Coaxial lines, description, 66
Commutation
   analysis, 156
   definition, 147
Conservation of energy, 196
Conversions systems, (definition), 147
Crossed field amplifiers, capabilities, 109

D
DC power supplies
   characteristics, steady state, 148
   dc-dc converter, analysis, 180
   regulators, definition, 179
   single phase, capacitor input filter, 154
   single phase bridge, analysis, 159
   single phase center tap, analysis, 149
   three phase bridge, analysis, 161
   three phase bridge, controlled, 185
   twelve pulse bridge, analysis, 169

Delta and wye network
 comparison, 32
 conversions, 33
Directional coupler, RF capabilities, 73
Distribution center, (definition), 9
Doppler, principle, 120

E

Energy
 conversions, 2
 sources, 2
 units, 3
Engine-generator sets, (definition), 5
Erickson, B.K., program author, 17
ESR, (definition), 209
Euler-Cauchy algorithm, second order, 135

F

Filters, dc power supply, 154
Fourier transform
 conversion expressions, 86
 definition, 79
 mathematical analysis, 81
 series, 80
 tabular analysis, 85
Frequency modulation, (definition), 103

G

Guillemin, E.A., system analyst, 239

H

Hard switches, (definition), 224
Hybrid, 3-dB RF, 118

I

Impedance, dc power supply source, 159
Improvement factors, (definition), 260
Inductance, (definition), 131
Inertia loads, acceleration, 188
Interfaces, (definition), 1
Inverse voltage diodes, characteristics, 251
Inversion systems
 ac-dc, analysis, 192
 dc-dc, analysis, 180

K

Klystron amplifiers, capabilities, 110

L

Leakage inductance, (definition), 11

M

Machines, dc
 generators, 190
 motors, 185
Magic T., capabilities, 76
Microstrip lines, capabilities, 68
Microwave radar horizon, (definition), 123
MOSFET power transistors
 capabilities, 117
 corporate RF structure, design, 118
Moving target indicator (MTI) radar, (definition), 120
Mutual inductance, (definition), 132

N

National Bureau of Standards units (NBS), 2
Network analysis elements, (definition), 130
Numerical integration algorithms
 first order, 134
 second order Euler-Cauchy, 134
 second order Runge-Kutta, 134

O

Operational amplifier, transient analysis, 142
Over-the-horizon backscatter (Oth-B) radar, (definition), 122

P

Parallel wire lines, capabilities, 64
Penstock, (definition), 2
Phase modulation, (definition), 103
Phased array radar, (definition), 122
Phasor diagram, application, 30
Power
 conversion systems, ac, 147
 definition, 3
 interface product, 8
 inversion systems, dc, 147
Power electronics, (definition), 5
Power factor
 correction, 16
 dc power supply, 161
 definition, 8
 service cost multiplier, 16
Power service cost multiplier, (definition), 16
Power transformer
 characteristics, 10
 percent R and X, definition, 12
Prime power system, (definition), 5
Pulse forming networks, analysis, 236

Pulse transformers, characteristics, 242
Pulsers, (definition), 206

R

Radar, (definition), 119
Reflection coefficient, (definition), 62, 236
Resistance, (definition), 130
RF power amplifiers, capabilities, 107

S

Shadow grids, (definition), 216
Snap-on, (definition), 152
Snubbers, dc power supply, 167
Soft switches, (definition), 226
Substations, (definition), 45
Switching power supply, (definition), 179
System analysis
    elements, 106
    functional diagrams, 105
    specification, 100
    statement of work, 101
    waveforms, 102
System International units (SI), 2
Systems of units, comparison, 3

T

Tetrode amplifiers
    capabilities, 112
    performance analysis, simplified, 113
Three phase circuits, capabilities, 28
Time variant elements, (definition), 133
Transients
    analysis approaches, 128
    definition, 4
Transmission lines
    auxiliary elements, 73

characteristic impedance, 56
circulators, isolators, 75
elements, 53
monitoring equipment,
    voltage standing wave ratio (VSWR), 73
    slotted line, 73
prime power line, definition, 22, 46, 59
propagation constant, 56
RF line, definition, 64
RF power division and combining, 76
Traveling-wave tube amplifiers, capabilities, 110
Triode amplifiers, capabilities, 112

U

Uninterruptible power source, analysis, 199

V

Volt-amperes
    definition, 8
    product, definition, 8

W

Waveforms
    amplitude modulation, 102
    frequency modulation, 103
    phase modulation, 103
    prime power, 8
    rectangular pulse, 81
    shaped pulse, 89
Waveguide lines, capabilities, 69
Wright, W.G., program author, 136
Wye and delta networks
    capability comparison, 32
    element conversion, 33